Replacement of Neanderthals by Modern Humans Series

Edited by

Takeru Akazawa
Research Institute, Kochi University of Technology
Kochi 782-8502, Japan
akazawa.takeru@kochi-tech.ac.jp

Ofer Bar-Yosef
Department of Anthropology, Harvard University
Cambridge, Massachusetts 02138, USA
obaryos@fas.harvard.edu

The planned series of volumes will report the results of a major research project entitled "Replacement of Neanderthals by Modern Humans: Testing Evolutionary Models of Learning", offering new perspectives on the process of replacement and on interactions between Neanderthals and modern humans and hence on the origins of prehistoric modern cultures. The projected volumes will present the diverse achievements of research activities, originally designed to implement the project's strategy, in the fields of archaeology, paleoanthropology, cultural anthropology, population biology, earth sciences, developmental psychology, biomechanics, and neuroscience. Comprehensive research models will be used to integrate the discipline-specific research outcomes from those various perspectives. The series, aimed mainly at providing a set of multidisciplinary perspectives united under the overarching concept of learning strategies, will include monographs and edited collections of papers focusing on specific problems related to the goals of the project, employing a variety of approaches to the analysis of the newly acquired data sets.

More information about this series at http://www.springer.com/series/11816

Emiliano Bruner • Naomichi Ogihara •
Hiroki C. Tanabe
Editors

Digital Endocasts

From Skulls to Brains

Springer

Editors
Emiliano Bruner
CENIEH
Burgos, Spain

Naomichi Ogihara
Keio University
Yokohama, Japan

Hiroki C. Tanabe
Nagoya University
Nagoya, Japan

Replacement of Neanderthals by Modern Humans Series
ISBN 978-4-431-56580-2 ISBN 978-4-431-56582-6 (eBook)
DOI 10.1007/978-4-431-56582-6

Library of Congress Control Number: 2017957848

Printed on acid-free paper

This Springer imprint is published by Springer Nature
The registered company is Springer Japan KK
The registered company address is: Chiyoda First Bldg. East, 8-1 Nishi-Kanda Chiyoda-ku, 101-0065 Tokyo, Japan

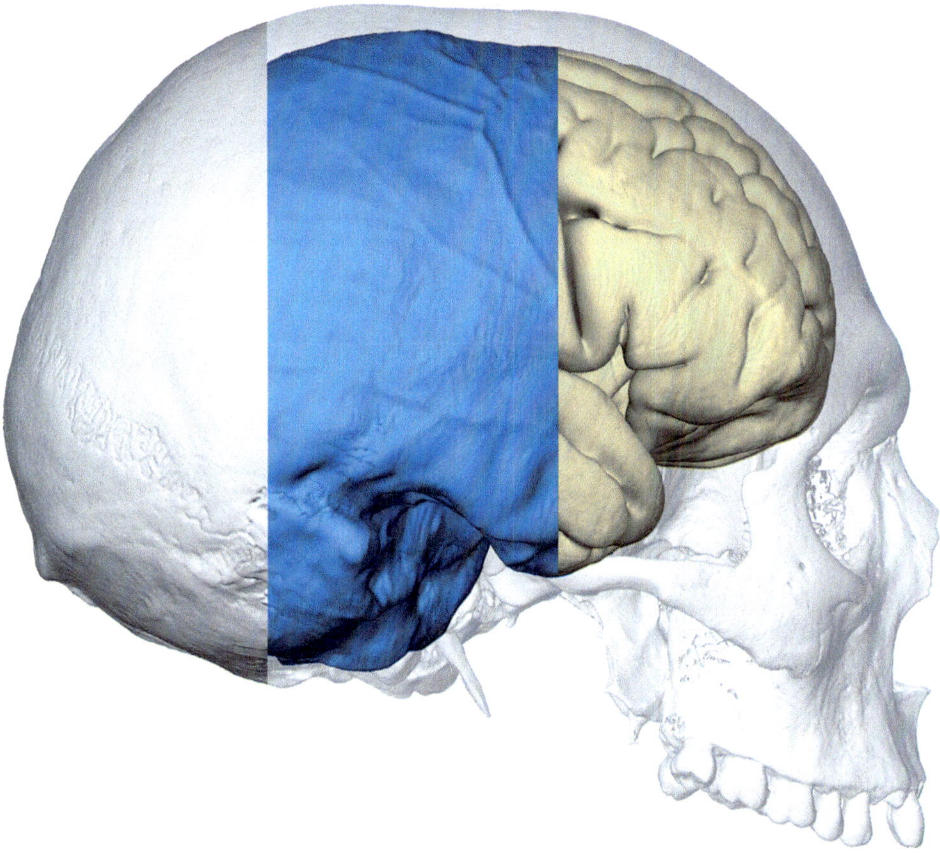

Brain, endocast, and skull digital reconstruction (courtesy of Simon Neubauer)

Preface

In recent years, computer-based techniques have led to a noticeable renaissance of most anatomical disciplines, involving new challenges and re-introducing old problems. Digital anatomy has represented a major advance in the visualization and exploration of anatomical elements, and computed morphometrics has supplied numerical and statistical tools for analyzing anatomical systems using proper quantitative approaches. Before this "pixel revolution," anatomy was often limited by reduced sample sizes and by methodological difficulties associated with physical dissections. Working with bodies, most of all when dealing with humans, implies a limited availability of individuals, difficulties in management and administration, and large and complex histological preparations. Furthermore, dissections only allow the study of the anatomical components outside of their functional conditions. Digital tools can be used to investigate large samples with an extreme resolution and within their biological context, preventing most of those limitations, which, decades ago, contributed to a sort of "freezing" of the anatomical fields, slowing down their development and often impeding the efficient dissemination of their achievements. Once the computed tools had become available on a large scale and many forgotten topics had been recovered from past literature, we realized that we still lacked much information regarding our own anatomy. In fact, we have spent the last decades principally investigating molecules and microscopic features, but we do not yet have a robust knowledge of our bones and vessels. For many macroanatomical traits, we still ignore the variations, influences, and developmental processes that generate the phenotypic variability of our species. Importantly, some of these anatomical traits may be crucial not only from an evolutionary perspective, but also from a medical point of view.

Physical dissections and other non-digital approaches are still mandatory and essential, but the complementary potentialities of these computed methods are outstanding. Nonetheless, as usual, power must be accompanied by adequate control of its capacities and limitations. Most of these methods are based on very complex and complicated technical and numerical assumptions and criteria that rely on elaborate programs, devices, and algebraic transformations, and they are based on an important background integrating electronics, informatics, and statistics. Therefore, the entangled numerical elaboration associated with these digital models requires competence and caution. Frequently, programs are sufficiently "user-friendly" to allow a basic manipulation of the data without any comprehensive knowledge of the processes involved. This usability further increases the possibility of a superficial use, interpretation, or understanding, of the actual outputs of a computerized analysis. Multidisciplinarity is, indeed, strictly required in such a complicated methodological context.

Most anatomical disciplines have taken advantage of these methodological changes, but one that probably has been particularly privileged by these digital approaches is neuroscience. Structural and functional imaging has induced a considerable revolution in all kinds of brain studies, including evolutionary neuroanatomy. This book is part of the 5-year (2010–2014) project "Replacement of Neanderthals by Modern Humans: Testing Evolutionary Models of Learning" (RNMH), funded by the Japanese Government (Ministry of Education, Culture, Sports, Science, and Technology, Grant-in-Aid for Scientific Research on Innovative Areas No. 22101001) and coordinated by Professor Takeru Akazawa. The project is based on a multidisciplinary approach, integrating cultural anthropology, biological sciences, and engineering, to investigate and compare cognitive and cultural capacities in modern humans and Neanderthals, and to make inferences on their respective learning abilities. This new volume of the RNMH Series is dedicated to brain evolution and paleoanthropology, focusing on recent advances in all those research areas investigating the brain form in extinct species. The book includes chapters on craniology, digital techniques, endocast reconstruction, craniovascular traits, surface analyses, landmarking, and on the relationships between the brain and the braincase. Furthermore, the volume includes chapters concerning the principal brain districts, and reviews the current knowledge regarding their evolution in humans and in nonhuman primates. The aim is to supply a comprehensive and updated reference on the challenges, advances, and limitations associated with the study of the brain form and functions in fossils, introducing the current state of the art and future directions of human paleoneurology.

Burgos, Spain Emiliano Bruner
Yokohama, Japan Naomichi Ogihara
Nagoya, Japan Hiroki C. Tanabe

Contents

On the Making of Endocasts: The New and the Old in Paleoneurology

1

Ralph L. Holloway

Abstract

Making endocasts with latex rubber has been around for many years. This chapter describes my methods which were not original and some of the experiences encountered. Other methods, using plaster of Paris, various silicon-based rubbers, and Admold (dental caulk), for sectioned crania are examined and their relative merits and problems compared, such as damage to original specimens, deterioration with time (especially with latex rubber), and tensile strength of silicon-based molds. The resolution is as good as it can get, compared to "virtual" endocasts. These older methods have largely been succeeded by the making of "virtual" endocasts through various scanning procedures, with numerous advantages such as being noninvasive of original fossil specimens, immediate coordinates for morphometric analyses, scan data sharing and replication, and production of actual virtual endocasts through 3-D printing.

Keywords

Brain endocasts • Latex • Silicon • Dental caulk • "Virtual" endocasts

1.1 Introduction

Emi Bruner's invitation to contribute an introductory chapter is a real challenge, particularly given his expressed desire for me to describe making endocasts with latex rubber. It might be useful to situate that process within a larger canvas of what is and has happened in paleoneurology regarding endocast studies and what is being studied and how (see Holloway 2014 for more extended discussion of paleoneurology).

From what I have gleaned from Tilly Edinger's (1975) massive (257 pages!) annotated bibliography, the earliest publication goes back to 1804. Pages 183–257 are devoted to the Hominidae, and there is a very fine forward by Professor Bryan Patterson which describes in great detail how the bibliography came to be. Clearly, paleoneurology has played an important role in the zoological sciences. When I wrote my dissertation (Holloway 1964), I had no idea that there was such a vast history and had only read papers devoted to questions of human brain evolution, although I was aware of and had admired Edinger's (1949) work on the evolution of the horse brain. Kotchetkova's book and endocasts were not available until Harry Jerison made it so. F. Symington, G.E. Smith, F. Weidenreich, C.U. Kappers, F. Tilney, C.J. Connolly, and G.H.W. Schepers were the fodder from which I came to the erroneous conclusion that endocasts were not of very much use in hominid evolution, as they seldom showed any reliable details, thanks to meningeal conspiracies and cisterna of cerebrospinal fluid covering areas where one needed details to be able to separate the cerebral lobes accurately. It was their volumes that were useful. Ironically, thanks to a lack of facilities at Columbia for doing histological work (Golgi-Cox) on primate brains ("if we do not know what is happening in the brains of Aplysia, the sea-slug, how can we possibly know what is

R.L. Holloway (✉)
Department of Anthropology, Columbia University, New York, NY 10027, USA
e-mail: rlh2@columbia.edu

E. Bruner et al. (eds.), *Digital Endocasts*, Replacement of Neanderthals by Modern Humans Series,
DOI 10.1007/978-4-431-56582-6_1

happening in primate brains..." from a Nobel laureate), my forays into dendritic branching were confined to rats (Holloway 1966). This led me to once again look at endocasts, and a semester's leave to work in P.V. Tobias' lab at Wits (University of Witwatersrand) in Johannesburg, South Africa, sealed my fate as a paleoneurologist. I knew the Taung endocast couldn't be some 500+ ml (cc), and I thus leapt into assessing its volume and morphology. I came back in 1971–1972 for a full year of research on the australopithecines and also worked with the Leakeys in Nairobi, Kenya, and on some of the Indonesian hominins in Teuku Jacob's lab in succeeding years, as well as the Solo specimens in Frankfurt, Germany. My goals at that time were finding accurate volumes and making endocast reconstruction that I thought were accurate.

Most of this history can be found in Holloway (2008) and Holloway et al. (2004), which I prefer not to recount here, particularly all of the controversial history with Dean Falk and Harry Jerison who are also leading experts in paleoneurology. In the late 1970s, I adopted the stereoplotting method described by Oyen and Walker (1977), which employed an apparatus that measured surface point length from a central homologous center in polar coordinates; some of the initial work is published in Holloway (1981). This was in the era of punched IBM cards and SPSS multivariate analyses. Fortunately, the equipment fell apart from so many measurements, and I was thus spared having to continue those studies (see Chap. 9 of this book).

The 1970s and 1980s were also an era in which I endocast close to 200+ ape endocasts from crania I borrowed from several museums, in addition to many of the australopithecines and early *Homo*. Close to 100 modern human endocasts were made from the lab collection of crania at Columbia and the American Museum of Natural History.

I can only hope that several hundred endocasts I have made survive the changing environments and will prove to be a useful collection for those wishing to pursue paleoneurological studies.

As I see it, there are about five ways of making endocasts:

1. I think the earliest attempts were to pour plaster of Paris directly into the crania (obviously not through the foramen magnum) probably first coating the internal table of the bone with shellac. The foramina of the cranial base would first be plugged, and delicate structures such as the clinoid processes and dorsal sellae, cribriform plate, as well as open cracks, or missing portions would be protected with plasticine or cotton wadding. These could only be done on the calva, cranial base portions, and not the whole, unless there were postmortem cracks or glue joints that could be separated and rejoined after the cast was extracted.
2. Similar to the above was the use of alginate, but in this case, the alginate cast formed a mold which could be covered with some other material which then could be a mold for a plaster endocast. As I recall from using it a few times, the material had no tear strength.
3. Liquid latex of varying consistencies became a standard in making endocasts. My earliest forays into this adventure were derived from what previous paleoneurologists (e.g., Len Radinsky 1967 and Tilly Edinger 1929, 1949) were using. The latex I used was called Admold, and it came from the Bronx, usually in gallon containers, with the consistency of a thin milk shake. I often added a small amount of red dye to effect a pink rather than crème complexion, which I thought made endocast details easier to see. To make the latex into "rubber" required a heat treatment, at about 100 °C and for about an hour. This was often done in various ovens, autoclaves, etc. This vulcanized the latex into a sheet with great tensile strength and flexibility, as the vulcanized product was extracted through the foramen magnum of the cranium. In addition to becoming something of an expert on making endocasts, I became an expert on handheld hair dryers that could be used in three different continents with different electrical voltages, outlets, etc. I was surely a host's pain in the neck for requesting such equipment and various stands (Bunsen burners a favorite) to hold the hair dryer so as to avoid the necessity of slave labor, etc. Nobody in their right mind would want to hold a hair dry in their hand for hours at a time! If dried and vulcanized properly, the extraction process could begin. This simply means getting the dried vulcanized endocast out of the skull, and that meant pulling it out through the foramen magnum for complete crania. I always used talcum or baby powder inside the endocast to prevent sticking when the endocast was collapsed. I would carefully release the endocast by using a finger (usually middle, but not with hylobatids, etc.) to initially detach the rubber from the foramen magnum and would apply some talcum power to that released interface as I worked the rubber into a completely collapsed state within the cranium. Now came the fun part: extracting the collapsed rubber endocast through the foramen magnum. This was done very gently mm by mm, collapsing the endocast as it peeled away from the bony surface and finally being rewarded with a pleasant-sounding "POP" (place the tip of the tongue on inner upper lip and flick forcefully forward and downward, and you will hear the sound of a latex rubber endocast emerging from the cranium). I then usually floated the endocast in water and filled it with liquid plaster to prevent distortion. After that, the foramen magnum area was capped with latex or plasticine, and the product was now ready for water displacement and various measurements with calipers and measuring tape.

Schoenemann et al. (2007) showed that this introduced only minor distortions, mostly confined to the basal region. Most of the early endocasts I made have undergone degeneration or caramelization (Fig. 1.1). I remember in particular the ones I made in Kenya and the Solo endocasts I made in Frankfurt while von Koenigswald was still alive. These were particularly difficult to make, as I recall it was during a very hot summer spell in Frankfurt, and I was working in my underwear. I made the layers too thin, also. These casts should be done again, but CT scanning is the way to go with such fragile specimens these days. The KNM-ER 1470 endocast (Fig. 1.2) was a special challenge. I wanted to stabilize the dimensions of the total latex, vulcanized in situ, so, much to Richard Leakey's temporary horror, I poured plaster into the latex-lined skull and told Richard to come back the next day. After the plaster had set, I simply dissolved the glued joints with acetone and, after the endocast was free, glued the cranial fragments back as they were. The Indonesian *Homo erectus* endocasts I made back in the 1970s were difficult, particularly Sangiran 10, 12, and 17. (See Holloway et al. 2004a for discussions, analysis, descriptions of fossil hominin endocasts.)

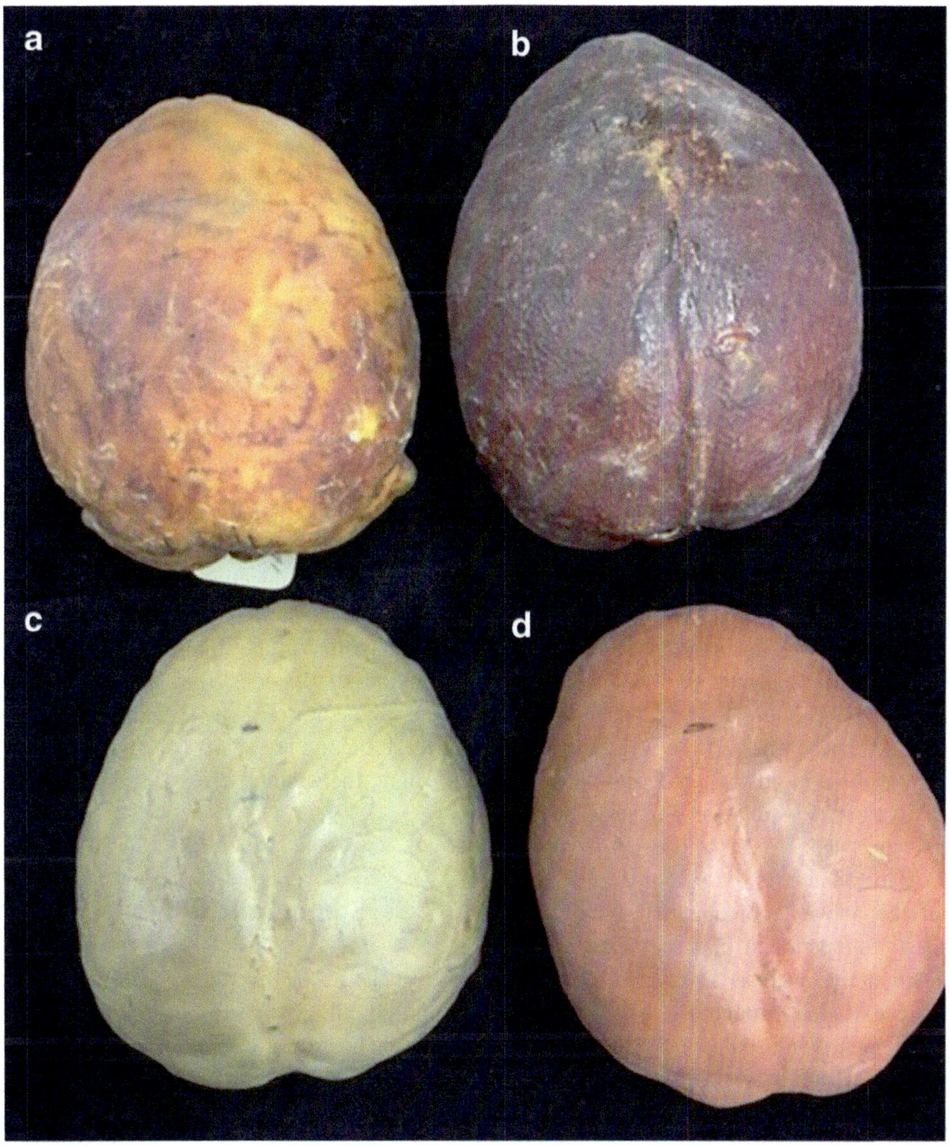

Fig. 1.1 Rubber latex endocasts showing various degrees of caramelization. (**a**, **b**) are gorillas; (**c**, **d**) are bonobos, the later showing pink coloration from adding red dye to Admold liquid latex

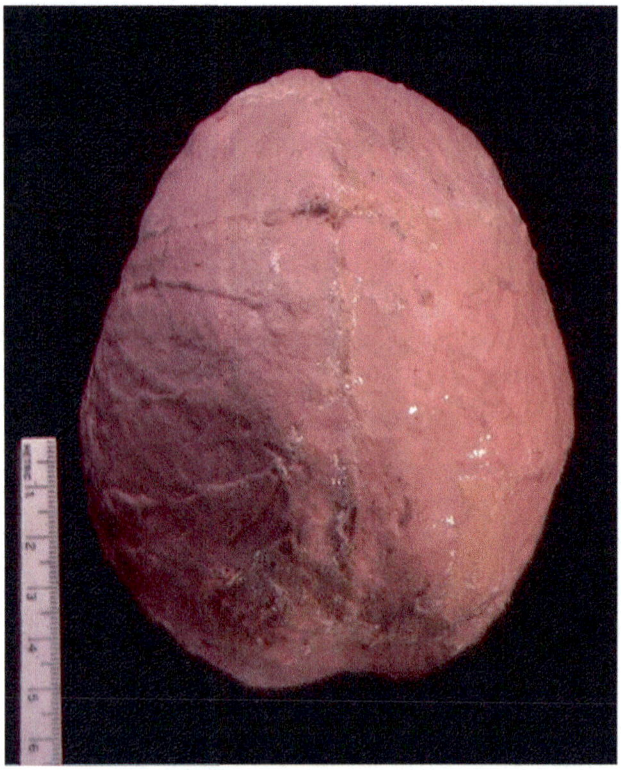

Fig. 1.2 The original KNM-ER 1470 *Homo rudolfensis* I made in Nairobi

Fig. 1.3 Ralph Holloway making an endocast using Dentsply

4. Various silicon casting products, e.g., Xantopren, became the standard way of making excellent casts of any of the bony elements of hominins during the 1960s and 1970s and are still used today. These molds were difficult to make and required considerable skill in making two halves tethered in plasticine and ending up with as small a flash line as possible. I used this method on a few of the australopithecines, such as Taung and SK 1585. The tensile strength was poor, compared to latex rubber, but the details were extremely fine as they were with latex. I still have some of those molds which do not deteriorate as does latex. Most of the wonderful Wenner-Gren casts were done this way, thanks to the skill of my friend and colleague, Dr. Alan Mann.
5. A variation of the above technique that I used when making endocasts on sectioned materials was to use a dental molding material such as Dentsply Aquasil LV Caulk, which was extruded through a gun that combined two compounds which would cure in 5 or less minutes (Fig. 1.3). This approach is wonderful on sectioned crania, or cranial fragments, but the casts have no tensile strength and, on modern human crania, require some skill in getting a thin flash line when the two halves are joined together and must be thick enough to avoid distortion. A small portion of two compounds, SmoothOn 320 A and B, mixed, is introduced through the foramen magnum, and the endocast rotated around so that the viscous compound would coat the entire endocast as it hardened while curing. The details of the endocranial surface are superb. The shelf life of these endocasts is unknown, but far longer than any of the latex rubber endocasts. Besides, they are usually a very pretty green or blue color (Fig. 1.4). Additional tools essential to such cast making are sharp scalpels to remove excess material along the flash line.

1.2 Some Concluding Remarks

All of the above five methods are "old," and each one has some potential to alter the bony surface, whether fossil or recent. The "new" refers, of course, to the use of CT, laser, and micro-CT scanning of the original fossil or specimen, and these methods are totally nondestructive. What results are a large number of scan sections, often at 0.5 mm intervals or lower. Obviously the quality of endocranial details will depend on the initial state of the fossil, the interval distance between slices, and the software package used to produce a "virtual" endocast. An immediate advantage is that not only

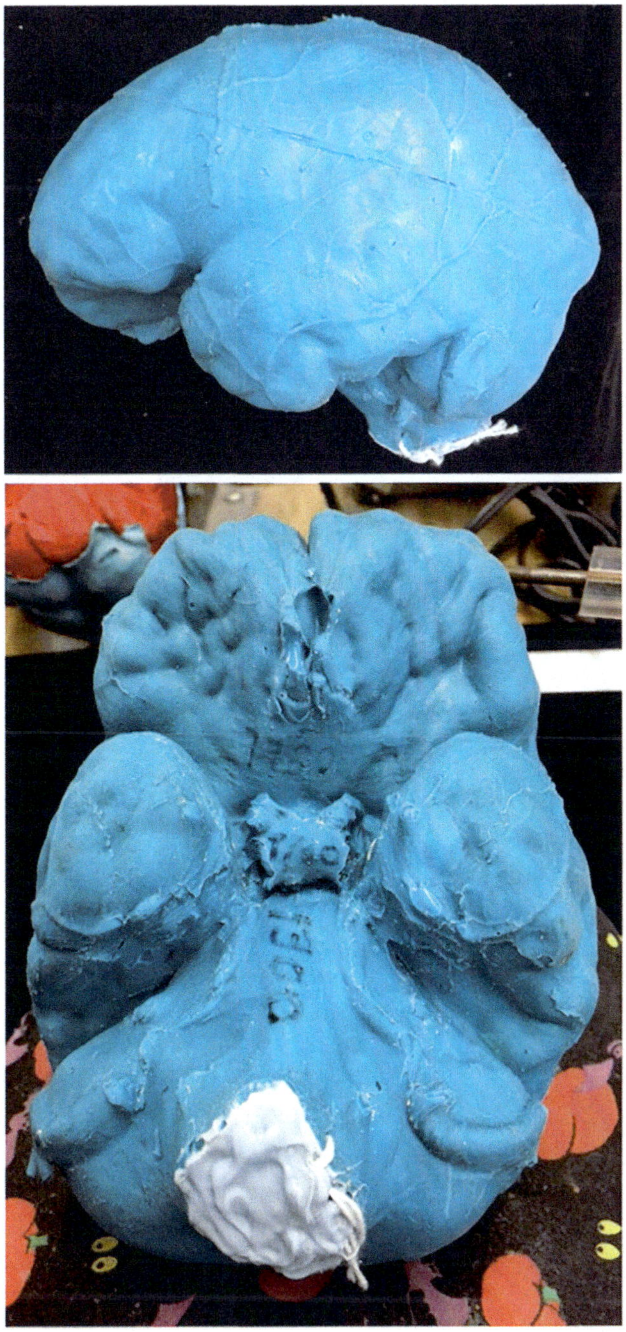

Fig. 1.4 *Above*: modern human endocast made with Dentsply (note flash line through the calva). *Below:* basal view of same modern human endocast

is the original fossil not damaged, nor mailed around various continents: the digital record is there, permanent, easy to upload and download to whoever might want to study the "virtual" result. The LB1 *Homo floresiensis* is a good example (Falk et al. 2005) or some of the recent works by Carlson et al. (2011), Neubauer et al. (2012), Gunz et al. (2009), Weber et al. (2012), Bruner and Manzi (2005), Zollikofer and Ponce de León (2013), etc. The *Homo naledi* fragments are available on MorphoSource.

Indeed, many of the endocasts made through methods 1–5 can themselves be scanned and be available as a repository of digital "virtual" endocasts. Almost all of the 200+ latex rubber endocasts of anthropoids I made during the 1970s have been scanned and are available through ORSA at the University of Pennsylvania, collected under the auspices of Drs. Janet Monge and Tom Schoenemann. This is an important process, as one of the problems with latex rubber is its gradual degradation or caramelization. The 80+ endocasts of modern humans (Figs. 1.5 and 1.6) I made during the last decade have also been scanned. Many of the plaster and plastic endocasts of hominins from my collection have also been scanned.

Interestingly, the various methods used to derive the volume of endocasts are not a constant. A remarkable collection of museum crania collected by Dr. Lynn Copes (2012) has been segmented by me, and the volumes derived (using Analyze 11) are often at variance with the recorded seed or shot volumes previously recorded, which are almost always higher in volume that those derived from either water displacement, CT, or laser scans, the latter two methods yielding volumes based on voxel counts. The last three methods yield only minor differences. Water displacement is tricky, in that the rubber or silicon has some degree of hydrophilia, and it is not uncommon for a rubber endocast to increase in volume by very small amounts as the number of immersions advances. This not always a good method for klutzes… Of course, while hitting that button in the software package saying "volume" is so convenient, it would be wise to remember that counting voxels is only as accurate as the initial segmentation that was done.

The new techniques of making virtual endocasts include algorithms for obtaining volumes, allow for measuring between points defined on the virtual surface, permit free rotation for both viewing and measurements, and also allow for correcting distortions, adding missing fragments, and reconstructing whole endocast portions based on sophisticated morphometric algorithms. One can even have some haptic experiences when one holds an actual endocast made from a 3-D printer using the CT scan data, but such experiences cannot match the haptic sensations with rubber or silicon, or even plaster endocasts, where the resolution is perfect.

Newer than "new" are my present experiences with working on the endocranial remains of *Homo naledi* (Berger et al. 2015). Here the authors have provided the entire world with the opportunity to freely download the CT sections of many of the remains for their analyses. When I did so (first I asked permission, after all I am "old school"), I sent the files over to my colleague Will Vanti in the library to print on their 3-D machine. What came back were brilliant red piece of plastic of both ecto- and endocranial surfaces (Fig. 1.7). Using Dentsply (see above), I made endocasts of DH1, DH3, and DH1. The first set, at 150 μm, didn't show any details I felt I

Fig. 1.5 Modern human crania and their endocasts

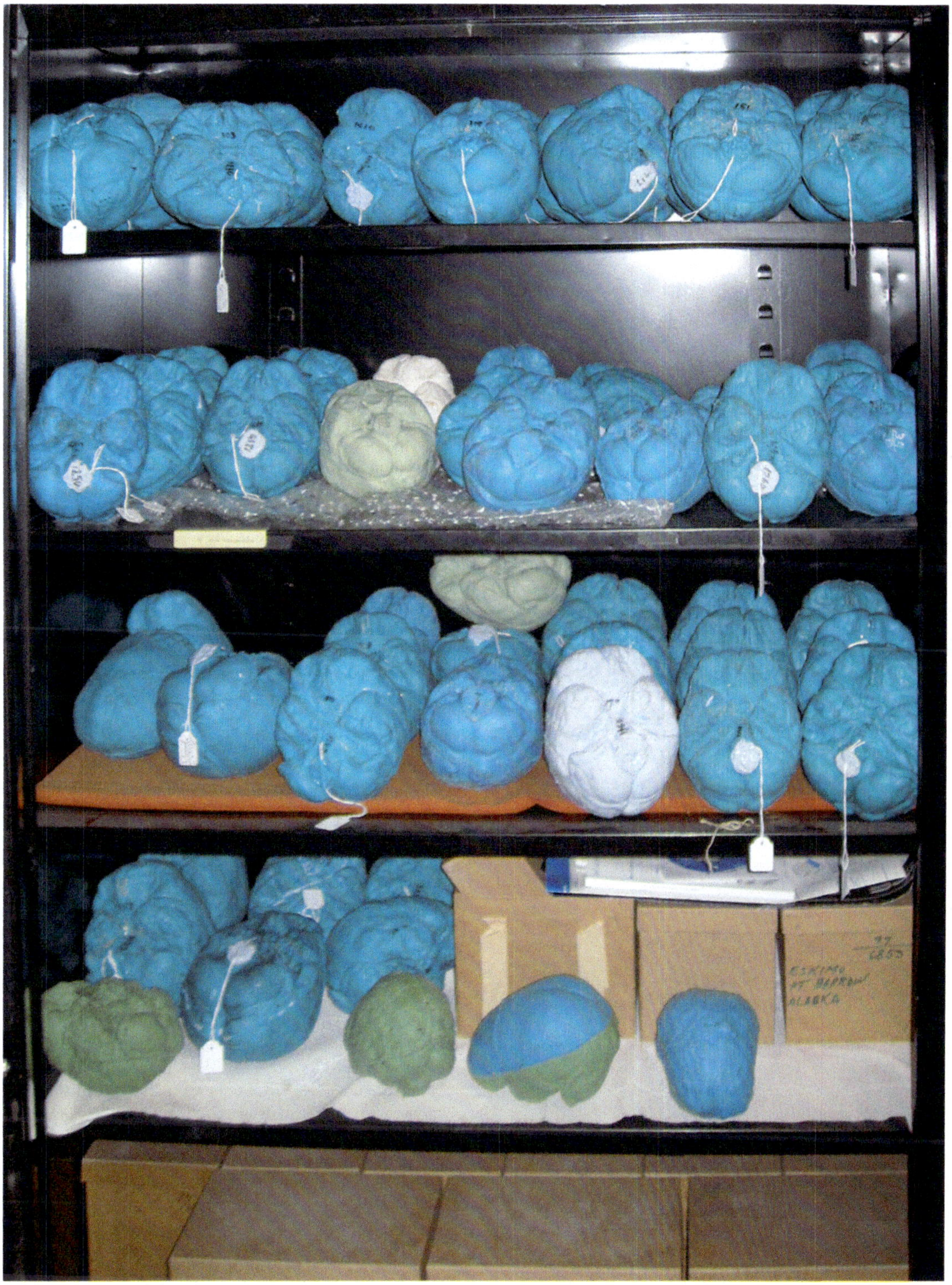

Fig. 1.6 File cabinet showing 80 modern human endocasts and microcephalics (*bottom row*)

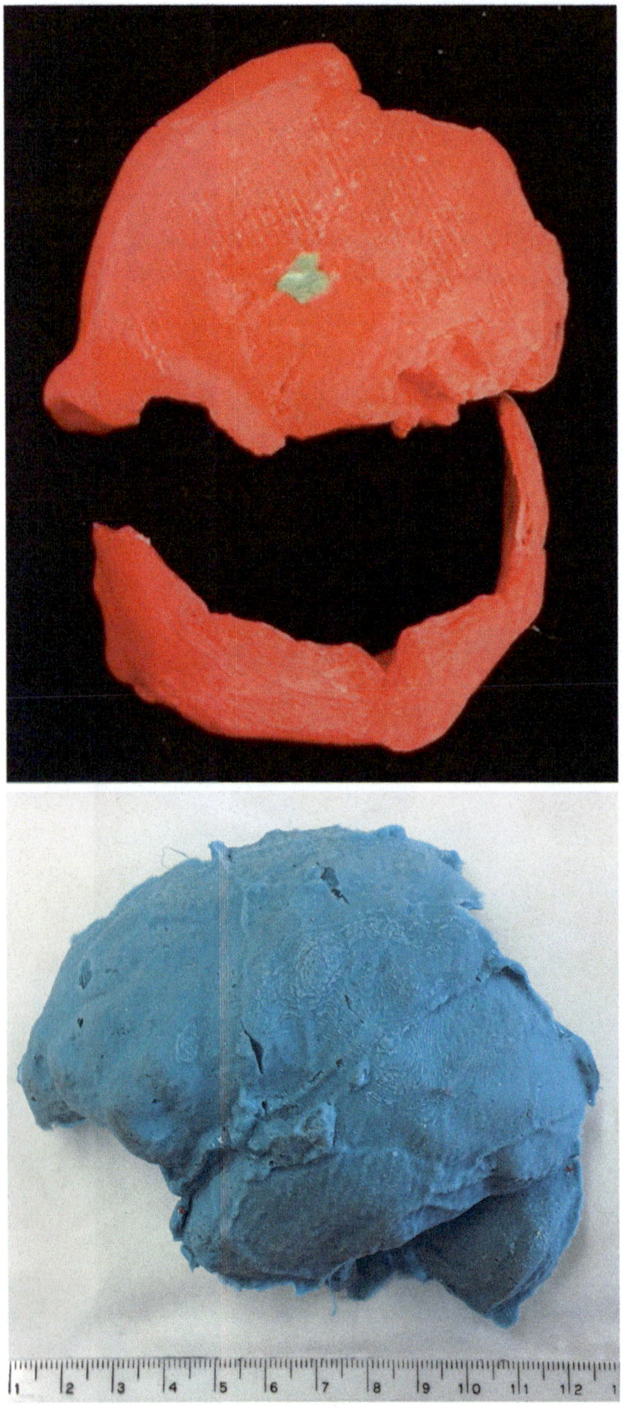

Fig. 1.7 *Above*: 3-D prints of some of the *Homo naledi* fragments from MorphoSource scans. *Upper* is frontal fragment, *bottom* is occipital fragment. *Below*: 3-D print of frontal endocranial surface of *Homo naledi*

could trust, so I asked Will if we could go to a finer resolution. Limited at 100 μm, the endocasts began to show some sulcal relief. I contacted the senior authors of the eLife paper and asked for help. Most importantly, I am working with another author, Dr. Heather Garvin, whose skills at illustrating this piece in different angles, with varied lighting, have made it possible to identify many endocranial features.

References

Berger LR, Hawks J, de Ruiter DJ, Churchill SE, Schmid P et al (2015) *Homo naledi*, a new species of the genus *Homo* from the Dinaledi Chamber, South Africa. elife 4:e09560

Bruner E, Manzi G (2005) CT-based description and phyletic evaluation of the archaic human calvarium from Ceprano, Italy. Anat Rec 285A:643–658

Carlson KJ, Stout D, Jashashvili T, de Ruiter DJ, Tafforeau P et al (2011) The endocast of of MH1, *Australopithecus sediba*. Science 333:1402–1407

Copes L (2012) Comparative and experimental investigations of cranial robusticity in Mid-Pleistocene Hominins. PhD dissertation, Arizona State University, 696 pages; 3505341

Edinger T (1929) Die fossilen Gehirne. Ergebneise Anat Entwicklungen Gesch 28:1–221

Edinger T (1949) Paleoneurology versus comparative brain anatomy. Confina Neurol Separatum 9:5–24

Edinger T (1975) Paleoneurology 1804–1966: an annotated bibliography. Springer-Verlag, Berlin

Falk D, Hildebolt C, Smith K, Morwood MJ, Sutikna T et al (2005) The brain of LB1, *Homo floresiensis*. Science 308:242–245

Gunz P, Mitteroecker P, Neubauer S, Weber GW, Bookstein FL (2009) Principles for the virtual reconstruction of hominin crania. J Hum Evol 57:48–62

Holloway RL (1964) Quantitative relations in the primate brain. PhD thesis. University of California, Berkeley, pp 174

Holloway RL (1966) Dendritic branching: some preliminary results of training and complexity in rat visual cortex. Brain Res 2:393–396

Holloway RL (1981) Exploring the dorsal surface of hominoid brain endocasts by stereoplotter and discriminant analysis. Phil Trans R Soc London B 292:155–166

Holloway RL (2008) The human brain evolving: a personal retrospective. Annu Rev Anthropol 37:1–19

Holloway RL (2014) Paleoneurology, resurgent! In: Bruner E (ed) Human paleoneurology. Springer Series in Bio/Neuroinformatics, Berlin, pp 1–11

Holloway RL, Broadfield DC, Yuan MS (2004) Brain endocasts: the paleoneurological evidence. Wiley, Hoboken

Neubauer S, Gunz P, Weber GW, Hublin JJ (2012) Endocranial volume of *Australopithecus africanus*: new CT-based estimates and the effects of missing data and small sample size. J Hum Evol 62: 498–510

Oyen OJ, Walker A (1977) Stereometric craniometry. Am J Phys Anthropol 46:177–182

Radinsky L (1967) The oldest primate endocast. Am J Phys Anthropol 27:385–388

Schoenemann PT, Gee J, Avants B, Holloway RL, Monge J, Lewis J (2007) Validation of plaster endocast morphology through 3D CT image analysis. Am J Phys Anthropol 132:188–192

Weber GW, Gunz P, Neubauer S, Mitteroecker P, Bookstein FL (2012) Digital South African fossils: morphological studies using reference-based reconstruction and electronic preparation. In: Reynolds SC, Gallagher A (eds) African genesis: perspectives on Hominin evolution. Cambridge, New York, pp 298–316

Zollikofer CPE, Ponce de León M (2013) Pandora's growing box: inferring the evolution and development of hominin brains from endocasts. Evol Anthropol 22:20–33

Digital Reconstruction of Neanderthal and Early *Homo sapiens* Endocasts

2

Naomichi Ogihara, Hideki Amano, Takeo Kikuchi, Yusuke Morita, Hiromasa Suzuki, and Osamu Kondo

Abstract

Endocranial morphology is currently the most useful source of information available for estimating the brain morphology and, hence, possible differences in cognitive ability in fossil hominins. Recently, computed tomography has been widely used to construct digital models of the endocranial cavity. With ongoing advances in computer-assisted morphological techniques, digital endocasts allow detailed analyses of morphological variability between hominin fossils and modern humans. This paper reviews digital reconstructions and morphological analyses of fossil endocasts and presents the digital reconstructions of complete endocasts of specimens of four Neanderthals and four early *Homo sapiens* based on CT scan data. Possible differences in the brain structure between Neanderthals and early *Homo sapiens* were identified based on a three-dimensional geometric morphometric analysis of the reconstructed endocasts. Our results demonstrated that ecto- and endocranial shapes are quantitatively different between Neanderthals and early *Homo sapiens*. The cranium of early *Homo sapiens* shows relative enlargement of the cerebellar region and relative expansion of the parietal area, possibly indicating that neuroanatomical organization is different between the two species.

Keywords

Fossil • Brain • Cerebellum • Geometric morphometrics

2.1 Introduction

Endocranial morphology is currently the most useful source of information available for estimating brain morphology and, hence, possible differences in cognitive ability in fossil hominins. Therefore, efforts have traditionally been made to construct casts from original fossil crania. Specifically, silicone rubber was poured onto the internal surface of fossil braincases through the foramen magnum to make a cast, and the extracted rubber cast was then filled with plaster to stabilize the shape of the cast. To analyze variation in morphology of the cranial cavity, linear dimensions were measured, and sulcus patterns were identified on the plaster endocasts (Holloway et al. 2004; Holloway 2008). However, although great care was taken to construct plaster endocasts, considerable deformation occurs, and errors of about 2 mm reportedly exist on the overall endocranial surfaces of plaster endocasts (Schoenemann et al. 2007).

Recently, the use of X-ray computed tomography (CT) for morphological analyses of fossil materials has become more widespread. This technique is now one of the most widely used methods to acquire and analyze the

N. Ogihara (✉) • H. Amano • T. Kikuchi • Y. Morita
Department of Mechanical Engineering, Faculty of Science and Technology, Keio University, Yokohama 223-8522, Japan
e-mail: ogihara@mech.keio.ac.jp; hideki_amano_0307@yahoo.co.jp

H. Suzuki
Department of Precision Engineering, Graduate School of Engineering, University of Tokyo, Tokyo 113-8656, Japan
e-mail: suzuki@den.t.u-tokyo.ac.jp

O. Kondo
Department of Biological Sciences, Graduate School of Science, University of Tokyo, Tokyo 113-0033, Japan
e-mail: kondo-o@bs.s.u-tokyo.ac.jp

E. Bruner et al. (eds.), *Digital Endocasts*, Replacement of Neanderthals by Modern Humans Series,
DOI 10.1007/978-4-431-56582-6_2

morphology of fossil specimens in the field of physical anthropology (Zollikofer and Ponce de León 2005; Gunz et al. 2009; Weber and Bookstein 2011; Ogihara et al. 2015). Using CT, the endocranial surface can also be determined, allowing construction of three-dimensional (3D) virtual models of the endocranial surfaces without damaging the original specimen. The spatial resolution of medical CT is about 0.3 mm, much smaller than the overall error of conventional plaster endocasts. Furthermore, using digital modeling, glue and plaster can be removed from the original specimen to separate the fragments constituting the fossil cranium, allowing reassembly of these fragments (Kikuchi and Ogihara 2013). Missing regions of the reassembled cranium can be geometrically or statistically interpolated. If the reconstruction is conducted using a digital model, deformations can be corrected based on geometric processing technologies, such as spatial warping techniques (Ogihara et al. 2006; Gunz et al. 2009). Therefore, digital endocasts hold great promise for more precise morphological comparisons of endocranial surfaces among different species in the human lineage.

The first morphological study of the hominin endocranium using digital endocasts was published in 1990, when Conroy et al. (1990) reported the endocranial capacity of *Australopithecus africanus* (MLD37/38) from 3D reconstructed digital endocasts. Since then, assessment of endocranial capacity and morphology based on digital endocasts has become increasingly common for working toward understanding the evolution of the human brain (Conroy et al. 1998, 2000a, b; Seidler et al. 1997; Recheis et al. 1999; Tobias 2001; Neubauer et al. 2004, 2012; Coqueugniot et al. 2004; Balzeau et al. 2005, 2013; Falk et al. 2005; Falk and Clarke 2007; Wu et al. 2008; Berger et al. 2010; Carlson et al. 2011; Kranioti et al. 2011; Kubo et al. 2011; Benazzi et al. 2011, 2014; Neubauer 2014; Amano et al. 2015). Furthermore, more detailed analyses of morphological variability in the endocranial shape have recently been carried out due to ongoing advances in geometric morphometric techniques (Neubauer et al. 2009, 2010).

The present paper reviews digital reconstructions and morphological analyses of fossil endocasts. We also present digital reconstructions of endocasts of four Neanderthal and four early modern human crania. We then describe possible differences in the brain structure between Neanderthals and early modern humans that were identified based on a 3D geometric morphometric analysis of the reconstructed endocasts of the fossil crania to infer possible differences in cognitive ability in fossil hominins.

2.2 Digital Reconstruction of Endocasts

An X-ray CT scanner is essentially a 3D shape digitizing device that captures both the external and internal structures comprising a biological specimen. Therefore, CT is an ideal tool for studying 3D morphology of endocasts. Figure 2.1 shows the process of constructing a digital endocast using a CT scanner. The first step is to obtain CT scan data of the original cranium. From a series of consecutive cross-sectional images of the specimen, the bony object region is segmented by thresholding, and its 3D isosurface is generated as a triangular mesh model using a computer graphics algorithm, such as the marching cubes algorithm (Fig. 2.1).

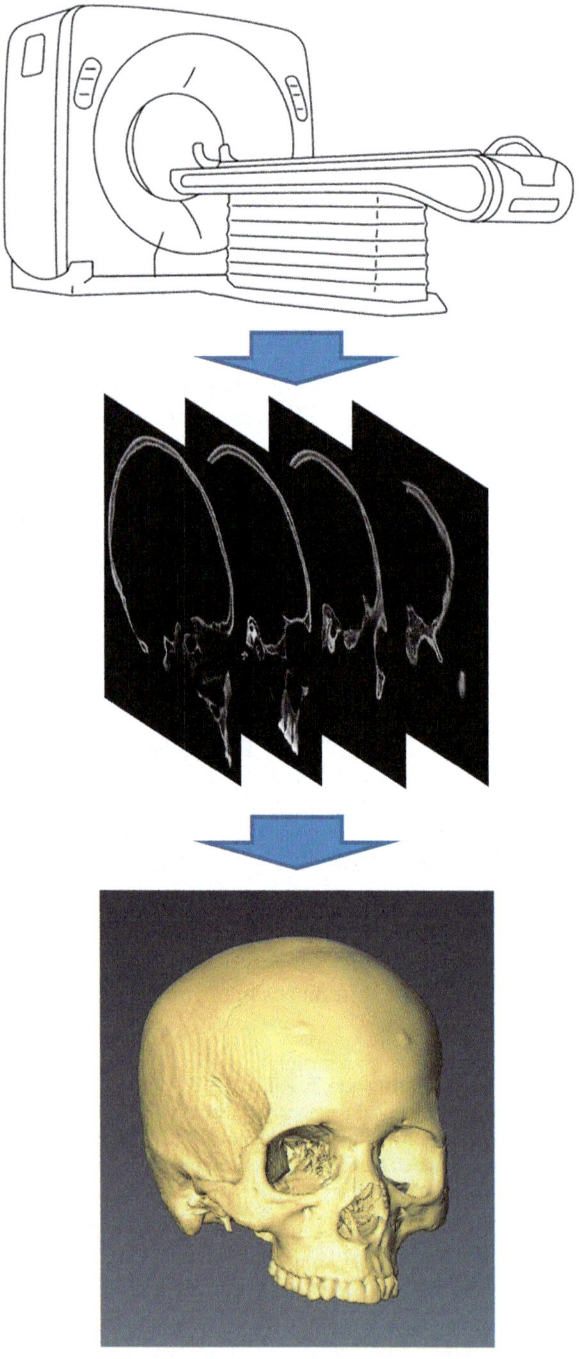

Fig. 2.1 Process of reconstructing a digital model of a cranium using an X-ray CT scanner (Ogihara et al. 2015) (Reprinted with the permission from the Anthropological Society of Nippon)

To create a digital endocast, the external surface of the cranium should be removed. For this, the external neurocranial surface is first selected using a paintbrush tool (Fig. 2.2a), and the selected surface is deleted. If the cranium is viewed from above, the internal surface of the cranial base is visible because the internal neurocranial surface facing inferiorly is invisible (transparent) because the surface facing inward is viewed from the back (Fig. 2.2b). Therefore, the internal surface including the basicranial surface can be entirely selected by the paintbrush tool. By selecting and deleting the inverse of the selected endocranial surface, the complete endocranial surface can be selected (Fig. 2.2c).

a b c d

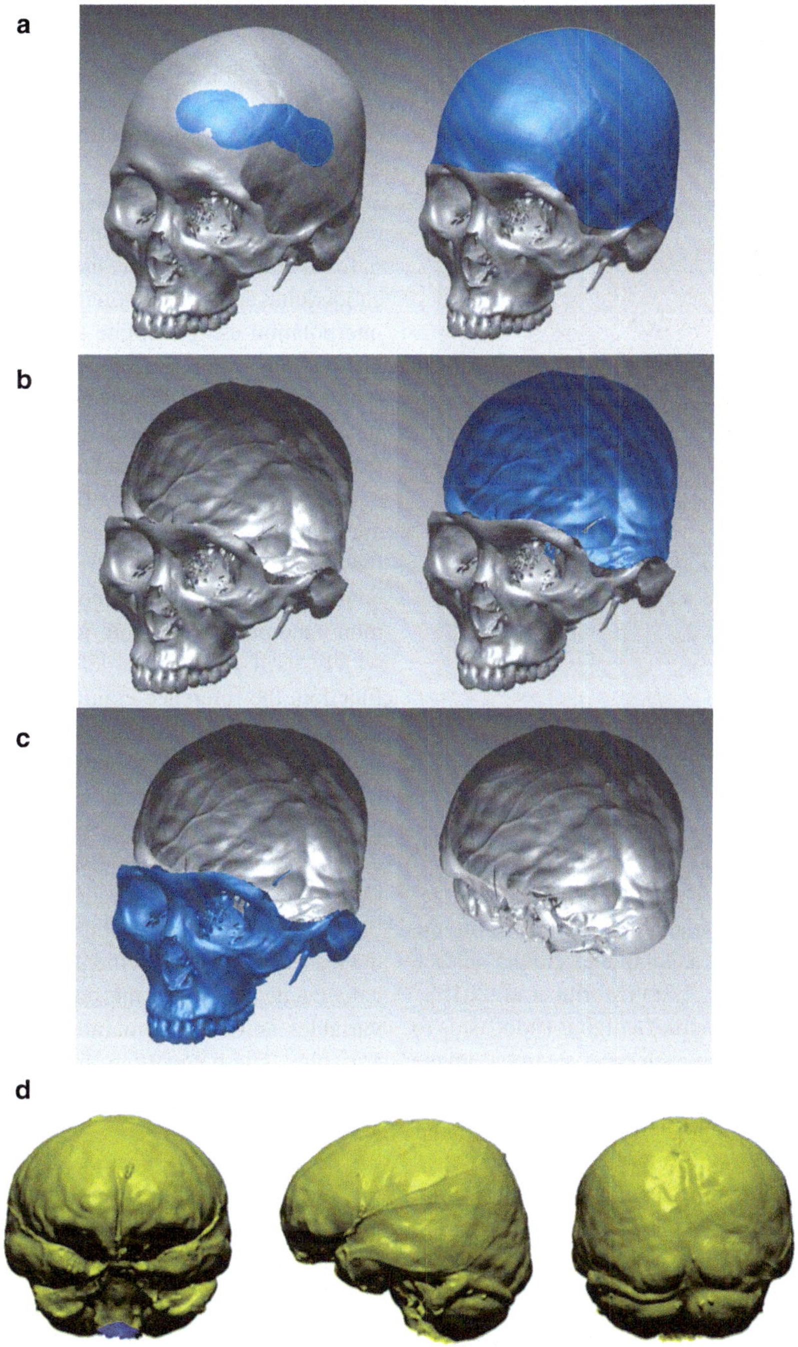

Fig. 2.2 Removal of the external surface of a cranium. (**a**) A paintbrush tool is used to select the external neurocranial surface. (**b**) The selected external surface is removed, and the entire internal surface is selected by the paintbrush tool. Note that the internal cranial surface is facing inward. Therefore, the concave surface of the occipital region is visible. (**c**) By selecting the inverse of the selected endocranial surface, the endocranial surface can be selected. (**d**) Holes on the surfaces such as the foramen magnum and neural foramina are filled using the fill-hole command, and normal vectors of the surface mesh triangles are flipped to the opposite direction to generate a closed surface model of the endocranium

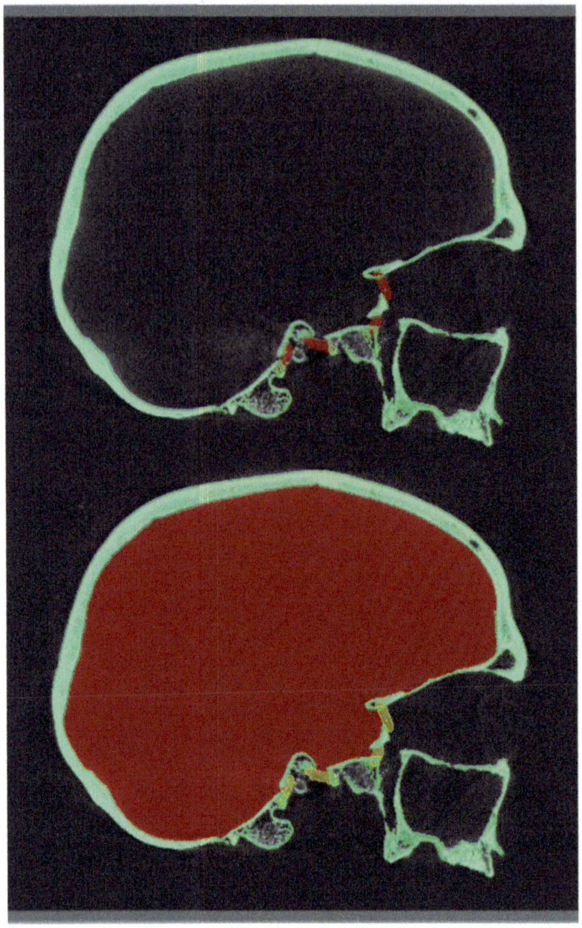

Fig. 2.3 Extraction of an endocranial cavity on a cross-sectional image using a region-growing algorithm

Lastly, holes on surfaces such as the foramen magnum and neural foramina are filled using a fill-hole command. Normal vectors of the surface mesh triangles are flipped to the opposite direction to generate a complete, closed surface model of the endocranium (Fig. 2.2d) (Morita et al. 2015).

Another possible way to construct a digital endocast is to extract the endocranial cavity on each cross-sectional image using the so-called region-growing algorithm (Fig. 2.3). Specifically, an initial seed is assigned in the endocranial cavity of each image. Then, the region is expanded until the region reaches the edge that is determined by thresholding prior to the region growing. To do so, openings due to foramina and nerve canals should be manually closed by drawing lines before beginning the digital reconstruction. This process is repeated for all consecutive cross-sectional images, and a 3D surface of the segmented volume is generated to create an endocast (Kubo et al. 2011).

These manual reconstructions of digital endocasts are, however, time-consuming and require patience. Therefore, efforts have also recently been made to computationally extract an endocast surface from a stack of CT images (Michikawa et al. 2017). In this extraction method, the seed is placed, and foramina and canals are closed automatically, with the assumption that the endocast is the largest cavity in the images. Although it takes hours to manually create a cranial endocast, the automatic method requires less than 10 min, hopefully facilitating morphological studies of endocasts. The automatically and manually constructed endocasts have been confirmed to be identical (Michikawa et al. 2017).

However, cranial fossils are usually fragile and only partially preserved. Accurate interpolation of missing parts in fossil crania is therefore essential for correct estimation of endocranial and, thus, brain morphology. For this, geometric interpolation using a spline function and statistical interpolation using multivariate regression have been proposed (Gunz et al. 2009). Geometric interpolation using a spline function interpolates a missing part based on data mapped from a complete reference specimen (Fig. 2.4). Specifically, common existing anatomical landmarks and semi-landmarks are digitized on the reference. Then, a deficient cranium and the deformation function from the reference to the target damaged cranium are defined based on the digitized common landmarks. The thin-plate spline (TPS) function is widely used for such a deformation function. Using this function, the reference cranium is matched to the damaged cranium to compensate for its missing parts. If many reference samples are used for interpolation, the degree of uncertainty in interpolation can also be evaluated (Gunz et al. 2009).

On the other hand, statistical interpolation is based on multivariate regression estimates of missing coordinates based on a sample of complete specimens as a reference database (Fig. 2.5). Specifically, multivariate regressions are calculated with the missing coordinates as dependent variables and other remaining coordinates as independent variables. These equations are then applied to predict missing cranial parts. For example, Amano et al. (2014) attempted to mathematically interpolate missing coordinates of crania based on a reference database of cranial morphology and successfully demonstrated the efficacy of the interpolation method (Fig. 2.5). However, estimation of missing landmarks on the basicranial region is reportedly difficult, possibly due to the low correlation between the shape of the basicranium and the rest of the cranium. See Gunz et al. (2009) and Ogihara et al. (2015) for more details about the interpolation methods.

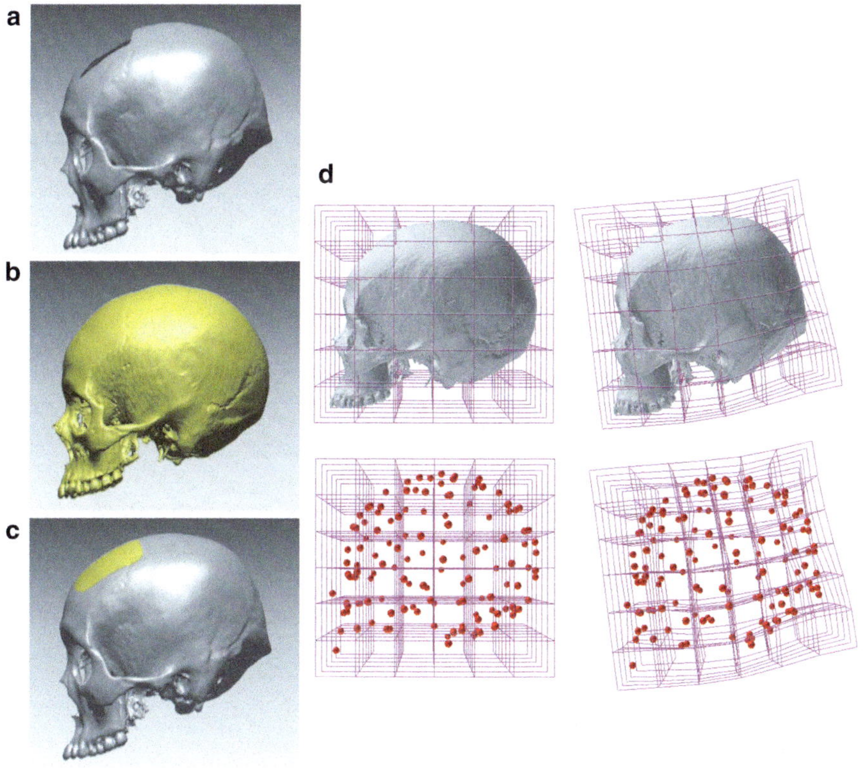

Fig. 2.4 Geometric interpolation using a thin-plate spline (TPS) function. (**a**) A deficient cranium with a missing region. (**b**) A complete reference cranium. (**c**) The missing portion of the deficient cranium is interpolated by warping the complete reference cranium. (**d**) The TPS function is widely used as a deformation function. Common existing anatomical landmarks and semi-landmarks are digitized on the reference and deficient cranium. The deformation function from the reference to the target damaged cranium is defined based on the digitized common landmarks

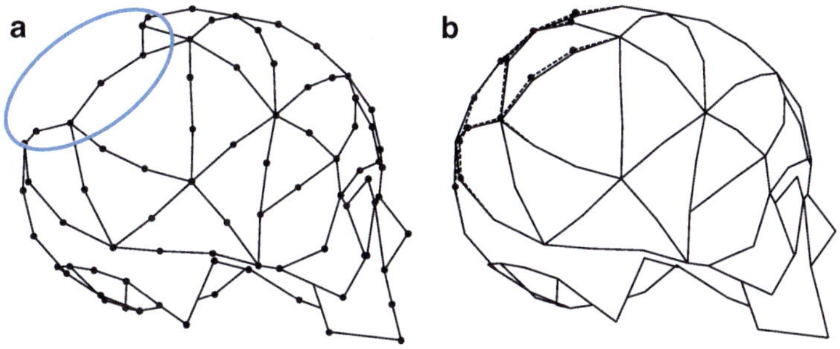

Fig. 2.5 Statistical interpolation based on multivariate regression. Coordinates of missing landmarks on a virtual deficient cranium (**a**) are estimated by calculating multivariate regressions with the missing coordinates as dependent variables and the other remaining coordinates as independent variables (**b**) (Amano et al. 2014)

2.3 Endocasts of Neanderthals and Early *Homo sapiens*

Using the above techniques, we performed digital reconstruction of digital endocasts of specimens of four Neanderthals and four early *Homo sapiens* as shown in Fig. 2.6. The four Neanderthals are Amud 1 (Suzuki and Takai 1970) (dated 50,000–70,000 years old; Valladas et al. 1999; Rink et al. 2001), Forbes' Quarry 1 (Busk 1865) (no dating information), La Chapelle-aux-Saints 1 (Boule 1908; Bouyssonie et al. 1909) (dated 47,000–56,000 years old; Grün and Stringer 1991), and La Ferrassie 1 (Capitan and Peyrony 1909) (dated 43,000–45,000 years old; Guerin et al. 2015). The four early *Homo sapiens* are Cro-Magnon 1 (Lartet 1868; Broca 1868) (dated 28,000 years old; Henry-Gambier 2002), Mladeč 1 (Szombathy 1925) (dated 31,000 years old; Wild et al. 2005), Qafzeh 9 (Vandermeersch

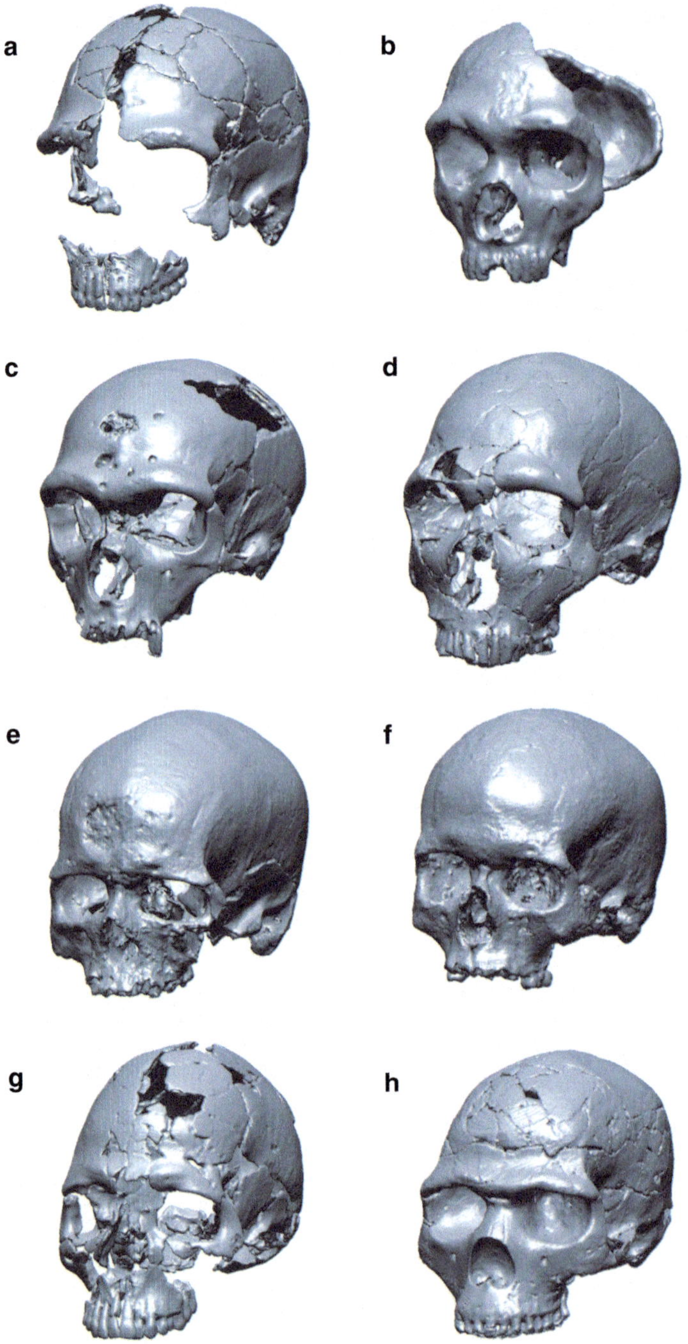

Fig. 2.6 Fossil crania of specimens of Neanderthals (**a–d**) and early *Homo sapiens* (**e–h**) used in the present study. (**a**) Amud 1, (**b**) Forbes' Quarry 1, (**c**) La Chapelle-aux-Saints, (**d**) La Ferrassie 1, (**e**) Cro-Magnon 1, (**f**) Mladeč 1, (**g**) Qafzeh 9, (**h**) Skhul 5

1981) (dated 90,000–120,000 years old; Valladas et al. 1988; Schwarcz et al. 1988; Grün and Stringer 1991), and Skhul 5 (McCown and Keith 1939) (dated 100,000–135,000 years old; Mercier et al. 1993; Grün et al. 2005) (Fig. 2.6).

For Amud 1, we first digitally removed the adhesive and plaster from the original CT data and isolated and disassembled the original cranial fragments comprising the fossil based on segmentation procedures such as thresholding and region-growing techniques (Fig. 2.7). These fragments were then mathematically reassembled in a virtual environment based on joint smoothness (Kikuchi and Ogihara 2013). The missing facial, basicranial, and endocranial regions were geometrically interpolated using a composite Neanderthal cranium (La Chapelle-aux-Saints 1 cranium whose missing central basicranial areas were interpolated by matching the Forbes' Quarry 1 cranium

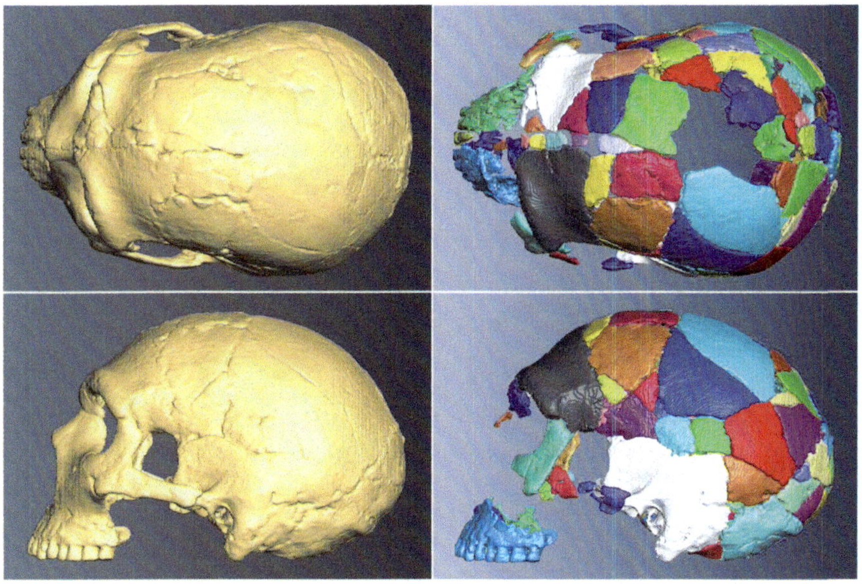

Fig. 2.7 Digital models of Amud 1 as originally reconstructed by Suzuki (1970) with (*left*) and without (*right*) plaster. The cranium is composed of numerous fragmented pieces, and substantial portions of the facial and basicranial regions are missing

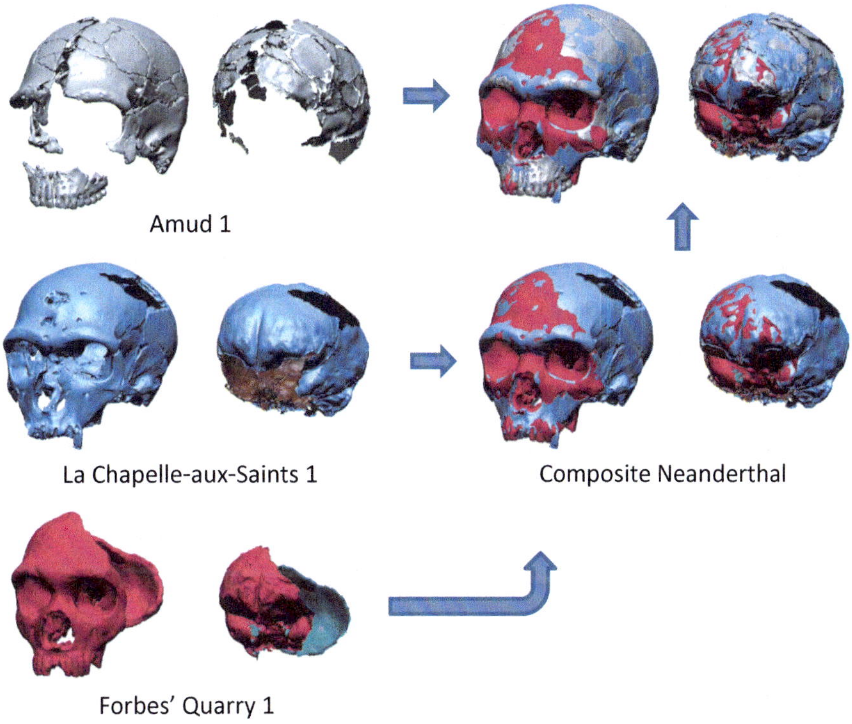

Fig. 2.8 Virtual reconstruction of the Amud 1 cranium (Amano et al. 2015) (Reprinted with the permission from John Wiley & Sons)

using a TPS function) as a reference cranium (Fig. 2.8). The remaining openings were compensated by matching a modern Japanese cranium (KUMA-554) using the TPS deformation, and the reconstruction was completed. Virtual reconstruction of the Amud 1 cranium is described in detail in Amano et al. (2015).

In the Forbes' Quarry 1 cranium, the basal region including the frontal lobe was preserved, but most of the left side was missing. The missing regions of the Forbes' Quarry 1 cranium were interpolated by warping the La Chapelle-aux-Saints 1 cranium (Fig. 2.9). The remaining openings were compensated by matching the modern Japanese cranium

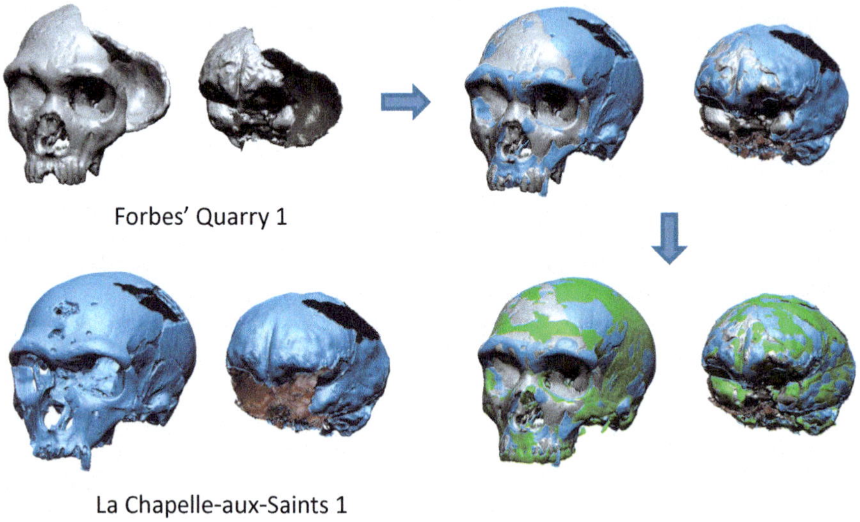

Fig. 2.9 Virtual reconstruction of the Forbes' Quarry 1 cranium

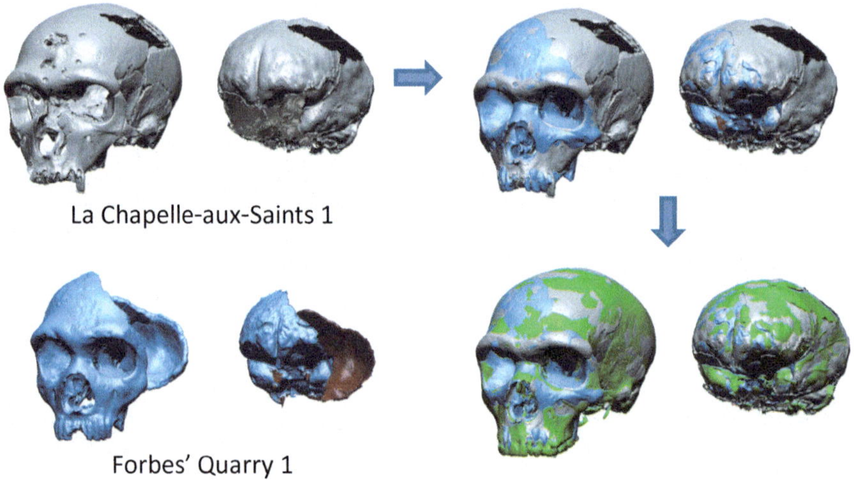

Fig. 2.10 Virtual reconstruction of the La Chapelle-aux-Saints 1 cranium

(KUMA-554) using the TPS deformation. The damaged portion of the skull was not reconstructed using the reflection of the opposite side because of possible cranial shape asymmetry.

The La Chapelle-aux-Saints 1 cranium was almost complete except for central basicranial areas. The missing basicranial region was interpolated by matching the Forbes' Quarry 1 cranium (Fig. 2.10). The remaining openings were compensated by matching the modern Japanese cranium (KUMA-554) using the TPS deformation.

In the La Ferrassie 1 cranium, the neurocranium and the occipital bone were preserved, but the anterior basal region was missing. The missing basicranial region was interpolated by matching the Forbes' Quarry 1 cranium (Fig. 2.11). The remaining openings were compensated by matching the modern Japanese cranium (KUMA-554) using the TPS deformation.

The fossil crania of the early *Homo sapiens* specimens were generally better preserved. We digitally removed the stone matrix and plaster where necessary and extracted well-preserved endocranial surfaces. For the Cro-Magnon 1, Qafzeh 9, and Skhul 5 crania, the modern Japanese cranium (KUMA-554) was matched onto the fossil endocasts to compensate for the missing surface areas to obtain complete endocranial surfaces (Fig. 2.12). The endocast of the Mladeč 1 is almost perfectly preserved except for a small deficit at the edge of the foramen magnum. We therefore did not use a reference cranium but rather used the fill-hole tool to compensate for the small missing surface.

To define a deformation function from one cranial specimen to another for interpolation, a set of homologous landmark coordinates that can be observed on both specimens must be obtained. For this, we acquired 62 anatomical

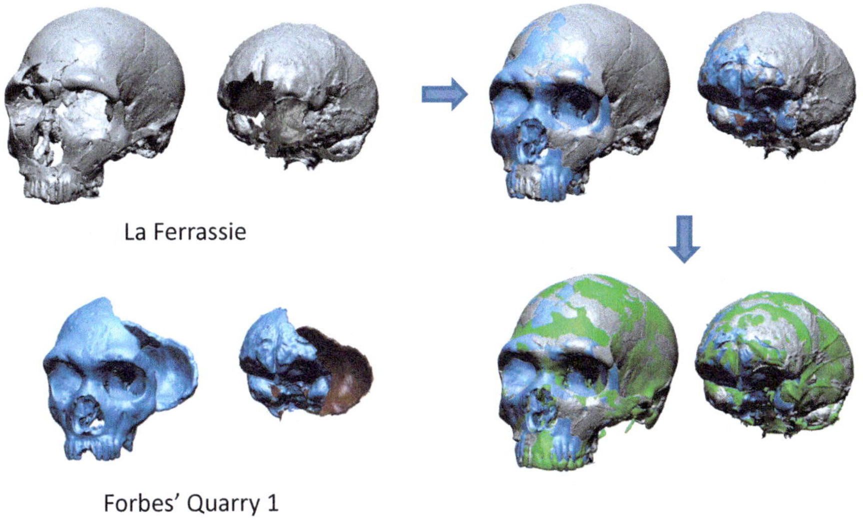

Fig. 2.11 Virtual reconstruction of the La Ferrassie 1 cranium

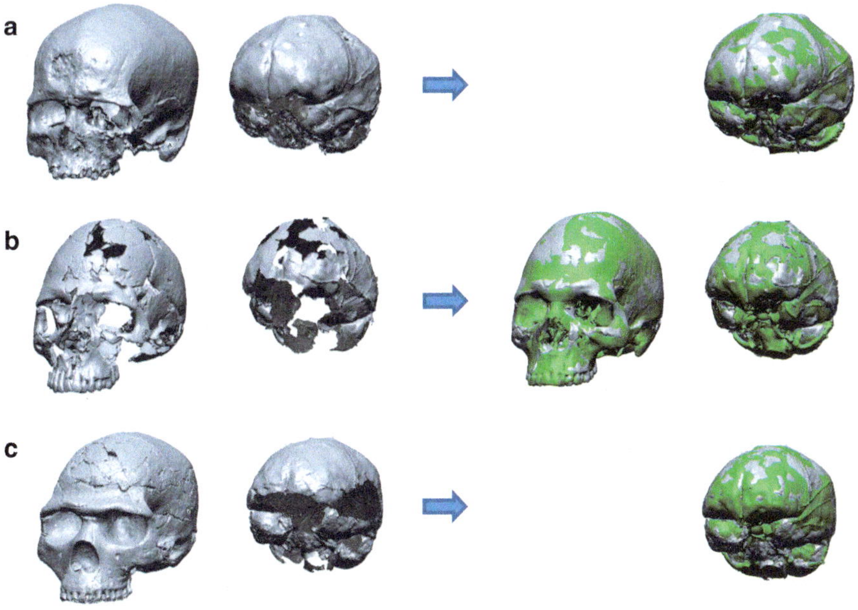

Fig. 2.12 Virtual reconstruction of the fossil crania of early *Homo sapiens*. (**a**) Cro-Magnon 1, (**b**) Qafzeh 9, (**c**) Skhul 5

landmarks on the external surface and 14 equally spaced points along curves approximated by Bazier functions (Morita et al. 2013) (Fig. 2.13). We also defined a total of 85 sliding semi-landmarks across the entire neurocranial surface based on the shortest paths between pairs of anatomical landmarks and equally spaced points along the curves (Morita et al. 2013). Similarly, we defined 30 anatomical landmarks on the endocranial surface and 22 equally spaced points along endocranial curves as well as 133 surface endocranial sliding semi-landmarks (Fig. 2.13). See Amano et al. (2015) for landmark definitions.

The 3D reconstructions of the digital endocasts are presented in Fig. 2.14 (see Appendix for the six-sided views). As shown in Fig. 2.14, endocranial surfaces of the four Neanderthal and four early *Homo sapiens* crania were successfully reconstructed in a virtual environment. The endocranial volumes of Neanderthals, Amud 1, Forbes' Quarry 1, La Chapelle-aux-Saints 1 and La Ferrassie 1, were 1736 cc, 1183 cc, 1512 cc and 1671 cc, respectively, and those of early Homo sapiens, Cro-Magnon 1, Mladeč 1, Qafzeh 9, and Skhul 5, were 1589 cc, 1596 cc, 1424 cc, and 1395 cc, respectively. Such virtual reconstruction of the complete geometry of the fossil

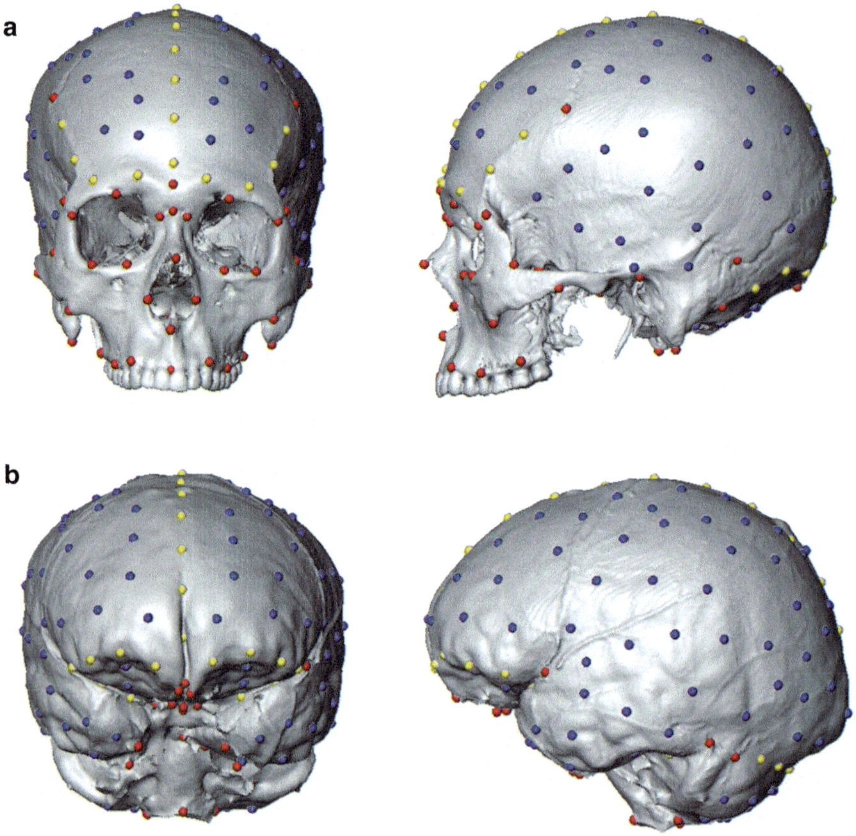

Fig. 2.13 Landmarks used to define thin-plate spline functions for geometric interpolation of the fossil crania (Amano et al. 2015). (**a**) Ectocranial landmarks. (**b**) Endocranial landmarks (Reprinted with the permission from John Wiley & Sons)

crania allows detailed comparative analysis of ecto- and endocranial morphology between the two species.

2.4 3D Morphometrics of Endocasts

Studies on endocasts have historically focused on endocranial volume (ECV), which can be used to approximate brain size (Falk 2012). Such studies have clearly demonstrated that the ECV of hominins has increased during the process of human evolution (Hublin et al. 2015). However, brain evolution and encephalization are not just a matter of size but also a matter of structure and organization. Therefore, researchers have tried to identify sulcus patterns on the extracted virtual endocasts (Holloway et al. 2004; Holloway 2008; Falk 2014). However, identifying cortical features such as imprints of sulci and gyri on the endocranial surface is actually very difficult. Although imprints of sulci and gyri extracted from crania are somewhat pronounced in nonhuman anthropoids, such as macaques (Kobayashi et al. 2014), and in human children (Zollikofer and Ponce de León 2013), such imprints are very subtle on the human adult cranium. Figure 2.15 shows a comparison of a modern human cranium and the brain enclosed in it. Here the CT and magnetic resonance images from one male participant were registered to each other to maximize mutual information between the CT and magnetic resonance images (Ogihara et al. 2015). Endocast and brain surfaces were then 3D reconstructed. As shown in Fig. 2.15, the sulcal patterns are generally not visible on the internal surface of the adult human cranium. The same is true for the adult chimpanzee cranium, although the imprints are quite prominent on the cranium of a juvenile chimpanzee (Fig. 2.16). Therefore, identification of cortical features and the relative size of brain regions from the fossil endocranial surfaces in Neanderthals and early *Homo sapiens* is currently quite difficult. However, the quality of imprints may be related to the spatial resolution of medical CT. Micro-CT may provide finer details about imprints than medical CT.

To quantitatively analyze the overall shape of the endocranial cavity, researchers have traditionally measured a set of linear metric variables taken from physical or virtual endocasts, such as maximum length, chords, and distances between two anatomical landmarks, and analyzed the difference in endocranial shape based on indices (ratios) or multivariate analyses (Falk et al. 2000, 2005; Broadfield et al. 2001; Balzeau et al. 2012, 2013). However, a set of linear measurements may have limited applicability in the analysis

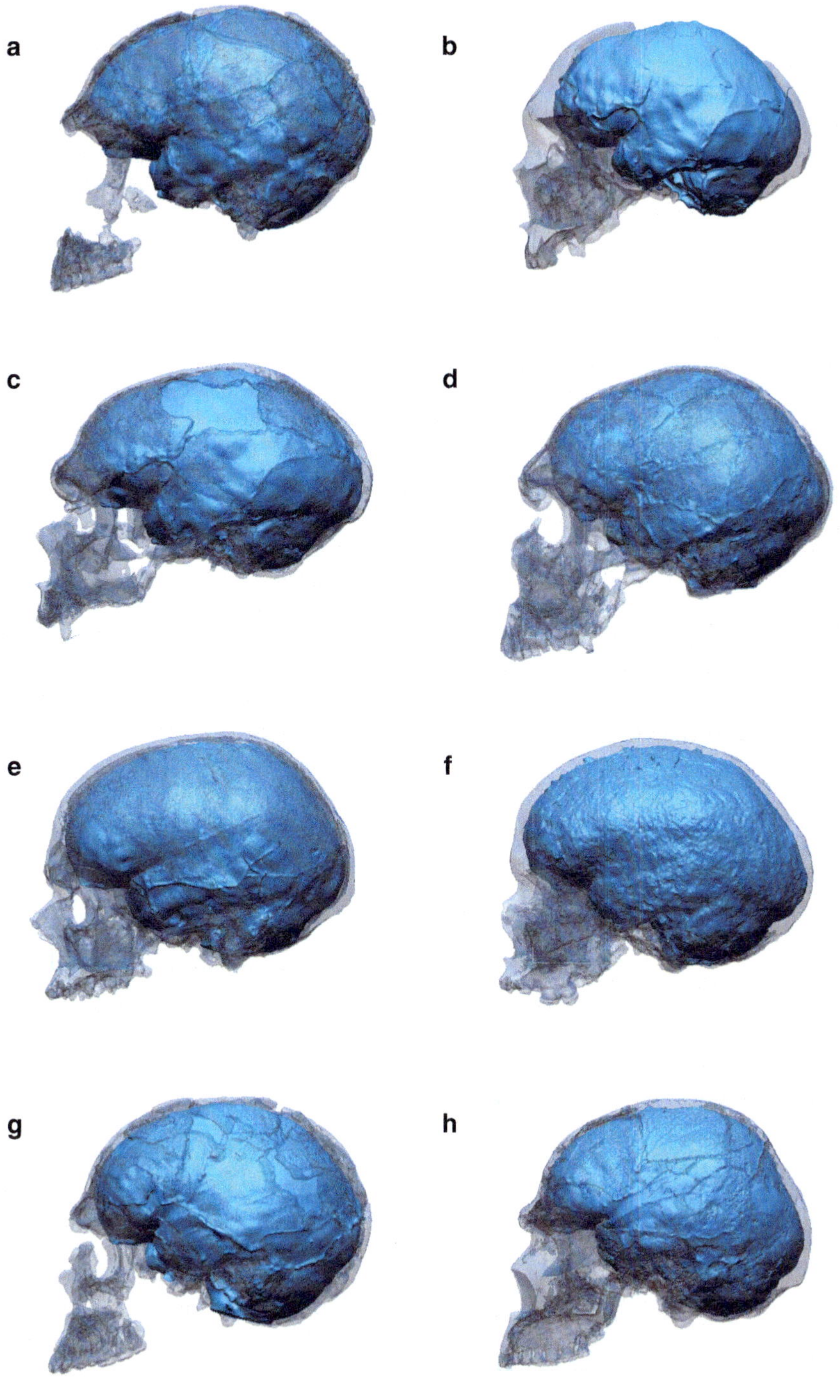

Fig. 2.14 Digital endocasts of specimens of Neanderthals (**a–d**) and early *Homo sapiens* (**e–h**). (**a**) Amud 1, (**b**) Forbes' Quarry 1, (**c**) La Chapelle-aux-Saints, (**d**) La Ferrassie 1, (**e**) Cro-Magnon 1, (**f**) Mladeč 1, (**g**) Qafzeh 9, (**h**) Skhul 5

of endocranial shape (Holloway 1981), because the overall spatial relationships of landmarks in each endocast are not preserved in the conventional multivariate analyses based on a set of linear measurements.

Thus, with a 3D geometric morphometric technique, a quantitative approach used to analyze shape variations based on landmark coordinates (Bookstein 1991; O'Higgins 2000; Adams et al. 2004; Slice 2005; Mitteroecker and Gunz 2009) was recently applied for quantitative comparisons of endocranial morphology. In these studies, homologous landmarks were digitized on the surface of each specimen, and landmark coordinates were normalized by centroid size for size-independent shape analysis. Landmark coordinates were then registered using the Procrustes method, and shape

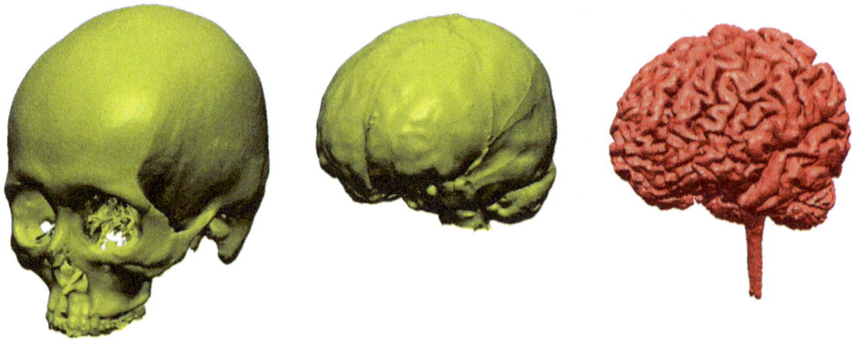

Fig. 2.15 Comparison of a human cranium and the brain enclosed in it (Ogihara et al. 2015). The CT and magnetic resonance images from one male participant were registered to each other to maximize mutual information between the CT and magnetic resonance images. Note that the sulcal patterns are not visible on the internal surface of the adult human cranium (Reprinted with the permission from the Anthropological Society of Nippon)

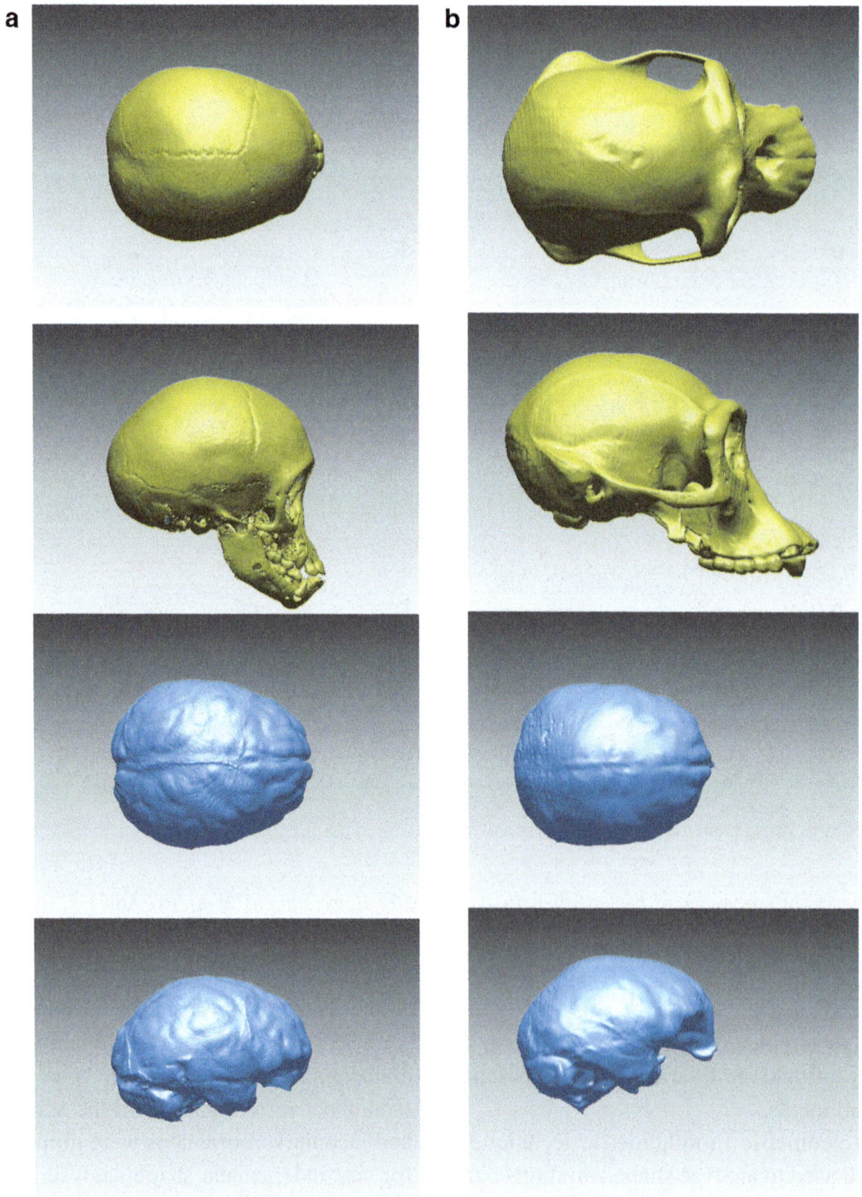

Fig. 2.16 Endocasts of a juvenile (**a**) and an adult female (**b**) chimpanzee. Note that imprints of sulci and gyri on the endocranial surface are somewhat prominent in the juvenile chimpanzee but not in the adult chimpanzee. The juvenile chimpanzee is a formalin-fixed specimen (JMC-3788), and the adult chimpanzee is a dry bone specimen (Musa), both housed at the Japan Monkey Center

variations among specimens were quantified based on differences (Procrustes residuals) in all specimens using multivariate statistical analyses such as principal component analysis (PCA). Using the geometric morphometric technique, Bruner et al. (2003) attempted for the first time to analyze the endocranial morphology of fossil hominins based on a set of landmark coordinates defined on the endocast. Morphological variations in sagittally projected endocasts (Bruner 2004) and 3D basicranial shape (anterior, middle, and posterior cranial fossae) (Bastir et al. 2008; Bruner and Ripani 2008) based on anatomical landmarks have also been investigated. Bienvenu et al. (2011) also analyzed 3D endocranial variations in humans and great apes based on 37 anatomical landmarks defined on the entire endocranium. However, only a few anatomical landmarks can be defined on the endocasts because the endocast has an ovoid shape. In addition, few characteristic morphological features necessary for defining landmarks can be found on its surface except for the basicranial surface. As a consequence, the landmarks used in those studies generally concentrated on the basicranium.

Therefore, the semi-landmark method has recently been used for analyses of endocranial vault morphology when limited definable landmarks are available (Bookstein 1997; Gunz et al. 2005, 2009). In this method, semi-landmarks on curves or surfaces are measured on a template specimen, projected onto all other specimens in the sample, and subsequently allowed to slide to minimize the TPS bending energy between each specimen and the mean shape. This method has recently been used to provide detailed descriptions of ontogenetic endocranial shape changes in humans (Neubauer et al. 2009) and to compare the growth trajectory of humans with that of Neanderthals (Gunz et al. 2010, 2012). The semi-landmark method was also applied by Bastir et al. (2011) to clarify detailed differences in basicranial morphology between Neanderthals and fossil and extant humans. However, the studies mentioned above are the only ones that have carried out detailed 3D shape analyses of the human endocranium using semi-landmarks to clarify the detailed morphological variability in extant or fossil hominins.

2.5 Cranial Shape Analysis of Neanderthals and Early *Homo sapiens*

Using the reconstructed fossil crania, we clarified the morphological variability of the ecto- and endocranial shapes using semi-landmark-based geometric morphometrics. For this, a total of 161 and 171 anatomical and semi-sliding landmarks were defined on the ecto- and endocranial surfaces of the fossil crania, respectively (Fig. 2.17).

On the ectocranial surface, a total of 62 anatomical landmarks were acquired. In addition, a total of 28 equally spaced points along the midsagittal curve, the nuchal line, the temporal line, and the supraorbital line were also defined as landmarks. Therefore, a total of 90 landmarks were identified as non-sliding landmarks. We defined semi-sliding landmarks on one specimen chosen as a template based on the shortest paths between pairs of anatomical landmarks (Morita et al. 2013). Along these paths, we obtained 71 equally spaced points, resulting in a total of 161 landmarks.

On the endocranial surface, a total of 16 anatomical landmarks were digitized. In addition, a total of 28 equally spaced points along the midsagittal curve, the anterior boundary of the anterior cranial fossa, and the lower border of the groove for the transverse sinus were also defined as landmarks (Morita et al. 2015). Therefore, a total of 44 landmarks were identified as non-sliding landmarks for each endocranium. We also defined a total of 127 semi-sliding landmarks on one specimen chosen as a template, resulting in a total of 171 landmarks.

The positions of all landmarks were symmetrized (Zollikofer and Ponce de León 2002) to eliminate possible asymmetric components of shape variation in the template specimen. In this study, Templand in the EVAN Toolbox (www.evan-society.org) was used to calculate the location of the sliding landmarks on each endocast.

The ecto- and endocranial shapes represented by the locations of the non-sliding and sliding landmarks were then analyzed independently using Morphologika geometric morphometric software version 2.5 (O'Higgins and Jones 1998). First, the landmark coordinates were normalized by centroid size for size-independent shape analysis, and the size-normalized endocasts were then superimposed using the method of least squares (generalized Procrustes analysis). Principal components (PCs) of endocranial shape variations among the specimens were then computed based on the variance-covariance matrix of the Procrustes residuals of all specimens used in the present study (O'Higgins and Jones 1998; O'Higgins 2000).

In addition to the four Neanderthal and four early *Homo sapiens* crania, we added to our investigation a total of 56 crania (23 female and 33 male) from the modern Japanese population housed at the Laboratory of Physical Anthropology, Kyoto University (Morita et al. 2014, 2015). We also added a total of 17 and 6 crania (sex unknown) from the modern European (France 2, Germany 4, and Czech 11) and Indian populations, respectively, housed at the University Museum, the University of Tokyo, for comparisons.

The results of PCA concerning morphological variability in the ectocranial surface of the Neanderthal and anatomically modern humans are presented in Fig. 2.18 as

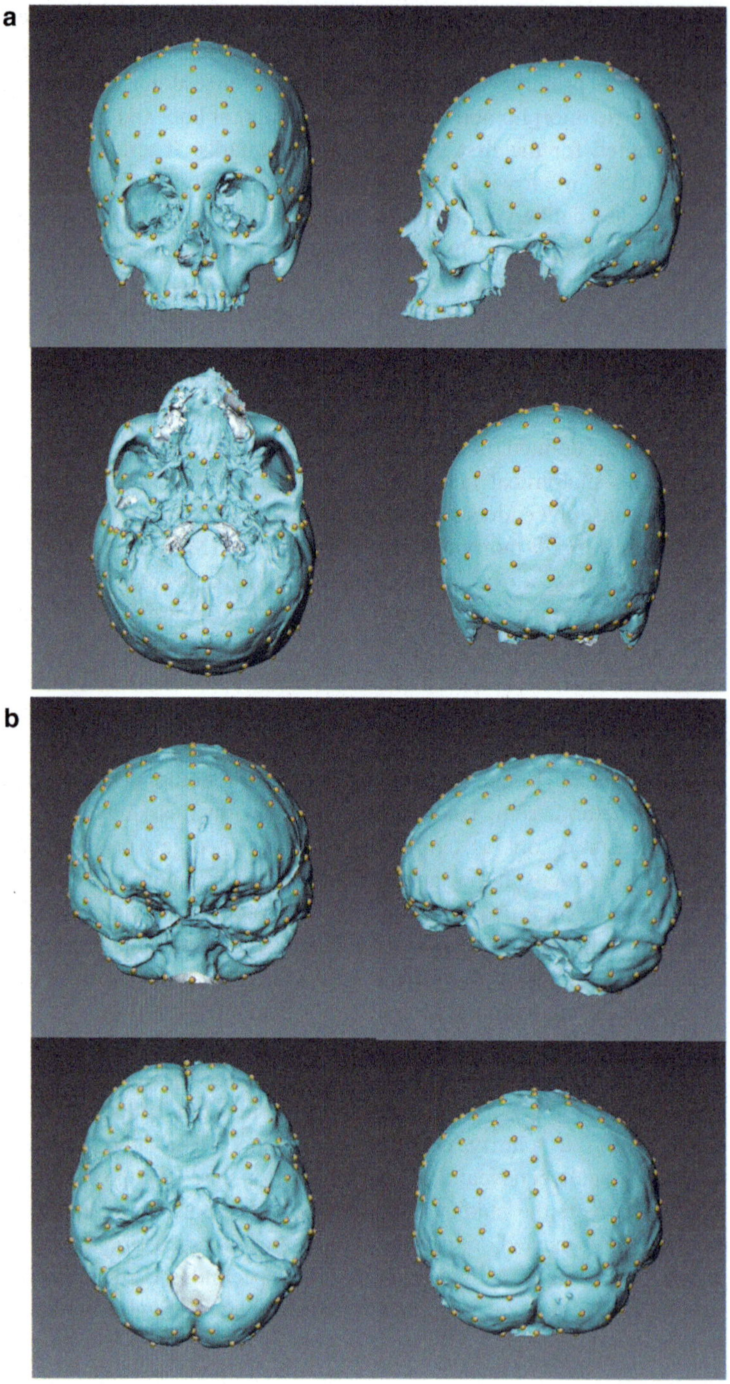

Fig. 2.17 Landmarks used for cranial shape analysis (**a**) Ectocranial landmarks. (**b**) Endocranial landmarks

a plot of the first principal component (PC1) versus the second principal component (PC2). The first two PCs accounted for 34.6% of the variation (21.3% and 13.3% for PC1 and PC2, respectively). In the present study, only the first two PCs were considered to be dominant because no clear separations among the Neanderthal, early *Homo sapiens*, modern Japanese, and modern European were observed in the remaining components.

The pattern of shape variations in Fig. 2.18 indicates that scores of PC1 were relatively higher in the Neanderthal (A, G, L, F) than in anatomically modern humans (C, M, Q, S). Thus, the fossil crania are clearly separated from the modern human crania along PC1.

Figure 2.19 shows the 3D ectocranial shape variabilities along PC1 by warping the cranial shape represented by the wireframe that connected the landmarks. With a decrease in

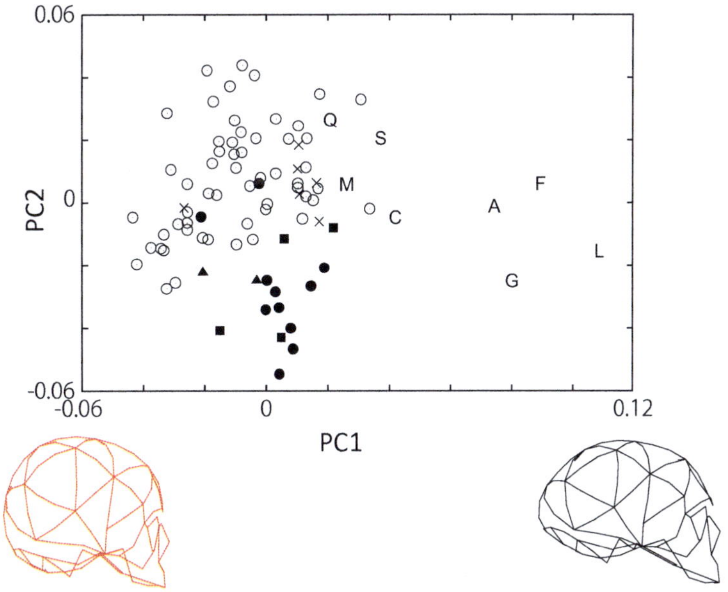

Fig. 2.18 The results of principal component (PC) analysis of ectocranial shape variation. PC1 (x-axis) versus PC2 (y-axis). *White circles* = modern Japanese; *black circles*, *squares*, and *triangles* = modern European (Czech, Germany, and France, respectively); crosses = modern Indian; *A* Amud 1, *G* Forbes' Quarry 1, *L* La Chapelle-aux-Saints, *F* La Ferrassie 1, *C* Cro-Magnon 1, *M* Mladeč 1; *Q* Qafzeh 9, *S* Skhul 5

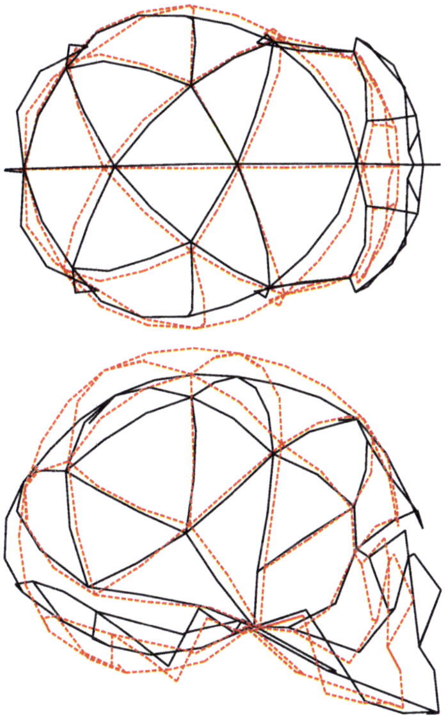

Fig. 2.19 Variations in ectocranial shape represented by the first PC. Variations are visualized with 3D deformation of the wireframe that connected the landmarks. *Solid line*: PC1 = 0.1. *Dotted line*: PC1 = −0.02

PC1, a relative contraction of the endocranial length, a relative elongation of the endocranial breadth, and a relative elongation of the endocranial height were observed (Fig. 2.19). These observations indicate that the cranium is more dolichocephalic if the PC1 score is larger, whereas it is more brachycephalic if the score is smaller. Furthermore, with a decrease in PC1, the parietal and cerebellar regions protruded more posterosuperiorly and posteroinferiorly, respectively.

The results of PCA concerning morphological variability in the endocasts of the Neanderthals and anatomically modern humans are presented in Fig. 2.20 as a plot of PC1 versus PC2. The first two PCs accounted for 35.0% of the variation (20.1% and 15.0% for PC1 and PC2, respectively).

The pattern of shape variations in Fig. 2.20 shows that scores on PC1 were relatively higher in the Neanderthal (A, G, L, F) and early *Homo sapiens* (C, M, Q, S) and vice versa in modern human populations (Japanese, European, and Indian populations), indicating that the PC1 roughly corresponds to the shape difference between the fossil hominins (Neanderthals and early *Homo sapiens*) and extant humans. However, the Neanderthal endocasts displayed relatively lower scores on PC2 and were separated from those of anatomically modern humans, suggesting that the shape variations along PC2 were possibly linked to the difference in endocranial morphology between the two species.

Figure 2.21 shows the 3D shape variabilities along PC1 and PC2 by warping the endocranial shape represented by the wireframe that connected the landmarks. With a decrease in PC1, a relative contraction of the endocranial length, a relative elongation of the endocranial breadth, and a relative elongation of the endocranial height were observed (Fig. 2.21). These observations indicate that the endocranium is more dolichocephalic if the PC1 score is

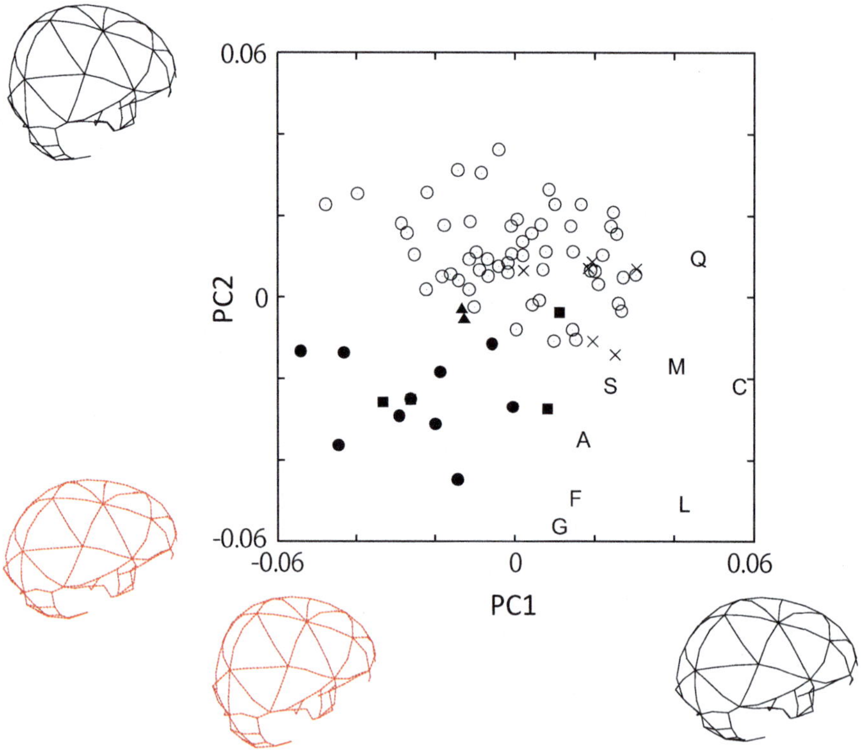

Fig. 2.20 The results of principal component (PC) analysis of endocranial shape variation. PC1 (x-axis) versus PC2 (y-axis). *White circles* = modern Japanese; *black circles*, *squares*, and *triangles* = modern European (Czech, Germany, and France, respectively); crosses = modern Indian; *A* Amud 1, *G* Forbes' Quarry 1, *L* La Chapelle-aux-Saints, *F* La Ferrassie 1, *C* Cro-Magnon 1, *M* Mladeč 1, *Q* Qafzeh 9, *S* Skhul 5

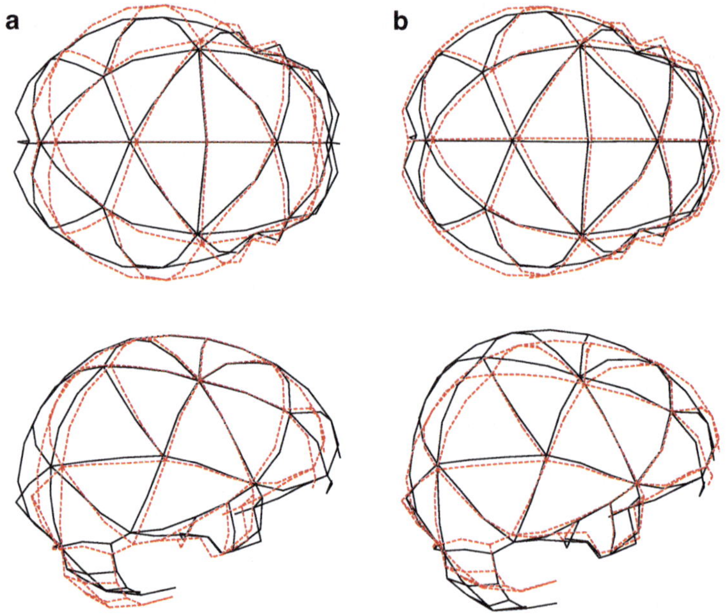

Fig. 2.21 Variations in endocranial shape represented by the first two PCs. Variations are visualized with 3D deformation of the wireframe that connected the landmarks. *Solid line*: PC1 = 0.04, PC2 = 0.04. *Dotted line*: PC1 = –0.04, PC2 = –0.04

larger as in fossil hominins, whereas it is more brachycephalic if the score is smaller as in extant humans. Furthermore, the frontotemporal region protruded less in the anterior direction with a decrease in PC1. However, the relative position of the internal occipital protuberance, corresponding to the boundary between the occipital lobe and the cerebellum, was virtually unchanged. Therefore, in a dolichocephalic cranium as in fossil hominins, the occipital lobe specifically protrudes backward, whereas in a brachycephalic cranium as in extant humans, the cerebellum is specifically enlarged.

With an increase in PC2, the parietal and cerebellar regions protruded more posterosuperiorly and posteroinferiorly, respectively, indicating a tendency for "globurization" along the PC2. On the other hand, with a decrease in PC2, the parietal region was relatively more elongated. These observations suggest that early *Homo sapiens* have a relatively larger cerebellum and parietal regions compared to Neanderthals.

Although skull and endocast shapes should be interrelated to each other, the pattern of morphological variation between the ecto- and endocranial shapes was considerably different. This may be because the ectocranial analysis included face regions. Neanderthals have relatively more protruding facial features than humans, as also extracted in our ectocranial analysis (Fig. 2.17). Therefore, Neanderthals were comparatively more separated from early and modern *Homo sapiens* in our ectocranial analysis than in our endocranial analysis.

2.6 Inferring Differences in Brain Functions

About 40,000 years ago, Neanderthals and early *Homo sapiens* coexisted in Europe for about 5000 years (Higham et al. 2014). However, Neanderthals disappeared, whereas early *Homo sapiens* have survived to this day. This disappearance of the Neanderthals and the expansion of modern humans have been explained by a number of hypotheses such as differences in the ability to adapt to a rapidly changing climate and environment (van Andel and Davies 2003; Finlayson and Carrion 2007); differences in technical, economic, and social systems (Adler et al. 2008; Shea 2008); differences in livelihood strategies (Bocherens et al. 2005; Richards and Trinkaus 2009); differences in language skills (Mellars 2004); assimilation between the two populations (Smith et al. 2005); and cognitive differences due to anatomical and functional differences in the brains of the two populations (Klein 2008). Nevertheless, details of the processes and reasons leading to the replacement remain unclear.

Our results demonstrated that endocranial shapes are quantitatively different between Neanderthals and early *Homo sapiens*. Specifically, our geometric morphometric analyses revealed that the modern human cranium shows relative enlargement of the cerebellar region and relative expansion of the parietal area, as suggested by other studies (Bruner et al. 2010, 2014; Gunz et al. 2010; Weaver 2005). Because the brain is molded in accordance with the endocranial cavity during development, these differences in the endocranial shape indicate that neuroanatomical organization may be different between the two species.

The cerebellum is traditionally considered to play an important role in motor control, particularly for coordinated fine control of the complex musculoskeletal system. However, recent studies on cognitive functions of the cerebellum suggest that the cerebellum has strong mutual connections with the cerebral cortex and plays important roles in planning, execution, and understanding of complex behavioral sequences, such as tool use and language (Imamizu et al. 2000; Barton 2012; Marien et al. 2014). Furthermore, the cerebellum may greatly contribute to efficient communication and social interactions in humans as a neural simulator, as the internal model, which is acquired in the cerebellum, is essential for predicting the mental state of another from communicative actions (Wolpert et al. 2003). Therefore, relative enlargement of the cerebellum may have enhanced the cognitive function of the brain in the modern human lineage (Middleton and Stick 1994).

The parietal lobe (excluding somatosensory areas) is considered an "association cortex" because of its integrative roles in multimodal inputs (Culham and Kanwisher 2001). The main functions of the parietal lobe have been documented as visuospatial integration and object manipulation, but some authors have suggested that the parietal lobe has more critically important, higher-order cognitive functions (Coolidge 2014), such as numerical processing (Dehaene et al. 2003) and interplay of egocentric/epicentric/autobiographical memory (Land 2014). Therefore, morphological differences in parietal lobes between Neanderthals and *Homo sapiens* may also be related to differences in cognitive function between the two species (Bruner 2010).

Pearce et al. (2013) suggested based on orbit size that Neanderthals had larger eyes and visual cortices than early *Homo sapiens*, but relatively smaller areas for other brain systems such as the parietal lobes (Bruner et al. 2010). Although the shape variation along the PC2 axis (which separates the Neanderthals from the early *Homo sapiens*) does not indicate a shape difference in the occipital region, this shape variation was observed along the PC1 in the present study. The reason for this discrepancy is currently unclear, but the present study suggested that a large occipital region may be a shared morphological characteristic between the Neanderthals and early *Homo sapiens*.

Recent morphological studies on the pattern of endocranial ontogeny of Neanderthals and modern humans suggested that relative expansion of the parietal and cerebellar regions occurs during the early postnatal period in modern humans so that the endocranium becomes more globular. However, this is not the case in the Neanderthal lineage (Gunz et al. 2010, 2012). Ontogenetic differences in the cranium and hence the brain between Neanderthals and early *Homo sapiens* may thus differentiate the developmental process of learning and social skills in early childhood (Hublin et al. 2015), possibly leading to critical differences in cognitive functions and learning capacity between the two populations.

Only a slight difference in the genetic and epigenetic sequences exists between Neanderthals and modern humans. However, such a subtle difference may become significant in terms of natural selection and may have led to differences between the two species in cognitive abilities for communication and social interactions (Meyer et al. 2012; Prüfer et al. 2014; Gokhman et al. 2014). The present study clearly demonstrated that subtle but innate morphological differences in the endocranial shape (and possibly the brain enclosed in the endocranial cavity) actually existed between Neanderthals and anatomically modern *Homo sapiens*. Although other hypotheses are certainly not excluded, such differences in endocranial morphology and hence brain structure and organization may have affected the fate of the two species. These differences may have provoked the disappearance of the Neanderthals and the successive expansion of modern humans until today. Here, we only compared the endocranial shape, which may not necessarily reflect actual neuronal differences between the two species. More studies are certainly necessary to clarify the background and direct causes of the replacement of Neanderthals by anatomically modern humans.

Acknowledgment The authors express their sincere gratitude to T. Akazawa of Kochi Institute of Technology for giving the opportunity to participate in the research project "Replacement of Neanderthals by Modern Humans: Testing Evolutionary Models of Learning" and for lending his continuous guidance and support throughout the course of the study. The authors also thank Y. Rak and I. Hershkovitz of Tel Aviv University and C. P. E. Zollikofer and M. Ponce de León of the University of Zurich for kindly allowing the use of CT scan data of the Amud 1 and Qafzeh 9; P. Mennecier and A. Froment of Muséum national d'Histoire naturelle for La Chapelle-aux-Saints 1, La Ferrassie 1, and Cro-Magnon 1; M. Bastir of Museo Nacional de Ciencias Naturales and C. Stringer of Natural History Museum for Forbes' Quarry 1; and D. Lieberman, O. Herschensohn, and M. Morgan of Harvard University for Skhul 5. The CT scan data of the Mladeč 1 were obtained from the digital archive of fossil hominoids, the University of Vienna. We also thank M. Nakatsukasa of Kyoto University, G. Suwa of the University of Tokyo, and T Takano of Japan Monkey Center for their kind permission to study the cranial materials under their care and D. Kubo of Hokkaido University for CT scanning of the crania housed at the University of Tokyo. This study was supported by a Grant-in-Aid for Scientific Research from the Japanese Ministry of Education, Culture, Sports, Science and Technology (#22101006).

Appendix

This appendix provides the plates of the six-sided views of the eight reconstructed endocasts. The blue surfaces are interpolated missing regions (Plates 2.1, 2.2, 2.3, 2.4, 2.5, 2.6, 2.7, and 2.8).

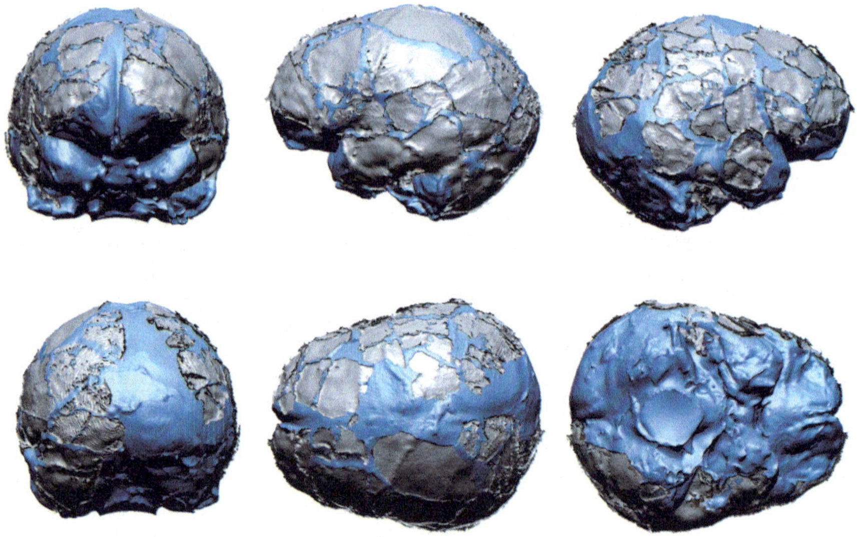

Plate 2.1 Amud 1

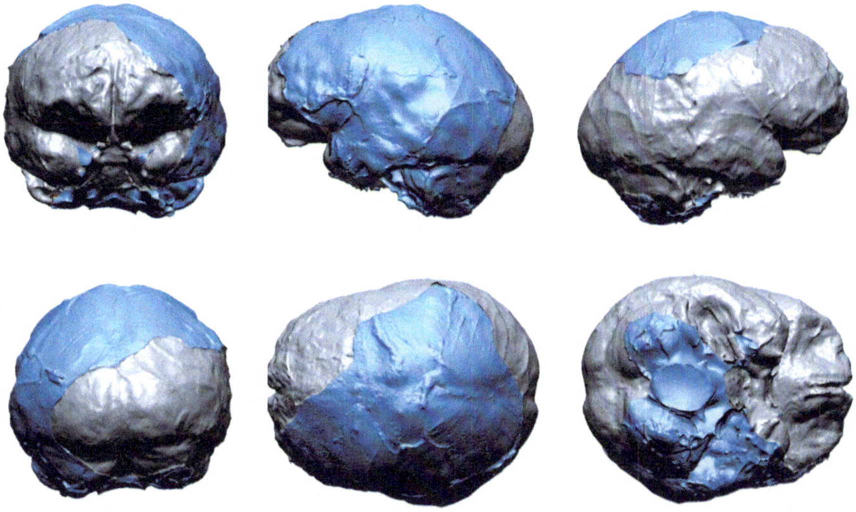

Plate 2.2 Forbes' Quarry 1

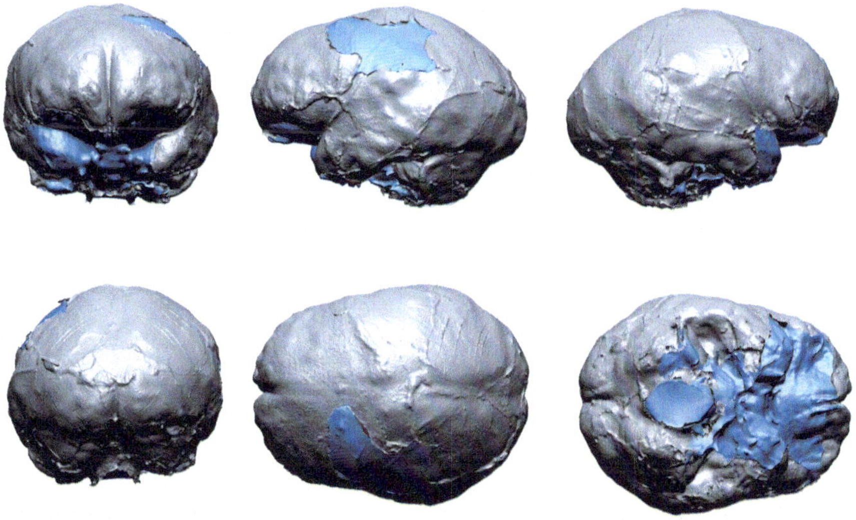

Plate 2.3 La Chapelle-aux-Saints 1

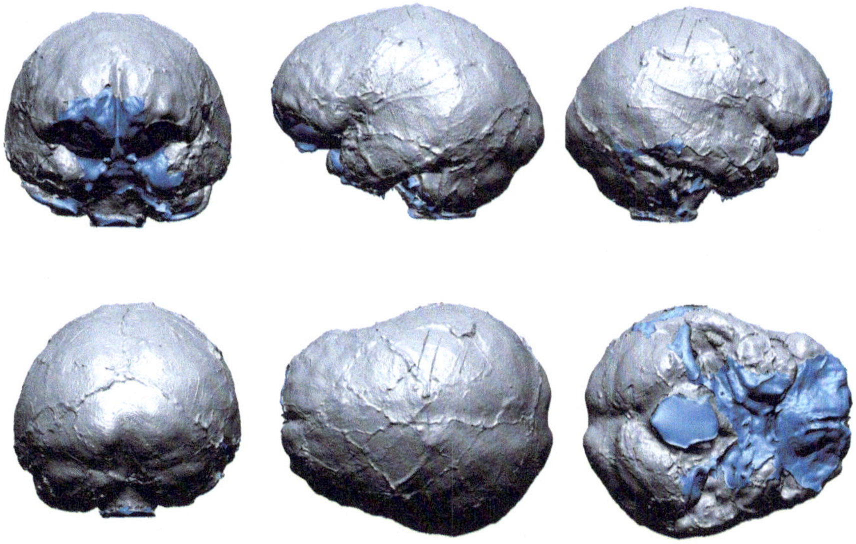

Plate 2.4 La Ferrassie 1

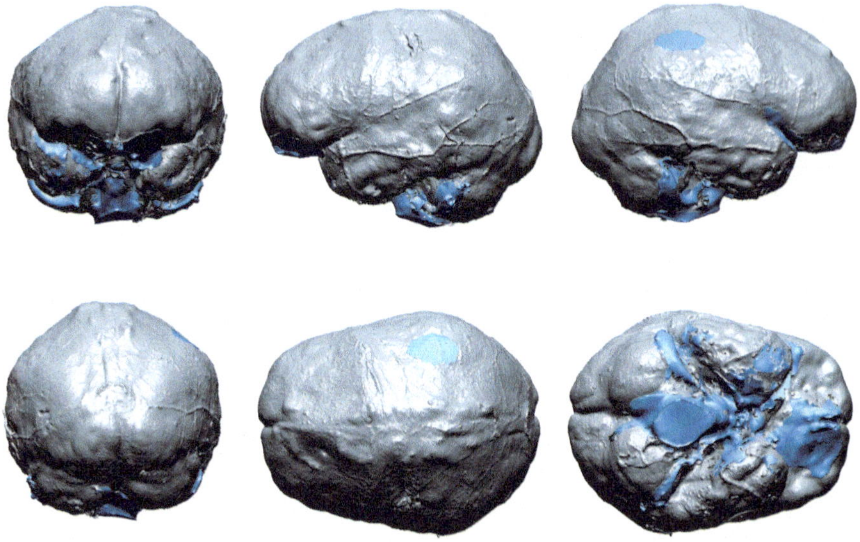

Plate 2.5 Cro-Magnon 1

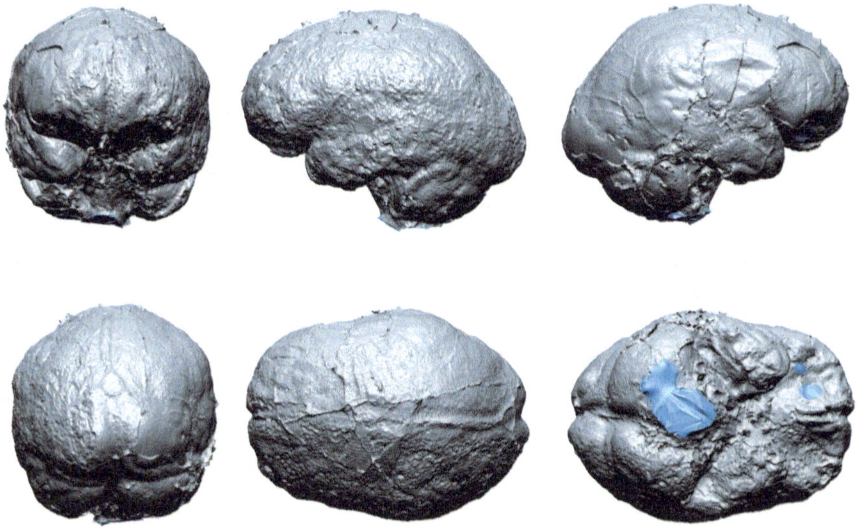

Plate 2.6 Mladeč 1

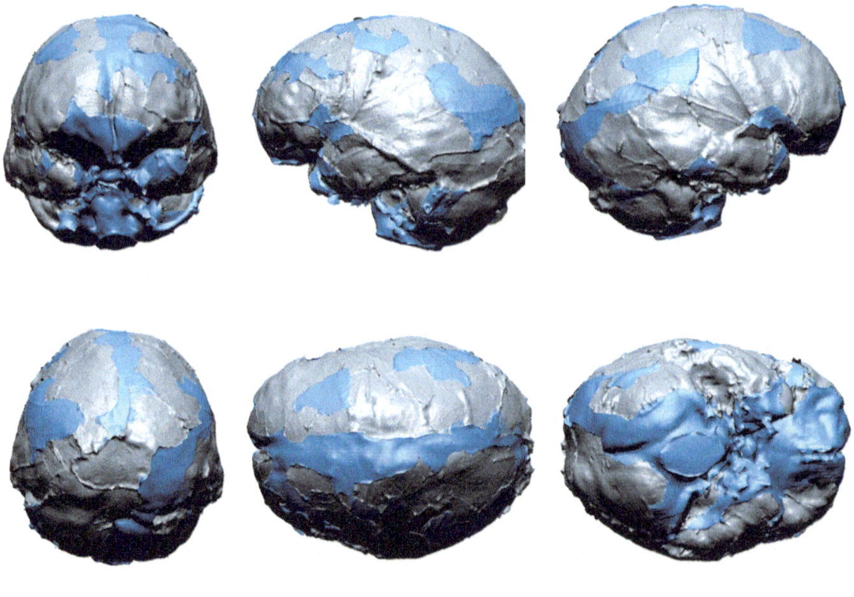

Plate 2.7 Qafzeh 9

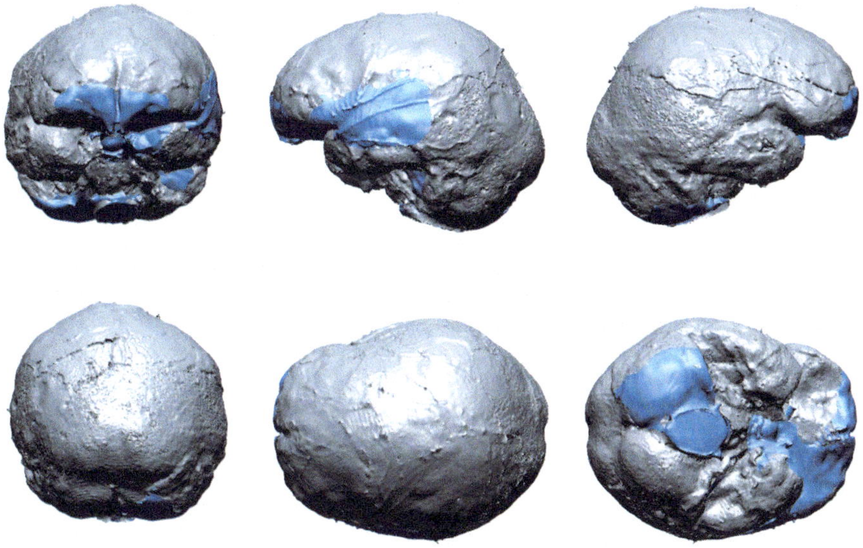

Plate 2.8 Skhul 5

References

Adams DC, Rohlf FJ, Slice DE (2004) Geometric morphometrics: ten years of progress following the 'revolution'. Ital J Zool 71:5–16

Adler DS, Bar-Yosef O, Belfer-Cohen A, Tushabramishvili N, Boaretto E, Mercier N, Valladas H, Rink WJ (2008) Dating the demise: neandertal extinction and the establishment of modern humans in the southern Caucasus. J Hum Evol 55:817–833

Amano H, Morita Y, Nagano H, Kondo O, Suzuki H, Nakatsukasa M, Ogihara N (2014) Statistical interpolation of missing parts in human crania using regularized multivariate linear regression analysis. In: Akazawa T, Ogihara N, Tanabe HC, Terashima H (eds) Dynamics of learning in Neanderthals and modern humans vol. 2 cognitive and physical perspectives. Springer, New York, pp 161–169

Amano H, Kikuchi T, Morita Y, Kondo O, Suzuki H, Ponce de León MS, Zollikofer CPE, Bastir M, Stringer C, Ogihara N (2015) Virtual reconstruction of the Neanderthal Amud 1 cranium. Am J Phys Anthropol 158:185–197

Balzeau A, Grimaud-Herve D, Jacob T (2005) Internal cranial features of the Mojokerto child fossil (East Java, Indonesia). J Hum Evol 48:535–553

Balzeau A, Holloway RL, Grimaud-Hervé D (2012) Variation and asymmetries in regional brain surface in the genus *Homo*. J Hum Evol 62:696–706

Balzeau A, Grimaud-Herve D, Detroit F, Hollowaoy RL, Combes B, Prima S (2013) First description of the Cro-Magnon 1 endocast and study of brain variation and evolution in anatomically modern Homo sapiens. Bull Mém Soc Anthropol Paris 25:1–18

Barton RA (2012) Embodied cognitive evolution and the cerebellum. Philos Trans R Soc B 367:2097–2107

Bastir M, Rosas A, Lieberman DE, O'Higgins P (2008) Middle cranial fossa anatomy and the origin of modern humans. Anat Rec 291:130–140

Bastir M, Rosas A, Gunz P, Pena-Melian A, Manzi G, Harvati K, Kruszynski R, Stringer C, Hublin JJ (2011) Evolution of the base of the brain in highly encephalized human species. Nat Commun 2:588

Benazzi S, Bookstein F, Strait D, Weber G (2011) A new OH5 reconstruction with an assessment of its uncertainty. J Hum Evol 61:75–88

Benazzi S, Gruppioni G, Strait DS, Hublin J-J (2014) Virtual reconstruction of KNM-ER 1813 *Homo habilis* cranium. Am J Phys Anthropol 153:154–160

Berger LR, de Ruiter DJ, Churchill SE, Schmid P, Carlson KJ, Dirks PH, Kibii JM (2010) *Australopithecus sediba*: a new species of homo-like australopith from South Africa. Science 328:195–204

Bienvenu T, Guy F, Coudyzer W, Gilissen E, Roualdès G, Vignaud P, Brunet M (2011) Assessing endocranial variations in great apes and humans using 3D data from virtual endocasts. Am J Phys Anthropol 145:231–246

Bocherens H, Drucker DG, Billiou D, Patou-Mathis M, Vandermeersch B (2005) Isotopic evidence for diet and subsistence pattern of the Saint-Cesaire I Neanderthal: review and use of a multi-source mixing model. J Hum Evol 49:71–87

Bookstein FL (1991) Morphometric tools for landmark data: geometry and biology. Cambridge University Press, Cambridge

Bookstein FL (1997) Landmark methods for forms without landmarks: localizing group differences in outline shape. Med Image Anal 1:225–243

Boule M (1908) L'homme fossile de la Chapelle-aux-Saints. C R Acad Sci Paris 147:1349–1352

Bouyssonie A, Bouyssonie J, Bardon L (1909) Découverte d'un squelette humain moustérien à La Bouffia de la Chapelle-aux-Saints. (Corréze). C R Hebd Soc Acad Sci Paris 147:1414–1415

Broadfield DC, Holloway RL, Mowbray K, Silvers A, Yuan MS, Marquez S (2001) Endocast of Sambungmacan 3 (Sm 3): a new homo erectus from Indonesia. Anat Rec 262:369–379

Broca P (1868) Sur les crânes et ossements des Eyzies. Bull Soc Anthropol Paris 3:350–392

Bruner E (2004) Geometric morphometrics and paleoneurology: brain shape evolution in the genus *Homo*. J Hum Evol 47:279–303

Bruner E (2010) Morphological differences in the parietal lobes within the human genus. Curr Anthropol 51:S77–S88

Bruner E, Ripani M (2008) A quantitative and descriptive approach to morphological variation of the endocranial base in modern humans. Am J Phys Anthropol 137:30–40

Bruner E, Manzi G, Arsuaga JL (2003) Encephalization and allometric trajectories in the genus Homo: evidence from the Neandertal and modern lineages. Proc Natl Acad Sci U S A 100:15335–15340

Bruner E, Martin-Loeches M, Colom R (2010) Human midsagittal brain shape variation: patterns, allometry and integration. J Anat 216:589–599

Bruner E, de la Cuetara JM, Masters M, Amano H, Ogihara N (2014) Functional craniology and brain evolution: from paleontology to biomedicine. Front Neuroanat 8:19

Busk G (1865) On a very ancient cranium from Gibraltar. Rep Br Assoc Adv Sci (Bath 1864) 91–92

Capitan L, Peyrony D (1909) Deux squelettes humains au milieu de foyers de l'époque moustérienne. C R Acad Inscrip Belles Lett Paris 53:797–806

Carlson KJ, Stout D, Jashashvili T, de Ruiter DJ, Tafforeau P, Carlson K, Berger LR (2011) The endocast of MH1, *Australopithecus sediba*. Science 333:1402–1407

Conroy GC, Vannier MW, Tobias PV (1990) Endocranial features of *Australopithecus africanus* revealed by 2- and 3-D computed tomography. Science 247:838–841

Conroy GC, Weber GW, Seidler H, Tobias PV, Kane A, Brunsden B (1998) Endocranial capacity in an early hominid cranium from Sterkfontein, South Africa. Science 280:1730–1731

Conroy GC, Falk D, Guyer J, Weber GW, Seidler H, Recheis W (2000a) Endocranial capacity in Sts 71 (*Australopithecus africanus*) by three-dimensional computed tomography. Anat Rec 258:391–396

Conroy GC, Weber GW, Seidler H, Recheis W, Zur Nedden D, Haile Mariam J (2000b) Endocranial capacity of the bodo cranium determined from three-dimensional computed tomography. Am J Phys Anthropol 113:111–118

Coolidge FL (2014) The exaptation of the parietal lobes in Homo sapiens. J Anthropol Sci 92:295–298

Coqueugniot H, Hublin JJ, Veillon F, Houet F, Jacob T (2004) Early brain growth in *Homo* erectus and implications for cognitive ability. Nature 431:299–302

Culham JC, Kanwisher NG (2001) Neuroimaging of cognitive functions in human parietal cortex. Curr Opin Neurobiol 11:157–163

Dehaene S, Piazza M, Pinel P, Cohen L (2003) Three parietal circuits for number processing. Cogn Neuropsychol 20:487–506

Falk D (2012) Hominin paleoneurology: where are we now? In: Hofman MA, Falk D (eds) Evolution of the primate brain: from neuron to behavior. Elsevier, London, pp 255–272

Falk D (2014) Interpreting sulci on hominin endocasts: old hypotheses and new findings. Front Hum Neurosci 8:134

Falk D, Clarke R (2007) Brief communication: new reconstruction of the Taung endocast. Am J Phys Anthropol 134:529–534

Falk D, Conroy GC, Recheis W, Weber GW (2000) Early hominid brain evolution_a new look at old endocasts. J Hum Evol 38:695–717

Falk D, Hildebolt C, Smith K, Morwood MJ, Sutikna T, Brown P, Jatmiko, Saptomo EW, Brunsden B, Prior F (2005) The brain of LB1, *Homo floresiensis*. Science 308:242–245

Finlayson C, Carrión JS (2007) Rapid ecological turnover and its impact on Neanderthal and other human populations. Trends Ecol Evol 22:213–222

Gokhman D, Lavi E, Prüfer K, Fraga MF, Riancho JA, Kelso J, Pääbo S, Meshorer E, Carmel J (2014) Reconstructing the DNA methylation maps of the Neandertal and the Denisovan. Science 344:523–527

Grün R, Stringer C (1991) Electron spin resonance dating and the evolution of modern humans. Archaeometry 33:153–199

Grün R, Stringer C, McDermott F, Nathan R, Porat N, Robertson S, Lois Taylor L, Mortimer G, Eggins S, McCulloch M (2005) U-series and ESR analyses of bones and teeth relating to the human burials from Skhul. J Hum Evol 49:316–334

Guerin G, Frouin M, Talamo S, Aldeias V, Bruxelles L, Chiotti L, Dibble HL, Goldberg P, Hubin JJ, Jain M, Lahaye C, Madelaine S, Maureille B, McPherron SJP, Mercier N, Murray AS, Sandgathe D, Steele TE, Thomsen KJ, Turq A (2015) A multi-method luminescence dating of the Palaeolithic sequence of La Ferrassie based on new excavations adjacent to the La Ferrassie 1 and 2 skeletons. J Archaeol Sci 58:147–166

Gunz P, Mitteroecker P, Bookstein FL (2005) Semilandmarks in three dimensions. In: Slice DE (ed) Modern morphometrics in physical anthropology. Kluwer Academic/Plenum Publishers, New York, pp 73–98

Gunz P, Mitteroecker P, Neubauer S, Weber GW, Bookstein FL (2009) Principles for the virtual reconstruction of hominin crania. J Hum Evol 57:48–62

Gunz P, Neubauer S, Maureille B, Hublin JJ (2010) Brain development after birth differs between Neandertals and modern humans. Curr Biol 20:R921–R922

Gunz P, Neubauer S, Golovanova L, Doronichev V, Maureille B, Hublin JJ (2012) A uniquely modern human pattern of endocranial development. Insights from a new cranial reconstruction of the Neandertal newborn from Mezmaiskaya. J Hum Evol 62:300–313

Henry-Gambier D (2002) Les fossiles de Cro-Magnon (Les Eyzies-de-Tayac, Dordogne)_nouvelles données sur leur position chronologique et leur attribution culturelle. Bull Mém Soc Anthropol Paris 14:89–112

Higham T, Douka K, Wood R, Ramsey CB, Brock F, Basell L, Camps M, Arrizabalaga A, Baena J, Barroso-Ruíz C, Bergman C, Boitard C, Boscato P, Caparrós M, Conard NJ, Draily C, Froment A, Galván B, Gambassini P, Garcia-Moreno A, Grimaldi S, Haesaerts P, Holt B, Iriarte-Chiapusso MJ, Jelinek A, Pardo JFJ, Maíllo-Fernández JM, Marom A, Maroto J, Menéndez M, Metz L, Morin E, Moroni A, Negrino F, Panagopoulou E, Peresani M, Pirson S, de la Rasilla M, Riel-Salvatore J, Ronchitelli A, Santamaria D, Semal P, Slimak L, Soler J, Soler N, Villaluenga A, Pinhasi R, Jacobi R (2014) The timing and spatiotemporal patterning of Neanderthal disappearance. Nature 512:306–309

Holloway RL (1981) Volumetric and asymmetry determinations on recent hominid endocasts: spy I and II, Djebel Ihroud I, and the sale Homo erectus specimens, with some notes on Neanderthal brain size. Am J Phys Anthropol 55:385–393

Holloway RL (2008) The human brain evolving: a personal retrospective. Annu Rev Anthropol 37:1–19

Holloway RL, Broadfield DC, Yuan MS (2004) The human fossil record: brain endocasts: the paleoneurological evidence. Wiley, Hoboken

Hublin JJ, Neubauer S, Gunz P (2015) Brain ontogeny and life history in Pleistocene hominins. Philos Trans R Soc B 370:20140062

Imamizu H, Miyauchi S, Tamada T, Sasaki Y, Takino R, Putz B, Yoshioka T, Kawato M (2000) Human cerebellar activity reflecting an acquired internal model of a new tool. Nature 403:192–195

Kikuchi T, Ogihara N (2013) Computerized assembly of neurocranial fragments based on surface extrapolation. Anthropol Sci 121:115–122

Klein RG (2008) Out of Africa and the evolution of human behavior. Evol Anthopol 17:267–281

Kobayashi Y, Matsui T, Haizuka Y, Ogihara N, Hirai N, Matsumura G (2014) Cerebral sulci and gyri observed on macaque endocasts. In: Akazawa T, Ogihara N, Tanabe HC, Terashima H (eds) Dynamics of learning in Neanderthals and modern humans, vol. 2: cognitive and physical perspectives. Springer, Japan, pp 131–137

Kranioti EF, Holloway R, Senck S, Ciprut T, Grigorescu D, Harvati K (2011) Virtual assessment of the endocranial morphology of the early modern European fossil calvaria from Cioclovina, Romania. Anat Rec 294:1083–1092

Kubo D, Kono RT, Suwa G (2011) A micro-CT based study of the endocranial morphology of the Minatogawa I cranium. Anthropol Sci 119:123–135

Land MF (2014) Do we have an internal model of the outside world? Philos Trans R Soc Lond B 369:1–6

Lartet L (1868) Une sépulture des troglodytes du Périgord (crânes des Eyzies). Bull Soc Anthropol Paris 3:335–349

Marien P, Ackermann H, Adamaszek M, Barwood CHS, Beaton A, Desmond J, De Witte E, Fawcett AJ, Hertrich I, Küper M, Leggio M, Marvel C, Molinari M, Murdoch BE, Nicolson RI, Schmahmann JD, Stoodley CJ, Thürling M, Timmann D, Wouters E, Ziegler W (2014) Consensus paper: language and the cerebellum: an ongoing enigma. Cerebellum 13:386–410

McCown TD, Keith SA (1939) The stone age of Mount Carmel: the fossil human remains from the Levallosiso-Mousterian. Clarendon Press, Oxford

Mellars P (2004) Neanderthals and the modern human colonization of Europe. Nature 432:461–465

Mercier N, Valladas H, Bar-Yosef B, Vandermeersch B, Stringer C, Joron JL (1993) Thermoluminescence date for the Mousterian burial site of ES-Skhul, Mt. Carmel. J Archaeol Sci 20:169–174

Meyer M, Kircher M, Gansauge MT, Li H, Racimo F, Mallick S, Schraiber JG, Jay F, Prüfer K, de Filippo C, Sudmant PH, Alkan C, Fu Q, Do R, Rohland N, Tandon A, Siebauer M, Green RE, Bryc K, Briggs AW, Stenzel U, Dabney J, Shendure J, Kitzman J, Hammer MF, Shunkov MV, Derevianko AP, Patterson N, Andrés AM, Eichler EE, Slatkin M, Reich D, Kelso J, Pääbo S (2012) A high-coverage genome sequence from an archaic Denisovan individual. Science 338:222–226

Michikawa T, Suzuki H, Moriguchi M, Ogihara N, Kondo O, Kobayashi Y (2017) Automatic extraction of endocast surfaces from CT images of crania. PLoS One 12(4):e0168516

Middleton FA, Stick PL (1994) Anatomical evidence for cerebellar and basal ganglia involvement in higher cognitive function. Science 266:458–461

Mitteroecker P, Gunz P (2009) Advances in geometric morphometrics. Evol Biol 36:235–247

Morita Y, Ogihara N, Kanai T, Suzuki H (2013) Quantification of neurocranial shape variation using shortest paths connecting pairs of anatomical landmarks. Am J Phys Anthropol 151:658–666

Morita Y, Amano H, Nakatsukasa M, Kondo O, Ogihara N (2014) A geometric morphometric study of neurocranial shape variations in the crania of modern Japanese. In: Akazawa T, Ogihara N, Tanabe H, Terashima H (eds) Dynamics of learning in Neanderthals and modern humans volume 2: cognitive and physical perspectives. Springer, Japan, pp 131–137

Morita Y, Amano H, Ogihara N (2015) Three-dimensional endocranial shape variation in the modern Japanese population. Anthropol Sci 123:185–191

Neubauer S (2014) Endocasts: possibilities and limitations for the interpretation of human brain evolution. Brain Behav Evol 84:117–134

Neubauer S, Gunz P, Mitteroecker P, Weber GW (2004) Three-dimensional digital imaging of the partial *Australopithecus africanus* endocranium MLD 37/38. Can Assoc Radiol J 55:271–278

Neubauer S, Gunz P, Hublin JJ (2009) The pattern of endocranial ontogenetic shape changes in humans. J Anat 215:240–255

Neubauer S, Gunz P, Hublin JJ (2010) Endocranial shape changes during growth in chimpanzees and humans: a morphometric analysis of unique and shared aspects. J Hum Evol 59:555–566

Neubauer S, Gunz P, Weber GW, Hublin JJ (2012) Endocranial volume of *Australopithecus africanus*: new CT-based estimates and the effects of missing data and small sample size. J Hum Evol 62:498–510

Ogihara N, Nakatsukasa M, Nakano Y, Ishida H (2006) Computerized restoration of nonhomogeneous deformation of a fossil cranium based on bilateral symmetry. Am J Phys Anthropol 130:1–9

Ogihara N, Amano H, Kikuchi T, Morita Y, Hasegawa K, Kochiyama T, Tanabe HC (2015) Towards digital reconstruction of fossil crania and brain morphology. Anthropol Sci 123:57–68

O'Higgins P (2000) The study of morphological variation in the hominid fossil record: biology, landmarks and geometry. J Anat 197:103–120

O'Higgins P, Jones N (1998) Facial growth in Cercocebus torquatus: an application of three-dimensional geometric morphometric techniques to the study of morphological variation. J Anat 193:251–272

Pearce E, Stringer C, Dunber RIM (2013) New insights into difference in brain organization between Neanderthals and anatomically modern humans. Proc R Soc B 280:20130168

Prüfer K, Racimo F, Patterson N, Jay F, Sankararaman S, Sawyer S, Heinze A, Renaud G, Sudmant PH, de Filippo C, Li H, Mallick S, Dannemann M, Fu Q, Kircher M, Kuhlwilm M, Lachmann M, Meyer M, Ongyerth M, Siebauer M, Theunert C, Tandon A, Moorjani P, Pickrell J, Mullikin JC, Vohr SH, Green RE, Hellmann I, Johnson PJF, Blanche H, Cann H, Kitzman JO, Shendure J, Eichler EE, Lein ES, Bakken TE, Golovanova LV, Doronichev VB, Shunkov MV, Derevianko AP, Viola B, Slatkin M, Reich D, Kelso J, Pääbo S (2014) The complete genome sequence of a Neanderthal from the Altai Mountains. Nature 505:43–49

Recheis W, Macchiarelli R, Seidler H, Weaver DS, Schäfer K, Bondioli L, Weber GW, zur Nedden D (1999) Reevaluation of the endocranial volume of the Guattari 1 Neandertal specimen (Monte Circeo). Coll Antropol 23:397–405

Richards MP, Trinkaus E (2009) Isotopic evidence for the diets of European Neanderthals and early modern humans. Proc Natl Acad Sci U S A 106:16034–16039

Rink WJ, Schwarcz HP, Lee HK, Rees-Jones J, Rabinovich R, Hovers E (2001) Electron spin resonance (ESR) and thermal ionization mass spectrometric (TIMS) 230Th/234U dating of teeth in middle paleolithic layers at amud cave, Israel. Geoarchaeology 16:701–717

Schoenemann PT, Gee J, Avants B, Holloway RL, Monge J, Lewis J (2007) Validation of plaster endocast morphology through 3D CT image analysis. Am J Phys Anthropol 132:183–192

Schwarcz HP, Grün R, Vandermeersch B, Bar-Yosef O, Valladas H, Tchernov E (1988) ESR dates for the hominid burial site of Qafzeh in Israel. J Hum Evol 17:733–737

Seidler H, Falk D, Stringer C, Wilfing H, Muller GB, zur Nedden D, Weber GW, Reicheis W, Arsuaga JL (1997) A comparative study of stereolithographically modelled skulls of Petralona and Broken Hill: implications for future studies of middle Pleistocene hominid evolution. J Hum Evol 33:691–703

Shea JJ (2008) Transitions or turnovers? Climatically-forced extinctions of Homo sapiens and Neanderthals in the east Mediterranean. Quat Sci Rev 27:2253–2270

Slice DE (2005) Modern morphometrics. In: Slice DE (ed) Modern morphometrics in physical anthropology. Kluwer Academic/Plenum Publishers, New York, pp 1–45

Smith FH, Jankovic I, Karavanic I (2005) The assimilation model, modern human origins in Europe, and the extinction of Neandertals. Q J Econ 137:7–19

Suzuki H (1970) The skull of the Amud man. In: Suzuki H, Takai F (eds) The Amud man and his cave site. Keigaku Publishing, Tokyo, pp 123–206

Suzuki H, Takai F (1970) The Amud man and his cave site. Keigaku Publishing, Tokyo

Szombathy J (1925) Die diluvialen Menschenreste aus der Fürst-Johann-Höhle bei Leutsch in Mähren. Die Eiszeit 2(1–34):73–95
Tobias PV (2001) Re-creating ancient hominid virtual endocasts by CT-scanning. Clin Anat 14:134–141
Valladas H, Reyss JL, Joron JL, Valladas G, Bar-Yosef O, Vandermeersch B (1988) Thermoluminescence dating of Mousterian 'Proto-Cro-Magnon' remains from Israel and the origin of modern man. Nature 331:614–616
Valladas H, Mercier N, Froget L, Hovers E, Joron JL, Kimbel WH, Rak Y (1999) TL dates for the neanderthal site of the amud cave, Israel. J Archaeol Sci 26:259–268
van Andel TH, Davies W (2003) Neanderthals and modern humans in the European landscape during the last glaciation: archaeological results of the stage 3 project. McDonald Institute for Archeological Research, Cambridge
Vandermeersch B (1981) Les hommes fossiles de Qafzeh (Israel). Editions du CNRS, Paris
Weaver AH (2005) Reciprocal evolution of the cerebellum and neocortex in fossil humans. Proc Natl Acad Sci U S A 102:3576–3580
Weber GW, Bookstein FL (2011) Virtual anthropology. Springer Verlag, New York
Wild EM, Teschler-Nicola M, Kutschera W, Steier P, Trinkaus E, Wanek W (2005) Direct dating of early upper palaeolithic human remains from Mladeč. Nature 435:332–335
Wolpert DM, Doya K, Kawato M (2003) A unifying computational framework for motor control and social interaction. Philos Trans R Soc Lond B 358:593–602
Wu XJ, Liu W, Dong W, Que JM, Wang YF (2008) The brain morphology of Homo Liujiang cranium fossil by three-dimensional computed tomography. Chin Sci Bull 53:2513–2519
Zollikofer CPE, Ponce de León MS (2002) Visualizing patterns of craniofacial shape variation in Homo sapiens. Proc R Soc Lond B Biol 269:801–807
Zollikofer CPE, Ponce de León MS (2005) Virtual reconstruction: a primer in computer-assisted paleontology and biomedicine. Wiley, Hoboken
Zollikofer CPE, Ponce de León MS (2013) Pandora's growing box: inferring the evolution and development of hominin brains from endocasts. Evol Anthropol 22:20–33

3 Inferring Cortical Subdivisions Based on Skull Morphology

Yasushi Kobayashi, Toshiyasu Matsui, and Naomichi Ogihara

Abstract

Reconstructing brains of fossil hominids is one of the most important issues in anthropology. It is of particular interest to know the extent of cortical subdivisions in those brains since differences may indicate the differences in cognitive capabilities between fossil hominines and modern humans. We evaluated two approaches to infer borders of cortical regions based on skull morphology. The first approach is to identify cerebral sulci and gyri based on the surface morphology of endocasts. The second approach is to infer the location of cerebral sulci and gyri using their spatial relationship with cranial sutures and other landmarks. We review the historical origin of these two approaches and evaluate their validity in fossil hominine studies.

Keywords

Endocast • Suture • Sulcus • Cortical areas • Prefrontal cortex • Primate

3.1 Introduction

Knowing the characteristics of cognitive functions in fossil hominids is one of the most essential targets, not only in studies on those extinct species but also in understanding the nature of modern humans. Cognitive functions are implemented in the nervous system, particularly in the brain. Today, neuroscience has a great variety of research tools that elucidate the structure and function of the nervous system and monitor their changes. Recent techniques using molecular biology and pharmacology can even manipulate specific functions of the nervous system. However, for studies on fossil hominids, we can only use remaining hard tissues and infer the structure and functions of the nervous system from them.

In this chapter, we first review the structures of the brain that may differentiate our cognitive functions from those of fossil hominids and present two approaches to infer the extent of some parts of the brain based on the skull morphology. We primarily focus on the differences between modern humans and Neanderthals but also refer to important studies in other fossil hominids, as well as extant primate species.

3.2 Classical Views in Primate Brain Evolution

The most outstanding feature of the primate brain is the highly developed cerebral cortex. The volume of the brain, and its subregions, has been quantitatively analyzed using allometry. Jerison defined the "encephalization quotient" (EQ) to quantify the relative development of brain volume to body size and clearly showed that extant mammals possess larger brains than fossil mammals and extant reptiles of similar body weight (Jerison 1973).

Y. Kobayashi (✉) • T. Matsui
Department of Anatomy and Neurobiology, National Defense Medical College, Tokorozawa, Saitama, Japan
e-mail: yasushi@ndmc.ac.jp; matsuto@ndmc.ac.jp

N. Ogihara
Department of Mechanical Engineering, Faculty of Science and Technology, Keio University, Yokohama, Kanagawa, Japan
e-mail: ogihara@mech.keio.ac.jp

E. Bruner et al. (eds.), *Digital Endocasts*, Replacement of Neanderthals by Modern Humans Series,
DOI 10.1007/978-4-431-56582-6_3

Although Jerison discussed the importance of EQ as an index of intelligence applicable to a wide variety of species, behavioral and cognitive studies in primates provided somewhat different views concerning brain organization and intelligence. Based on these findings, absolute brain size and body size are considered to better correlate with the mental performance of nonhuman primates than EQ (see review by Gibson et al. 2001).

The proportions of the volume of different subdivisions of the brain are also informative when we evaluate different aspects of neural functions. Stephan focused on the "ascending primate scale"—a series of different classes of species comprising basic insectivores, progressive insectivores, prosimians, and simians—and defined the "encephalization index," the relative brain volume of a primate species compared to a basic insectivore of equal body weight (Stephan and Andy 1969). His data showed that brains of simians are larger than those of prosimians, which are in turn larger than those of insectivores. He also calculated the progression indices, a value expressing the degree of enlargement of a structure in one species in comparison with that of a typical basal insectivore of equal body weight. The simian neocortex exhibited by far the highest progression index, followed by the striatum, the diencephalon, and the cerebellum.

The neocortex is a division of the cerebral cortex, which underwent overwhelmingly rapid development during primate evolution. It was originally defined as the part of the cerebral cortex that does not receive direct or indirect inputs of olfactory information (Ariëns Kappers 1909). It roughly corresponds to the isocortex, which typically shows six-layered organization either in the adult or during development (Brodmann 1906). The isocortex did not expand uniformly throughout primate evolution but subdivided into relatively stable primary sensory- and motor-related areas and rapidly expanding association areas.

The association areas are regions of the cerebral cortex that are myelinated later than primary sensory and motor-related areas during development (Flechsig 1920). The association areas receive sensory information from primary sensory areas, integrate different sensory modalities, identify and locate objects, judge the surrounding environment, store and retrieve long-term memories, and plan and execute behaviors. They, thus, play a pivotal role in higher cognitive functions, especially in primates.

The association areas are roughly subdivided into frontal, parietal, occipital, and temporal. The frontal association areas (prefrontal areas) are bordered caudally by motor-related areas, mostly the premotor area. In the primate, the border on the lateral surface largely corresponds to the precentral sulcus. The parietal association areas are bordered rostrally by the postcentral sulcus, separating them from the primary somatosensory area, and the occipital association areas are separated from the primary visual area (V1) by the lunate sulcus running caudally. However, borders between the parietal, occipital, and temporal association areas are not macroscopically obvious, except for the parieto-occipital sulcus on the medial surface and the preoccipital notch on the latero-inferior margin. Accordingly, these association areas are often referred to together as the parieto-temporo-occipital association areas.

To demonstrate the evolutional changes of the primate association areas, Brodmann (1912) measured the surface area of the frontal association areas "regio frontalis," motor-related areas "regio precentralis," and total frontal lobe "Frontallappen" (Table 3.1). The data clearly showed that frontal association areas (prefrontal areas) expanded during primate evolution, particularly in greater apes and the modern human. The proportion of the motor-related areas remained rather stable from old-world monkey to greater apes, but markedly reduced in the modern human, probably due to the expansion of association areas, not only in the frontal association areas but also in the parieto-temporo-occipital association areas.

Blinkov and Glezer (1968) also measured the surface area of the cerebral cortex and showed that the proportion of the human frontal cortex in the whole brain was by far the largest in primate species (32.8% in human, 22.1% in chimpanzee, 21.3% in orangutan, 21.2% in gibbons).

The cerebral cortex did not expand alone during evolution. In addition to association and commissural connections between different cortical areas, the cortex has robust projections to and from different subcortical structures, which also developed, keeping step with cortical changes during evolution. The portion of the white matter that interconnects these structures expanded as well.

Table 3.1 Proportion of the surface area of the prefrontal and frontal cortices to that of the total cerebral cortex according to Brodmann (1912)

	Regions		
Species	"Regio frontalis" (prefrontal areas)	"Regio precentralis" (areas 4 and 6)	"Frontallappen" (frontal lobe)
Rabbit	2.2%		
Cat	3.4%		
Dog	6.9%		
Lemur	7.2–8.3%		
Marmoset	8.9%		
Capuchin	9.2%	13.3%	22.5%
Guenon	11.1%	13.5%	24.6%
Macaque	11.3%	11.9%	23.2%
Hylobates	11.3%	10.1%	21.4%
Chimpanzee	16.9%	13.6%	30.5%
Human	29.0%	7.3%	36.3%

3.3 Updated Information on Primate Brain Evolution

Advances in modern neuroanatomical and imaging studies also provided abundant quantitative data on the organization of the brain in extant primate species, including humans. Recent studies by Herculano-Houzel provided a totally new approach to estimate the magnitude of development of the mammalian brains including primates. They homogenized tissue and counted the numbers of neurons and glial cells with minimum bias that is inherent in the counting procedures using conventional histology (Herculano-Houzel and Lent 2005). Their findings showed that primate brains have a larger number of neurons than rodent brains of similar size (Herculano-Houzel et al. 2007). They also revealed that the prefrontal region of both human and nonhuman primates holds about 8% of cortical neurons and the human prefrontal cortex is enlarged along the same allometric trajectory as for other primates (Gabi et al. 2016).

In volumetric studies, X-ray computed tomography (CT) provided accurate measurements of the skull, not only in extant species but also in fossils. Endocranial volumes thus could be analyzed with much greater accuracy than with classical methods. For soft tissue analysis, magnetic resonance imaging (MRI) can demarcate nervous tissues from cerebrospinal fluid, gray matter from white matter, and enables volume analysis of different modules of the nervous system, each involved in different cognitive functions.

Concerning the volumetric evaluation of the extant primate species, including humans, Semendeferi et al. (2002) conducted a series of studies that demonstrated striking commonalities among humans and other great apes. When gray matter volumes were compared, the human had the largest frontal lobes only in absolute terms, while the proportion of the frontal lobes to the total cortex is very similar in humans (37.7%) and other great apes (35.4% in chimpanzee). Gibbons (29.4%) and monkeys (30.6% in rhesus, 29.6% and 31.5% in capuchin) had significantly smaller frontal cortical volume than the great apes, but the difference was smaller than that in Brodmann's data (Brodmann 1912). Similarly, the proportion of the parieto-occipital sector in the human brain was not noteworthy. In contrast, the proportion of the temporal cortex was greater in humans than in other apes.

The similarity of the frontal proportion in great apes and humans seems to contradict Brodmann's findings. Semendeferi attributed the discrepancy to the small sample size of previous studies; however, one important difference needs to be pointed out. Brodmann measured the surface area of the cortex, while Semendeferi analyzed cortical volumes. Brodmann's data showed that the proportion of areas 4 and 6 dropped quite markedly in humans compared to the chimpanzee. Areas 4 and 6 are the thickest cortical areas in the primate brain, which contribute a lot in the volumetric comparisons, whereas the prefrontal areas are markedly thinner than areas 4 and 6 (Fig. 3.1). This means that in the volumetric analyses of the

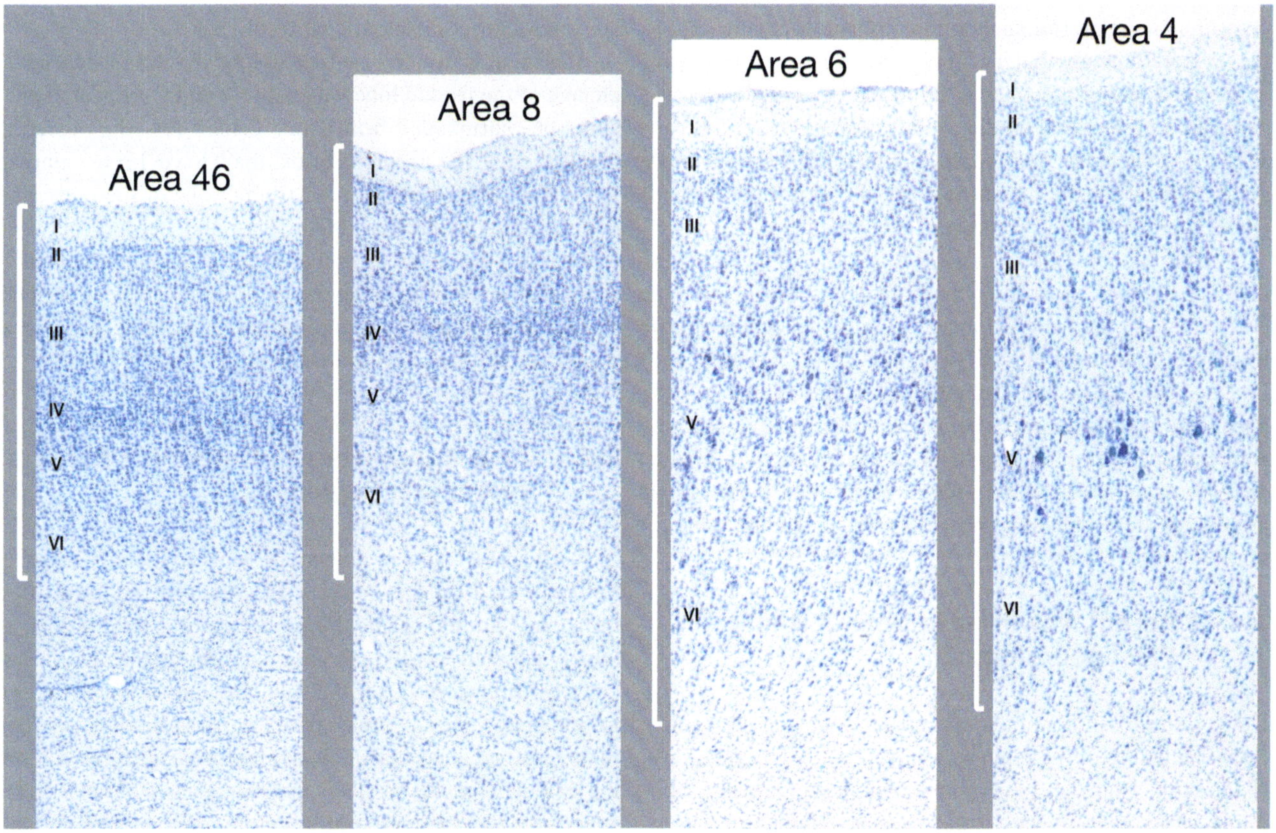

Fig. 3.1 Cortical structure of frontal areas in the macaque monkey. Sections were stained using the Nissl method for cytoarchitecture. Prefrontal areas (areas 46 and 8) are much thinner than premotor and primary motor areas (areas 6 and 4). *Roman numerals* represent the numbers of layers

frontal cortex, the expansion of the prefrontal areas may be partly masked by the relative decrease of areas 4 and 6 when we compare the proportion to the total cortical volume. The precentral cortical volume showed indeed a larger difference between humans and other great apes (28.8–33% in humans, 25.5–29.7% in other great apes) than the entire frontal cortical volume (Semendeferi et al. 2002), although the difference in the volumetric analysis remains smaller than in surface area analysis. A more recent study examining the prefrontal cortical volume in comparison with that of the primary visual cortex (striate cortex) demonstrated that the prefrontal/striate proportion in human is markedly enlarged compared to chimpanzee, while chimpanzee has also a larger value than macaque monkeys (Passingham and Smaers 2014).

Evolution of the nervous tissue does not affect the gray matter alone. The white matter that connects various regions of the cortex and subcortical nuclei also changes during the evolution. Schoenemann et al. (2005) differentiated volumes of gray matter and white matter using MRI and showed that the largest difference between human and nonhuman primates was the prefrontal white matter volume in proportion to the total cerebral white matter. The prefrontal white matter volume represented 10.9% of the total cerebral white matter in humans and 7.7% in the other great apes, whereas the proportion of prefrontal gray matter to the total cerebral gray matter showed a much smaller difference between humans and the other great apes (14.4% in human, 13.4% in great apes). These values need further refinement since the definition of the prefrontal sector in this study is not based on the cytoarchitecture but an approximation in which they define the level of the anterior tip of the genu of the corpus callosum as the caudal border of the prefrontal cortex. It can safely be concluded that the development of the prefrontal cortex comprises the increased cortical connections to and from other parts of the cortex and subcortical structures.

For this line of research, a study by Glasser and van Essen (2011) may open a possibility to more precise delineation of cortical areas based on MRI images. Using the ratio of signal intensities obtained in T1-weighted and T2-weighted images, they estimated the myelin content of the cortex and illustrated the borders between motor and somatosensory areas. In combination with this method, volumetric analysis using MRI will provide more detailed information of cortical areas of extant primate species including human.

3.4 Recent Advances in Neanderthal Paleoneurology

Concerning fossil hominids, we have to focus on the evolutionary changes of the brain that may affect the morphology of the skull. There is a long history of studies of the outer and inner structures of the skull. Recent computer-assisted reconstruction of the fossil skulls and imaging techniques, including CT and MRI, have brought about less-biased and statistically sophisticated tools in this field of research.

In the evolutionary differences between Neanderthals and modern humans, Bruner and his colleagues analyzed landmarks on the computer-reconstructed virtual endocasts of anatomically modern humans, Neanderthals, and more archaic hominines and demonstrated "parietal expansion" in modern humans compared with the Neanderthals and the other hominines (Bruner et al. 2003; Bruner 2004). The findings imply the possibility that modern humans obtained additional capacity in cognitive functions implemented in the parietal lobe, for example, visuospatial coordination and integration.

For the moment, it is reasonable to assume that parietal expansion resulted from increased cortical and white matter volume in the parietal lobe. However, there remains a possibility that expansion of other regions of the brain caused a secondary shift in the location of the parietal lobe. The net changes of the parietal volume therefore cannot be determined unless we (1) locate the border of the parietal lobe and (2) quantify the development of subcortical structures such as the basal ganglia and the diencephalon. The second factor is not a particular issue because of the relative stability of subcortical structures during hominid evolution. The first factor, however, needs to be examined given that a shift of the cortical borders has often occurred during evolution, for example, a caudal shift of the border of the primary visual area V1 (lunate sulcus on the lateral surface) has occurred since hominines differentiated from apes.

The same is true for the frontal lobe. Even if a significant change of the frontal lobe was not proven in terms of expansion or shrinkage from the center of the cerebral hemispheres, the frontal cortex may have been enlarged when the caudal border of the frontal lobe or the prefrontal cortex shifted caudally. Identifying the borders between cortical regions is thus a prerequisite for quantitative analysis of the proportional changes of different cortical regions. We have evaluated two approaches in tackling this issue, which we will describe in this chapter.

3.5 Two Approaches to Determine Borders of Cortical Regions

In fossil species, where it is not possible to identify regional borders based on the internal structures of the cortex, we have only limited means in approximating the borders. A widely adopted method is to use cerebral sulci as proxies of cortical borders. Some of the sulci are known to correspond to borders of cortical areas, particularly to borders between major cortical areas that are well preserved during evolution: the Sylvian fissure between the frontal and parietal lobes superiorly and the temporal lobe inferiorly, the central sulcus between the primary motor area (M1) and the primary somatosensory area (SI), the postcentral sulcus between SI and the posterior parietal lobe, the intraparietal sulcus between the superior and inferior parietal lobules, and the lunate sulcus demarcating the rostral border of the primary visual area (V1). Other sulci are also useful to identify the location of some cortical areas because they contain those areas, for example, the calcarine sulcus for V1, the inferior frontal sulcus or principal sulcus for area 46, and the intraparietal sulcus for anterior, lateral, medial, and ventral intraparietal areas (areas AIP, LIP, MIP, and VIP).

Recent advances in geometric morphometrics provided a statistically secure approach to evaluate the morphology of fossil skulls. On the other hand, we cannot determine the extent of cortical areas directly from skulls in fossil species. We need to know the relationship between skull landmarks and the cortical borders before inferring the extent of an area.

The abovementioned sulci often leave ridges, and adjacent gyri leave imprints or depressions on the inner surface of the skull. On the endocast of the skull, therefore, we can observe the convolutional patterns of the cortical surface. In cases where we can observe those patterns, we can identify the major sulci and gyri on the endocast without depending on indirect inferences. The major difficulty in this approach is that the convolutional patterns are not obvious in all primate species.

Another approach is to infer the locations of sulci using skull landmarks such as sutures, glabella, bregma, lambda, inion, pterion, and asterion. Although causal relationships between landmarks and cerebral structures are not biologically confirmed, these landmarks are robust and can be identified in many fossil skulls.

3.6 Locating Cerebral Sulci Based on Endocast Surface Morphology

This first approach has been used for fossil skulls since the early ages of fossil hominine research. Boule and Anthony (1911) illustrated almost the entire extent of the lateral sulcus, portions of the orbital, superior and middle frontal, postcentral, external parieto-occipital, superior temporal, lunate, and calcarine sulci on the endocast of La Chapelle-aux-Saints. Anthony (1913) described imprints of major sulci including some portions of the lateral sulcus; the superior, middle, and inferior frontal sulci; the postcentral sulcus; the intraparietal sulcus; and the parieto-occipital and lunate sulci on the endocast of La Quina.

In terms of the validity of those inferences, Symington (1916) criticized the simple assumption of the correspondence of endocranial morphology to cerebral convolutions and stated that "the simplicity or complexity of the cerebral fissures and convolutions cannot be determined with any degree of accuracy from endocranial casts." Le Gros Clarke et al. (1936) compared endocasts of chimpanzees with the brains derived from the same individual and pointed out the risk in using endocranial depressions to identify cerebral sulci, particularly in the parietal area. Ogawa et al. (1970) confirmed Symington's concerns in their study on the Amud endocast. Smith-Agreda (1955) examined more than 300 modern human skulls and reported that impressions representing cerebral gyri were observed typically in the anterior and middle cranial fossa, and the correspondence was less secure on the inner surface of the lateral wall of the skull. Only in a few abnormal cases were the convolutional patterns clearly observed up to the vertex. These findings indicate that the inference of the cerebral gyri and sulci using the skull may be reliable in the basal portion of the skull but is increasingly difficult toward the vertex.

On the other hand, endocasts of smaller primate skulls usually show marked convolutional patterns that apparently correspond to cerebral sulci and gyri. Le Gros Clarke (1945) estimated the convolutional pattern of the cerebral cortex using endocasts in fossil lemuroids. Radinsky (1972) reviewed the taxonomic characteristics of endocasts of monkeys and presented some data on the sulci observed on the endocasts. A recent study on fossil cercopithecoid skulls also showed marked imprints on the endocasts that apparently corresponded to the major sulci on the brain (Beaudet et al. 2016). Even in hominines, an *Australopithecus*, Taung Child, exhibited imprints that closely resembled cerebral sulci and gyri (Dart 1925, 1940). However, identification of sulci is not always unambiguous (for discussion concerning the lunate sulcus, see papers by Falk (1980, 1983) and Holloway (Holloway 1980)). The most problematic issue is that it is impossible to evaluate the reliability of the inference in fossil species.

To evaluate this approach, we first compared macaque monkey skulls with brains derived from the same animals (Kobayashi et al. 2014a). Figure 3.2 shows the skull and brain of a crab-eating macaque. In all the examined endocasts, we clearly identified depressions corresponding to the principal, arcuate, central, intraparietal, lunate, lateral, superior temporal, anterior middle temporal, medial, and lateral orbital sulci. We also observed very shallow indentations, for instance, the superior precentral and postcentral dimples. Even individual differences in the course of some sulci were confirmed on the endocast, for example, in the lower end of the central sulcus and connections between the medial and lateral orbital sulci.

We next analyzed dry skull specimens that were stored at the National Defense Medical College for educational purposes (Table 3.2). We scanned the skulls using an Asterion CT scanner (Toshiba; Tokyo, Japan) and obtained full three-dimensional image stacks consisting of 253–281 contiguous, 0.5-mm-thick slices. The images consisted of a 512 × 512 pixel matrix, with a pixel size of 0.351 × 0.351 mm. CT

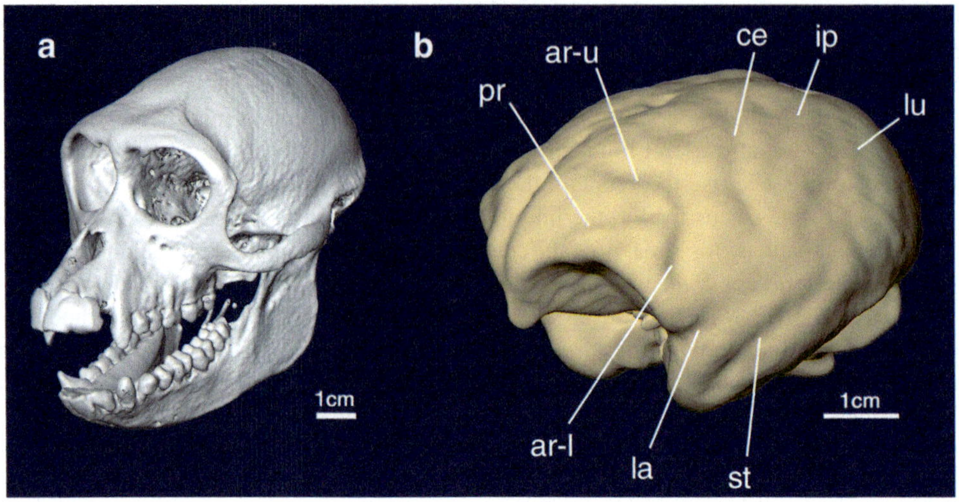

Fig. 3.2 A skull (**a**) and a virtual endocast (**b**) of a crab-eating monkey specimen reconstructed from CT data. *ar-l*, *ar-u* lower/upper limb of arcuate sulcus, *ce* central sulcus, *ip* intraparietal sulcus, *la* lateral sulcus, *lu* lunate sulcus, *pr* principal sulcus, *st* superior temporal sulcus

Table 3.2 Sulci identified on endocasts of dry skull specimen

	pr	ar	sf	fo	ipc	spc/spd	ce	ip	lu	la	st	amt	ot	o	lo	mo	CS
Lemur catta	○						○	○		○	○		○				—
Alouatta	○						○	○	○	○	○	○		○			+
Presbytis cristata 1	○	○					○	○		○	○	○			○	○	—
Presbytis cristata 2	○	○					○	○	○	○	○	○			○	○	+
Presbytis cristata 3	○	○					○	○		○	○	○			○	○	—
Lagothrix 1	○		○		○	○	○	○	○	○	○	○			○	○	+
Lagothrix 2	○		○		○	○	○	○	○	○	○	○			○	○	+
Pygathrix nemaeus	○	○				○	○	○	○	○	○	○			○	○	+
Macaca arctoides 1	○	○				○	○	○	○	○	○	○			○	○	+
Macaca arctoides 2	○	○				○	○		○	○	○	○			○	○	+
Macaca fascicularis 1	○	○				○	○	○	○	○	○	○			○	○	+
Macaca fascicularis 2	○	○				○	○	○	○	○	○	○			○	○	±
Macaca fuscata 1	○	○				○	○		○	○	○	○			○	○	+
Macaca fuscata 2	○	○				○	○		○	○	○	○			○	○	+
Hylobates	○		○	○	○	○	○	○	○	○	○		○		○	○	+
Pan troglodytes										○					○	○	+

images were analyzed using the Amira 5.4 software package (Visage Imaging; Berlin, Germany) on a Z620 workstation (Hewlett-Packard Japan; Tokyo, Japan) on a Mac Pro computer (Apple; Cupertino, CA, USA). To create virtual endocasts, we selected pixels inside the skull in each slice based on their density and reconstructed the surfaces. Cerebral sulci were identified based on our macroscopic samples of monkey brains for crab-eating macaques, Japanese macaques, and chimpanzees, as well as using descriptions in previous reports (Connolly 1936; Connolly 1950; Paxinos et al. 2000; Bailey et al. 1950).

Table 3.2 also shows the cerebral sulci identified on the endocasts. The skull of the chimpanzee was opened, so that we could get an incomplete reconstruction of the endocast. In monkeys and a gibbon, we observed the major cerebral sulci that faced the inner surface of the skull (Fig. 3.3). As for the sulci demarcating cerebral lobes, the lateral sulcus (la) and the central sulcus (ce; Rolando fissure) were clearly observed in all the skulls. In the frontal lobe, sulci observed in all cases included the principal sulcus (pr), which is homologous to the inferior frontal sulcus in human, and the upper and lower limbs of the arcuate sulcus (ar), which correspond to the superior frontal sulcus (sf) and the inferior precentral sulcus (ipc), respectively. The medial and lateral orbital sulci (mo, lo) were seen in a gibbon and monkeys, except for the lemur and the howler monkey, the latter having only a single orbital sulcus (o). In the parietal lobe, the intraparietal sulcus (ip) was identified in all the skulls except for two Japanese macaque specimens, which had the largest skulls among the monkeys we examined. In the Japanese macaques, the parietal lobe was so smooth that hardly any trace of sulci was detected; the central sulcus was

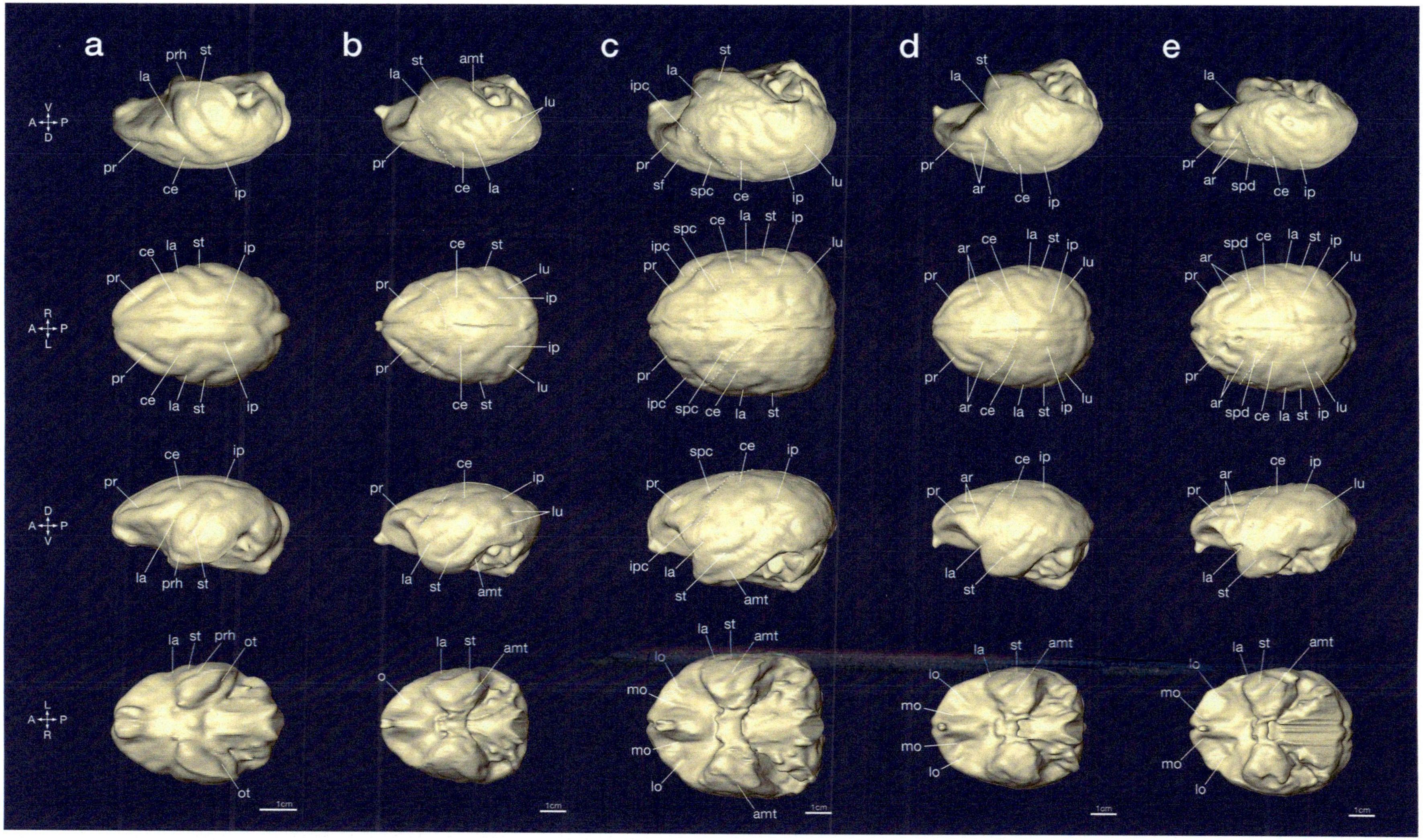

Fig. 3.3 Virtual endocasts reconstructed from dry skull specimens. (**a**) Lemur (*Lemur catta*), (**b**) howler monkey (*Alouatta*), (**c**) woolly monkey (*Lagothrix*) 2, (**d**) silvered leaf monkey (*Presbytis cristata*) 2, (**e**) douc (*Pygathrix nemaeus*), (**f**) *red*-faced stump-tailed macaque (*Macaca arctoides*) 1, (**g**) crab-eating macaque (*Macaca fascicularis*) 2, (**h**, **i**) Japanese macaque (*Macaca fuscata*) 1–2, (**j**) gibbon (*Hylobates*). *Top*, *middle-top*, *middle-bottom*, and *bottom* panels show the *right lateral*, *top*, *left lateral*, and *bottom* sides of the endocasts, respectively. A scale bar represents 1 cm for each specimen. *amt* anterior middle temporal sulcus, *ar* arcuate sulcus, *ce* central sulcus, *fo* fronto-orbital sulcus, *ip* intraparietal sulcus, *ipc* inferior precentral sulcus, *la* lateral sulcus, *lo* lateral orbital sulcus, *lu* lunate sulcus, *mo* medial orbital sulcus, *o* orbital sulcus, *ot* occipitotemporal sulcus, *pr* principal sulcus, *prh* posterior rhinal sulcus, *sf* superior frontal sulcus, *spc* superior precentral sulcus, *spd* superior precentral dimple, *st* superior temporal sulcus

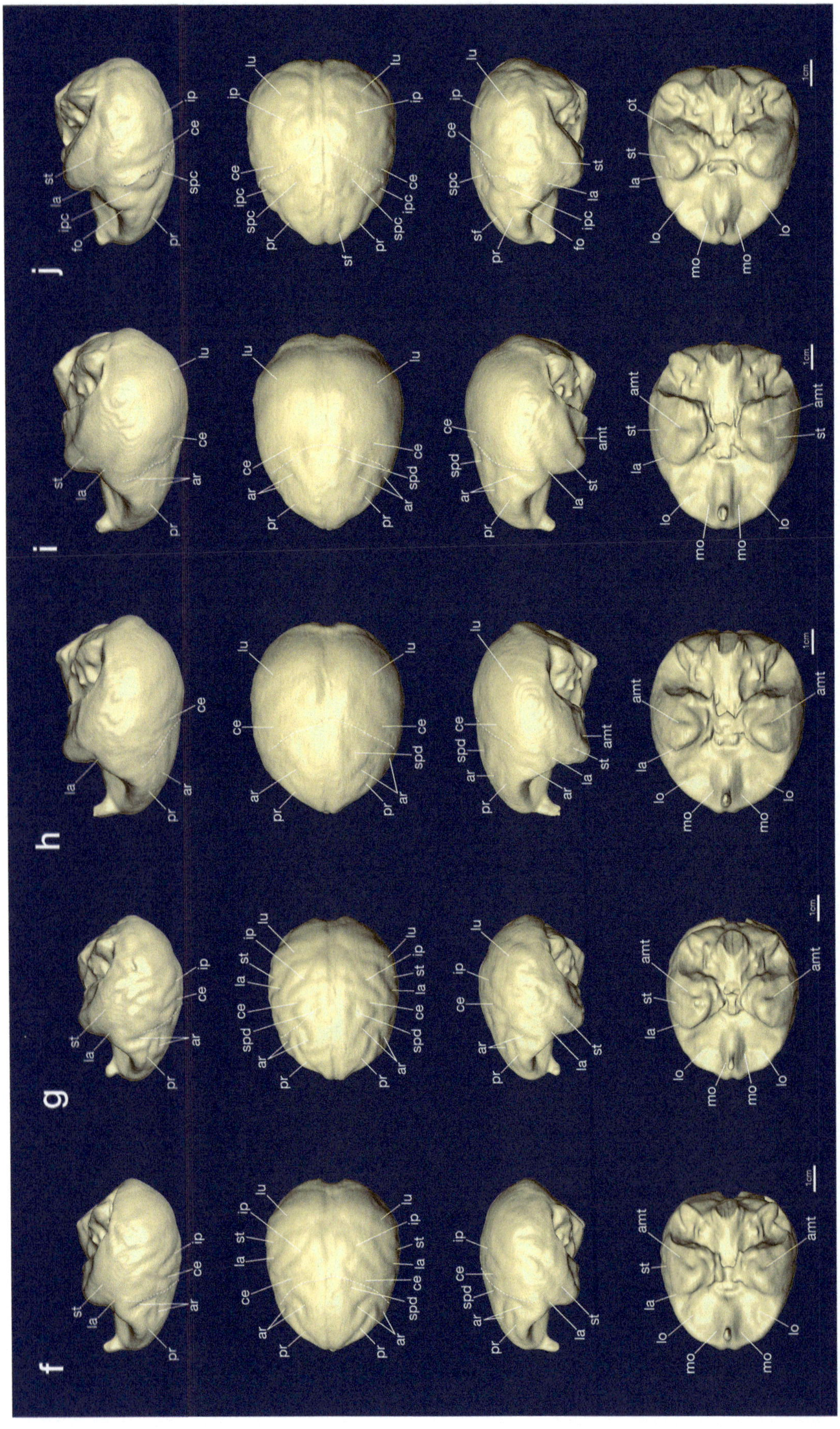

Fig. 3.3 (continued)

less distinct than in other monkeys. In the occipital lobe, the lunate sulcus (lu) was observed in most of the skulls, although very faintly in the Japanese macaques. In the temporal lobe, the superior temporal sulcus (st) was most distinct in its anterior portion but obscured posteriorly in most of the endocasts. The anterior middle temporal sulcus (amt) was also found in most of the cases. We cannot deduce a general rule from the limited number of cases analyzed in the study above, but imprints on the endocasts will be highly informative in monkeys and lesser apes.

In humans and the other great apes, the inference of the location of cerebral sulci is more problematic. Our preliminary study in human infants showed convolutional patterns not only on the basal part but also on the vault of the endocast, although correspondence to cerebral sulci was not always clear (Matsui and Kobayashi 2012). The connective tissues and blood vessels on the brain surface are thinner in young animals. The major factors that seem to obscure the course of sulci are (1) the thick connective tissue and cistern around the anterior portion of the lateral sulcus and (2) the superior, inferior, and superficial middle cerebral veins that are often located on cerebral sulci. In addition, developing brains may exert more influence on the adjacent skull to make extra space in which to grow. Infant skulls may provide a clue in locating cerebral sulci on endocasts in species with large brains and skulls. Several major cerebral sulci were actually identified on the endocast of a child skull of Mojokerto (Balzeau et al. 2005). Developmental studies will be necessary to further evaluate this approach.

Another important issue is the deformation and fragmentation of skulls during fossilization process. Virtual reconstruction of the skulls on computers provides practical and useful tools for endocast analysis (for review, see Zollikofer 2002; Gunz et al. 2009; Ogihara et al. 2015). For example, in our project "Replacement of Neanderthals by Modern Humans" (http://www.koutaigeki.org/eng/index.html), we have also carried out research on the statistical interpolation of the missing parts of the skull (Kikuchi and Ogihara 2013; Amano et al. 2014, 2015) and on the assessment of the left-right asymmetry after elimination of the deformity of the original skull (Kondo et al. 2014). These methods will provide a less-biased reconstruction of skulls and contribute not only to the better estimation of the overall dimensions of endocasts but also to the more accurate reproduction of their surface morphology.

3.7 Inferring Cerebral Sulci Based on Skull Landmarks

This approach was not originally adopted for studies on fossil species but for neurological and neurosurgical needs before the development of radiological imaging techniques. A large number of studies were carried out since Broca's description on the spatial relationship between the central sulcus and the coronal suture (Broca 1861) (see review by Broca (1876) and Anderson and Makins (1889b)). For example, Horsley (1892) measured the location of the upper end of the central sulcus in the sagittal arch from glabella to inion; the central sulcus was located 12.5 mm posterior from the midpoint of the sagittal arch, which corresponded to 55.7% of the glabella – inion distance. Anderson and Makins (1889a) also reported that the upper end of the central sulcus fell between the midsagittal point and 19 mm posterior from it. Cunningham and Horsley (1892) conducted a more elaborated analysis and stated that the central and precentral sulci were "remarkably constant in its relative position to the rest of the hemisphere."

Interestingly, Cunningham and Horsley (1892) showed illustrations of the dissected heads of humans and nonhuman primates, in which the frontal and parietal bones, as well as the squamous parts of the temporal and occipital bones, were largely removed except for the portions adjacent to the sutures (Fig. 3.4). In nonhuman primates (rhesus, baboon, cebus, orangutan, and chimpanzee), the inferior precentral sulcus, or its homologue, the lower limb of the arcuate sulcus, was largely hidden underneath the bones comprising the coronal suture, while in humans the inferior precentral sulcus was located posteriorly, some distance from the suture. Flatau and Jacobsohn (1899) also illustrated the cerebral convolutional pattern with the skull and certain sutures in a lemur, macaque monkey, and chimpanzee (Fig. 3.5). The lower portion of the coronal suture fell over the position of the inferior precentral sulcus in the chimpanzee and the lower arcuate sulcus in the macaque.

These findings prompted us to reevaluate the relationship between the sulci and sutures. We first utilized the skulls and brains of five *Macaca fascicularis* monkeys. We examined and compared the locations of the coronal suture and the lower limb of the arcuate sulcus (Kobayashi et al. 2014b). The coronal suture was identified on CT images as curved low-density lines extending laterally from the bregma. The arcuate sulcus was traced on lateral photographic images of the brain and was superimposed on the CT images of the skull. In this study, we defined the plane through the frontal and occipital poles as horizontal and measured the horizontal distance of the suture and the sulcus from the frontal pole at different dorsoventral levels (from level 0 at the fronto-occipital line through level 10 at the vertex of the brain (Fig. 3.6). The distances were normalized by using their proportions to the fronto-occipital length of the endocast. The data showed that the lower limb of the arcuate sulcus was located slightly anterior to the lower half of the coronal suture within a very limited area: the average distance ± S.D. was 0.0–1.4% ± 1.1–3.0% of the distance between the frontal and occipital poles.

We next analyzed the dry skull specimens used for the evaluation of the first approach (examples are shown in Fig. 3.3). The suture was fused completely and left no trace of the coronal suture in the lemur and two of the silvered leaf monkey specimens and was partially obscured in a crab-

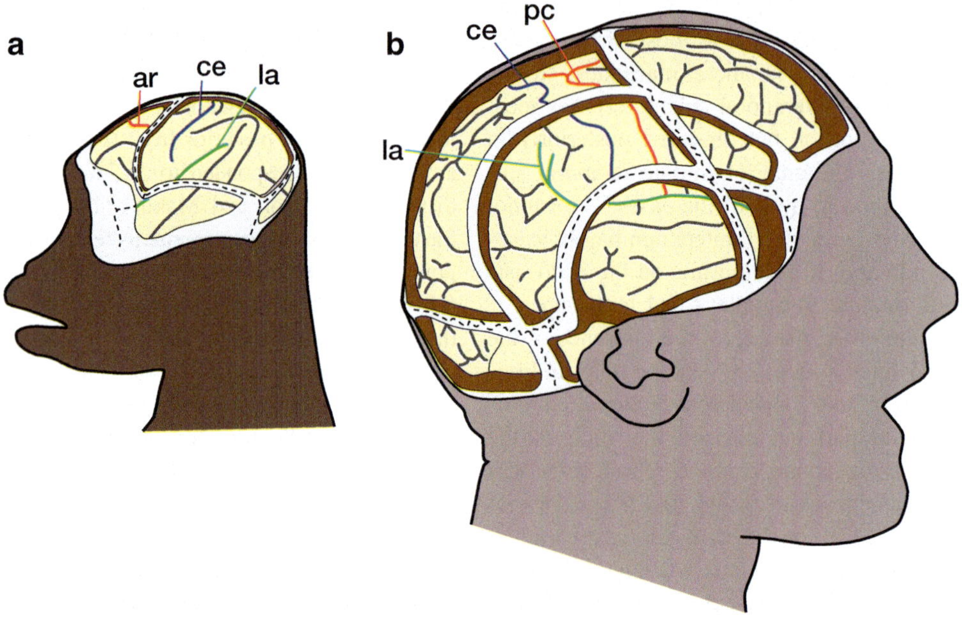

Fig. 3.4 Spatial relationship between sutures and cerebral sulci in rhesus monkey (**a**) and human (**b**) illustrated by Cunningham (Redrawn from Cunningham and Horsley 1892). *ar* arcuate sulcus (upper limb), *ce* central sulcus, *la* lateral sulcus, *pc* precentral sulcus. Note that bones adjacent to the temporal ridge remained in (**a**)

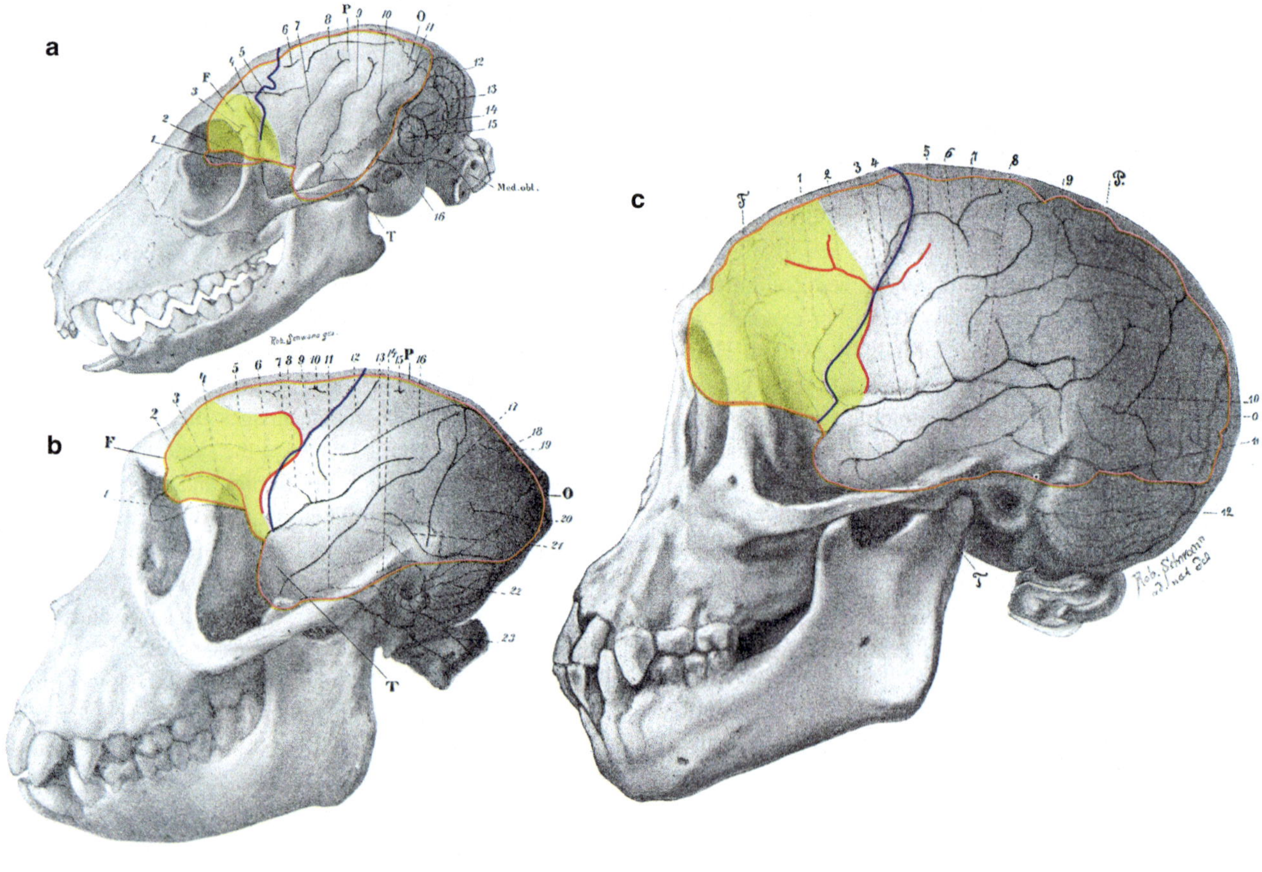

Fig. 3.5 Spatial relationship between sutures and cerebral sulci in lemur (**a**), rhesus monkey (**b**), and chimpanzee (**c**), illustrated by Flatau and Jacobsohn (1899). Color was added to the original picture. *Blue lines* and *red lines* represent the coronal suture and the arcuate sulcus/inferior precentral sulcus, respectively. Areas painted in yellow depict the prefrontal cortex determined in later studies

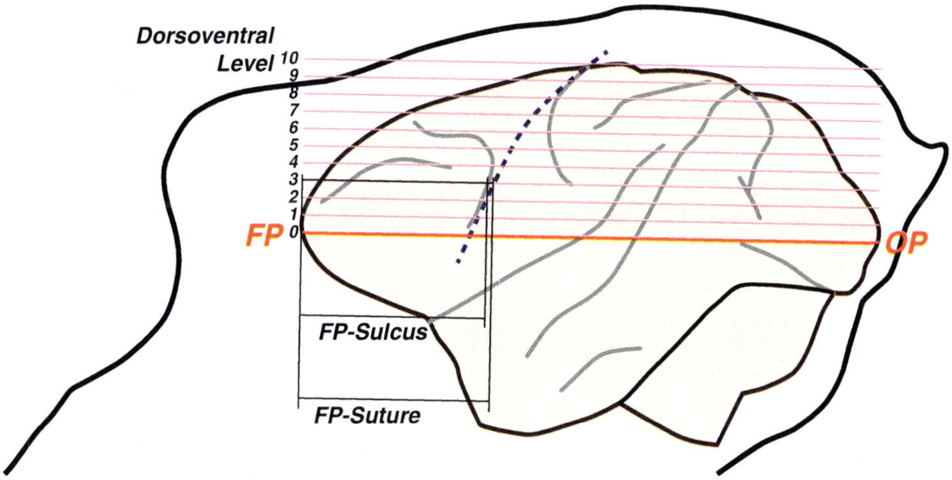

Fig. 3.6 Measurements of the locations of the coronal suture and the arcuate sulcus. We defined the line connecting the frontal and occipital poles (FP, OP) as the horizontal line. The space between the horizontal line and the vertex of the brain was divided by nine lines with equal intervals. We determined the relative dorsoventral levels: from DV0 on the horizontal line to DV10 on the line through the vertex and measured the horizontal distance of the suture and the sulcus from the frontal pole. To normalize between species and individual differences of the skull and brain size, we used the ratio of the distance to the fronto-occipital length (FP-OP)

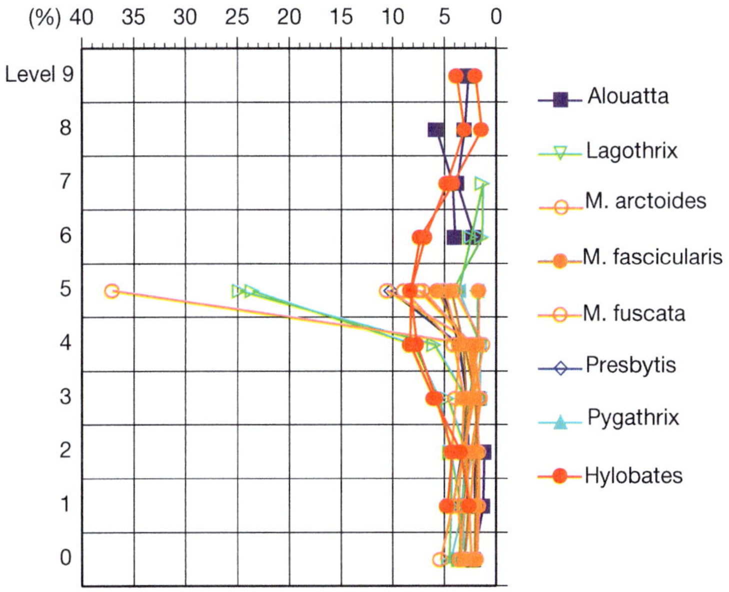

Fig. 3.7 Distance of the arcuate or precentral sulcus from the coronal suture. In dry skulls, sutures were easily identified as sharp, thin ridges on the endocasts in contrast to smooth, wide bumps corresponding to cerebral gyri. The *horizontal axis* of the graph is for the distance from the coronal suture in proportion to the fronto-occipital length (in percentage), and the vertical axis is for the dorsoventral levels as indicated in Fig. 3.6. The *right edge* indicates the location of the coronal suture, and the *left side* is anterior. *Colored lines* represent the locations of the sulci; different species are shown with different markers

eating macaque specimen. In other skulls, however, the coronal suture was easily identified, both on CT images and on the reconstructed endocasts. At dorsoventral levels 0–1, the arcuate sulcus was often absent or obscured due to the proximity of the lateral sulcus. At levels 2–4, the suture and sulcus were clearly and most frequently observed on the endocasts. At level 5, the arcuate sulcus made a sharp curve anteriorly to form its upper limb and ended between levels 5 and 6 except in the howler monkey and the gibbon. Even though the samples were derived from a wide variety of taxa, the lower limb of the arcuate sulcus or the inferior precentral sulcus was generally located slightly anterior to the lower half of the coronal suture with an amazingly small variance at levels below 4 (Fig. 3.7). The location of the lower half of the coronal suture and the lower limb of the arcuate sulcus exhibited strong correlation at levels 2–3 (Fig. 3.8a).

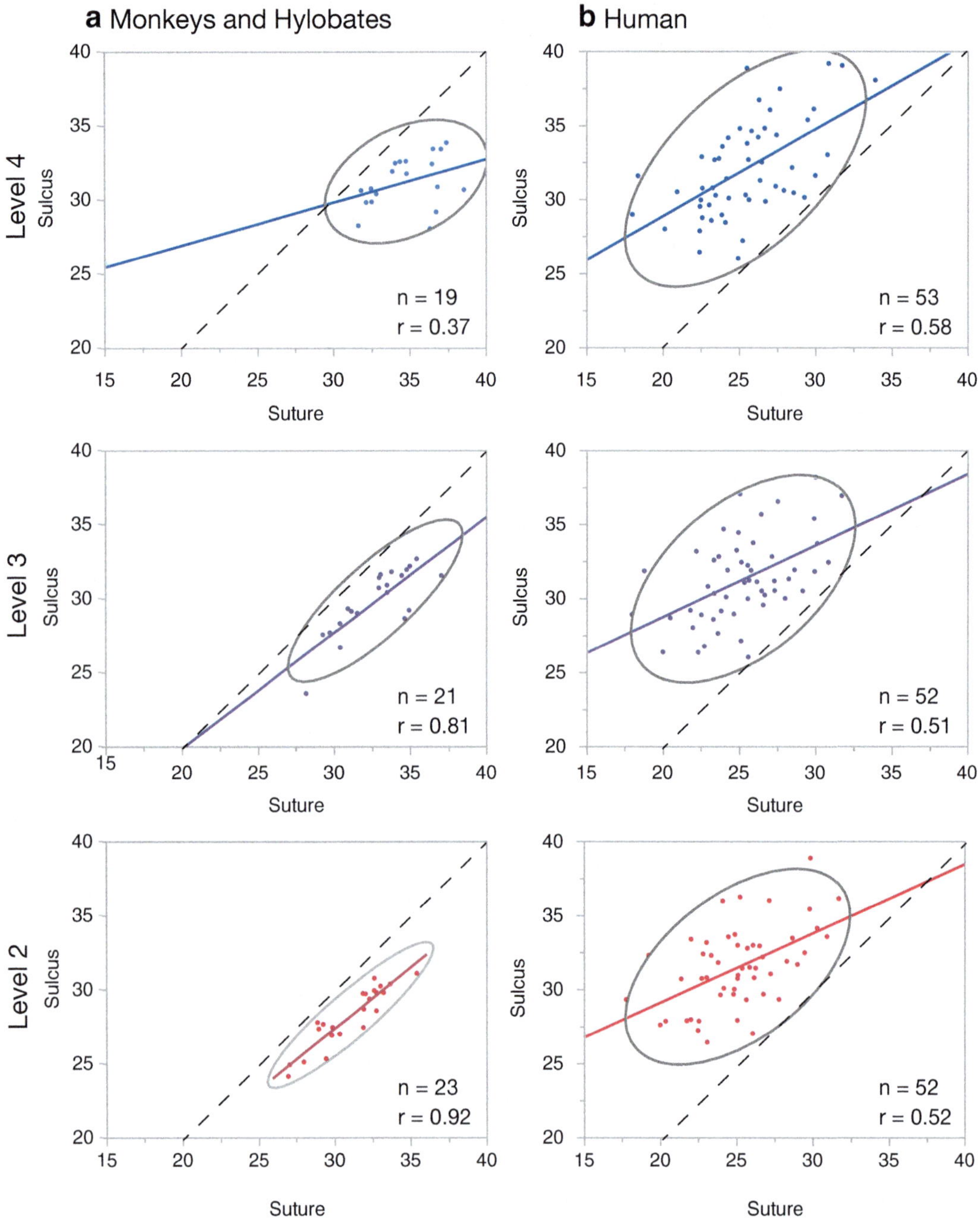

Fig. 3.8 Correlation of the locations of the coronal suture and the arcuate sulcus/inferior precentral sulcus. In dry skull specimens of monkeys and a gibbon (**a**), the locations of the suture and arcuate sulcus show strong correlation at dorsoventral levels 2–3 and moderately strong correlation at level 4, where the sulcus bends anteriorly in several samples. In human cadavers (**b**), the location of the suture exhibited moderately strong correlations at levels 2–4. The regression line and the 95% probability ellipse are drawn in each scatter plot

A similar correlation was also observed in the human, in which we measured the location of the suture on the inner surface of the skull and that of the sulcus on the brain in Japanese cadavers. It should be noted that the individual differences were even larger in humans than in primate samples from different taxa (Fig. 3.8b). However, the location of the coronal suture and the inferior precentral sulcus showed moderately strong correlations, which indicate that, solely based on a skull, we may be able to infer the location of the precentral sulcus of the same individual. We plan to increase the number of the analyzed cases to establish a statistically verified method to infer the location of the inferior precentral sulcus and apply it to Neanderthal skulls.

Recently, the extent of the precuneus, which is highly variable in the human, was statistically analyzed using MRI (Bruner et al. 2015b) and compared with the extent of the parietal bone (Bruner et al. 2015a). The results showed low correlation between the parietal lobe and parietal bone lengths ($r = 0.27$ for chords and 0.32 for arcs) and between the precuneus length and the parietal bone length ($r = 0.20$ for chords and 0.24 for arcs). Both correlations are considerably lower than those observed between the locations of the coronal suture and the inferior precentral sulcus in our study. The reason of this discrepancy is not clear, but one possible factor is that bregma and lambda are formed by the fusion of the concentric ossification of the frontal, parietal, and occipital bones long after birth, and their positions can be modified by differences in ossification speed between adjacent bones. The lateral aspect of the coronal suture is fused earlier and might be more stable than the locations of bregma and lambda. A comprehensive analysis of the individual differences of subdivisions of the brain and the skull will be necessary to evaluate the validity of the use of skull landmarks as reference points to infer borders of brain regions.

3.8 Concluding Remarks

In this article, we have summarized classical and recent views concerning the cortical evolution in primates including human and also presented two approaches we examined recently to infer the extent of cortical areas based on the skull morphology.

The method using the endocast surface morphology is more straightforward and useful in monkeys and gibbons which have relatively smaller skulls and may be also applicable to very young individuals in greater apes and humans. However, probably due to the thickness of connective tissue and the vascularity, it is not suitable in adult humans except for the lower portion of the cortex, such as the orbitofrontal and inferior temporal cortices.

The other method using the coronal suture as a landmark to infer the location of the precentral sulcus can be applied in any skull unless the suture is fused by ossification. This approach is, however, based on an assumption that the relationship between the locations of the suture and the sulcus is stable in different species. Because the relationship in the modern human is different from that in monkeys and gibbons, further studies in the extant great apes will be necessary before we interpolate it in fossil hominines.

Acknowledgment The authors thank Ms. Mayumi Watanabe, Mr. Tohru Tamai, and Mr. Hiroshi Sasaki at the Department of Anatomy and Neurobiology, National Defense Medical College, for the assistance in histology and cadaver analysis. This study was supported by Grant-in-Aid for Scientific Research on Innovative Areas (Grant No. 23101509, 25101711) from the Japanese Ministry of Education, Science, Culture, and Technology.

References

Amano H, Morita Y, Nagano H, Kondo O, Suzuki H, Nakatsukasa M, Ogihara N (2014) Statistical interpolation of missing parts in human crania using regularized multivariate linear regression analysis. In: Dynamics of learning in neanderthals and modern humans, vol 2. Springer, pp 161–169

Amano H, Kikuchi T, Morita Y, Kondo O, Suzuki H, Ponce de León MS, Zollikofer CP, Bastir M, Stringer C, Ogihara N (2015) Virtual reconstruction of the Neanderthal Amud 1 cranium. Am J Phys Anthropol 158(2):185–197

Anderson W, Makins GH (1889a) Experiments in Cranio-cerebral topography. J Anat Physiol 23(Pt 3):455–465

Anderson W, Makins GH (1889b) Experiments in Cranio-cerebral topography. Lancet 2:61–64

Anthony R (1913) L'encéphale de l'homme fossile de La Quina. Bull Mém Soc d'Anthropol Paris:117–194

Ariëns Kappers CU (1909) The phylogenesis of the palaeo-cortex and archi-cortex compared with the evolution of the visual neo-cortex. Arch Neurol Psychiatr 4:161–173

Bailey P, von Bonin G, McCulloch WS (1950) The isocortex of the chimpanzee. The University of Illinois Press, Urbana

Balzeau A, Grimaud-Herve D, Jacob T (2005) Internal cranial features of the Mojokerto child fossil (East Java, Indonesia). J Hum Evol 48 (6):535–553

Beaudet A, Dumoncel J, de Beer F, Duployer B, Durrleman S, Gilissen E, Hoffman J, Tenailleau C, Thackeray JF, Braga J (2016) Morphoarchitectural variation in South African fossil cercopithecoid endocasts. J Hum Evol 101:65–78

Blinkov SM, Glezer IaI (1968) The human brain in figures and tables: a quantitative handbook. Basic Books, New York

Boule M, Anthony R (1911) L'encéphale de l'homme fossile de la Chapelle-aux-Saints. Anthropologie 22:129–196

Broca P (1861) Remarques sur le siége de la faculté du langage articulé, suivies d'une observation d'aphémie (perte de la parole). Bull Soc Anat 2e Ser 6:330–357

Broca P (1876) Sur la topographie cranio-cérébrale ou sur les rapports anatomiques du crane et du cerveau. Revue d'Anthroplogique:193–248

Brodmann K (1906) Beiträge zur histologischen Lokalisation der Grosshirnrinde. Fünfte Mitteilung: Über den allgemeinen Bauplan des Cortex pallii bei den Mammaliern und zwei homologe Rindenfelder im besonderen. Zugleich ein Beitrag zur Furchenlehre. J Psychol Neurol 6(Ergänzungsheft):275

Brodmann K (1912) Neue Ergebnisse über die vergleichende histologische Lokalisation der Grosshirnrinde mit besonderer Berücksichtigung des Stirnhirns. Anat Anz 41:157–216

Bruner E (2004) Geometric morphometrics and paleoneurology: brain shape evolution in the genus homo. J Hum Evol 47(5):279–303

Bruner E, Manzi G, Arsuaga JL (2003) Encephalization and allometric trajectories in the genus homo: evidence from the Neandertal and modern lineages. Proc Natl Acad Sci U S A 100(26):15335–15340

Bruner E, Amano H, de la Cuetara JM, Ogihara N (2015a) The brain and the braincase: a spatial analysis on the midsagittal profile in adult humans. J Anat 227(3):268–276

Bruner E, Roman FJ, de la Cuetara JM, Martin-Loeches M, Colom R (2015b) Cortical surface area and cortical thickness in the precuneus of adult humans. Neuroscience 286:345–352

Connolly CJ (1936) The fissural pattern of the primate brain. Am J Phys Anthropol 21(3):301–422

Connolly CJ (1950) External morphology of the primate brain. Thomas, Springfield

Cunningham DJ, Horsley V (1892) Contribution to the surface anatomy of the cerebral hemispheres: with a chapter upon Cranio-cerebral topography. Academy House, Dublin

Dart RA (1925) Australopithecus africanus: the man-ape of South Africa. Nature 115(2884):195–199

Dart RA (1940) The status of Australopithecus. Am J Phys Anthropol 26(1):167–186

Falk D (1980) A reanalysis of the South African australopithecine natural endocasts. Am J Phys Anthropol 53(4):525–539

Falk D (1983) The Taung endocast: a reply to Holloway. Am J Phys Anthropol 60(4):479–489

Flatau E, Jacobsohn L (1899) Handbuch der Anatomie und vergleichenden Anatomie des Centralnervensystem der Säugetiere. 1. Makroskopischer Teil. Karger, Berlin

Flechsig P (1920) Anatomie des menschlichen Gehirns und Rückenmarks auf myelogenetischer Grundlage. Thieme, Leipzig

Gabi M, Neves K, Masseron C, Ribeiro PF, Ventura-Antunes L, Torres L, Mota B, Kaas JH, Herculano-Houzel S (2016) No relative expansion of the number of prefrontal neurons in primate and human evolution. Proc Natl Acad Sci U S A 113(34):9617–9622

Gibson KR, Rumbaugh D, Beran M (2001) Bigger is better: primate brain size in relationship to cognition. In: Evolutionary anatomy of the primate cerebral cortex. Cambridge University Press, Cambridge, pp 79–97

Glasser MF, Van Essen DC (2011) Mapping human cortical areas in vivo based on myelin content as revealed by T1- and T2-weighted MRI. J Neurosci 31(32):11597–11616

Gunz P, Mitteroecker P, Neubauer S, Weber GW, Bookstein FL (2009) Principles for the virtual reconstruction of hominin crania. J Hum Evol 57(1):48–62

Herculano-Houzel S, Lent R (2005) Isotropic fractionator: a simple, rapid method for the quantification of total cell and neuron numbers in the brain. J Neurosci 25(10):2518–2521

Herculano-Houzel S, Collins CE, Wong P, Kaas JH (2007) Cellular scaling rules for primate brains. Proc Natl Acad Sci U S A 104 (9):3562–3567

Holloway RL (1980) Revisiting the south African Taung australopithecine endocast: the position of the lunate sulcus as determined by the stereoplotting technique. Am J Phys Anthropol 56(1):43–58

Horsley V (1892) On the topographical relations of the cranium and surface of the cerebrum. In: Topography CttSAotCHWaCUC-c (ed) Contribution to the surface anatomy of the cerebral hemispheres: with a chapter upon Cranio-cerebral topography. Academy house, Dublin, pp 306–358

Jerison HJ (1973) Evolution of the brain and intelligence. Academic, New York

Kikuchi T, Ogihara N (2013) Computerized assembly of neurocranial fragments based on surface extrapolation. Anthropol Sci 121 (2):115–122

Kobayashi Y, Matsui T, Haizuka Y, Ogihara N, Hirai N, Matsumura G (2014a) Cerebral sulci and gyri observed on macaque endocasts. In: Dynamics of learning in neanderthals and modern humans, vol 2. Springer, pp 131–137

Kobayashi Y, Matsui T, Haizuka Y, Ogihara N, Hirai N, Matsumura G (2014b) The coronal suture as an indicator of the caudal border of the macaque monkey prefrontal cortex. In: Dynamics of learning in neanderthals and modern humans, vol 2. Springer, pp 139–143

Kondo O, Kubo D, Suzuki H, Ogihara N (2014) Virtual endocast of Qafzeh 9: a preliminary assessment of right-left asymmetry. In: Dynamics of learning in neanderthals and modern humans, vol 2. Springer, pp 183–190

Le Gros Clark WE (1945) Note on the palaeontology of the lemuroid brain. J Anat 79(Pt 3):123–126

Le Gros Clark WE, Cooper DM, Zuckerman S (1936) The endocranial cast of the chimpanzee. J R Anthropol Inst G B Irel 66:249–268

Matsui T, Kobayashi Y (2012) Developing cranial parameters that delineate subdivisions of the brain – skulls of human infants. Paper presented at the The 5th conference on replacement of Neanderthals by modern humans: testing evolutionary models of learning, Tokyo, April 14–16, 2012

Ogawa T, Kamiya T, Sakai S, Hosokawa H (1970) Some observation on the endocranial cast of the Amud man. In: Suzuki H, Takai F (eds) The Amud man and his cave site. The University of Tokyo Press, Tokyo, pp 411–424

Ogihara N, Amano H, Kikuchi T, Morita Y, Hasegawa K, Kochiyama T, Tanabe HC (2015) Towards digital reconstruction of fossil crania and brain morphology. Anthropol Sci 123 (1):57–68

Passingham RE, Smaers JB (2014) Is the prefrontal cortex especially enlarged in the human brain allometric relations and remapping factors. Brain Behav Evol 84(2):156–166

Paxinos G, Huang XF, Toga AW (2000) The rhesus monkey brain in stereotaxic coordinates. Academic, San Diego

Radinsky L (1972) Endocasts and studies of primate brain evolution. The functional and evolutionary biology of primates, ed R Tuttle Aldine[DF]

Schoenemann PT, Sheehan MJ, Glotzer LD (2005) Prefrontal white matter volume is disproportionately larger in humans than in other primates. Nat Neurosci 8(2):242–252

Semendeferi K, Lu A, Schenker N, Damasio H (2002) Humans and great apes share a large frontal cortex. Nat Neurosci 5(3):272–276

Smith-Agreda V (1955) The distribution of the impressions gyrorum at the inner side of the human cranium; with utilization of endocranial casts. Dtsch Z Nervenheilkd 173(1):37–68

Stephan H, Andy OJ (1969) Quantitative comparative neuroanatomy of primates: an attempt at a phylogenetic interpretation. Ann N Y Acad Sci 167(1):370–387

Symington J (1916) Endocranial casts and brain form: a criticism of some recent speculations. J Anat Physiol 50(Pt 2):111–130

Zollikofer CP (2002) A computational approach to paleoanthropology. Evol Anthropol: Issues News Rev 11(S1):64–67

Fossil Primate Endocasts: Perspectives from Advanced Imaging Techniques

4

Amélie Beaudet and Emmanuel Gilissen

Abstract

Compared to their putative insectivore-like ancestors, extant primates show an enlarged brain relative to body weight, a larger neocortex and proportionally decreased olfactory bulbs. Besides hypotheses based on the comparative neuroanatomy of extant taxa, the only direct evidence documenting such long-term evolutionary history is provided by fossil endocasts. However, due to the unpredictable yet unavoidable impact of taphonomic processes, the reliability of data from the fossil record is complicated by the nature of the investigated structures themselves. Nonetheless, palaeoneurology has recently enlarged its traditional investigative toolkit by integrating descriptive morphology with advanced methods of high-resolution 3D imaging and computing. In addition to the development of digital restoration techniques, the introduction of analytical methods for investigating topographic differences in morphostructural organization and quantitatively characterizing intra- and interspecific variation patterns provides new possibilities for the study of the primate fossil record, especially for assessing brain evolutionary tracks.

Keywords

Palaeoneurology • Primate brain evolution • Deformation-based models • Semi-automatic sulci detection • Computer-assisted reconstruction

4.1 A Review of Non-human Primate Palaeoneurology

4.1.1 Evidence from Primate Palaeoneurology

One of the most fascinating unsolved problems in palaeoneurology centres on unravelling the selective pressures that were responsible for both the increase in relative brain size and the cerebral reorganization in primates (Radinsky 1975). Investigating the evolutionary changes of the brain across various species is critical to characterize phylogenetic specializations and provide insights into the interaction between an organism and its environment (Barton and Harvey 2000; de Winter and Oxnard 2001). As stated by Le Gros Clark (1971: 227–228): "[Thus] the progressive elaboration and differentiation of the cortex in the evolving Primates have led to increasing powers of apprehending the nature of

A. Beaudet (✉)
School of Geography, Archaeology and Environmental Studies, University of the Witwatersrand, Johannesburg, South Africa

Department of Anatomy, University of Pretoria, Pretoria, South Africa
e-mail: beaudet.amelie@gmail.com

E. Gilissen
Department of African Zoology, Royal Museum for Central Africa, Tervuren, Belgium

Laboratory of Histology and Neuropathology, Université Libre de Bruxelles, Brussels, Belgium
e-mail: emmanuel.gilissen@africamuseum.be

E. Bruner et al. (eds.), *Digital Endocasts*, Replacement of Neanderthals by Modern Humans Series,
DOI 10.1007/978-4-431-56582-6_4

external stimuli, a greater capacity for a wider range of adjustments to any environmental change, and an enhancement of the neural mechanisms for effecting more delicately co-ordinated reactions".

The relative importance and the timing of two critical processes in the evolution of the primate brain, i.e. cortical reorganization and size increase, has been largely questioned (Gonzales et al. 2015). Recent evidence that brain size and level of gyrification are controlled by different genes in extant catarrhine primates corroborates the often observed lack of correlation between these two structural parameters (Welker 1990) and suggests that encephalization and cerebral complexity could have evolved independently (Rogers et al. 2010). Despite the relative scarcity of direct fossil evidence, palaeontological studies provide valuable evidence, notably by revealing that cerebral complexity preceded enlarged brain size in particular primate lineages (Gonzales et al. 2015).

For a better understanding of primate brain evolution, it is then crucial to integrate and combine different approaches as well as various sources of data. However, this prerequisite represents a critical challenge since the data based either directly on the brain or indirectly on endocasts are by nature very different (Neubauer 2014). Our knowledge of the evolution of the primate brain primarily relies on the interpretation of palaeoneurological evidence and on comparative information from extant species, in which the brain and behaviour can be investigated directly (Armstrong and Falk 1982). Fossil endocasts are replicas of the internal table of the bony braincase and provide the only direct evidence of brain evolution. When the neurocranium is filled with sediment during fossilization, morphological information about the external brain surface may be preserved as a natural endocast, as remarkably illustrated by the South African primate fossil records (e.g. the hominin specimens Taung, Sts 60, SK 1585 or the cercopithecoid specimens MP3a, MP 36, STS 538, STS 564, STS 565; Brain 1981; Holloway et al. 2004). Endocasts thus constitute a proxy for investigating and quantifying variations in brain size, global brain shape and neocortical surface morphology (if the dura mater was thin enough in the living animal), including imprints of cerebral convolutions (i.e. gyri and sulci) (Holloway 1978; Holloway et al. 2004; Falk 2014; Neubauer 2014). The imprinted cerebral surface corresponds to much of the forebrain, especially the neocortex, which includes visual, auditory, somatosensory, motor as well as association areas (Radinsky 1975). However, the information the neocortical surface provides is limited to the external morphology, and little can be said about the evolution of subcortical elements.

In 1978, Holloway defined at least six levels of useful evidence to be gleaned from endocasts, depending on their completeness and the replicability of neocortical details: (1) the gross brain size, which corresponds to the volume of neural mass; (2) the areal determination, which is the surface of the endocast divided into major lobar regions; (3) the major sulcal and gyral identifications and blood vessel patterns; (4) the identification of secondary and tertiary sulcal and gyral convolutions and their functional correspondence; and (5), to some extent, the subcortical relationships of these parameters.

Two levels received particular interest in palaeoneurology. Firstly, brain size has been extensively explored in the primate fossil record, even used as a criterion to define taxa at some points (e.g. definition of the genus *Homo* by Leakey et al. 1964). Secondly, the pattern of cerebral fissures apparent on the endocasts can suggest the location of cortical functional divisions (Kaas 2006). Based on cortical maps determined by neurophysiologists, the coincidence between architectonic and functional entities has been used to interpret fossil endocast (Radinsky 1975; Falk 1981, 1982). Moreover, sulcal variation has been demonstrated to be a reasonable predictor of cytoarchitecture for primary and secondary regions, such as visual, somatosensory and motor areas (Fischl et al. 2008). However, it has been shown that the cerebral areas delimitated by sulci on the external cortical surface do not systematically coincide with functional areas (Amunts et al. 1999).

4.1.2 General Evolutionary Trends and Comparative Studies of Living Mammals

Increase in size and development of the fissuration pattern are two key processes in primate brain evolution. In 1971, Le Gros Clark pointed out that "undoubtedly the most distinctive trait of the Primates, wherein this order contrasts with all other mammalian orders in its evolutionary history, is the tendency towards the development of a brain which is large in proportion to the total body weight, and which is particularly characterized by a relatively extensive and often richly convoluted cerebral cortex" (Le Gros Clark 1971: 227). In his review of primate brain evolution based on living mammal and fossil endocasts, Radinsky (1975) added supplementary fundamental neuronal changes: "To summarize, evolutionary trends suggested by comparative studies of the brains of living insectivores and primates include: increase in size of brain relative to body weight, increase in amount of neocortex (beyond what one would expect from the increase in brain size); decrease in relative size of olfactory bulbs; increase in amount of visual cortex and size of lateral geniculate body (possibly accounted for by the overall increase in brain size); development of a central sulcus in anthropoids rather than the coronal sulcus seen in prosimians" (Radinsky 1975: 659).

Several lines of evidence highlight various evolutionary trends described by Le Gros Clark and Radinsky, which could be summarized as follows: (1) relative increase in brain size, (2) relative increase of the neocortical surface, (3) reorganization of the sensory system and (4) fissuration pattern reorganization. The four lines are briefly detailed here.

4.1.2.1 Brain Size

On average, compared to other mammalian orders, living primates have larger brains relative to body size. Based on Radinsky's estimations (1975), relative brain size in living strepsirhines exceeds from 2.4 to 7.0 times what would be expected in a basal insectivores (sensu Stephan et al., 1991) of comparable weight. The interpretation of the increase over time in primate brain size and the earliest evidence of this evolutionary trend were largely debated in the literature (for a review, see Radinsky 1982). Because several features of brain morphology scale to absolute brain size (e.g. volume of the frontal cortex, degree of gyrification), variation in the primate brain size is critical for understanding brain evolution and evolutionary process (Zilles et al. 1988; Semendeferi et al. 2002).

4.1.2.2 Neocortical Surface

Since its emergence in a mammalian ancestor, the neocortex has expanded in both relative and absolute size, particularly in anthropoid primates, in which the neocortex comprises up to 80% of the brain mass (Rakic and Kornack 2001). Insectivore brains are relatively small and have little neocortex with few (if any) neocortical folds. Indeed, folds appeared on the primate brain with the evolution of larger brains containing more neocortex (Radinsky 1975). The causes and mechanisms of cortical folding are largely debated in terms of either selective pressure or developmental and structural processes (Van Essen 1997; Hilgetag and Barbas 2005; Toro 2012; Zilles et al. 2013; Bayly et al. 2014; Ronan and Fletcher 2015; Tallinen et al. 2016).

4.1.2.3 Sensory System

Visual areas represent about 50% of the entire primate neocortex, and the neocortical expansion is associated with variance in functionally specific parts of the visual system (Barton 1998). Comparative studies between neural structures in extant primates demonstrate a greatly modified visual system in comparison with their earliest relatives, with an increased number of visual cortical areas enlarged and specialized for detailed vision (Felleman and Van Essen 1991; Kaas 2006). Besides the development of the visual cortex, the other remarkable change in early primate neocortex was the reorganization of the motor and premotor cortex. In contrast with their mammalian relatives, both strepsirhines and haplorhines have a large and well-defined primary motor field (M1) (Kaas 2006).

4.1.2.4 Fissuration Pattern

As described by Radinsky (1975), a longitudinally oriented fissure (i.e. the coronal sulcus) separates the representation of the head from that of the forelimb in the primary motor and somatosensory cortex in strepsirhines and in most mammals with a comparable degree of neocortical folding. On the contrary, in anthropoids, no major fissure exists in that position, and the primary motor cortex is instead separated from the primary somatosensory cortex by a transverse fissure (i.e. the central sulcus) (Radinsky 1975).

4.1.3 Reconstructing the Primate Brain Evolution from Extant Diversity and Fossil Endocasts

4.1.3.1 Evidence from Extant Diversity

All stages experienced in the course of primate brain evolution, from the simple smooth hemisphere to a highly convoluted one, are suggested to be represented among living primate taxa (Hill 1972). Based on previous work by Le Gros Clark (1971), Hill (1972) proposed an evolutionary model that included living mammals as representatives of each evolutionary stage, from fundamental to a more complex pattern. Accordingly, the elephant shrew *Macroscelides*, now included in the order Afrotheria, represents the basal brain morphology.

The diurnal tree shrew *Tupaia* was regarded as an example of the early reduction of the olfactory lobe and the development of visual areas in the primate lineage (Hill 1972). However, it now seems that many of the primate-like features observed in the visual system of tree shrews are actually evolutionary convergences that arose independently from those observed in primates (Kaas 2002). Such parallel evolution appears to also have occurred during early primate evolution: for instance, late Palaeocene to middle Eocene microsyopids probably expanded their cerebral cortex and evolved an improved visual sense independently from euprimates (Silcox et al. 2010).

Changes in cortical regions are characterized by different reorganizational tempos. Comparison of the myelo- and cytoarchitecture of the granular frontal cortex in the strepsirhine *Galago* and in the anthropoid *Macaca* suggests that considerable changes occurred in this cortical region during primate evolution, including the addition of new areas in anthropoids (Preuss and Goldman-Rakic 1991a). In contrast, many, although possibly not all, of the parietal and temporal association areas present in *Macaca* appear to have evolved early in primate history, prior to the divergence of the lineages that led to strepsirhines and haplorhines (Preuss and Goldman-Rakic 1991b).

The neopallium expansion in both strepsirhines and haplorhines is accompanied by a complex cortical folding and by a definite pattern of sulci (e.g. deep Sylvian fissure, calcarine complex, superior temporal sulcus, rectus sulcus, intraparietal sulcus). In Cebidae, Cercopithecidae and Colobidae, the hemispheres are enlarged, and the neocortical surface is richly convoluted, with the occurrence of a well-marked central sulcus and of various fissures on the occipital lobe, including the lunate sulcus. The degree of neocortical fissuration is particularly high in hominoids, where the visual neocortex is shifted further back, around the occipital pole on the mesial aspect of the cortex (von Bonin and Bailey 1961).

4.1.3.2 Evidence from Fossil Endocasts

Despite the paucity of available hard evidence, Radinsky pointed out that several of the known fossil specimens provide relevant information for understanding primate brain evolution, notably the tempo of the emergence of the extant primate condition, and published most of the earliest critical descriptions (Radinsky 1967, 1970, 1973, 1974). The fossil record suggests that the expansion of the visual cortex and the reduction in size of the olfactory bulbs that distinguish strepsirhines from basal insectivores had already appeared at the beginning of the second major primate evolutionary radiation, 55 million years ago. Indeed, from early Eocene deposits of North America, the euprimate *Tetonius homunculus* shows a remarkably advanced brain compared to the condition typical of basal insectivores, with enlarged occipital and temporal lobes and reduced olfactory bulbs. However, brain morphology remains less derived than later primates because of relatively small frontal lobes (Radinsky 1967, 1974). The features observed on the *Tetonius*' endocast surface corroborate the results of the cytoarchitectonic studies published by Preuss and Goldman-Rakic (1991a, b). In addition, with the unique exception of a visible Sylvian fissure, in this fossil representative, the neocortical surface is featureless (Radinsky 1967, 1970), which is a typical feature of euprimates (Orliac et al. 2014), although also missing in another early euprimate, *Smilodectes gracilis*, a middle Eocene basal adapoid (Gazin 1965).

In early euprimates such as *Tetonius*, but also in *Smilodectes gracilis*, the presence of enlarged occipital and temporal cortical areas suggests that the expansion of these regions involved in optic and auditory functions, respectively, may have been one among the critical adaptations responsible for the early Eocene primate radiation (Radinsky 1967, 1970, 1974).

According to Radinsky, the anthropoid record suggests that, 25–30 million years ago, the visual cortex and the olfactory bulb relative size had reached the condition observed in extant taxa. The transversely oriented central sulcus, distinguishing modern anthropoids from most strepsirhines, was also present in at least some among the earliest anthropoids, such as *Aegyptopithecus* from the Oligocene deposits of Fayum, Egypt (Radinsky 1973, 1974).

4.2 Extracting, Reconstructing and Characterizing Fossil Endocasts

4.2.1 Extracting and Reconstructing Fossil Endocasts

When the neurocranium is filled with sediment during the fossilization process, information about brain morphology and organization may be recorded as a natural endocast. For fossil specimens not preserving a natural endocast, it is possible to generate an imprint of the endocranial surface by filling the braincase with casting material (Holloway et al. 2004). However, thanks to recent developments in imaging techniques, virtual endocasts may now be generated from computed (micro)tomography (CT and μCT) data (Zollikofer and Ponce de León 2005; Weber and Bookstein 2011; Neubauer 2014).

Dealing with the fossil record nearly invariably implies facing the problem of fragmentation and distortion of the remains. These factors can be partially balanced by using advanced virtual imaging techniques (Zollikofer et al. 1998; Zollikofer 2002; Gunz et al. 2009; Weber and Bookstein 2011). To illustrate the valuable contribution provided by virtual imaging to primate palaeoneurology, here we present three selected cases of fossil cercopithecoids from Plio-Pleistocene South African deposits: (1) the female individual MP 224 from the site of Makapansgat, attributed to *Parapapio broomi* by Freedman (1976); (2) the specimen STS 564 that corresponds to the holotype of *Parapapio broomi*, from Sterkfontein Member 4 (Freedman 1957); and (3) the female *Parapapio antiquus* TP 8 from Taung (formerly Tvl. 639), originally described by Freedman (1957, 1961). The specimens are currently housed at the University of the Witwatersrand, Johannesburg, and at the Ditsong National Museum of Natural History, Pretoria, and have been detailed by microtomography at the South African Nuclear Corporation (Necsa) of Pelindaba and at the Palaeosciences Centre of the University of the Witwatersrand, Johannesburg (spatial resolution ranging from 0.072 to 0.082 mm).

In principle, sediments can be removed digitally, and the endocranial cavity can be virtually separated from the braincase using semi-automatic threshold-based segmentation (Spoor et al. 1993; see also the region-based segmentation approach that relies on topographic concepts and which is known as the "watershed transform" in Meyer and Beucher 1990; Roerdink and Meijster 2001) (Fig. 4.1). Lately,

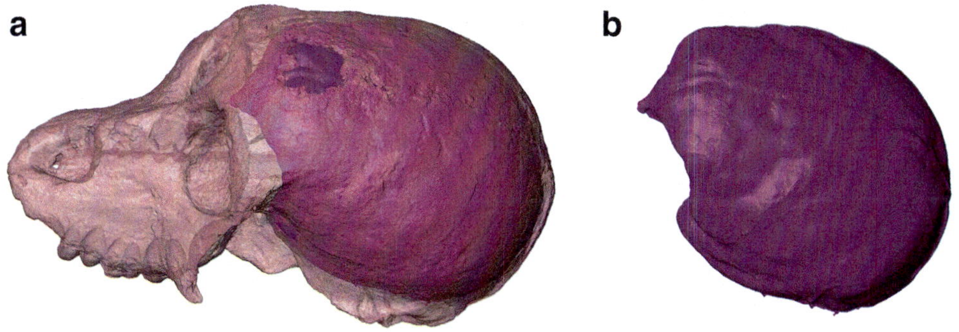

Fig. 4.1 Microtomography-based virtual reconstruction (**a**) and extraction (**b**) of the endocast from the fossil cercopithecoid specimen MP 224 (*Parapapio broomi*) from Makapansgat, South Africa. The braincase is rendered semi-transparent. Images not to scale

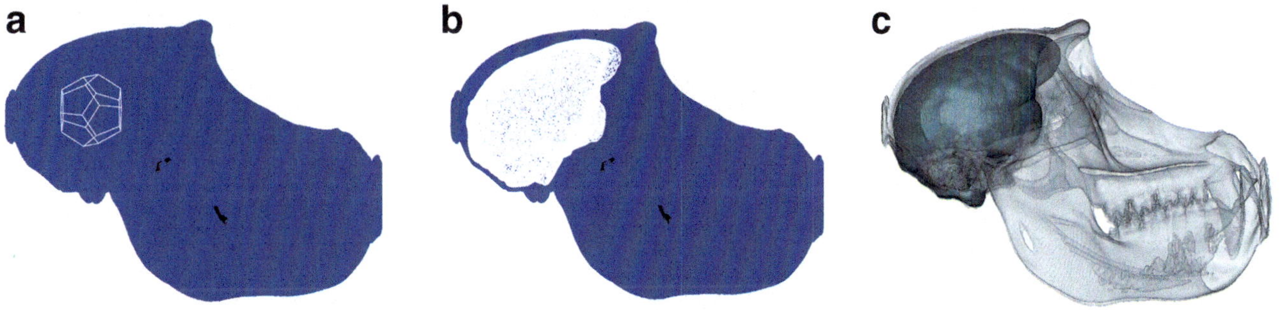

Fig. 4.2 Schematic rendering of the successive steps performed by the software Endex for the automatic extraction of the endocast in a specimen of the extant *Theropithecus gelada*. The initial mesh inside the braincase (**a**) spatially grows (**b**) and finally fits the inner cranial surface (**c**)

additional methods were developed for the automatic extraction of endocasts. Among the most effective, the Endex software (Subsol et al. 2010; http://liris.cnrs.fr/gilles.gesquiere/wiki/doku.php?id=endex) is particularly suitable for specimens preserving an empty endocranial cavity. A mesh represented by a sphere is positioned inside the endocranial space and deformed along the three orthogonal directions (Fig. 4.2). Through this deformation process, the mesh moulds the inner cranial surface and produces a virtual endocast. Both regularity and constrained smoothing contribute to deform the mesh within the limits imposed by the braincase (e.g. the virtual endocast could not spread out of natural openings such as the *foramen magnum*).

The type of damage most commonly observed on fossil specimens results from post-depositional taphonomic dynamics and can be considered within three main categories: (1) missing parts (e.g. MP 224, STS 564), (2) fragment assembling (e.g. TP 8) and (3) plastic deformation (not specifically illustrated here, but see comments below).

4.2.1.1 Missing Parts

The fossil specimen may lack some endocranial areas, as in MP 224, where part of the right temporal region is missing, and in STS 564, lacking a significant portion of the left hemisphere (Fig. 4.3). When topographically limited, the missing parts can be estimated and reconstructed based on the remaining bony parts by using classical landmark-based geometric morphometric tools and algorithmic approaches (i.e. statistical and geometric reconstructions; Gunz et al. 2004, 2009). These methods rely on the fact that information captured by shape coordinates is typically redundant, especially when the measurement points are closely spaced (Gunz et al. 2009). The non-preserved regions can now be digitally and automatically closed by digitizing a curve network around the margin of the missing area and by creating a NURBS (non-uniform rational basis spline) surface that matches the points along the curves through, for instance, the Rhino v5.0 software (R. McNeel & Associates) (Benazzi et al. 2011). Here, the result of the application of the automatic closure process to the MP 224 *Parapapio* specimen from Makapansgat is illustrated in Fig. 4.3. When the missing regions affect a larger endocranial area (e.g. part of the cranium is not preserved), because of the bilateral symmetry of the brain, bony portions that are preserved on one side can be mirrored in order to reconstruct the missing side (Gunz et al. 2009). Here, the right hemisphere of the specimen STS 564 was mirrored to reconstruct the left counterpart (Fig. 4.3).

4.2.1.2 Fragment Assembling

Virtual reconstructions have also focused on assembling fragments. In some cases, pieces of skulls physically restored do not perfectly articulate, and this impacts on, and biases, the endocast reconstruction. However, once

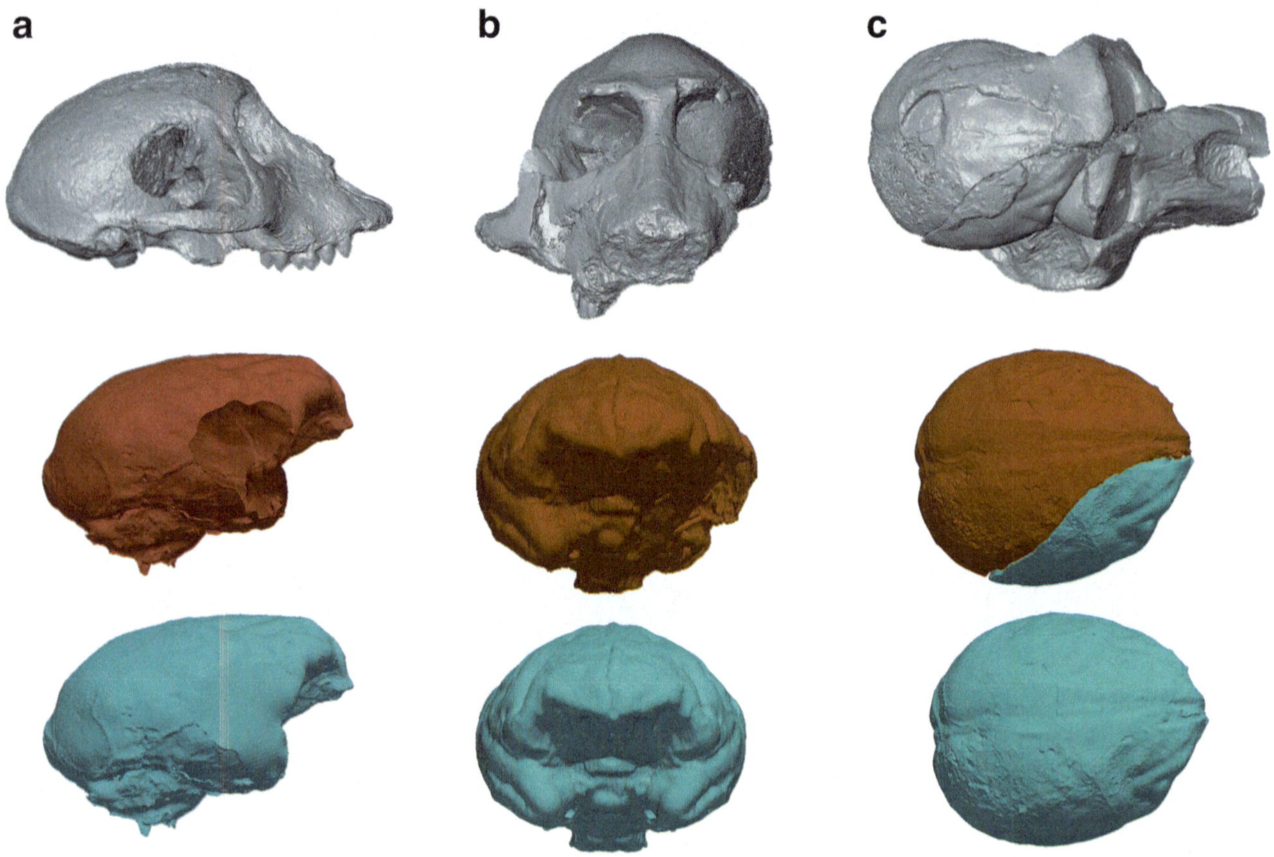

Fig. 4.3 Virtual rendering of the South African fossil cercopithecoid specimens MP 224 (**a** *Parapapio broomi* from Makapansgat), STS 564 (**b** *Parapapio broomi* from Sterkfontein), and TP 8 (**c** from *Parapapio antiquus* from Taung) (*top row*) and digital extraction and reconstruction of the corresponding endocasts (*middle and bottom rows*) (Beaudet 2015; Beaudet et al. 2016). Surfaces in red and green represent the original and the corrected volumes, respectively. The missing temporal area in MP 224 has been digitally filled. The left hemisphere in STS 564 has been rendered by mirroring the right counterpart. The fissure running through the TP 8 right hemisphere has been corrected by digitally reassembling the two halves. Images not to scale

segmented out of volumetric images, virtual bones can be readily reassembled manually on the computer. Natural endocasts affected by breakages and separated into several fragments, such as in the case of the *Parapapio antiquus* endocast TP 8 from Taung divided in two halves because of an anteroposterior fracture, can be corrected by virtually reassembling the two parts (Fig. 4.3).

4.2.1.3 Crushing and Deformation

Plastic distortions affecting the bony remains or the natural endocasts are difficult to identify and assess, especially when the morphological inter- and intra-individual variability is unknown, which is nearly always the case for extinct species. The operation of reconstructing an *ante-mortem* shape from a deformed specimen is called "retrodeformation" (Tallman et al. 2014). Assuming a low degree of developmentally related normal deviation from symmetry, one can correct it by a symmetrization process (e.g. reflection and averaging method in Gunz et al. 2009 or algorithmic symmetrization in Tallman et al. 2014). As stated by Gunz et al. (2009), any reconstruction is based on assumptions about functional constraints, integration, symmetry as well as sex variation, species affinity and taphonomic distortion, regardless of whether it relies on plaster material handling or computer processing.

4.2.2 Characterizing Fossil Endocasts

In 1978, Holloway noted that "Most of the research done in primate (and other animal) paleoneurology is necessarily of a qualitative nature which can, and often does, incorporate each of the first five levels of description mentioned above-that is, whether gyrus *x* or sulcus *y* can be seen, whether the frontal lobe is relatively small or large (or smaller or larger than *z*), whether the primary visual striate cortex is expanded or not, or whether or not the olfactory bulbs are reduced" (Holloway 1978: 187). However, palaeoneurology has recently enlarged its traditional investigative toolkit by integrating methods of high-resolution imaging, 3D modelling and statistical analyses granting a higher degree of reliability of the quantitative and qualitative estimates. In addition to basic descriptions and traditional metrical analyses, recent incorporation and validation of computer-based techniques for reconstructing and comparing endocranial casts have substantially improved the quality of data delivered by endocasts (Zollikofer et al. 1998; Zollikofer 2002; Gunz et al. 2009; Weber and Bookstein 2011). Accordingly, the advanced methods in virtual palaeoanthropology contribute to the characterization of the organization and morphology of lobes and to the identification of gyral and sulcal pattern, corresponding, respectively, to levels (ii) and (iii) in Holloway's description (Holloway 1978).

4.2.2.1 Deformation-Based Models: Interspecific Variation

Digital data make quantitative analysis of overall endocranial shape possible, notably through the landmark-/semilandmark-based geometric morphometrics (Bruner et al. 2009; Bruner 2004; Neubauer et al. 2009, 2010; Gunz 2015), or via the registration of surfaces from the correspondence of anatomical landmarks (Specht et al. 2007). However, even if efficient in compartmentalizing the endocranial cavity, the use of methodological toolkits based on landmarks and semilandmarks positioned on surfaces or extracted from curves captures little information about the brain itself and its subdivisions (Neubauer 2014). One potential alternative is to combine a detailed analysis of the sulcal pattern together with the characterization of the overall endocranial shape via the deformation-based models (Durrleman et al. 2012a, b; Beaudet et al. 2016; Dumoncel et al. 2014; Beaudet 2015).

The deformation-based model has been demonstrated to be a relevant tool for the registration of morphoarchitectural variations in primate endocranial ontogenetic trajectories, depicting both global and local changes (Durrleman et al. 2012a), and also valuable for taxonomic and evolutionary studies (Dumoncel et al. 2014; Beaudet 2015; Beaudet et al. 2016). Through the deformation computation process, both global and local changes could be rendered, compared and discussed. This mathematical approach relies on the construction of group-average surface models (i.e. global mean shape) and their deformation to the investigated surfaces (i.e. specimens). Contrary to landmark-based geometric morphometric analyses, comparison between surfaces does not assume a point-to-point correspondence between samples (Glaunès and Joshi 2006; Durrleman et al. 2012a, b; Dumoncel et al. 2014). The magnitude of displacements from the reference to specimens is rendered by colour maps from, for instance, dark blue (lowest values) to red (highest values) onto the surfaces. Figure 4.4 illustrates the displacements from a global mean shape to the cercopithecoid fossil specimens MP 224, STS 564 and TP 8 (see also Beaudet et al. 2016). When compared to the mean shape, MP 224 is antero-posteriorly elongated and the temporal lobes are convergent, while in STS 564 the parietal area is elevated and the volume of the temporal lobes is reduced. The large vectors recorded inferiorly for STS 564 and TP 8 may indicate taphonomic distortion and/or approximation in the segmentation process. It is noteworthy that the reconstructed temporal area in MP 224 does not affect the overall deformation results.

4.2.2.2 Deformation-Based Models: Intraspecific Variation

For an accurate evaluation of interspecific variation, intraspecific variability should be considered and quantified. Intraspecific studies could help one to elucidate whether different regions of the brain are more or less prone to evolve in a coordinated or independent way (Gómez-Robles et al. 2014). In the absence of large fossil assemblages, living populations have to be considered as reliable proxies for the assessment of reference variation patterns. Previous landmark-based studies attempted to characterize the primate endocranial and brain morphological variation (e.g. Gómez-Robles et al. 2013). The variability of primate brains, including those of humans, has been also explored and mapped using coarse spatial matching (e.g. Gilissen 2001; Zilles et al. 2001). In this respect, deformation-based models are particularly suitable (Durrleman 2010). Intragroup variability may be explored by performing a principal component analysis using the deformation momenta from the mean shape to specimens (Durrleman 2010). Deformations along the axis illustrate how the mean shape varies within the population from the mean to the standard deviation (σ) (i.e. the square root of variance, the corresponding eigenvalue). The extent of morphological variation revealed by a sample of eight endocasts representing five extant *Papio* taxa (*anubis* $n = 2$, *cynocephalus* $n = 1$, *cynocephalus kindae* $n = 2$, *hamadryas* $n = 1$ and *ursinus* $n = 2$) is shown in Fig. 4.5. In this case, a mean shape was computed from the *Papio* sample and deformed along the two axes of the principal component analysis. Along both PC1 and PC2, the observed variation mainly affects the temporal lobes and the superior part of the parietal area and, in some extent, the frontal pole, indicating that these regions are relatively variable within this taxon.

4.2.2.3 Semi-automatic Sulci Detection

Because of variable degrees of preservation, individual variation and ambiguous homology, identification of cortical convolutions from endocast may be problematic (Conolly 1950; Holloway et al. 2004; Falk 2009). However, sulcal patterns reproduced on endocast provide important information for studying the evolution of the primate cerebral cortex,

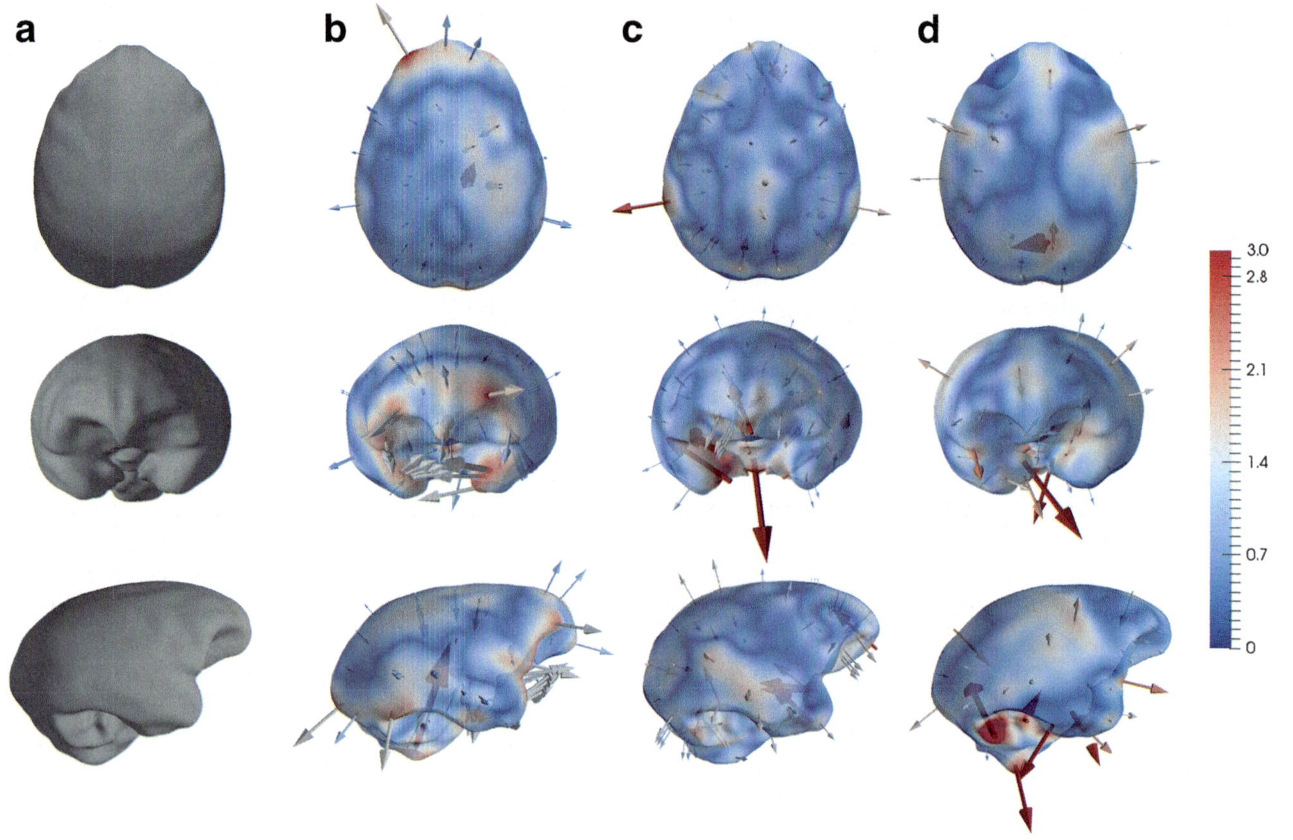

Fig. 4.4 Comparative maps of morphological deformations from the global mean shape (**a**) to the South African fossil cercopithecoid specimens MP 224 (**b** *Parapapio broomi* from Makapansgat), STS 564 (**c** *Parapapio broomi* from Sterkfontein) and TP 8 (**d** *Parapapio antiquus* from Taung) (Modified from Beaudet et al. 2016). Cumulative displacement variations (in mm) are rendered by a pseudo-colour scale ranging from *dark blue* (*lowest* values) to *red* (*highest* values). Vectors represent the magnitude and the orientation of the deformations from the mean shape. Images not to scale

and significant effort should be performed in order to extract as much details from any available fossil specimen (Falk 2014). The sulcal pattern is usually identified and described by visual inspection of the endocranial surface (e.g. Holloway et al. 2004; Falk 2014). Based on previous similar studies (Subsol 1995, 1998), an automatic method for the detection of the neocortical reliefs based on the algorithm introduced by Yoshizawa and co-workers (2007, 2008) in 3D meshes of topographical variation (i.e. crest lines) has been recently developed (Beaudet 2015; Beaudet et al. 2016). Following the automatic detection process, manual corrections can be performed by removing the non-anatomical features using references from extant primates (e.g. Primate Brain Bank; Netherlands Institute for Neuroscience; the Netherlands, www.primatebrainbank.org; Comparative Mammalian Brain Collections; the University of Wisconsin; Michigan State University; and the National Museum of Health and Medicine, http://brainmuseum.org/).

Results of the automatic detection and manual correction computed on the three fossil specimens used as examples in this contribution are illustrated in Fig. 4.6, where they have been compared to the condition of a living baboon individual (*Papio hamadryas*) derived from the online database Primate Brain Bank (see also Beaudet et al. 2016). In STS 564, only the right hemisphere is represented. The three specimens preserve the arcuate (arc) and principal (p) sulci, as well as part of the Sylvian fissure(s). Imprints of the lateral calcarine (lc) and inferior occipital sulcus (oci) are visible in STS 564, while the central (c), intraparietal (ip) and temporal superior (ts) sulci are detectable in MP 224 and STS 564. These results convincingly demonstrate that, in the case of the fossil record, the variable degree of preservation of the investigated specimens, the quality of the imaging system and the efficiency of the semi-automatic detection all influence the degree of reliability of the sulcal pattern assessment based on fossil endocasts (Fig. 4.6). Moreover, the correspondence of endocranial impressions to cerebral sulci and gyri has been largely questioned in primate palaeoneurology (e.g. Le Gros Clark et al. 1936), and further studies are needed to assess the reliability of cerebral cranial imprints for inferring fossil neuroanatomical evidence (e.g. Kobayashi et al. 2014).

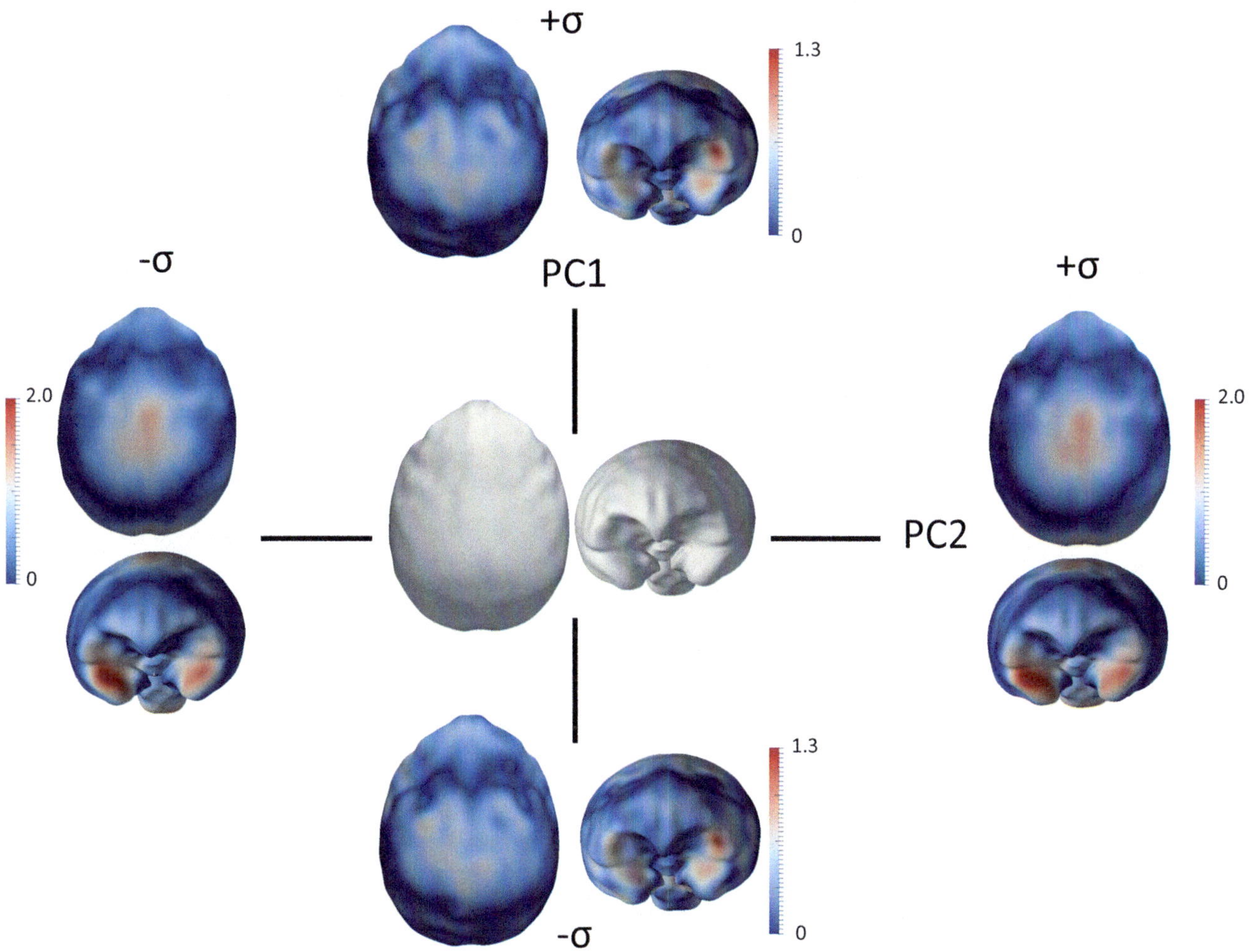

Fig. 4.5 Morphological variability of the taxon mean shape (in *grey*) computed for the extant *Papio* endocasts along the first (PC1) and second (PC2) axes of the principal component (Beaudet 2015). Deformations illustrate how the mean shape varies within the sample from the mean to the standard deviation (+/−σ). Cumulative displacement variations (in mm) are rendered by a pseudo-colour scale ranging from *dark blue* (*lowest values*) to *red* (*highest values*). Images not to scale

4.3 Conclusions and Perspectives

Although the primate fossil record had steadily increased, subtle aspects concerning the tempo and mode of some of the most critical changes occurred during brain evolution – i.e. increase in brain size and expansion of the neocortical surface and reorganization of the sensory system and of the fissuration pattern – remain to be elucidated. Endocasts, the only direct evidence of extinct neuroanatomical conditions, preserve two fundamental bodies of information: the gross cerebral morphology and the sulcal pattern (Holloway 1978). Recent advances in the field of palaeoneurological research convincingly demonstrate that, despite some unavoidable limitations intimately related to fossilization dynamics, such information can be virtually extracted through automatic segmentation methods (e.g. Endex software, watershed transform) and confidently reconstructed in cases of missing data, fragment assembling or plastic deformation via mirroring processes or automatic closure applications. Combining the quantitative description of the overall endocranial conformation and the sulcal pattern through newly developed analytical approaches offers the unique opportunity to track some of the finest cortical changes that occurred in primate neuroanatomy evolution. Also, because the deformation-based models do not assume a point-to-point correspondence between samples, as it is the case of classical landmark-based geometric morphometrics, and concern the overall endocranial surface, this methodology represents a powerful tool for the comparative assessment of most neuroanatomical features, especially given the potential modular organization of the brain and the associated mosaic evolutionary patterns described for some taxa (Preuss and Goldman-Rakic 1991a, b; Gómez-Robles et al. 2014). Since sulcal variation has been accepted as a reasonable predictor of cytoarchitecture for primary and secondary regions such as the visual, somatosensory and motor areas (Fischl et al. 2008), an accurate description of the

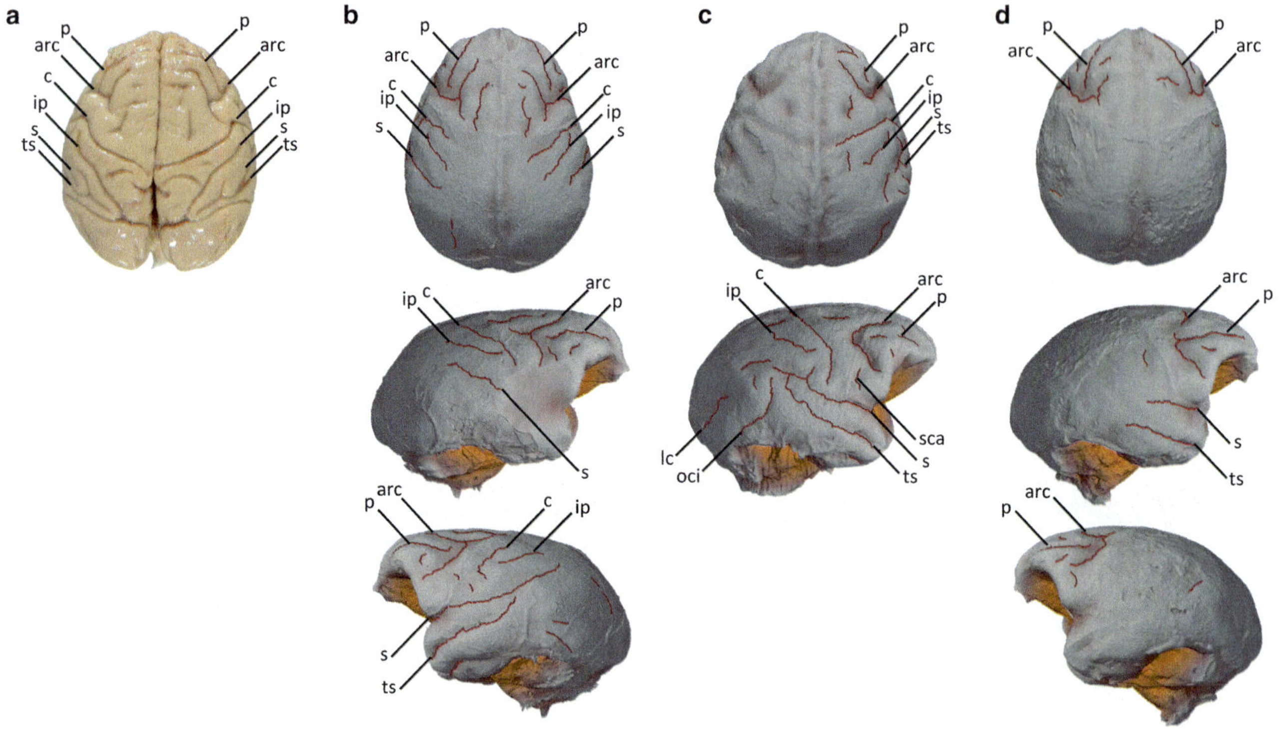

Fig. 4.6 Superior view of the brain of a specimen of the extant *Papio hamadryas* (**a** image courtesy of the Primate Brain Bank, Netherlands Institute for Neuroscience, the Netherlands) compared with virtual reconstructions of the endocasts extracted from the South African fossil cercopithecoid specimens MP 224 (**b** *Parapapio broomi* from Makapansgat), STS 564 (**c** *Parapapio broomi* from Sterkfontein) and TP 8 (**d** *Parapapio antiquus* from Taung), also rendered with their sulcal impressions (in upper and lateral views) (Modified from Beaudet et al. 2016). Because in TP 8 the left hemisphere was reconstructed as mirrored surface of the right counterpart, only the right side is annotated. *Arc* arcuate sulcus, *ip* intraparietal sulcus, *lc* lateral calcarine sulcus, *oci* inferior occipital sulcus, *p* principal sulcus, *s* Sylvian (lateral) fissure, *sca* subcentral anterior sulcus, *scp* subcentral posterior sulcus, *ts* superior temporal sulcus. Only the sulci mentioned in the text are labelled here. Images not to scale

neocortical sulci topographic distribution and arrangement is necessary for detecting the functional determinants of cerebral changes and for challenging ongoing evolutionary hypotheses. In this perspective, a further investigative step will be represented by the development of analytical protocols granting the automatic sulci recognition and labelling in fossil endocasts, as already occurs for human brains (e.g. BrainVISA, http://brainvisa.info/index_f.html).

Acknowledgements We are grateful to the editors E. Bruner, N. Ogihara and H. Tanabe for their kind invitation to contribute this volume. We are indebted to J. Cuisin (Paris), G. Fleury (Toulouse), S. Potze (Pretoria), W. Wendelen (Tervuren) and B. Zipfel (Johannesburg) for having granted access to fossil and comparative materials under their care from the Ditsong National Museum of Natural History (Pretoria), the Musée d'Histoire naturelle in Toulouse, the Musée national d'Histoire naturelle (Paris), the Royal Museum for Central Africa (Tervuren) and the University of the Witwatersrand (Johannesburg). We also thank K. Carlson and T. Jashashvili (Johannesburg), G. Clément and M. Garcia-Sanz (Paris), B. Duployer and C. Tenailleau (Toulouse), L. Bam, F. de Beer and J. Hoffman (Pretoria) for microtomographic acquisitions performed at the Accès Scientifique à la Tomographie à Rayons-X (AST-RX) imagery platform set at the Musée national d'Histoire naturelle (Paris), at the French Research Federation FERMaT (Toulouse), at the Palaeosciences Centre of the University of the Witwatersrand (Johannesburg) and at the South African Nuclear Corporation (Pelindaba). For scientific contribution and/or discussion and comments to the results summarized in this study, we are especially grateful to J. Braga (Toulouse), L. Bruxelles (Toulouse), M. Cazenave (Pretoria), E. Delson (New York), J. Dumoncel (Toulouse), S. Durrleman (Paris), D. Ginibriere (Toulouse), J. Heaton (Birmingham), R. Holgate (Pretoria), N. Jablonski (University Park), J.P. Jessel (Toulouse), O. Kullmer (Frankfurt), R. Macchiarelli (Poitiers & Paris), M. Nakatsukasa (Kyoto), L. Pan (Toulouse), G. Subsol (Montpellier), D. Stratford (Johannesburg), J.F. Thackeray (Johannesburg) and C. Zanolli (Toulouse). The French research federation FERMaT (FR3089), the National Research Foundation (NRF) and Department of Science and Technology (DST) of South Africa are acknowledged for providing micro-X-ray tomography laboratory facilities. This work was granted access to the HPC resources of CALMIP supercomputing centre under the allocation 2015-[P1440] attributed to the laboratory AMIS (Toulouse). Research is supported by the Centre of Research and Higher Education (PRES) of Toulouse, the Midi-Pyrénées Region and the French Ministry of Foreign Affairs.

References

Amunts K, Schleicher A, Bürgel U, Mohlberg H, Uylings HBM, Zilles K (1999) Broca's region revisited: cytoarchitecture and intersubject variability. J Comp Neurol 412:319–341

Armstrong E, Falk D (1982) Primate brain evolution: methods and concepts. Plenum Press, New York
Barton RA (1998) Visual specialization and brain evolution in primates. Proc Biol Sci 265:1933–1937
Barton RA, Harvey PH (2000) Mosaic evolution of brain structure in mammals. Nature 405:1055–1058
Bayly P, Taber L, Kroenke C (2014) Mechanical forces in cerebral cortical folding: a review of measurements and models. J Mech Behav Biomed Mater 29:568–581
Beaudet A (2015) Caractérisation des Structures Crânio-Dentaires Internes des Cercopithécoïdes et Etude Diachronique de leurs Variations Morphologiques dans la Séquence Plio-Pléistocène Sud-Africaine. PhD dissertation, Université de Toulouse
Beaudet A, Dumoncel J, de Beer F, Duployer B, Durrleman S, Gilissen E, Hoffman J, Tenailleau C, Thackeray JF, Braga J (2016) Morphoarchitectural variation in South African fossil cercopithecoid endocasts. J Hum Evol 101:65–78
Benazzi S, Bookstein FL, Strait DS, Weber GW (2011) A new OH5 reconstruction with an assessment of its uncertainty. J Hum Evol 61:75–88
Brain CK (1981) The hunters of the hunted? An introduction to African cave Taphonomy. University of Chicago Press, Chicago
Bruner E (2004) Geometric morphometrics and paleoneurology: brain shape evolution in the genus *Homo*. J Hum Evol 47: 279–303
Bruner E, Mantini S, Ripani M (2009) Landmark-based analysis of the morphological relationship between endocranial shape and traces of the middle meningeal vessels. Anat Rec 292:518–527
Connolly CJ (1950) External morphology of the primate brain. C.C. Thomas, Springfield
de Winter W, Oxnard CE (2001) Evolutionary radiations and convergences in the structural organization of the mammalian brain. Nature 409:710–714
Dumoncel J, Durrleman S, Braga J, Jessel J-P, Subsol G (2014) Landmark-free 3D method for comparison of fossil hominins and hominids based on endocranium and EDJ shapes. Am J Phys Anthropol 153(suppl 56):110 (abstract)
Durrleman S (2010) Statistical models of currents for measuring the variability of anatomical curves, surfaces and their evolution. PhD dissertation, Université Nice-Sophia Antipolis
Durrleman S, Pennec X, Trouvé A, Ayache N, Braga J (2012a) Comparison of the endocranial ontogenies between chimpanzees and bonobos via temporal regression and spatiotemporal registration. J Hum Evol 62:74–88
Durrleman S, Prastawa M, Korenberg JR, Joshi S, Trouvé A, Gerig G (2012b) Topology preserving atlas construction from shape data without correspondence using sparse parameters. In: Ayache N, Delingette H, Golland P, Mori K (eds) MICCAI 2012, Part III. LNCS, vol 7512. Springer, Heidelberg, pp 223–230
Falk D (1981) Sulcal patterns of fossil *Theropithecus* baboons: phylogenetic and functional implications. Int J Primatol 2:57–69
Falk D (1982) Mapping fossil endocasts. In: Armstrong E, Falk D (eds) Primate brain evolution: methods and concepts. Plenum Publishing Company, New York, pp 217–226
Falk D (2009) The natural endocast of Taung (*Australopithecus africanus*): insights from the unpublished papers of Raymond Arthur Dart. Am J Phys Anthropol 49:49–65
Falk D (2014) Interpreting sulci on hominin endocasts: old hypotheses and new findings. Front Hum Neurosci 8:134
Felleman DJ, Van Essen DC (1991) Distributed hierarchical processing in the primate cerebral cortex. Cereb Cortex 1:1–47
Fischl B, Rajendran N, Busa E, Augustinack J, Hinds O, Yeo BT, Mohlberg H, Amunts K, Zilles K (2008) Cortical folding patterns and predicting cytoarchitecture. Cereb Cortex 18:1973–1980
Freedman L (1957) The fossil Cercopithecoidea of South Africa. Ann Transv Mus 23:121–262
Freedman L (1961) New cercopithecoid fossils, including a new species, from Taung, Cape Province, South Africa. Ann S Afr Mus 46: 1–14
Freedman L (1976) South African fossil Cercopithecoidea: a re-assessment including a description of new material from Makapansgat, Sterkfontein and Taung. J Hum Evol 5:297–315
Gazin CL (1965) An endocranial cast of the Bridger middle Eocene primate *Smilodectes gracilis*. Smithson Misc Coll 149:1–14
Gilissen E (2001) Structural symmetries and asymmetries in human and chimpanzee brains. In: Falk D, Gibson KR (eds) Evolutionary anatomy of the primate cerebral cortex. Cambridge University Press, Cambridge, pp 187–215
Glaunès JA, Joshi S (2006) Template estimation from unlabeled point set data and surfaces for computational anatomy. In: Pennec X, Joshi S (eds) Proceedings of the international workshop on the Mathematical Foundations of Computational Anatomy. Copenhagen, pp 29–39
Gómez-Robles A, Hopkins D, Sherwood C (2013) Increased morphological asymmetry, evolvability and plasticity in human brain evolution. Proc R Soc B Biol Sci 280:20130575
Gómez-Robles A, Hopkins D, Sherwood CC (2014) Modular structure facilitates mosaic evolution of the brain in chimpanzees and humans. Nat Comm 5:4469
Gonzales LA, Benefit BR, McCrossin ML, Spoor F (2015) Cerebral complexity preceded enlarged brain size and reduced olfactory bulbs in old world monkeys. Nat Comm 6:7580
Gunz P (2015) Computed tools for paleoneurology. In: Bruner E (ed) Human paleoneurology. Springer, Zurich, pp 39–55
Gunz P, Mitteroecker P, Bookstein, FL, Weber GW (2004) Computer-aided reconstruction of incomplete human crania using statistical and geometrical estimation methods. Enter the past: computer applications and quantitative methods in archaeology. BAR International Series, Oxford, pp 96–98
Gunz P, Mitteroecker P, Neubauer S, Weber GW, Bookstein FL (2009) Principles for the virtual reconstruction of hominin crania. J Hum Evol 57:48–62
Hilgetag CC, Barbas H (2005) Developmental mechanics of the primate cerebral cortex. Anat Embryol 210:411–417
Hill WCO (1972) Evolutionary biology of the primates. Academic, London
Holloway RL (1978) The relevance of endocasts for studying primate brain evolution. In: Noback CR (ed) Sensory systems of primates. Plenum Press, New York, pp 181–200
Holloway RL, Broadfield DC, Yuan MS (2004) The human fossil record: brain endocasts – the paleoneurological evidence. Wiley-Liss, New York
Kaas JH (2002) Convergences in the modular and areal organization of the forebrain of mammals: implications for the reconstruction of forebrain evolution. Brain Behav Evol 59:262–272
Kaas JH (2006) Evolution of the neocortex. Curr Biol 16:910–914
Kobayashi Y, Matsui T, Haizuka Y, Ogihara N, Hirai N, Matsumura G (2014) Cerebral sulci and gyri observed on macaque endocasts. In: Akazawa T, Ogihara N, Tanabe HC, Terashima H (eds) Dynamics of learning in Neanderthals and modern humans, vol 2. Springer, Japan, pp 131–137
Le Gros Clark WE (1971) The antecedents of man: an introduction to the evolution of the primates. Edinburgh University Press, Edinburgh
Le Gros Clark WE, Cooper DM, Zuckerman S (1936) The endocranial cast of the chimpanzee. J Roy Anthropol Inst 66:249–268
Leakey LSB, Tobias PV, Napier JR (1964) A new species of genus *Homo* from Olduvai Gorge. Nature 202:7–9
Meyer F, Beucher S (1990) Morphological segmentation. J Vis Commun Image Represent 1:21–46
Neubauer S (2014) Endocasts: possibilities and limitations for the interpretation of human brain evolution. Brain Behav Evol 84:117–134

Neubauer S, Gunz P, Hublin J-J (2009) The pattern of endocranial ontogenetic shape changes in humans. J Anat 215:240–255
Neubauer S, Gunz P, Hublin J-J (2010) Endocranial shape changes during growth in chimpanzees and humans: a morphometric analysis of unique and shared aspects. J Hum Evol 59:555–566
Orliac MJ, Ladevèze S, Gingerich PD, Lebrun R, Smith T (2014) Endocranial morphology of Palaeocene *Plesiadapis tricuspidens* and evolution of the early primate brain. Proc R Soc B 281: 20132792
Preuss TM, Goldman-Rakic PS (1991a) Myelo- and cytoarchitecture of the granular frontal cortex and surrounding regions in the strepsirhine primate *Galago* and the anthropoid primate *Macaca*. J Comp Neurol 310:429–474
Preuss TM, Goldman-Rakic PS (1991b) Architectonics of the parietal and temporal association cortex in the strepsirhine primate *Galago* compared to the anthropoid primate *Macaca*. J Comp Neurol 310:475–506
Radinsky L (1967) The oldest primate endocast. Am J Phys Anthropol 27:385–388
Radinsky L (1970) The fossil evidence of prosimian brain evolution. In: Noback CR, Montagna W (eds) Primate brain. Appleton-Century-Croft, New York, pp 209–224
Radinsky L (1973) *Aegyptopithecus* endocasts: oldest record of a pongid brain. Am J Phys Anthropol 39:239–247
Radinsky L (1974) The fossil evidence of anthropoid brain evolution. Am J Phys Anthropol 41:15–28
Radinsky L (1975) Primate brain evolution. Am Sci 63:656–663
Radinsky L (1982) Some cautionary notes on making inferences about relative brain size. In: Armstrong E, Falk D (eds) Primate brain evolution: methods and concepts. Plenum, New York, pp 29–37
Rakic P, Kornack DR (2001) Neocortical expansion and elaboration during primate evolution: a view from neuroembryology. In: Falk D, Gibson KR (eds) Evolutionary anatomy of the primate cerebral cortex. Cambridge University Press, Cambridge, pp 30–56
Roerdink JBTM, Meijster A (2001) The watershed transform: definitions, algorithms and parallelization strategies. Fund Inform 41:187–228
Rogers J, Kochunov P, Zilles K, Shelledy W, Lancaster J, Thompson P, Duggirala R, Blangero J, Fox PT, Glahn DC (2010) On the genetic architecture of cortical folding and brain volume in primates. NeuroImage 53:1103–1108
Ronan L, Fletcher PC (2015) From genes to folds: a review of cortical gyrification theory. Brain Struct Funct 220:2475–2483
Semendeferi K, Lu A, Schenker N, Damasio H (2002) Humans and great apes share a large frontal cortex. Nat Neurosci 5:272–276
Silcox MT, Benham AE, Bloch JI (2010) Endocasts of *Microsyops* (Microsyopidae, primates) and the evolution of the brain in primitive primates. J Hum Evol 58:505–521
Specht M, Lebrun R, Zollikofer CPE (2007) Visualizing shape transformation between chimpanzee and human braincases. Vis Comput 23:743–751
Spoor F, Zonneveld F, Macho GA (1993) Linear measurements of cortical bone and dental enamel by computed tomography: applications and problems. Am J Phys Anthropol 91:469–484
Stephan H, Baron G, Frahm HD (1991) Comparative brain research in mammals volume 1. Insectivora. Springer, New York
Subsol G (1995) Construction Automatique d'Atlas Anatomiques Morphométriques à Partir d'Images Médicales Tridimensionnelles. PhD dissertation, Ecole Centrale de Paris
Subsol G (1998) Crest lines for curve based warping. In: Toga A (ed) Brain warping. Academic Press, San Diego, pp 241–262
Subsol G, Gesquière G, Braga J, Thackeray F (2010) 3D automatic methods to segment "virtual" endocasts: state of the art and future directions. Am J Phys Anthropol 141(suppl 50):226–227
Tallinen T, Chung JY, Rousseau F, Girard N, Lefèvre J, Mahadevan L (2016) On the growth and form of cortical convolutions. Nat Phys. doi:10.1038/nphys3632
Tallman M, Amenta N, Delson E, Frost SR, Ghosh D, Klukkert ZS, Morrow A, Sawyer GJ (2014) Evaluation of a new method of fossil retrodeformation by algorithmic symmetrization: crania of papionins (primates, Cercopithecidae) as a test case. PLoS One 9(7): e100833
Toro R (2012) On the possible shapes of the brain. Evol Biol 39: 600–612
Van Essen DC (1997) A tension-based theory of morphogenesis and compact wiring in the central nervous system. Nature 385:313–318
von Bonin G, Bailey P (1961) Pattern of the cerebral isocortex. Primatologia II/2. Karger, New York
Weber GW, Bookstein FL (2011) Virtual anthropology: a guide to a new interdisciplinary field. Springer, London
Welker W (1990) Why does cerebral cortex fissure and fold? A review of determinants of gyri and sulci. In: Jones EG, Peters A (eds) Cerebral cortex. Vol 8b. Comparative structure and evolution of cerebral cortex, Part II. Plenum Press, New York, pp 3–136
Yoshizawa S, Belyaev A, Yokota H, Seidel HP (2007) Fast and faithful geometric algorithm for detecting crest lines on meshes. Proceedings of the 15th Pacific conference on computer graphics and applications, pp 231–237
Yoshizawa S, Belyaev A, Yokota H, Seidel HP (2008) Fast, robust, and faithful methods for detecting crest lines on meshes. Comput Aided Geom D 25:545–560
Zilles K, Armstrong E, Schleicher A, Kretschmann HJ (1988) The human pattern of gyrification in the cerebral cortex. Anat Embryol 179:173–179
Zilles K, Kawashima R, Dabringhaus A, Fukuda H, Schormann T (2001) Hemispheric shape of European and Japanese brains: H 3-D MRI analysis of intersubject variability, ethnical, and gender differences. NeuroImage 13:262–271
Zilles K, Palomero-Gallagher N, Amunts K (2013) Development of cortical folding during evolution and ontogeny. Trends Neurosci 36:275–284
Zollikofer CPE (2002) A computational approach to paleoanthropology. Evol Anthropol 11:64–67
Zollikofer CPE, Ponce de León MS (2005) Virtual reconstruction: a primer in computer-assisted paleontology and biomedicine. Wiley Interscience, Hoboken
Zollikofer CPE, Ponce de León MS, Martin RD (1998) Computer-assisted paleoanthropology. Evol Anthropol 6:41–54

The Evolution of Avian Intelligence and Sensory Capabilities: The Fossil Evidence

5

Stig A. Walsh and Fabien Knoll

Abstract

Crocodiles and birds are the only living representatives of Archosauria, a once diverse clade of vertebrates that mastered terrestrial, aerial and aquatic environments during the Mesozoic. Because the braincases of archosaurs are largely ossified, the group has particularly benefited from advances in non-destructive visualisation of endocranial structures over the past two decades. Here, we focus on the neurosensory evolution in the avian lineage of the Archosauria, a group in which the Bauplan of most representatives is optimised to accommodate the functional demands of flight. Neurosensory evolution in birds included a trend towards an enlargement of the telencephalon relative to the rest of the brain, an increased vestibular system sensitivity and probably also a widening of auditory frequency range and an increased reliance on visual stimuli. Despite a relatively smooth surface, bird endocasts provide crucial information on the evolution of a critical structure, the Wulst, which underwent significant enlargement during the Cenozoic and is found with highly variable form in all extant birds. With our increasing awareness of avian cognitive capacity and neural structure, the evolution of the brain in the sauropsid lineage represents an increasingly useful comparative tool against which the development of the synapsid lineage brain of primates can be assessed. Current refinements in quantification of brain structures in extant birds are improving the reliability of the information derived from the external surface of endocasts. This, in turn, should result in a better understanding of the palaeoneurology of extinct birds and other dinosaurs.

Keywords

Bird • Dinosaur • Neurosensory evolution • Wulst • Flocculus

5.1 Introduction

Fossil cranial specimens referable to hominids are clearly of crucial importance to studies of the human brain evolutionary pathway. However, knowledge of neurosensory evolution in non-human animals is also important for understanding the context and drivers of vertebrate brain evolution as a whole and represents a backdrop against which key stages in human brain evolution may be viewed. Many, if not all, of the key events in the story of vertebrate life on Earth would have required concomitant changes to neurosensory capabilities and brain development. These

S.A. Walsh
Department of Natural Sciences, National Museums Scotland, Chambers Street, Edinburgh EH1 1JF, UK
e-mail: s.walsh@nms.ac.uk

F. Knoll (✉)
Fundación Conjunto Paleontológico de Teruel-Dinópolis, Avda. Sagunto s/n, 44002 Teruel, Spain

School of Earth and Environmental Sciences, University of Manchester, Oxford Road, Manchester M13 9PL, UK
e-mail: knoll@fundaciondinopolis.org; fabien.knoll@manchester.ac.uk

E. Bruner et al. (eds.), *Digital Endocasts*, Replacement of Neanderthals by Modern Humans Series,
DOI 10.1007/978-4-431-56582-6_5

evolutionary events include the appearance of jaws in gnathostomes, which would have involved the evolution and development of the trigeminal branches and Gasserian ganglion. The tetrapod invasion of land would have led to the loss of the lateral line system and the modification of the vestibular system to compensate for the loss of support from water and, indeed, an adaptation of other sense organs for perceiving airborne stimuli. Mastery of aerial locomotion in amniotes would have involved further enhancement of vestibular, proprioceptive and visual pathways and has been the subject of debate for well over a century (Walsh and Milner 2011a). Added to these pioneer events are the many instances of adaptive reversals over evolutionary time, such as the secondary adaptation to life in water or loss of flight, all of which required significant modifications to existing neurosensory structures optimised for specific environments (Walsh et al. 2014).

It is important to note that a number of evolutionary adaptive events, including acquisition of bipedality, flight capability and secondary adaptation to life in water, have occurred in our own mammal (synapsid) amniote lineage. However, all of these events are restricted to crown-group mammals, and none is known to have occurred before the Cretaceous-Palaeogene extinction event and subsequent mammalian adaptive radiation. Similar adaptive events are documented in the non-mammalian (sauropsid) lineage, but in sauropsids the adaptations occurred several times in sometimes distantly related clades (e.g. secondary adaptation to life in water in chelonians, squamates, ichthyosaurs, sauropterygians, crocodyliforms and birds).

Both sauropsids and synapsids appear to have inherited a basic neural architecture from a common stem-amniote ancestor that lived before the two lineages diverged around 330 million years ago (Jarvis et al. 2005). Nonetheless, a number of their characteristic features must have evolved separately. For instance, a basic telencephalic cortical structure appears to be common in amniotes, but the characteristic mammalian six-layered neocortex cytoarchitecture is restricted to crown-group mammals (Reiner et al. 2005). By comparison, the avian pallium has a nuclear rather than a layered structure (Jarvis et al. 2005). This indicates that the telencephalic neural reorganisation that accompanied major adaptive events such as the parallel evolution of avian and mammalian flight was probably achieved through different sequences of change in the two lineages. In other words, synapsids and sauropsids reached the same functional neurosensory adaptation, but via different routes (Jarvis et al. 2005). Knowledge of brain evolution in the sauropsid lineage is therefore clearly relevant as a comparative case study for the synapsid neural evolutionary trajectory (Lefebvre et al. 2004; Walsh and Milner 2011a).

The sauropsid clade Archosauria, which includes *inter alia* crocodiles, non-avian dinosaurs and birds, appears to demonstrate a trend towards brain enlargement similar to that recognised long ago for mammals (e.g. Marsh 1886; Jerison 1973). A branch of this clade, the birds, has achieved relatively large brains with respect to their body size and even displays cognitive abilities that rival and sometimes exceed those of primates (Emery and Clayton 2004). These abilities include complex vocalisation (Petkov and Jarvis 2012), social learning (Tebbich et al. 2001), Machiavellian intelligence (Clayton et al. 2007), cooperative hunting (Yosef and Yosef 2010), self-recognition (Prior et al. 2008) and tool manufacture, use and preference (Hunt and Gray 2007). Consequently, their neurosensory evolution is particularly relevant as a comparative case study for hominids.

As with hominid palaeoneurology, our knowledge of brain development in extinct sauropsid animals hinges on the availability of suitable fossil skull specimens. Although the field has advanced greatly since the advent of non-destructive X-ray computed tomography (CT), the rarity of suitable material remains a fundamental impediment to our understanding of brain evolution in the clade. Computed tomographic analysis of fossil archosaurs is also often problematic in ways that may be less likely to be encountered when investigating hominoid fossils. At one end of the archosaur fossil spectrum are large skulls (e.g. *Tyrannosaurus rex*: Brochu 2003) that are too large to fit in conventional scanners or achieve adequate X-ray penetration, while at the other end are exceptionally small and fragile skeletons that are not normally preserved in three dimensions (e.g. flattened birds from Chinese Konservat Lagerstätten: Zhou 2004). A further problem is that Archosauromorpha includes many early forms that are not closely related to living taxa and which may also have possessed novel or unusual neural adaptations that are either not present or have been coapted in derived extant representatives.

Despite these difficulties, knowledge of archosaur palaeoneurology has improved considerably over the past two decades. Here, we provide an overview of recent advances in this field with special reference to the evolution of the avian brain and that of birds' close relatives, the non-avian dinosaurs and crocodylians.

5.2 Historical Background

Archosaur palaeoneurology has a long history of research. In fact, the earliest mention of any brain cavity endocast was that of a pterosaur (an extinct clade of flying reptiles; Oken 1819). The first mention of an avian endocast and, thereby, the birth of palaeoneurological investigations on birds is to be found a few years later in a discrete *nota bene* in the second edition of Cuvier's (1822) *Recherches sur les Ossemens fossiles*.

As well as providing the earliest description of an archosaur endocast, Owen (1842, 1871, 1879) was also the first to figure the endocast of an extinct bird, earlier than the first physical endocast from a non-avian saurischian was made (Marsh 1884). Both Owen and Marsh had developed a strong interest in vertebrate brain evolution (Walsh and Knoll 2011), but Gratiolet (1858) seems to have been the first to realise the potential of artificial endocasts in the study of mammal and bird brain evolution.

In the last five decades, there has been a slow revival of interest for the field (e.g. Dechaseaux 1970; Hoch 1975; Mlíkovský 1980; Elżanowski and Galton 1991). The first digital reconstructions of extant and extinct bird endocasts were realised at the beginning of the twenty-first century (Franzosa 2004; Domínguez Alonso et al. 2004), a few years later than the very first digital reconstruction of a saurischian endocast (Knoll 1997; Knoll et al. 1999). This development of digital 3D reconstruction has led to a boom in interest in avian palaeoneurology (e.g. Milner and Walsh 2009; Walsh and Milner 2011b; Ksepka et al. 2012; Proffitt et al. 2016), and the potential to apply quantitative methods to the analysis of endocast composition through large libraries of datasets of extant and fossil taxa (e.g. Balanoff et al. 2013; Walsh et al. 2013).

5.3 Archosaur Digital Endocasts: Quality of Data

Today, nearly all archosaur palaeoneurological work focusses on high-resolution 'digital' endocasts, which generally preserve more detail than natural or synthetic physical endocasts because they allow visualisation of fragile endocranial structures, such as the pituitary and some vascular loops in birds (Walsh and Knoll 2011). Some encouraging indication that traces of original external brain morphology may sometimes be detectable in exceptionally preserved fossils using phase contrast X-ray tomography has been found (Pradel et al. 2009). No such discovery has been made so far in a fossil archosaur, but evidence for fossilised intracranial soft tissues has been put forward recently in an ornithischian dinosaur (Brasier et al. 2016). Apart from this single exception, our knowledge of fossil archosaur brain form remains centred on analysis of the shape and overall size of the brain *cavity*.

As with mammals, the endocranial surface of the archosaur brain cavity provides some information about brain shape and maximal brain dimensions, but the degree of fidelity offered by this surface varies greatly between clades. This is due to the varying thickness of the meningeal envelopes (e.g. dura mater), which typically enfold not only a cerebrospinal fluid-filled space but also blood vessels and sinuses that may also account for a significant proportion of the endocranial space (Witmer et al. 2008). As such, the brain cavity cast is never a representation of the brain itself (Hurlburt et al. 2013), and in some taxa this poor fidelity can make recognition even of the position of major brain regions, such as the telencephalon, problematic (e.g. some fossil crocodiles; Fig. 5.1a). For this reason, care should be taken not to confuse endocast anatomical nomenclature with terminology used for true neural structures. A thorough review and standardisation of palaeoneurological nomenclature for use across vertebrates as a whole is probably overdue.

At the extreme opposite end of the spectrum, many birds exhibit a relationship between the brain, dural envelope and skull roof so tight that the shape of the dorsal surface of the skull actually mimics the shape of the dorsal surface of the underlying brain. Even in those bird and dinosaur taxa with less intimate brain-to-skull relationships, clear impressions of the passage of blood vessels are often visible on the endocranial surface (e.g. Osmólska 2004; Picasso et al. 2009). However, although the avian cerebellum is folded, impressions of these folia are not visible on the walls of the cerebellar fossa in all taxa due to the presence of other tissues overlying the brain in this region. An impression of the pineal (situated on the dorsal brain surface in the angle between the caudal extent of the two telencephalic lobes and cerebellum) has been suggested to be present in dinosaurs (e.g. Kundrát 2007). In many sauropods, the presence of a parietal foramen is likely to have been associated with a pineal organ (e.g. Knoll et al. 2012), although it is often too large to relate to the diencephalic neural structure alone. In these, the structure situated where the pineal would be expected may instead represent an expansion of the dural envelope, essentially due to a venous sinus system (Witmer et al. 2008). A diminutive pineal was reported in the oldest known avialan, *Archaeopteryx* (Domínguez Alonso et al. 2004), but does not appear to be present in more recent reconstructions of this taxon (Balanoff et al. 2013). Other Mesozoic birds also do not appear to have a pineal trace, and none is apparent in digital endocasts generated from extant birds (Walsh et al. 2016).

The archosaur telencephalon is lissencephalic, and only the position of major structures such as the olfactory lobes and interhemispheric fissure can be determined on the surface of the endocast. Consequently, functional areas within the telencephalon are mostly impossible to index on digital endocasts, unlike endocasts of gyrencephalic mammals, which may exhibit function-related sulci and gyri (but see Bruner 2003). An exception to this rule can be found in birds, where a feature of the dorsal telencephalon known as the 'Wulst' can be traced easily in most species. The Wulst appears to be restricted to crown-group Aves (Neornithes), and its significance will be discussed in more detail later.

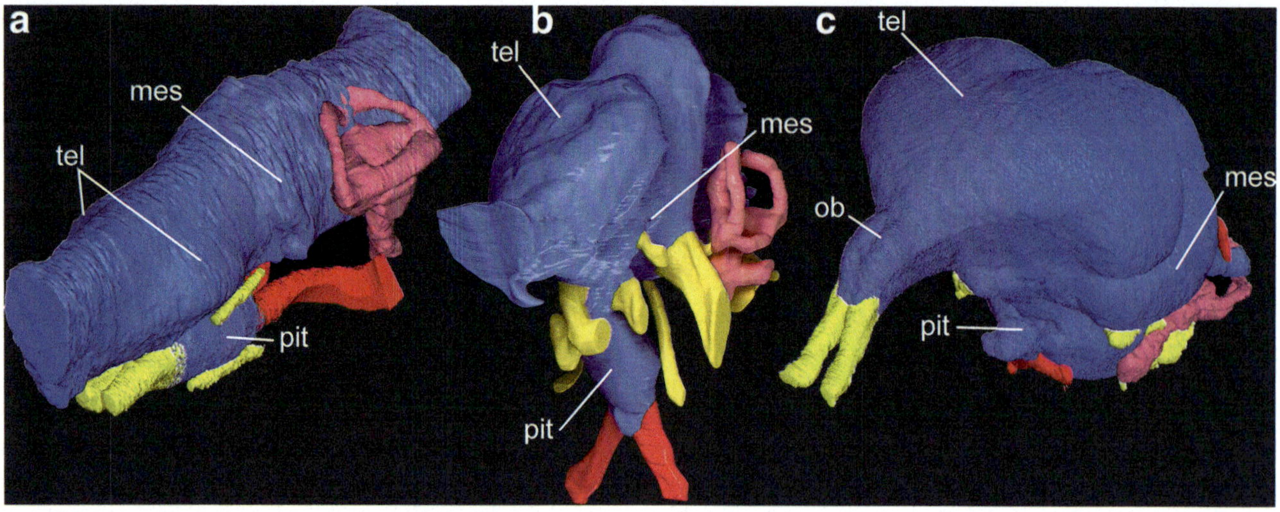

Fig. 5.1 Examples of fossil archosaur digital endocasts in oblique view. (**a**) Marine crocodile (*Steneosaurus* cf. *gracilirostris*) from the Jurassic of the United Kingdom (for full description, see Brusatte et al. 2016) exhibiting tube-like form. (**b**) Sauropod dinosaur endocast (Lithostrotia indet.) from the Cretaceous of Spain (see Knoll et al. 2015) showing a more contracted morphology. (**c**) Avian endocast (*Cerebavis cenomanica*) from the Cretaceous of Russia with a strongly involute shape (see Walsh et al. 2016). Colour code: brain cavity endocast, *blue*; inner-ear labyrinth endocast, *pink*; nerve canal endocasts, *yellow*; arterial canal endocasts, *red*. Abbreviations: *mes* mesencephalon cast, *ob* olfactory bulb cast, *pit* pituitary region cast, *tel* telencephalon. Not to scale

Issues with variable endocast fidelity aside, brain cavity reconstructions can often be further complicated by incomplete ossification of the brain cavity (Walsh et al. 2014; Sobral et al. 2016). The extent of this problem does not seem to have been extensively surveyed across Archosauria, although ossification of some parts of the neurocranium is well known to be synapomorphic for archosaurs (Bhullar and Bever 2009). Incomplete ossification is typical in avian braincases in the caudal area and roof of the orbit, where the optic nerve foramina may be replaced by a single large fenestra (Hall et al. 2009), and the base of the fossa and foramina housing the olfactory bulb and nerve (Ali et al. 2008; SAAW personal observation). In these instances, the morphology of the exposed regions of the diencephalon (optic chiasma) and ventral olfactory bulb region may be difficult to resolve.

5.4 What the Digital Endocasts Show

The plesiomorphic archosaur brain Bauplan (Fig. 5.2) is a little different from that of other diapsid reptiles: situated along an approximately straight brain axis is a relatively simple cerebellum positioned dorsal to an elongate medulla, paired hemispheres of the mesencephalon are situated rostral to the cerebellum, and the two halves of the telencephalon (pallium and olfactory lobes) are situated further rostral to the mesencephalon. This arrangement results in an elongate

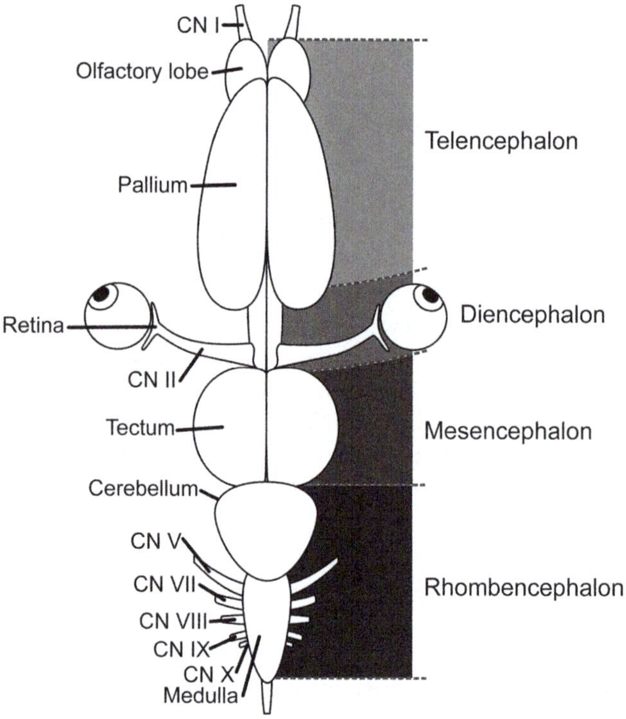

Fig. 5.2 Schematic representation of the composition of a primitive sauropsid brain (dorsal view). This basic brain Bauplan demonstrates an unflexed brain axis and illustrates the major brain regions (telencephalon, diencephalon, mesencephalon and rhombencephalon) together with some of the regional divisions (tectum, medulla, etc.) and cranial nerves originating from the medulla oblongata

brain and is the basic brain form found in living crocodiles, alligators, caimans and gharials (Witmer et al. 2008; Brusatte et al. 2016) and also in the endocasts of many dinosaurs and their archosaur ancestors (Walsh and Knoll 2011). As mentioned above, taxa with large proportions of dural envelope surrounding the brain may reveal little of the original external brain morphology. In some elongate archosaur brains, this may result in a poorly informative tube-like endocast (Fig. 5.1a).

Birds represent a derived condition with respect to brain morphology: the brain axis deviates from the main axis of the spinal cord, producing a noticeably flexed form in lateral view. Much of this flexion occurs within the region of the mesencephalon (Pearson 1972), with a strong angle often developed between the long axis of the telencephalon and that of the brain as a whole (Dubbeldam 1989). The angle of intersection between the main axis of the brain and the bill also varies from a near parallel orientation to the main axis of the bill (e.g. *Phalacrocorax*) to the brain axis being oriented almost vertically in the skull with a large angle between bill and main brain axis (e.g. *Capella*; Portmann and Stingelin 1961). Lateral enlargement of the telencephalon and flexion of the main axis of the brain during evolution has resulted in the more caudal brain regions being essentially folded under the telencephalon in many species. In birds and non-avian bird-like theropods, the endocast is largely brain-like, indicating a fairly close brain-to-skull relationship and relatively small dural envelope volume. Endocast volume in these taxa is therefore likely to be a reasonable approximation of brain volume (Iwaniuk and Nelson 2002).

In terms of brain size relative to body size (Encephalisation Quotient), it has long been recognised that bird brains are large compared with those of living reptiles (Jerison 1973). The problem in avian palaeoneurology is that fossil bird skulls that are sufficiently complete and three dimensionally preserved to allow for an estimation of endocast volume are very rare. Moreover, in order to calculate encephalisation quotient (EQ) values, some estimate of body mass must be made, and most currently accepted methods to achieve this require particular postcranial bones to be preserved with the skull. Such specimens in which a well-preserved braincase is found associated with convenient elements of the postcranial skeleton are exceedingly rare. In fact, our knowledge of Mesozoic avian brain development centres on just a handful of taxa (Walsh and Milner 2011a), only two of which are complete enough for a reasonable estimate of brain volume to be made: *Archaeopteryx lithographica* (Domínguez Alonso et al. 2004) and *Cerebavis cenomanica* (Walsh et al. 2016). Of those two, only *Archaeopteryx* includes postcrania. A similar situation persists throughout the Cenozoic fossil record (Jerison 1973; Walsh and Milner 2011a). Added to this paucity of hard data is the questionable reliability of body mass estimates in general (Christiansen and Fariña 2004). Body weight in living bird species is known to vary greatly depending on when measurements are made (e.g. pre- or post-breeding or migration season; Dunning 2008). All of these factors affect the reliability of EQ calculation.

Another problem for assessing trends in brain size in early birds is that a number of non-avian theropod dinosaurs closely related to birds appear to have been capable of flight and also had brains that were at least as bird like in form and relative volume (EQ) as that of *Archaeopteryx* (Balanoff et al. 2013). As such, the trend towards modern bird-like EQ values actually appears to have occurred multiple times during the evolution of the ancestors of modern birds. This blurring of perceived definitions begs the question of whether it is possible to definitively recognise a stem 'bird' from fossil remains during this time (Balanoff et al. 2014). By the Cretaceous, the lineage that would eventually lead to modern birds, the Neornithes, shared the skies with other true avian clades, such as the Enantiornithes. These clades are well defined as 'avian', but only the Neornithes survived into the Cenozoic and present day. The record of Cenozoic neornithine endocasts is relatively good compared with that of birds (or even paravians as a whole) during the Mesozoic. Unfortunately, virtually nothing is currently known about brain form and EQ in the non-neornithine avian lineages (Walsh and Milner 2011a). Consequently, although we know that the highly encephalised and complex brain of living bird species must have evolved from a small and simple crocodile-like brain in the avian ancestral lineage, the timing and rate of change are presently far from clear.

Because variation in external brain shape reflects differences in internal brain structure to some extent, studies of endocasts can provide useful information about brain structure evolution over time. Enlargement (relative to the proportions of the brain as a whole) of particular brain regions dealing primarily with specific processing tasks has been suggested to correlate with the relative importance to the organism of those processing tasks through variations in neuronal packing (Principal of Proper Mass; Jerison 1973). This general concept has long been applied to bird and dinosaur endocast study (e.g. Edinger 1929), albeit mostly in a qualitative framework. Some support for the approach in birds has come from quantitative analysis of volume composition of brain tissue in these main functional regions, which found strong correlations between brain composition (relative proportions of brain regions) and ecology and behaviour ('cerebrotype analysis': Iwaniuk and Hurd 2005; Fuchs et al. 2014). Iwaniuk and Hurd (2005) recovered five main cerebrotypes, mostly centred on relative volumes of divisions of the rhombencephalon (e.g. cerebellum) and divisions of the telencephalon (nidopallium and hyperpallium). Locomotor style was one behavioural aspect that correlated with brain region composition. Fuchs et al. (2014) were able to characterise migratory bird brains based on compositional geometry, generated from brain sections. Both studies suggest that further brain characterisation is possible, although this may depend on the regions measured.

Could such approaches make it possible to characterise a 'flying brain'? Identification of such a flight-capable brain (sensu Witmer and Ridgely 2007) has become something of a Holy Grail in bird, dinosaur and pterosaur palaeoneurology. It is clear that the overall enlargement of the modern avian brain is centred on relative expansion of the cerebellum, mesencephalon (optic tectum) and telencephalon (Dubbeldam 1998), all of which have been suggested to be important for flight (see Walsh and Milner 2011a for a review). For instance, increases in processing capacity in the cerebellum would be expected because of the role of the cerebellum in maintaining balance through vestibular and visual sensory input. The expansion of the vestibular system early in avian evolution (Domínguez Alonso et al. 2004) is consistent with the importance of such sensory processing, although enlargement and increased complexification of the course of the semicircular canals have never been demonstrated to correlate directly with flying ability in birds. Birds also have the largest eyes relative to body size of any terrestrial vertebrate (Martin 1985), and enlargement of the optic tectum would be expected for primary integration of the visual signal. However, despite the role of centres of somatosensory control in the telencephalon, the relevance of its enlargement to flight control during evolution is presently unclear. Lesion studies have shown that birds with destroyed telencephala can fly and land if thrown in the air (Salzen and Parker 1975), and the reasons for the expansion of this region may be more complex than generally assumed.

The difficulty of applying a cerebrotype-like approach to characterising avian endocasts is obvious: cerebrotype analysis quantifies regional volume through measurement of region area in serial-sectioned brain tissue, but only the external expression of these boundaries at best can be determined on the surface of an endocast. While it would be possible to dissect digital endocasts into regional components using these external boundaries, either as voxel or mesh model volumes (Walsh and Milner 2011b), attempts to do so (Balanoff et al. 2013; Walsh et al. 2013) always involve a great deal of uncertainty due to the unknown internal extent of those boundaries. Cross validation of such digital dissection approaches with volume data derived from dissections of actual brain tissue is clearly needed. Preliminary tests on small samples of avian taxa indicate that there is no significant difference between digital and wet dissection region volumes as a percentage of the whole brain/endocast for *some* regions such as the telencephalon (Walsh and Jones unpublished data), but sample size in those tests was small, and the effect on results of interoperator variation has yet to be tested.

Some subregions of the avian brain are known to be important for flight control and are ostensibly feasible to isolate in digital endocasts for measurement of volume. One of these, the cerebellar flocculus, protrudes from the side of the cerebellum through the rostral semicircular canal and is responsible for processing the vestibuloocular reflex (VOR) and vestibulocollic reflex (VCR). The VOR is likely to be particularly important for flight because it acts to stabilise gaze on an object when both the object and viewer may be moving rapidly through 3D space. The flocculus is not externally well developed in most quadrupedal archosaurs (including extant crocodilians), but is noticeably enlarged in some bipedal archosaurs (Sobral et al. 2016), theropod dinosaurs (Witmer et al. 2008), pterosaurs (Witmer et al. 2003) and birds (Milner and Walsh 2009). Enlargement of the flocculus in the endocasts of extinct archosaurs would be predicted to indicate enhanced VOR/VCR processing, which had been suggested to relate to flight control (Domínguez Alonso et al. 2004; Milner and Walsh 2009). However, empirical testing of this using extant avian species with known flight capabilities found no correlation between flocculus fossa size and aerial manoeuvrability; a weak negative association was even detectable (Walsh et al. 2013). The reasons for this result may have been due to other tissues sharing the floccular fossa with neural tissue or the effects of neural plasticity or a combination of these and other factors. Nonetheless, the result highlights that attempting to decompose digital endocasts into component regions is not straightforward.

If characterisation of archosaur endocasts using volumes and surface areas is problematic, it seems surprising that brain surface shape characterisation using 3D geometric morphometrics has so far not been used to any great extent. To our knowledge, only one study (Kawabe et al. 2013) has investigated variations in brain shape between species of birds. This analysis showed that orbit size, and hence eye size, affects avian brain shape, a finding that is interesting because requirements for flight probably limit the range of skull morphological variation in birds to a greater extent than in mammals. Simply put, the need to retain a small and relatively light skull with a thin interobital septum means that the brain, eyes and ears must be packed into a small space. An extreme instance of this can be seen in the barn owl (*Tyto alba*), in which the large, almost forward-facing eyes have probably been displaced by the need to enlarge the auditory apparatus for detecting high-frequency vocalisations of their small mammal prey (Martin 2009). Besides, large eyes are important for light gathering in these nocturnal predators, and the eyes have become so large that their sclerotic rings have become fused to orbital margin, necessitating modification of head rotation in place of eye movement. The result of this eye displacement into a more forward-facing situation is partial stereopsis (Martin 2009), an ability shared with other avian taxa, all of which also possess a stereopsis-related modification of the brain – an enlarged Wulst (Iwaniuk et al. 2008).

The Wulst is a dorsal expansion of the telencephalon (hyperpallium), normally demarcated from the rest of the telencephalon by a groove called the vallecula (not visible on avian endocasts; Fig. 5.3). The Wulst has attracted considerable attention because it appears to achieve mammalian neocortex-like function via a different cytoarchitecture; rather than the six layers of mammalian neocortex, the Wulst exhibits modality-specific nuclei with interregion connections similar to those found between neocortical layers (Reiner 2009). Although the Wulst is largest in birds with visual specialisations like stereopsis (Iwaniuk et al. 2008), the structure is also involved with many other tasks such as navigation during migration (Mouritsen et al. 2005) and higher functions such as tool use (Mehlhorn et al. 2010).

The Wulst has not been identified with certainty in any Mesozoic bird (but see Chatterjee 1991; Balanoff et al. 2013), and the oldest known definite Wulst development is from the Lower Eocene London Clay Formation of England (55 ma; Milner and Walsh 2009; Walsh and Milner 2011b). At this time, Wulst development was already similar to that of living birds, suggesting the structure was already present in some, if not all, Mesozoic Neornithes (Walsh and Milner 2011b). As mentioned, its presence in other Mesozoic avian groups is presently unknown. The ornithurine (the clade to which Neornithes belongs) *Cerebavis cenomanica* lacks a Wulst (Fig. 5.1C), indicating that the structure had not evolved to any degree that would be noticeable on an endocast by 93 ma, unless it appeared in an ornithurine more closely related to living neornithines (Walsh et al. 2016).

Although the timing and distribution of Wulst appearance is currently uncertain, the evolution of the structure may represent a further pulse in telencephalon expansion that probably occurred after the initial paravian expansion of the region. If so, current evidence would suggest the 40 million years or so between the Cenomanian and Eocene was an important period for avian brain evolution. Although *Halcyornis toliapicus* from the Eocene of England possessed a reasonably large Wulst (Walsh and Milner 2011b), the Wulsts of all other Paleogene birds (Milner and Walsh 2009; Tambussi et al. 2015) are within the smaller range found in living birds and are narrow and confined to the rostral region of the telencephalon. This condition conforms to Stingelin's (1957) predicted ancestral form, from which the wide variation of Wulst shapes and positions on the telencephalon seen in living birds presumably evolved. This variation today includes Wulsts positioned rostrally or caudally on the dorsal telencephalon or in some cases covering the entire dorsal (and parts of the ventral) surface of the telencephalon (Stingelin 1957; Fig. 5.3). Some support for a Cenozoic trend towards Wulst enlargement from a rostrally positioned form comes from penguin endocasts. In these, the Wulst had noticeably extended caudally and laterally between 34 ma (Tambussi et al. 2015) and 23 ma (Ksepka et al. 2012). Although larger Wulst developments may one day be found in Paleogene birds, current evidence would point to continued enlargement during the Cenozoic that culminated in the large Wulst volumes seen in owls, corvids and others today.

In addition to Wulst enlargement, a general increase in telencephalon relative size may also be detectable from the earliest paravians and throughout the Cenozoic. However, the patchy taxonomic and temporal sampling during this interval, as well as difficulty in reliably isolating the telencephalon for volume measurements, makes this difficult to test. If the trend is real, what is its underlying cause? Certainly, the avian clades that exhibit the largest telencephalon volumes (crows and parrots) also tend to have the highest levels of observed intelligence (Lefebvre et al. 2004), but did this intelligence arise as a consequence of telencephalon expansion for some other purpose, or was a trend towards problem-solving abilities already present?

It seems safe to assume that any initial telencephalic expansion required for flight would probably have ceased after the full evolution of powered flight at the base of the avian line. However, the telencephala of Eocene neornithines are proportionately larger than those of any known Mesozoic birds (Walsh and Milner 2011a), so it does appear that telencephalon expansion had not simply stopped after birds became proficient fliers. Large brains and powered flight are both energetically expensive (Isler and van Schaik 2006), so continued increases in brain size require explanation as to what drives these increases. Milner and Walsh (2009) suggested advantages in neural size and complexity in Neornithes may have given the group a competitive advantage over other Mesozoic avian clades (e.g. Enantiornithes) through behavioural flexibility at the K-Pg mass extinction. Certainly, a large and energetically expensive brain would be expected to select *against* a particular taxon during times of ecological collapse, unless it in some way provides an advantage. If this suggestion has any basis, it may be that living birds are descended from survivor ancestors that already possessed the basic neural architecture to allow them to achieve the primate-like levels of intelligence today seen in crows and parrots.

Even if living birds are continuing this trend towards telencephalon enlargement, there are, of course, limits to how large bird brains can eventually grow. Firstly, large brains require large heads to house them, so increases in body size would be required to counter the 'stone in the paper airplane' effect. This would be less important in secondarily flightless birds, and loss of flight might also reduce energy expenditure. However, brains in flightless birds actually appear either to reduce (Bennett and Harvey 1985;

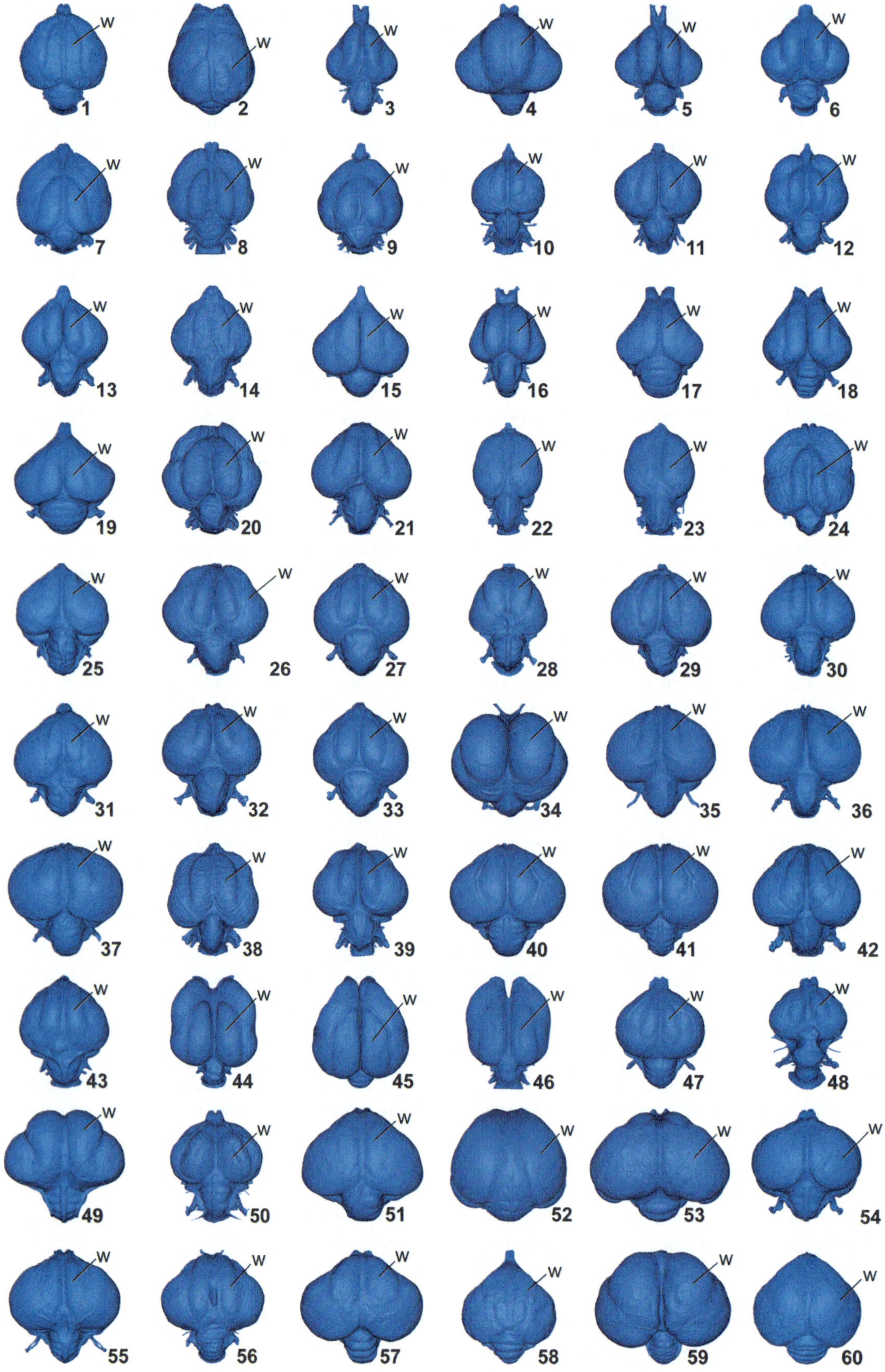

Fig. 5.3 Examples of variation in Wulst shape and position in 60 avian digital endocasts (in dorsal view; rostral to the top of the page). *W* indicates the position of the Wulst. *1 Rhynchotus rufescens*; *2 Apteryx haastii*; *3 Casuarius casuarius*; *4 Struthio camelus*;

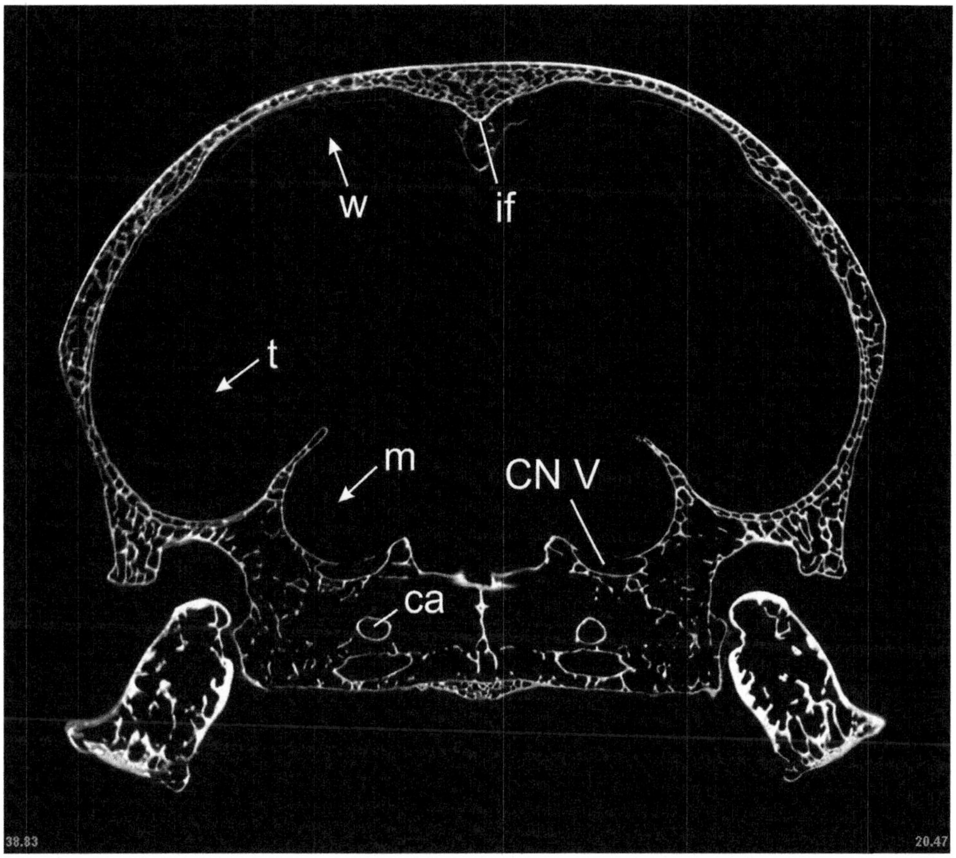

Fig. 5.4 Coronal tomography of *Corvus corax* (common raven) illustrating the wrapping effect of the telencephalon over the mesencephalon. *Ca* canal of the carotid artery, *CN V* foramen for the trigeminal nerve, *if* interhemispheric fissure, *m* fossa for the mesencephalon, *t* fossa for the telencephalon, *w* fossa for the Wulst

Corfield et al. 2008) or not change markedly in relative size (Iwaniuk et al. 2004). Presumably, evolutionary drivers to maintain or increase brain size in terrestrial environments have been less powerful than those for the evolution of flight, but the explanation is likely to be more complex than simple energetics. In crows, growth of the pallium is so dramatic that the structure partially wraps the mesencephalon (Fig. 5.4), a condition approaching that noted in primates. The observation that neuron density in the telencephalon of corvids and parrots is higher than in other birds and primates provides an explanation for how these birds are able to match primate intelligence with a smaller brain (Olkowicz et al. 2016). It also suggests that the brains of flying birds may actually be nearing the physical outer limits of capacity within the braincase.

5.5 Conclusions

The brains of birds (particularly the telencephalon) show the derived condition of having a large size (compared to other archosaurs and reptiles). Birds also possess remarkably complex brains compared with other archosaurs, and some are capable of feats of intelligence and behavioural innovation

Fig. 5.3 (Continued) *5 Dromaius novaehollandiae*; *6 Rhea americana*; *7 Aythya fuligula*; *8 Cygnus olor*; *9 Tachyeres brachypterus*; *10 Gallus gallus*; *11 Phasianus colchicus*; *12 Grus grus*; *13 Gavia immer*; *14 Podiceps cristatus*; *15 Eudyptula* sp.; *16 Diomedea exulans*; *17 Pelagodroma marina*; *18 Fulmarus glacialis*; *19 Pelecanoides urinatrix*; *20 Pelecanus erythrorhynchos*; *21 Fregata magnificens*; *22 Phalacrocorax carbo*; *23 Phalacrocorax harrisi*; *24 Threskiornis aethiopicus*; *25 Phaethon lepturus*; *26 Ciconia ciconia*; *27 Ardea cinerea*; *28 Rhynchops niger*; *29 Larus argentatus*; *30 Creagrus furcatus*; *31 Gelochelidon nilotica*; *32 Stercorarius skua*; *33 Alca torda*; *34 Tyto alba*; *35 Buteo buteo*; *36 Aquila chrysaetos*; *37 Circus cyaneus*; *38 Vultur gryphus*; *39 Sagittarius serpentarius*; *40 Falco tinnunculus*; *41 Falco subbuteo*; *42 Pandion haliaetus*; *43 Opisthocomus hoatzin*; *44 Ara macao*; *45 Amazona aestiva*; *46 Strigops habroptila*; *47 Columba livia*; *48 Pezophaps solitaria*; *49 Podargus strigoides*; *50 Steatornis caripensis*; *51 Apus apus*; *52 Selasphorus rufus*; *53 Trogon curucui*; *54 Alcedo atthis*; *55 Coracias garrulous*; *56 Ramphastos dicolorus*; *57 Tyrannus tyrannus*; *58 Hirundo rustica*; *59 Corvus corax*; *60 Acanthorhynchus superciliosus*. Not to scale that rival primates. Our current knowledge of archosaur brain evolution provides some clues as to how this neural complexity arose, but throws up more questions than can be answered with the fossil evidence currently available, even using X-ray tomographic techniques. The paucity of bird skulls suitable for scanning, particularly in the Mesozoic but also in important intervals in the Cenozoic (e.g. sampling of Miocene to Recent corvids), is frustrating efforts to answer key questions in avian brain evolution. Probably the most important of these questions is why the avian telencephalon relative size has continued to increase after birds acquired effective flight.

Acknowledgements We thank S. Brusatte for kindly making available images of the endocast of *Steneosaurus* cf. *gracilirostris* segmented by A. Muir (Edinburgh University) as well as R. Ridgely (Ohio University) for the digital endocast of the indeterminate Spanish lithostrotian. Two anonymous reviewers provided insightful comments. FK's research was funded by the European Union (PIEFGA-2013-624969); SAW's research was supported by NERC grant NE/H012176/1.

References

Ali F, Zelenitsky D, Therrien F, Weishampel D (2008) Homology of the "ethmoid complex" of tyrannosaurids and its implications for the reconstruction of the olfactory apparatus of non-avian theropods. J Vertebr Paleontol 28:123–133

Balanoff AM, Bever GS, Norell MA (2014) Reconsidering the avian nature of the oviraptorosaur brain (Dinosauria: Theropoda). PLoS One 9(12):e113559

Balanoff AM, Bever GS, Rowe TB, Norell MA (2013) Evolutionary origins of the avian brain. Nature 7465:93–96

Bennett PM, Harvey PH (1985) Relative brain size and ecology in birds. J Zool 207:151–169

Bhullar BAS, Bever GS (2009) An archosaur-like laterosphenoid in early turtles (Reptilia: Pantestudines). Breviora 518:1–11

Brasier MD, Norman DB, Liu AG, Cotton LJ, Hiscocks JEH, Garwood RJ, Antcliffe JB, Wacey D (2016) Remarkable preservation of brain tissues in an early cretaceous iguanodontian dinosaur. Geol Soc Lond Spec Publ 448:SP448.3

Brochu CA (2003) Osteology of *Tyrannosaurus rex*: insights from a nearly complete skeleton and high-resolution computed tomographic analysis of the skull. J Vertebr Paleontol 22(suppl 4):1–138

Bruner E (2003) Fossil traces of human thought: palaeoneurology and the evolution of the genus. Homo J Anthropol Sci 81:29–56

Brusatte SL, Muir A, Young MT, Walsh SA, Steel L, Witmer LM (2016) The braincase and neurosensory anatomy of an early Jurassic marine crocodylomorph: implications for crocodilian sinus evolution and sensory transitions. Anat Rec 299:1511–1530

Chatterjee S (1991) Cranial anatomy and relationships of a new Triassic bird from Texas. Philos Trans R Soc Lond B 332:277–342

Christiansen P, Fariña RA (2004) Mass prediction in theropod dinosaurs. Hist Biol 16:85–92

Clayton NS, Dally JM, Emery NJ (2007) Social cognition by food-caching corvids: the western scrub-jay as a natural psychologist. Philos Trans R Soc Lond B 362:507–522

Corfield JR, Wild JM, Hauber ME, Parsons S, Kubke MF (2008) Evolution of brain size in the Palaeognath lineage, with an emphasis on New Zealand ratites. Brain Behav Evol 71:87–99

Cuvier G (1822) Recherches sur les Ossemens fossiles, vol 3, 2nd edn. G. Dufour & E. d'Ocagne, Paris

Dechaseaux C (1970) Moulages endocraniens d'oiseaux de l'Éocène Supérieur du Bassin de Paris. Ann Paléontol 56:69–72

Domínguez Alonso P, Milner AC, Ketcham RA, Cookson MJ, Rowe TB (2004) The avian nature of the brain and inner ear of *Archaeopteryx*. Nature 430:666–669

Dubbeldam JL (1989) Shape and structure of the avian brain, an old problem revisited. Acta Morphol Neerl Scand 27:33–43

Dubbeldam JL (1998) Birds. In: Nieuwenhuys R, Ten Donkelaar HJ, Nicholson C (eds) The central nervous system of vertebrates, vol 3. Springer, Berlin, pp 1525–1636

Dunning JB (2008) CRC handbook of avian body masses, 2nd edn. CRC Press, Boca Raton

Edinger T (1929) Die fossilen Gehirne. Ergeb Anat Entwicklungsgesch 28:1–249

Elżanowski A, Galton PM (1991) Braincase of Enaliornis, an early Cretaceous bird from England. J Vertebr Paleontol 11:90–107

Emery NJ, Clayton NS (2004) The mentality of crows: convergent evolution of intelligence in corvids and apes. Science 306: 1903–1907

Franzosa JW (2004) Evolution of the brain in theropoda (Dinosauria). PhD dissertation, The University of Texas at Austin, Austin

Fuchs R, Winkler H, Bingman VP, Ross JD, Bernroider G (2014) Brain geometry and its relation to migratory behavior in birds. J Adv Neuro Res 1:1–9

Gratiolet P (1858) Sur l'encéphale du *Caïnotherium commune*, Brav. Extraits p v séances, Soc philom Paris 23:19–23

Hall MI, Iwaniuk AN, Gutiérrez-Ibáñez C (2009) Optic foramen morphology and activity pattern in birds. Anat Rec 292:1827–1845

Hoch E (1975) Amniote remnants from the eastern part of the lower Eocene North Sea basin. Colloq Int CNRS 218:543–562

Hunt G, Gray RD (2007) Parallel tool industries in New Caledonian crows. Biol Lett 3:173–175

Hurlburt GR, Ridgely RC, Witmer LM (2013) Relative size of brain and cerebrum in tyrannosaurid dinosaurs: an analysis using brain-endocast quantitative relationships in extant alligators. In: Parrish JM, Molnar RE, Currie PJ, Koppelhus EB (eds) Tyrannosaurid paleobiology. Indiana University Press, Bloomington, pp 134–155

Isler K, van Schaik C (2006) Costs of encephalisation: the energy trade-off hypothesis tested on birds. J Hum Evol 51:228–243

Iwaniuk AN, Heesy CP, Hall MI, Wylie DRW (2008) Relative Wulst volume is correlated with orbit orientation and binocular visual field in birds. J Comp Physiol A 194:267–282

Iwaniuk AN, Hurd PL (2005) The evolution of cerebrotypes in birds. Brain Behav Evol 65:215–230

Iwaniuk AN, Nelson J (2002) Can endocranial volume be used as an estimate of brain size in birds? Can J Zool 80:16–23

Iwaniuk AN, Nelson JE, James HF, Olson SL (2004) A comparative test of the correlated evolution of flightlessness and relative brain size in birds. J Zool 263:317–327

Jarvis ED, Güntürkün O, Bruce L, Csillag A, Karten H, Kuenzel W, Medina L, Paxinos G, Perkel DJ, Shimizu T, Striedter G, Wild JM, Ball GF, Dugas-Ford J, Durand SE, Hough GE, Husband S, Kubikova L, Lee DW, Mello CV, Powers A, Siang C, Smulders TV, Wada K, White SA, Yamamoto K, Yu J, Reiner A, Butler AB (2005) Avian brains and a new understanding of vertebrate brain evolution. Nat Rev Neurosci 6:151–159

Jerison HJ (1973) Evolution of the brain and intelligence. Academic, London

Kawabe S, Shimokawa T, Miki H, Matsuda S, Endo H (2013) Variation in avian brain shape: relationship with size and orbital shape. J Anat 223:495–508

Knoll F (1997) La boîte crânienne d'un théropode (Saurischia) du Jurassique des Vaches Noires: ostéologie et paléoneurologie. DEA dissertation, Université des Sciences et Techniques du Languedoc, Montpellier

Knoll F, Buffetaut E, Bülow M (1999) A theropod braincase from the Jurassic of the Vaches Noires cliffs (Normandy, France): osteology and palaeoneurology. Bull Soc Géol Fr 170:103–109

Knoll F, Witmer LM, Ortega F, Ridgely RC, Schwarz-Wings D (2012) The braincase of the basal sauropod dinosaur *Spinophorosaurus* and 3D reconstructions of the cranial endocast and inner ear. PLoS One 7(1):e30060

Knoll F, Witmer LM, Ridgely RC, Ortega F, Sanz JL (2015) A new titanosaurian braincase from the cretaceous "Lo Hueco" locality in Spain sheds light on neuroanatomical evolution within Titanosauria. PLoS One 10(10):e0138233

Ksepka DT, Balanoff AM, Walsh S, Revan A, Ho A (2012) Evolution of the brain and sensory organs in Sphenisciformes: new data from the stem penguin *Paraptenodytes antarcticus*. Zool J Linnean Soc 166:202–219

Kundrát M (2007) Avian-like attributes of a virtual brain model of the oviraptorid theropod *Conchoraptor gracilis*. Naturwissenschaften 94:499–504

Lefebvre L, Reader SM, Sol D (2004) Brains, innovations and evolution in birds and primates. Brain Behav Evol 63:233–246

Marsh OC (1884) Principal characters of American Jurassic dinosaurs: part VIII. Am J Sci 27:329–340

Marsh OC (1886) Dinocerata. Monogr US Geol Surv 10:1–243

Martin GR (1985) Eye. In: King AS, McLelland J (eds) Form and function in birds. Academic, New York, pp 311–373

Martin GR (2009) What is binocular vision for?: a birds' eye view. J Vis 9:1–19

Mehlhorn J, Hunt GR, Gray RD, Rehkämper G, Güntürkün O (2010) Tool-making new Caledonian crows have large associative brain areas. Brain Behav Evol 75:63–70

Milner AC, Walsh SA (2009) Avian brain evolution: new data from Palaeogene birds (Lower Eocene) from England. Zool J Linnean Soc 155:198–219

Mlíkovský J (1980) Zwei Vogelgehirne aus dem Miozän Böhmens. Čas Miner Geol 25:409–413

Mouritsen H, Feenders G, Liedvogel M, Wada K, Jarvis ED (2005) Night-vision brain area in migratory songbirds. Proc Natl Acad Sci U S A 102:8339–8344

Oken L (1819) Pterodactylus longi- et brevirostris. Isis 2:1788–1798

Olkowicz S, Kocourek M, Lučan RK, Porteš M, Fitch WT, Herculano-Houzel S, Němec P (2016) Birds have primate-like numbers of neurons in the forebrain. Proc Natl Acad Sci U S A 113:7255–7260

Osmólska H (2004) Evidence on relation of brain to endocranial cavity in oviraptorid dinosaurs. Acta Palaeontol Pol 49:321–324

Owen R (1842) Report on British fossil reptiles: Pt II. Rep Brit Assoc Adv Sci 11:60–204

Owen R (1871) On *Dinornis* (Part XVI.): containing notices of the internal organs of some species, with a description of the brain and some nerves and muscles of the head of the *Apteryx australis*. Trans Zool Soc Lond 7:381–396

Owen R (1879) Memoirs on the extinct wingless birds of New Zealand, with an appendix on those of England, Australia, Newfoundland, Mauritius, and Rodriguez, vol 2. J van Voorst, London

Pearson R (1972) The avian brain. Academic, London

Petkov CI, Jarvis ED (2012) Birds, primates, and spoken language origins: behavioural phenotypes and neurobiological substrates. Front Evol Neurosci 4:1–24

Picasso MBJ, Tambussi C, Dozo MT (2009) Neurocranial and brain anatomy of a late Miocene eagle (Aves, Accipitridae) from Patagonia. J Vertebr Paleontol 29:831–836

Portmann A, Stingelin W (1961) The central nervous system. In: Marshall AJ (ed) The biology and comparative physiology of birds, vol 2. Academic, New York, pp 1–36

Pradel A, Langer M, Maisey JG, Geffard-Kuriyama D, Cloetens P, Janvier P, Tafforeau P (2009) Skull and brain of a 300-million-year-old chimaeroid fish revealed by synchrotron holotomography. Proc Natl Acad Sci U S A 106:5224–5228

Prior H, Schwarz A, Güntürkün O (2008) Mirror-induced behaviour in the magpie (*Pica pica*): evidence of self-recognition. PLoS Biol 6(8):e202

Proffitt JV, Clarke JA, Scofield RP (2016) Novel insights into early neuroanatomical evolution in penguins from the oldest described penguin brain endocast. J Anat 229:228–238

Reiner A (2009) Avian evolution: from Darwin's finches to a new way of thinking about avian forebrain organization and behavioural capabilities. Biol Lett 5:122–124

Reiner A, Yamamoto K, Karten HJ (2005) Organization and evolution of the avian forebrain. Anat Rec A 287:1080–1102

Salzen EA, Parker DM (1975) Arousal and orientation functions of the avian telencephalon. In: Wright P, Caryl PG, Vowles DM (eds) Neural and endocrine aspects of behavior in birds. Elsevier, Amsterdam, pp 205–242

Sobral G, Sookias RB, Bhullar BAS, Smith R, Butler RJ, Müller J (2016) New information on the braincase and inner ear of Euparkeria capensis broom: implications for diapsid and archosaur evolution. R Soc Open Sci 3:160072

Stingelin W (1957) Vergleichend morphologische untersuchungen am Vorderhirn der Vögel auf cytologischer und cytoarchitektonischer Grundlage. Helbing and Lichtenhahn, Basel

Tambussi CP, Degrange FJ, Ksepka DT (2015) Endocranial anatomy of Antarctic Eocene stem penguins: implications for sensory system evolution in Sphenisciformes (Aves). J Vertebr Paleontol 35:e981635

Tebbich S, Taborsky M, Fessl B, Blomqvist M (2001) Do woodpecker finches acquire tool-use by social learning? Proc R Soc Lond B 268: 2189–2193

Walsh SA, Iwaniuk AN, Knoll MA, Bourdon E, Barrett PM, Milner AC, Nudds R, Abel RL, Dello Sterpaio P (2013) Avian cerebellar floccular fossa size is not a proxy for flying ability in birds. PLoS One 8(6):e67176

Walsh SA, Knoll MA (2011) Directions in palaeoneurology. Spec Pap Palaeontol 86:263–279

Walsh SA, Milner AC (2011a) Evolution of the avian brain and senses. In: Dyke G, Kaiser G (eds) Living dinosaurs: the evolutionary history of modern birds. Wiley, Chichester, pp 282–305

Walsh SA, Milner AC (2011b) *Halcyornis toliapicus* (Aves: Lower Eocene, England) indicates advanced neuromorphology in Mesozoic Neornithes. J Syst Palaeontol 9:173–181

Walsh SA, Milner AC, Bourdon E (2016) A reappraisal of *Cerebavis cenomanica* (Aves, Ornithurae), from Melovatka, Russia. J Anat 229:215–227

Walsh SA, Zhe-Xi L, Barrett P (2014) Modern imaging techniques as a window to prehistoric auditory worlds. In: Köppl C, Manley G (eds) Insights from comparative hearing research. Springer, New York, pp 227–261

Witmer LM, Chatterjee S, Franzosa J, Rowe T (2003) Neuroanatomy of flying reptiles and implications for flight, posture and behavior. Nature 425:950–953

Witmer LM, Ridgely RC (2007) Evolving an on-board flight computer: brains, ears, and exaptation in the evolution of birds and other theropod dinosaurs. J Morphol 268:1150

Witmer LM, Ridgely RC, Dufeau DL, Semones MC (2008) Using CT to peer into the past: 3D visualisation of the brain and ear regions of birds, crocodiles and nonavian dinosaurs. In: Endo H, Frey R (eds) Anatomical imaging: towards a new morphology. Springer, Tokyo, pp 67–87

Yosef R, Yosef N (2010) Cooperative hunting in brown-necked raven (*Corvus rufficollis*) on Egyptian mastigure (*Uromastyx aegyptius*). J Ethol 28:385–388

Zhou Z (2004) The origin and early evolution of birds: discoveries, disputes, and perspectives from fossil evidence. Naturwissenschaften 91:455–471

The Endocranial Vascular System: Tracing Vessels 6

Gizéh Rangel de Lázaro, Stanislava Eisová, Hana Píšová, and Emiliano Bruner

Abstract

The vascular system is distributed throughout the cerebral, connective, and bony elements of the braincase, and it supplies an anatomical connection between these three components of the endocranial morphology. The imprints and traces left by arteries and veins in the bone thickness and surface can be useful in the analysis of vascular features in fossil specimens and archaeological samples. These traits can provide indirect physiological or morphogenetic information associated with evolutionary changes, demographic relationships, or individual life history. Digital anatomy and computed morphometrics have represented a major advance in the study of these craniovascular characters, for which there is still limited knowledge available regarding their variability, functions, and development. In this chapter, we present and discuss current evidence on the imprints of middle meningeal vessels, diploic veins, dural venous sinuses, and emissary veins. We review the morphological and functional information about these craniovascular features and their applications in paleontology, medicine, bioarchaeology, and forensic science.

Keywords

Craniovascular morphology • Middle meningeal artery • Diploic veins • Venous sinuses • Paleoneurology • Paleoangiology • Digital anatomy

6.1 Introduction

The craniovascular system consists of four major districts, namely, the pericranial, diploic, meningeal, and cerebral networks (Falk 1986; Saban 1995; Zenker and Kubik 1996; Hershkovitz et al. 1999; Skrzat et al. 2004; Bruner et al. 2005; Jivraj et al. 2009; Louis et al. 2009; Adeeb et al. 2012). They are mainly distributed in three anatomical levels: ectocranial, bone, and endocranial (Fig. 6.1). In the ectocranial level, there are vessels which drain the scalp and adjacent soft tissues (Patel 2009). In the bone level, the emissary veins can be found (Louis et al. 2009), which bridge the ectocranial and endocranial areas, and the diploic channels, which run within the bone matrix (García-González et al. 2009; Jivraj et al. 2009). The endocranial volume includes two networks: meningeal and cerebral. The first one includes the meningeal vessels (Bruner and Sherkat 2008; Patel and Kirmi 2009) and

G. Rangel de Lázaro (✉)
Área de Prehistoria, Universitat Rovira i Virgili, Avinguda Catalunya 35, 43002 Tarragona, Spain

Institut Català de Paleoecologia Humana i Evolució Social, Zona Educacional 4, Campus Sescelades URV (Edifici W3), 43007 Tarragona, Spain
e-mail: gizeh.rangel@urv.cat

S. Eisová • H. Píšová
Katedra Antropologie a Genetiky Cloveka, Univerzita Karlova, Viničná 7, 12843 Prague, Czech Republic
e-mail: stanaeisova@gmail.com; hana.pisova@natur.cuni.cz

E. Bruner
Programa de Paleobiología, Centro Nacional de Investigación sobre la Evolución Humana, Paseo Sierra de Atapuerca 3, 09002 Burgos, Spain
e-mail: emiliano.bruner@cenieh.es

E. Bruner et al. (eds.), *Digital Endocasts*, Replacement of Neanderthals by Modern Humans Series,
DOI 10.1007/978-4-431-56582-6_6

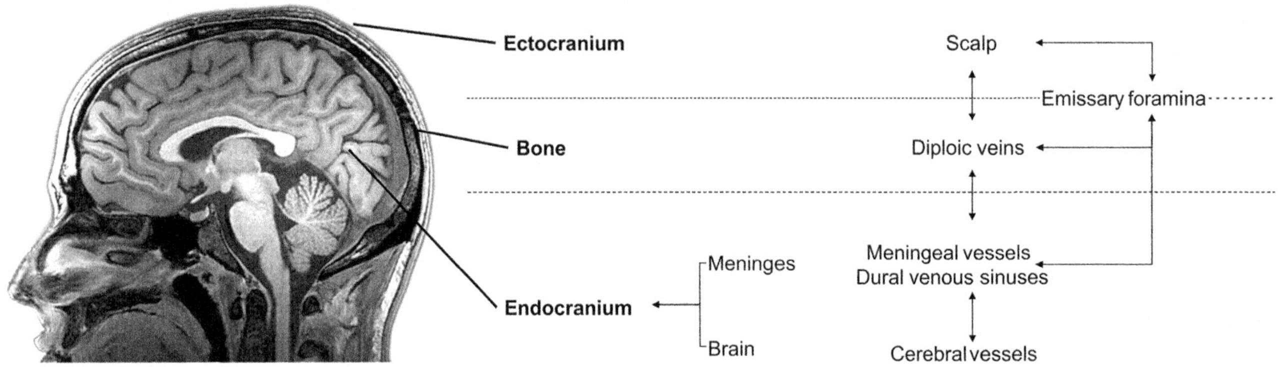

Fig. 6.1 Schematic representation of the craniovascular system, which comprises four major networks distributed in three layers

the dural venous sinuses, which form the major drainage pathways from the endocranium to the internal jugular veins (Rea 2015). The second one is formed by the cerebral veins and arteries, running within the brain mass (Touzani and MacKenzie 2007; Woolsey et al. 2008).

The veins supplying the endocranial space are thin walled and valveless, which facilitate blood flow in both directions (Gray 1913; Hartmann and Hoyer 2012; Schapira 2007; Schmidt and Thews 2013). The distinction among these vascular networks is more a matter of nomenclature and topology than functional roles, because all their vessels are actually integrated and connected as a single system, and their embryological (Adeeb et al. 2012; Robert 2004) and functional boundaries (Schapira 2007; Mortazavi et al. 2013; Schmidt and Thews 2013; Rea 2015) are not really separate in terms of homology or development. Early anatomical classifications and descriptions of these traits were carried out on cadaveric dissections and histological preparations in the nineteenth and early twentieth centuries (Breschet 1829; Raciborski 1841; Gray 1913), and several methodological limits hampered a proper development of a reliable terminology. As a consequence, anatomical terms and definitions are in many cases variable and subjective and still based on a nomenclature which is general and descriptive.

In the last two decades, digital anatomy and computed morphometrics have represented a major advance in macroanatomical studies (Bookstein 1991; Bruner 2004; Zelditch et al. 2004; Slice 2005; Zollikofer and Ponce de León 2005; Rilling 2008, 2014), revealing the necessity of a more careful and efficient terminological approach. Currently, the anatomical studies of the endocranial vascular system combine new biomedical technologies and computed morphometrics, allowing digital reconstruction of the vascular traces and quantitative analyses based on their variation and covariation (Hershkovitz et al. 1999; Rangel de Lázaro et al. 2016; Eisová et al. 2016).

The relationship between skull and endocranial vascular system is a key matter in evolutionary anthropology, paleoanthropology, and human anatomy. In terms of evolution, the endocranial vascular system provides evidence of phylogenetic differences (Grimaud-Hervé 1997; Bruner et al. 2005; Bruner and Sherkat 2008), possibly associated with encephalization and brain thermoregulation (Bruner et al. 2011a, b). The management of blood flow may play a crucial thermal role in physiological (Caputa 2004; Wysocki 2008; Hartmann and Hoyer 2012; Mortazavi et al. 2013) and evolutionary terms (Falk 1990; Cabanac 1995; Falk and Gage 1997; Bruner et al. 2011a, b, 2012). Vascular traits are also considered in fields such as bioarchaeology, forensic science, and medicine. In bioarchaeology, craniovascular features are used as individual or group morphological markers (Hauser and De Stefano 1985; Hershkovitz et al. 1999). In forensic science, the vascular networks can supply information regarding individual life history and recognition (Lee et al. 2014; Short et al. 2014). This same information is also relevant in medicine, clinics, and surgery, being associated with infective diseases, the distribution of oxygen-rich blood to the brain, and endocranial heat management (Ribas et al. 2006; Bruner and Sherkat 2008; García-González et al. 2009; Louis et al. 2009; Patel 2009; Bertolizio et al. 2011). The mechanisms associated with the endocranial blood flow are still largely uncertain, and the information concerning vascular morphogenesis and variability are rather scarce. In this chapter, we present and discuss the information available on the craniovascular features used in anthropology to evaluate the anatomy of the vascular system from cranial remains.

6.2 The Craniovascular System

6.2.1 Morphogenesis: Bone and Vessels

Cranial morphogenesis is a mixture of genetic, biomechanical, and biochemical factors susceptible to environmental and physiological responses (Moss and Young 1960; Enlow 1990; Sperber 2001; Bastir et al. 2006; Bastir 2008; Neubauer et al. 2010; Bruner et al. 2014; Bruner 2015). Therefore, both functional and structural factors can influence the relationship between cranial bones and vessels. Functional factors are based on shared biochemical signals or shared metabolic processes, while structural factors are based on spatial constraints and biomechanical interactions. Growth and development indeed play a major role in generating this biological background (Lieberman 1996; Eichmann et al. 2005). Age-related changes associated with vascular growth can affect bone morphology, the rate of mineralization, or bone maturation (Percival and Richtsmeier 2013). Research focused on the endocranial vascular system must hence necessarily take into account morphogenesis and use information from embryology and histology.

In multiphasic anatomical systems formed by different elements and different tissues such as the head, genetic programs will be oriented by epigenetic factors, channeling morphological variability according to alternative pathways of phenotypic expression (Nanney 1958; Waddington 2012; Müller et al. 2013). It must be noted that in molecular biology, "epigenetic" refers to environmental modifications of DNA which can be inherited or at least influence the genetic expression between generations (Burian 2004). However, in embryology and anatomy, the term has referred to all nongenetic factors that can influence the phenotype, leading to the expression of some characters in terms of presence or degree. In this sense, most craniovascular traits are "epigenetic" because they may be sensitive to environmental or developmental factors, although often showing a genetic background (Hauser and De Stefano 1989; Manzi et al. 1996; Moore 2015).

In terms of transmission, epigenetic traits can be heritable through direct genetic effects (genes coding for traits) or by indirect genetic influence (genes influencing general developmental processes which induce the expression of the traits). In the latter case, the expression of the character is not strictly genetic, but it should be interpreted as a secondary consequence of a general morphogenetic balance.

The morphogenesis can be, at least theoretically, divided into *growth* (size changes) and *development* (shape changes) (Moss and Young 1960; Lieberman 1996; Eichmann et al. 2005; Percival and Richtsmeier 2013). Shape and size changes between the brain and braincase during morphogenesis must be properly balanced to provide a functional phenotype. Veronika Kochetkova (1978) was a pioneer in paleoneurology, hypothesizing structural influences among sulci, gyri, cranial base, cerebral pressure, meninges, vascular system, and cerebrospinal fluid. In 1960, Moss and Young suggested that changes in the size of the braincase (growth) are mainly due to the pressure applied by the brain during growth. At the same time, changes in neurocranial shape (development) are mostly due to redistribution and reorientation of the growth pressures and strains through the main connective layers, which act as biomechanical tensors. According to this model, the *falx cerebri* and the *tentorium cerebelli* may be the major connective structural sheet, positioned between the cerebral hemispheres and between the brain and cerebellum, respectively. The meningeal layers separate and physically anchor the brain and the braincase and are at the same time an important support for the endocranial vessels (middle meningeal artery and venous sinuses). Therefore, the same elements hypothesized to be relevant for the structural organization of the morphogenetic processes also represent the shared physical interface among cranial, cerebral, and vascular systems. Nonetheless, many hypotheses in functional craniology have yet to be verified through proper quantitative analyses, due to the difficulties in handling these factors and variables in a strict experimental framework.

Whatever the relationship, the influence between soft and hard tissue is reciprocal, and many vascular elements leave their traces on the cranial bones supplying information on physiological functions related to blood flow, oxygenation, and thermoregulation (Richtsmeier et al. 2006; Mortazavi et al. 2013; Bruner 2015). The presence and degree of expression of these bony imprints will depend on physical variables, such as the size of the vessel, the meningeal thickness, the blood pressure, the brain pressure, and the pressure of the cerebrospinal fluid. Bruner and Sherkat (2008) reported a good correspondence between meningeal arteries and their traces in surgical cases. However, in general, the absence of an imprint should not be interpreted automatically as the absence of the corresponding vascular elements, and there may be some differences between the actual vascular morphology and its imprints left on the bone surface (O'Loughlin 1996).

During early ontogenetic stages, the imprints from vessels and circumvolutions on the endocranial surface are more pronounced, while throughout adulthood, they smoothen gradually (Neubauer et al. 2009, 2010; Zollikofer and Ponce de León 2013).

The cranial morphogenetic process is based on a regulated balance between bone deposition and bone resorption achieved by osteoblasts and osteoclasts, respectively (Lieberman et al. 2000; Martínez-Maza et al. 2006). Any epigenetic alteration between size and shape variations can lead to an excess (*hyperostosis*) or defect (*hypostosis*) of ossification, so generating an unbalanced phenotype (Manzi and Vienna 1997; Manzi et al. 2000). In the case of hyperostosis, osteoblasts deposit more bone than required, or osteoclasts do not remove the bone in excess. In contrast, during hypostosis, the ossification pattern does not remain on a par with the morphological changes, generating isolated ectopic areas of intramembranous ossifications (known as Wormian or supernumerary bones) and numerous or unfused sutures (O'Loughlin 2004). The dynamics and factors associated with these epigenetic traits are largely unknown, but these features are generally used in archaeology, population biology, and forensic anthropology, as individual or group osteological markers (Hauser and De Stefano 1989). Extreme degrees of hyperostosis and hypostosis may generate bone diseases caused by various systemic factors and may be associated with specific lifestyle and environmental stresses. In this sense, the vascular traces of the skull represent hyperostotic crests and hypostotic grooves formed after structural relationships between the bone and the contacting tissue (Eichmann et al. 2005), and their morphology must be interpreted according to the underlying balance of bone remodeling.

Some traits associated with early ontogenetic stages (e.g., the emissary foramina) are less influenced by secondary effects expressed in later maturation processes and hence are more sensitive to a direct genetic component (Del Papa and Perez 2007). Other characters will show a larger sensitivity to biomechanical and structural effects induced by the physical interactions between soft and hard tissues. For example, the development of the middle meningeal vascular patterns and the course of dural venous sinuses can be influenced by cranial deformations, due to artificial cultural practice or to pathological and subpathological craniosynostosis (O'Loughlin 1996). The vessels associated with the deformed surface are generally shallow and flat, while the vessels running along the undistorted areas compensate that change, being wider and deeper. Furthermore, morphological asymmetries of the skull can induce morphological asymmetries of the vessels, stressing further the importance of the environment on their morphogenesis. The resulting phenotype is the balance between a network of forces, in which genes often give only general "commands" associated with cell proliferation and regulation. Such commands need to be conveniently coordinated by inner (integration) and outer (environmental) factors (Dawkins 1989; Van Speybroeck 2002). Even though some experimental studies support the functional matrix hypothesis (Moss 1968; Kyrkanides et al. 2011), the specific mechanisms behind these processes are still unknown.

6.2.2 Vault Thickness and Bone Histology

In paleontology and archaeology, vascular analysis is limited to the imprints of the vessels, left on the bony surface, and within the bony matrix. Therefore, a proper knowledge of the bone biology is mandatory in order to interpret the morphology of the traces in terms of any possible structural and functional interaction.

The human vault bones are divided into three layers: an external table in contact with the scalp and ectocranial soft tissues, an intermediate cancellous bone (diploe), and an internal table in contact with the endocranial space. Cranial vault thickness (CVT) is computed as the distance between endocranial and ectocranial boundaries, and it undergoes major morphological changes in early ontogeny, with some minor alterations in adulthood (Hatipoglu et al. 2008; Scheuer and Black 2000; Li et al. 2011; Marsh 2013; Tsutsumi et al. 2013; Anzelmo et al. 2014; García Gil et al. 2015). The thickness of cranial vault bones differs not only among hominids (e.g., Kennedy 1991; Nawrocki 1991; Gauld 1996; Balzeau 2013; Curnoe and Green 2013; Copes and Kimbel 2016) but also varies in modern populations (Skrzat et al. 2004; Marsh 2013; Voie et al. 2014; Boruah et al. 2015). Multiple systemic and local stimuli influence the thickness of the cranial vault (Lieberman 1996; Moreira-Gonzalez et al. 2006; Baab et al. 2010; Menegaz et al. 2010). It has been suggested that the three skull layers are functionally independent, with the diploe responding to red-blood-cell levels and cortical bone more sensitive to mineral-ion level changes (Kennedy 1991). Furthermore, the inner layer is more susceptible to brain growth and the outer layer more susceptible to muscular strains (Moss and Young 1960).

The morphology and distribution of the diploe seem to be influenced by age-related changes (Koenig et al. 1995; Hershkovitz et al. 1999; García Gil et al. 2015). In young individuals, the diploe is homogenous and uniform, while with aging, the trabecular bone becomes more porous, sclerotic, and scattered (Skrzat et al. 2004). At the age of 5, the vault is 75–80 % of its final thickness, and 3 years later, the cranial cavity reaches adult size (Hatipoglu et al. 2008; Anzelmo et al. 2014). Vault thickness gradually increases without sex differences until the age of 20, and then it undergoes minor thinning through the rest of the life (Roche 1953; Li et al. 2011; Tsutsumi et al. 2013; Marsh 2013).

In early ontogenetic stages, CVT is linked to brain protection during growth and development (Koenig et al. 1995; Hershkovitz et al. 1999; Anzelmo et al. 2014), mechanical forces (muscles attachments), suture formation, and vessel enlargement (Moss and Young 1960; Enlow 1968; Goldsmith 1972). Conversely, growth hormones, physical activity, nutrition (Lieberman 1996; Skrzat et al. 2004; Baab et al. 2010), and local mechanisms (such as different types of muscular exercise and other kinds of mechanical demands) could partially alter CVT in later ontogeny (Menegaz et al. 2010; Copes 2012). Physical activity causes the release of growth hormone, which leads to an increase of the bone mass and vault thickness, by inducing and promoting osteogenesis (Lieberman 1996; Baab et al. 2010; Menegaz et al. 2010).

A quantitative mapping of the human cranium shows the parietal, occipital, and frontal bones form the thicker regions of the skull (Voie et al. 2014; Boruah et al. 2015; Lillie et al. 2015). Some differences in the diploe density of younger and older adults due to a reduction in calcium concentration have been observed (Ross et al. 1998; Skrzat et al. 2004; Lynnerup 2001; Lynnerup et al. 2005; Torres-Lagares et al. 2010; Sabancıoğulları et al. 2012; Anzelmo et al. 2014). Modern human cranial vault thickness ranges from 1.96 to 10.6 mm (Brown 1994; Hwang et al. 1999; Lynnerup 2001; Voie et al. 2014; Boruah et al. 2015; Lillie et al. 2015; Copes and Kimbel 2016). In general, no significant sexual differences have been described in CVT (Ross et al. 1998; Lynnerup et al. 2005; Torres-Lagares et al. 2010; Sabancıoğulları et al. 2012; Anzelmo et al. 2014; Eisová et al. 2016).

Modern humans are characterized by large parietal bones (Bruner et al. 2011a) with complex vascular networks (Bruner et al. 2005; Rangel de Lázaro et al. 2016), when compared with other human species. We recently published a survey on modern adult parietal bone, reporting a mean thickness value of 6.02 ± 1.54 mm for the total thickness, 1.52 ± 0.50 mm for the inner table, 2.84 ± 1.44 mm for the diploe, and 1.70 ± 0.48 mm for the outer table (Eisová et al. 2016; Fig. 6.2). Our results also suggested that the largest variation for the parietal thickness is associated with the variation of the diploe, since the outer and inner table tends to be more constant over all the bone. The outer table is in general thicker, denser, and stiffer (Peterson and Dechow 2002), while the inner table shows more homogeneous values. The upper areas of the parietal bone are thicker than the lower ones (Peterson and Dechow 2002, 2003; Anzelmo et al. 2014).

Bone thickness and vessel size (measured in terms of lumen size) can be hypothesized to have a proportional correlation if bone and vessels are sensitive to shared growth factors or, alternatively, they may display an inverse size correlation if involved in structural and spatial competition. Nonetheless, we found no correlation between skull size, parietal bone thickness, and vascular dimensions suggesting that, at least in terms of size, bones and vessels are influenced by independent factors (Eisová et al. 2016).

Also bone histology offers different possibilities of analysis. According to García Gil et al. (2015), it is possible to identify three different histological phases of cranial growth. Child vault bones present dense lamellar bone, large secondary osteons, and resorption cavities. In contrast, the adolescent bones show a larger extension of mineralized regions (highly remodeled areas) and smaller secondary osteons. In the adult, cancellous bone is largely expanded. The sealing of the cranial bone surfaces helps to minimize the bone porosity while increasing bone expansion (during childhood) and thickness (during youth). This "sealing process" could play an important role controlling head thermoregulation until the brain finishes its maturation.

6.3 Tracing the Vascular Networks

6.3.1 Physical Methods

Physical inspection, anatomical dissections, and corrosion cast techniques have been the traditional methods to study the morphology of the craniovascular system (Kaplan et al. 1972; Louis et al. 2009; Chen et al. 2011; Tubbs et al. 2015). In dry skulls, visual inspection of the endocranial cavity can be performed with endoscopes and dental mirrors. In complete crania, middle meningeal and dural venous imprints can be visualized through the foramen magnum (Boyd 1930; Saban 1995). Foramina of emissary vessels can be generally examined ectocranially and measured with steel wires or lead shots (Wysocki et al. 2006; Wysocki 2008; Murlimanju et al. 2011). The physical inspection of diploic channels is more difficult, as they are enclosed in the cranial bone thickness.

Cadaveric head dissection is one of the best available methods for understanding human vascular structures (Kresimir et al. 2001; Adeeb et al. 2012). Techniques based on silicon latex injection are used to dissect cadaveric heads (García-González et al. 2009). However, the identification of direct communication between vascular structures, such as diploic vessels and dural venous sinuses, can be tricky, because of the difficulty of filling low flow in specific points. In that case, the corrosion cast-inject solution may be a better method to preserve and provide a three-dimensional reconstruction of even the smallest vascular pathways (Zenker and Kubik 1996; San Millán Ruíz et al. 2004; Johnston et al. 2007).

Transilluminiscence is another method that has been used in anthropology as a complement to radiography and other imaging techniques (Barozzino and Sgro 2002; Horner 1962). In its basic form, a bright light source is introduced into the endocranial cavity, and the light shines across the bone thickness showing the diploic patterns on the ectocranial surface. This method can be ineffective in thicker skulls or when the diploic thickness is reduced.

In recent years, the combinations of digital methods with dissection and microscopic techniques are commonly applied in paleoanthropology (Rosas et al. 2008; Peña-Melián et al. 2011), human anatomy (Das et al. 2008; Tubbs et al. 2011), and medicine (San Millán Ruíz et al. 2002; Kim et al. 2014). Unfortunately, traditional techniques do not preserve the information about the bone counterparts effectively, which is the principal source of information in the anthropological field. To date, there are still few studies of the craniovascular system and their relation to bone tissue.

6.3.2 Biomedical Imaging Techniques

Biomedical imaging technologies encompass applied mathematics, engineering, physics, and chemistry and can be used to develop methods for processing and analyzing quantitative data from molecular/cellular images to large anatomical structures and physiological processes. Visual representations of bones, organs, tissues, and blood flow are obtained through ultrasound, radiopharmaceutical materials, magnetic fields, and X-rays. Imaging techniques are also applied in paleoanthropological research to investigate the overall morphological variability among individuals and among species found in the fossil record (Spoor et al. 2000; Sutton 2008; Wua and Schepartzb 2009; Weber 2015). Integrating results from paleoneurology and comparative neuroanatomy is useful to develop hypotheses about brain and craniovascular evolution, taking into consideration the anatomical changes that have characterized the genus *Homo* in the last two million years (Wind 1984; Bruner and Manzi 2005, 2006; Gunz et al. 2009).

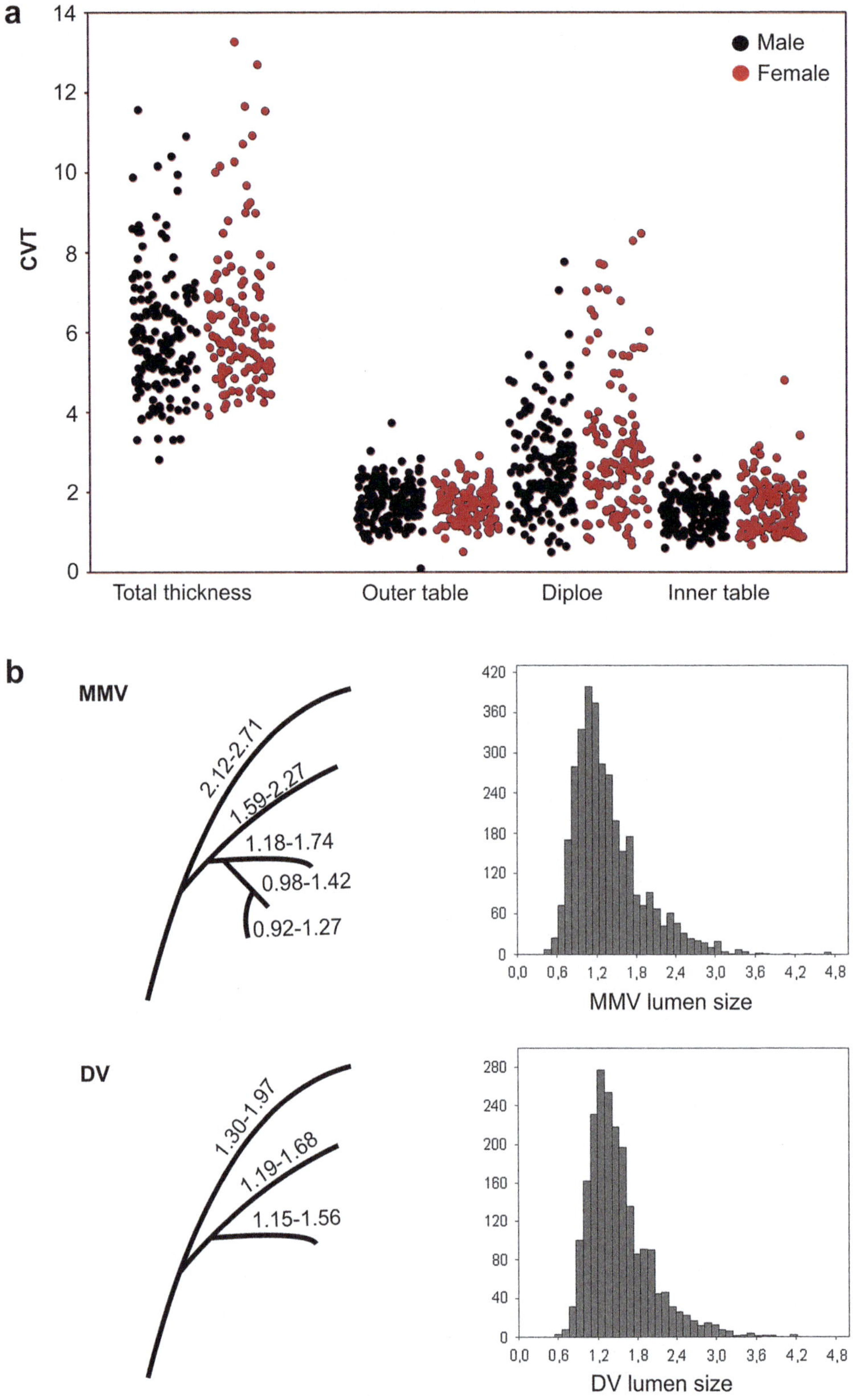

Fig. 6.2 (**a**) Cranial vault thickness (CVT) in modern human males and females. Jitter plot of the thickness measurements of 16 males and 13 females. Each individual was measured in five points in both parietal bones. No significant sex or side differences were detected. The thickness of the diploe is the most variable value, determining the major differences among individuals in the overall bone *thickness*. (**b**) On the parietal bone, five orders of branches can be generally recognized for the middle meningeal vessels (MMV) and three for the diploic vessels (DV). Lumen size is reported for each order, showing the 25th and 75th percentiles. The histograms show an overall distribution of the lumen diameter (Data after Eisová et al. 2016)

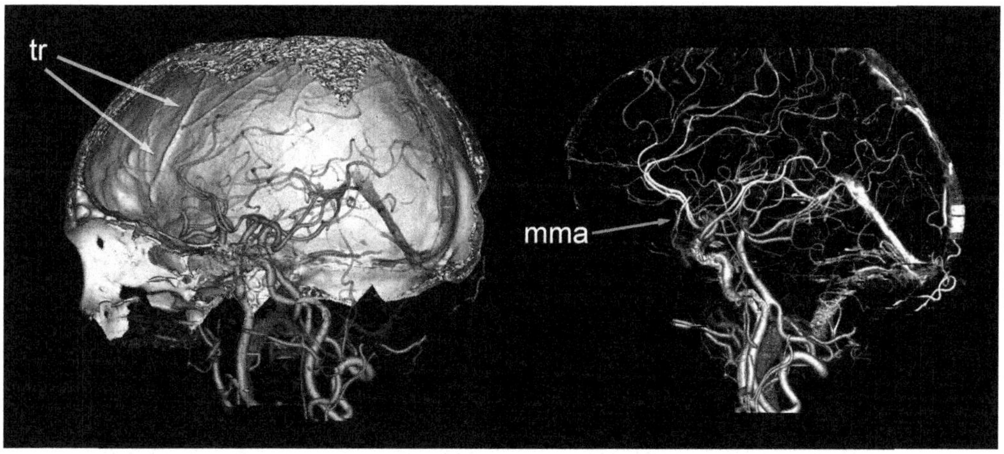

Fig. 6.3 Angiotomography can be used for the three-dimensional reconstruction of the craniovascular system. The middle meningeal artery is visible through the endocranial trace (tr), but blood is detected only in its basal portion (mma), suggesting that at rest the flow is generally interrupted (Modified after Bruner et al. 2011a, b)

Vascular morphology and blood metabolism are generally analyzed in vivo by using ultrasounds (Hajdú et al. 2008; Patel 2009); positron emission tomography (PET) with radioactive tracer to measure metabolic changes, blood flow, and oxygen use (Bremmer et al. 2011); and magnetic resonance imaging (MRI and functional MRI) through magnetic field and radiofrequency pulses (Sotero and Iturria-Medina 2011; Chatterjee et al. 2015; MacDonald and Frayne 2015). MR angiography and venography are used to generate high-contrast images of the vascular system (Connor and Jarosz 2002; Hu et al. 2007; Zonoobi et al. 2009; Yang and Guo 2015). Computed tomography (CT) is instead the most efficient method to study craniovascular features in living individuals, dry skulls, and fossil specimens. CT scans combine both X-rays and computer technology to acquire and reconstruct images of the body using multiple projections to create cross-sectional pictures (slices) showing the distribution of density values (Goldman 2007; Seeram 2015). CT scans are detailed in terms of contrast resolution and image quality, allowing better visualization of small density differences (Goebbel 2013). The density values are coded following a standard grayscale chromatic map (Hounsfield units) (Hounsfield 1980). The computer calculates the voxel-wise density values according to the capacity of attenuation of the energy signal, where each voxel is a three-dimensional element represented by a pixel and the thickness of the slice (Mahesh 2002). CT is used to model and virtually reconstruct the craniovascular characters (Rangel de Lázaro et al. 2016; Eisová et al. 2016; Píšová et al. 2017) because it is a nondestructive technique that allows fast image processing, accurate data, high image resolution, and the preservation of the specimens, avoiding any direct manipulation of the cranial remains. Angiotomography is another X-ray-based method used in vivo to visualize blood drainage and distribution by injecting high-density liquid contrasts during the radiographic scan (Curé et al. 1994; Patel 2009). Angiotomography showed that, in adults, the blood flow in the middle meningeal artery is scarce or even absent (Fig. 6.3), suggesting that blood flow in these vessels must be activated by specific physical, physiological, or environmental conditions (Bruner et al. 2011a, b).

Computed tomography, angiotomography, and other biomedical imaging techniques used in digital anatomy are largely contributing to the renaissance of new studies in vascular biology (Weinstein et al. 1977; Curé et al. 1994; Bruner et al. 2011a, b).

6.4 Anatomy and Morphology

The middle meningeal vessels, the diploic veins, and the venous sinuses are the main vascular elements when dealing with cranial samples. These structures leave their traces in the cranial bones, revealing the vascular patterns in extinct species or past populations. The anatomical organization of these vascular systems appears to be more developed in humans than in nonhuman primates, suggesting a role in evolutionary biology.

6.4.1 Middle Meningeal Artery

Middle meningeal vessels (MMV) are part of an endocranial craniovascular system interlocked between the layers of dura mater and cranial bones (Luttenberg 1959; Saban 1995; Patel and Kirmi 2009). Although most authors refer to this vascular structure as the middle meningeal *artery*, we prefer to use the more general term *vessels*, because bone imprints are probably left by both arteries and veins (Jones 1912; Falk and Nicholls 1992; Bruner and Sherkat 2008). The vessels are impressed on the endocranial walls, generally originating from the middle cranial fossa and ascending mostly to parietal surface (Fig. 6.4). Commonly, the middle meningeal artery arises from the maxillary branch of the external carotid artery and enters the skull through the foramen spinosum, rarely from the foramen ovale or from the petrosphenoidal fissure. Less frequently, the middle meningeal artery originates from distinct segments of the internal carotid artery (e.g., persistent stapedial artery, ophthalmic artery, or lacrimal artery) usually followed by the absence or reduction of the foramen spinosum (Royle and Motson 1973; McLennan et al. 1974; Dilenge and Ascherl 1980; Silbergleit et al. 2000; Manjunath 2001; Kimball et al. 2015). In such cases, the vessels pass through the superior orbital fissure or through the cranio-orbital foramen (also called meningo-orbital foramen) (Diamond 1991; Georgiou and Cassell 1992; Lui and Rhoton 2001; Tubbs et al. 2015). After entering the endocranial space, the basal stem of the MMV gives rise to two principal branches: anterior (bregmatic) and posterior (lambdatic). This separation is most frequently located within the middle cranial fossa. As the two branches ascend the parietal walls, one of them or both give rise to a third (obelic) branch. The middle meningeal network anastomoses with other vascular networks (e.g., diploic veins, dural venous sinuses, venous lacuna, occipital arteries, and pericranial arteries) (Louis et al. 2009; Hartmann and Hoyer 2012; Mortazavi et al. 2013). The branching pattern generates pseudo-fractal geometry (Zamir 1999, 2001; Bruner et al. 2005).

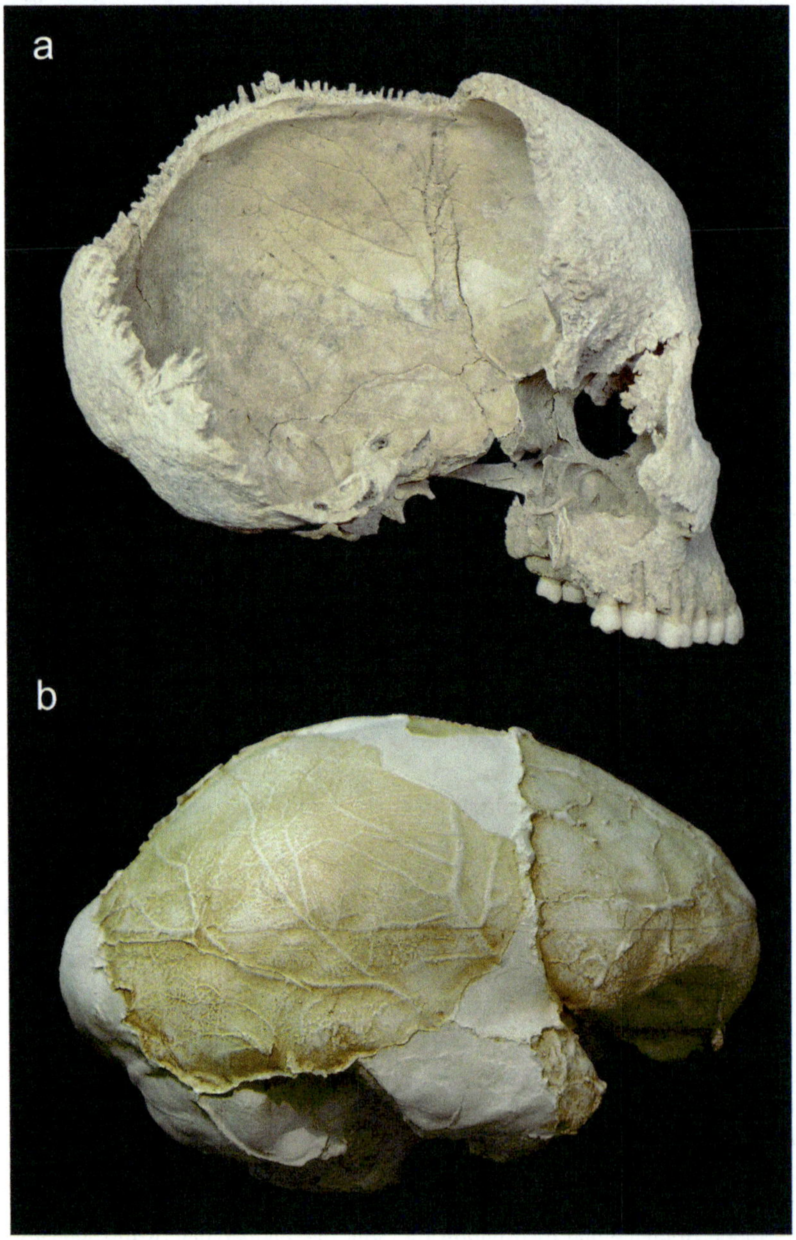

Fig. 6.4 (**a**) A modern human skull from an archaeological context showing, in the inner endocranial surface, the traces of the middle meningeal vessels. (**b**) The endocast of Arago (France, about 450,000 years before present), showing the imprints of the middle meningeal vessels on the mold of the parietal bone (Modified after Bruner 2010; Píšová et al. 2017)

The morphogenetic variables implicated in shaping the MMV are far from being understood (Rothman 1937; Falk 1993; Wu et al. 2006; Bruner and Sherkat 2008). The MMV can be observed from the early ontogenetic stages and associated with detailed endocranial imprints, although with aging the traces smoothen gradually (Saban 1995). The bone tissue is constantly molded by the tension and pressure exerted by the anatomical structures inside the endocranial space (Moss and Young 1960; Zollikofer and Ponce de León 2013). The degree of expression of the traces can be therefore due to different factors such as vessel size, blood and brain pressure, meningeal thickness, or the pressure of the cerebrospinal fluid.

Since the beginning of the twentieth century, different kinds of classifications and nomenclature have been proposed in both humans (Giuffrida-Ruggeri 1904; Derezinski 1934; Rothman 1937; Diamond 1991; Grimaud-Hervé 1997; Bruner et al. 2003) and nonhuman primates (Falk and Nicholls 1992; Falk 1993; Bruner and Sherkat 2008). The patterns of distribution of the MMV are described according to classification systems like those developed by Giuffrida-Ruggeri (1912) and Adachi (1928), following the branching schemes and topography of the vessels. Giuffrida-Ruggeri (1912) used the points of bifurcation of four specific branches as characters. The Adachi method is based on the origin of the middle branch (Adachi's types; Adachi 1928), ramifying from the anterior bregmatic branch (Type I), from the posterior lambdatic branch (type II), or from both branches (type III) (Marcozzi 1942; Fig. 6.5). Nonetheless, classes for such complex traits are extremely sensitive to different operational choices. General classifications (like Adachi's) are reliable but scarcely informative, while more specific classifications (based on more detailed anatomical variations) are more informative but also more subjective. In both cases, statistical inferences are difficult to support according to robust quantitative approaches.

The traces left by the MMV allow an indirect method to consider vascular patterns in modern humans but also in fossil specimens (Saban 1980, 1983, 1995; Grimaud-Hervé 1997). In modern humans, an increase in vascular complexity has been observed, when compared with other hominids (Bruner and Sherkat 2008). Vascular complexity seems not to be correlated with brain size or encephalization (Bruner et al. 2005). In adult modern humans, there is no evidence suggesting that the morphology of the vascular traces is influenced by cranial geometry (Bruner et al. 2009). Nevertheless, the vascular schemes are sensitive to the neurocranial form in pathologically and artificially deformed skulls, probably because of changes in the distribution of the endocranial pressure, which can influence the spatial arrangement of veins and arteries (O'Loughlin 1996).

Taking into account the scarce blood flow associated with the middle meningeal vessels in adult resting states, three hypotheses have been suggested to interpret their increased complexity in modern humans (Bruner et al. 2011a, b): (1) this system may play a key role in specific ontogenetic (possibly early) stages, when the brain demands higher metabolic/morphogenetic activity; (2) MMV blood flow may be mostly activated during excessive heat loading (physical exercise, emergency state, or pathological conditions); and (3) the vascular system may have biomechanical/protection functions associated with hydrodynamic pressure and not with heat or oxygenation. Nonetheless, it is also possible that the increase in complexity of the MMV in our species could be a secondary result of a general increase in the cerebral vascular system, whose networks are regulated by common growth factors (Bruner 2015).

6.4.2 Diploic Vessels

The vault bones are composed of two cortical layers (external and internal) separated by the diploe (Breschet 1829; Zenker and Kubik 1996; Toriumi et al. 2011; Hershkovitz et al. 1999). The diploic channels are pathways left by valveless diploic veins, running through the cancellous bone and interconnected through a network of microscopic channels (Rangel de Lázaro et al. 2016; Fig. 6.6). They are protected by a single endothelial layer (Zenker and Kubik 1996). The diploic veins connect extracranially with emissary veins (Píšová et al. 2017) and intracranially with middle meningeal vessels (Eisová et al. 2016) and dural sinuses (García-González et al. 2009; Patel and Kirmi 2009; Tsutsumi et al. 2013). The diploic veins are a blood pathway that bridges the endocranial and ectocranial vascular systems (Fig. 6.7). Therefore, the analysis of the diploic vessels may be relevant in anthropology, medicine, and paleontology, given their possible involvement in brain thermoregulation (Falk and Gage 1997; Caputa 2004; Wysocki 2008).

Several techniques have been used to study the diploic veins and their channels, such as corrosion casts (Testut 1893; Jefferson and Stewart 1928; Zenker and Kubik 1996), cadaveric head dissection (García-González et al. 2009), medical visualization through radiological methods (X-ray, angiography, computed tomography) (Hershkovitz et al. 1999; Skrzat et al. 2004; Patel 2009; Rangel de Lázaro et al. 2016; Eisová et al. 2016), magnetic resonance imaging technique (Hatipoglu et al. 2008; Jivraj et al. 2009; Tsutsumi et al. 2013), and experimental studies (Toriumi et al. 2011).

Hard bones often protect the diploe from postdepositional and taphonomical alterations, which facilitate its study in archaeology, paleoanthropology, and forensic contexts (Curnoe and Green 2013; Bruner et al. 2014; Lee et al.

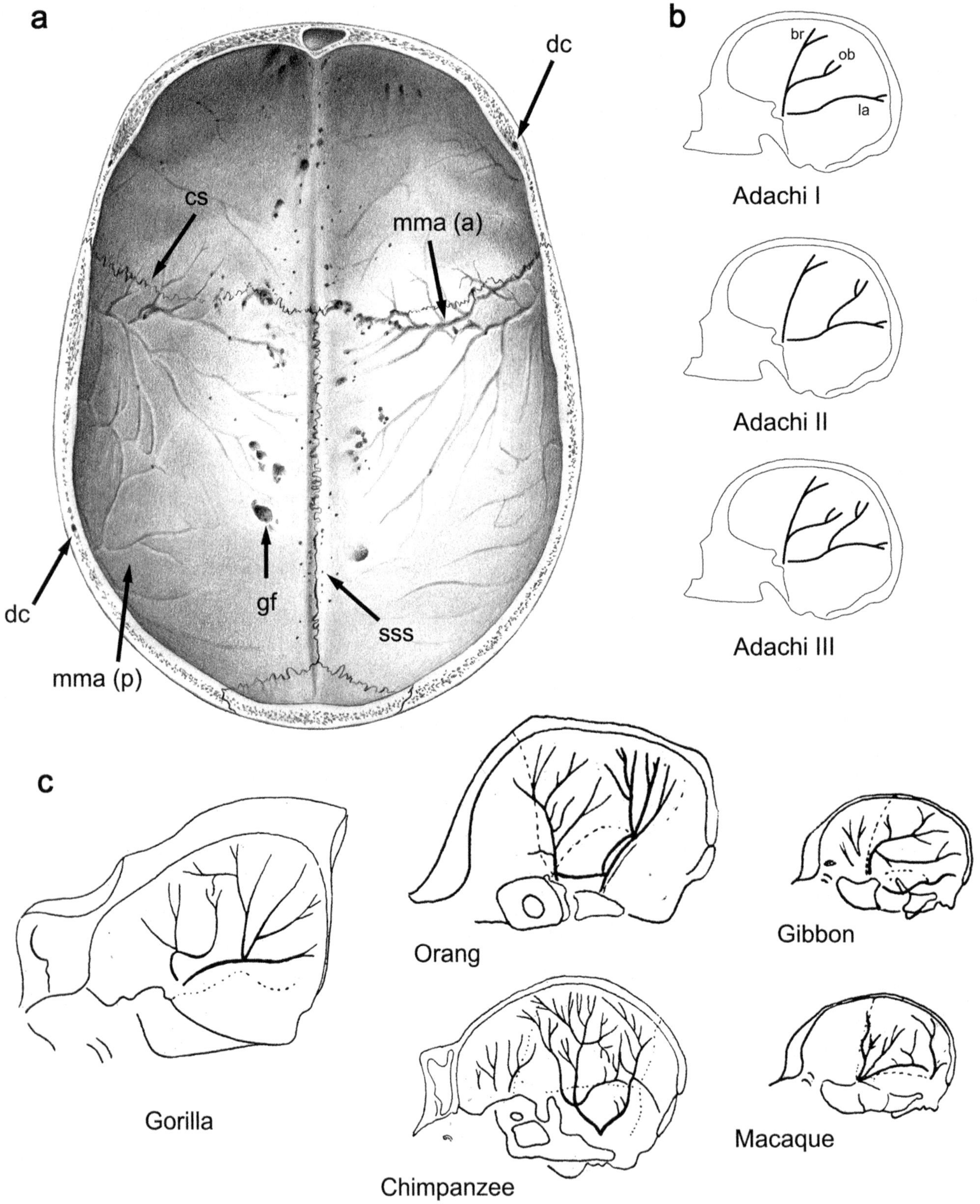

Fig. 6.5 (**a**) Drawing of the endocranial vault surface in modern humans shows the details of the vascular traces: *cs* coronal suture; *mma* (a) and (p), middle meningeal artery anterior and posterior; *dc* diploic channel; *gf* granular fovea; *sss* superior sagittal suture. (**b**) Adachi classes are a descriptive categorization applied to MMV distribution. The middle obelic branch can ramify from anterior bregmatic branch (type I), posterior lambdatic branch (type II), or both branches (type III). (**c**) Schematic patterns of the middle meningeal vessels in nonhuman primates (Original drawing of Marcozzi V, modified after Bruner and Sherkat 2008; Bruner et al. 2009)

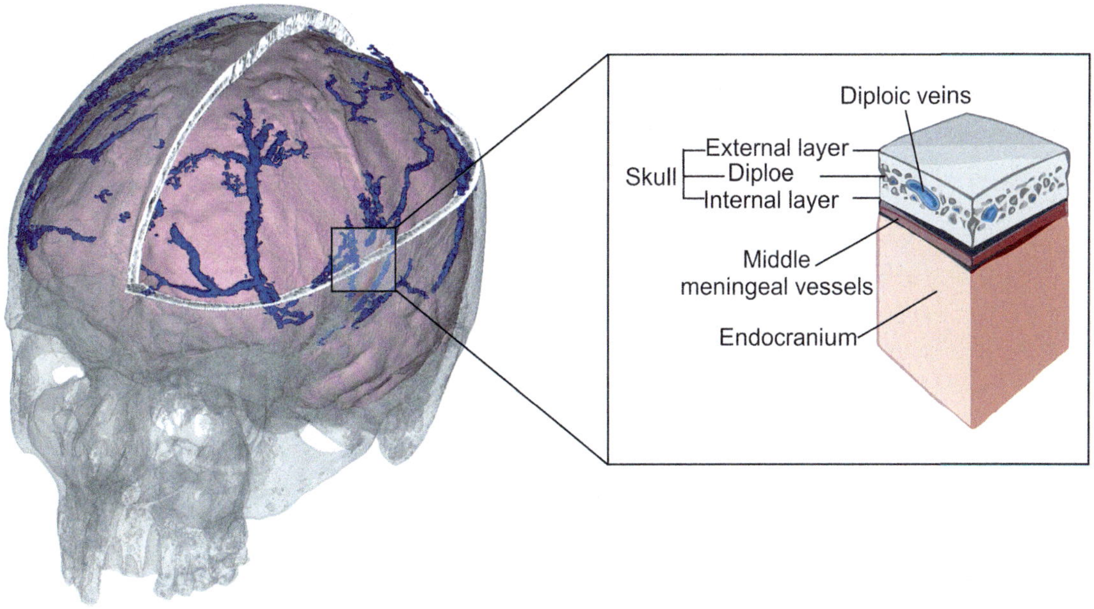

Fig. 6.6 Diploic vessels (*blue*) in a modern human skull and its relationship with the adjacent anatomical structures, namely, the middle meningeal vessels (*red*), skull layers (*gray*), and endocranium (*pink*)

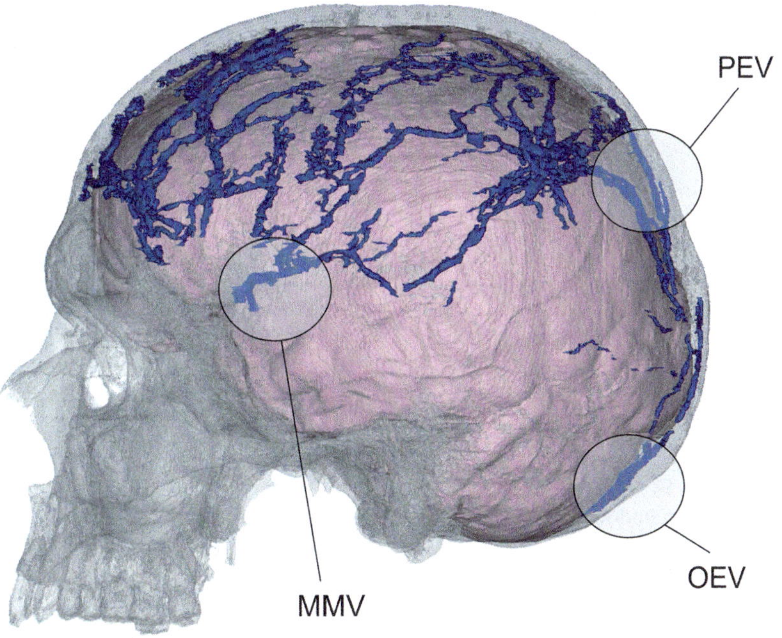

Fig. 6.7 Main communication areas between the diploic veins and meningeal vessels (MMV), parietal emissary veins (PEV), and occipital emissary vein (OEV)

2014; Weber 2014; Copes and Kimbel 2016; Rangel de Lázaro et al. 2016).

The diploic vein network appears to be more developed in humans than in nonhuman primates, suggesting phylogenetic changes (Hershkovitz et al. 1999). Also Neanderthals, despite their large cranial capacity, display a minor degree of vascularization when compared with modern humans (Rangel de Lázaro et al. 2016; Fig. 6.8). These differences in the diploic vascular complexity may parallel the ones described for the middle meningeal artery, more developed in modern humans and especially at the parietal bone. The emissary veins together with the diploic vessels could be part of a thermoregulatory network, refreshing the brain under conditions of hyperthermia (Cabanac 1995).

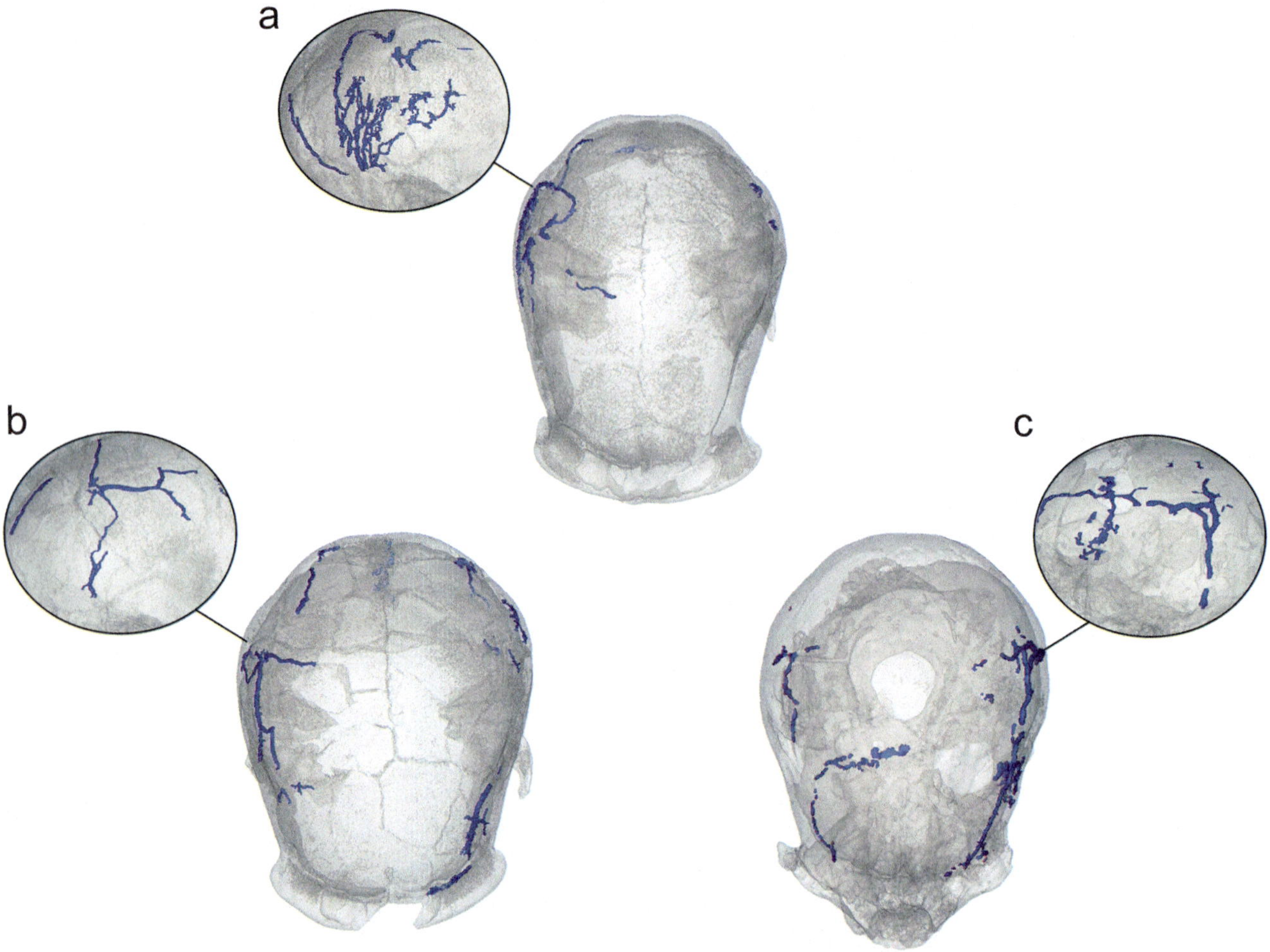

Fig. 6.8 Diploic channels in Neanderthals. (**a**) In Spy 1, the channels were observed mainly on the *right side*, with seven vertical branches going up to the superior parietal bone and subdivided into thin branches directed to the anterior parietal region and occipital bone. (**b**) In Spy 2, there is an incomplete central lacuna with three stems along the parietal area. (**c**) In Saccopastore 1, diploic vessels were localized mainly on the left side with one major vertical trunk extended superiorly and subdivided into two branches continuing to the frontal bone (Data after Rangel de Lázaro et al. 2016)

6.4.3 Dural Venous Sinuses

The dural venous sinuses drain the endocranial blood (Gray 1913; Curé et al. 1994). They are valveless venous channels, lined by endothelium, placed between the two layers of dura mater and within its folds (Knott 1881; Mancall and Brock 2010). Dural venous sinuses are divided into two interdependent groups. One encompasses the upper posterior sinuses (superior network), and the second includes the sinuses of the cranial base (inferior network) (Gray and Carter 1858; Curé et al. 1994). The sinuses run into internal jugular veins, at the posterior endocranial fossa.

The superior sagittal sinus begins at the *crista galli* and runs midsagittally alongside the metopic and sagittal suture into the *torcular Herophili* (confluence of the sinuses), a connection area of the upper posterior group of sinuses, positioned on the internal occipital protuberance. At the confluence, the superior sagittal sinus runs into the right transverse sinus in most cases and optionally in the left transverse sinus (Fig. 6.9). Two deep sinuses run between the hemispheres without contacting any bone surface: the inferior sagittal and the straight sinuses. The inferior sagittal sinus runs along the lower margin of the *falx cerebri* and continues into the straight sinus which runs, most commonly but not exclusively, into the left transverse sinus. The two transverse sinuses further continue into the sigmoid sinuses, at the jugular fossa of the cranial base. As an alternative route, the blood flow can continue in one (unilateral) or two (bilateral) small occipital sinuses in the medial lower occipital bones, running into marginal sinuses lateral to the foramen magnum and draining into the jugular fossae or into the vertebral venous system (Curé et al. 1994; Drake et al.

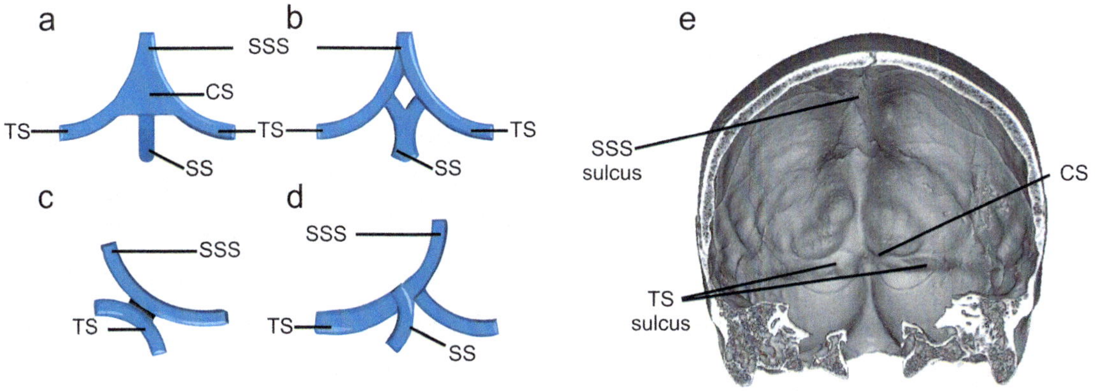

Fig. 6.9 Four main patterns of superior sagittal sinus (SSS) drainage into transverse sinus (TS) have been described: (**a**) The SSS reaches the confluence of the sinuses (CS), around the internal occipital protuberance, where it splits into both TS, after joining a large vascular area which receives one single straight sinus (SS); (**b**) Both SSS and SS split into two symmetrical channels; (**c**) Although the separation is not complete, SSS and SS largely run into one TS; (**d**) SSS deviates from the midline and continues into one (generally the *right*) TS, while SS follows the opposite scheme; (**e**) Imprints of the venous sinus pattern in the skull, namely, the SSS sulcus, the TS sulcus, and the CS

2013). This occipito-marginal (O/M) system can occasionally substitute or coexist with the transverse-sigmoid (T/S) system.

The pattern at the confluence of the sinuses is very variable, with variable schemes of connections and asymmetries (Matiegka 1923; Kaplan et al. 1972; Dora and Zileli 1980; Kimbel 1984; Falk 1986; Ayanzen et al. 2000; Singh et al. 2004; Kopuz et al. 2010). Neanderthals display more asymmetric patterns and fewer admixtures between the right and left flows (Peña-Melián et al. 2011). This might be a consequence of their larger endocranial breadth (Bruner and Holloway 2010), inducing a more pronounced separation of the sinus from the midsagittal area. However, similar patterns are also found in many anatomically modern humans (Knott 1881; Matiegka 1923; Curé et al. 1994).

On the endocranial base, grooves can be left by the superior and inferior petrosal sinuses and optionally by the petrosquamous and sphenoparietal sinus when present. The superior petrosal sinuses leave traces on the petrous part of the temporal bone, at the attachment of the *tentorium cerebelli*. The inferior petrosal sinuses form an imprint on the basilar part of occipital bone. Petrosal sinuses connect cavernous and lateral sinuses. Petrosquamous sinus courses alongside the petrosquamous suture and connects the transverse sinuses with the middle meningeal veins (Butler 1957; Saban 1995). It usually disappears during early ontogenesis, although the sinuses often fail to involutes and persist into adulthood (Marsot-Dupuch et al. 2001). The trace of the petrosquamous sinus in the bones is rarely detected.

The sphenoparietal sinus is not consistently defined in paleoanthropology and human anatomy (Grimaud-Hervé 2004; San Millán Ruíz et al. 2004). This vascular structure is situated below the lesser sphenoid wing, and it has been defined as a combination of lesser sphenoid wing sinus and anterior portion of middle meningeal veins (San Millán Ruíz et al. 2004). In paleoanthropology, the groove positioned behind the coronal suture is generally interpreted as the trace of the sphenoparietal sinus (Grimaud-Hervé 2004). Neanderthals showed higher prevalence (reaching 100 %) of the sphenoparietal sinus groove behind the coronal suture, when compared with anatomically modern humans (14%) (Grimaud-Hervé 2004).

6.4.4 Emissary Veins

Emissary veins are auxiliary valveless vessels that create the communications between intracranial dural venous sinuses and extracranial venous networks. These veins are already developed in early embryological stages when the extra- and intracranial venous networks are still not topographically separated by cranial bones (Butler 1957). After the ossification of the cranial bones, the veins enclosed by osseous tissue form foramina and canals that pass through the bones in recurrent and specific positions. Although the emissary veins are generally small in diameter, occasionally they can take over the role of an aberrant sinus and constitute an important part of the venous drainage system (Boyd 1930; Shapiro and Robinson 1967; Anderson et al. 1997; Marsot-Dupuch et al. 2001). In the bones, the emissary veins leave foramina and curved channels of variable number, size, and position, with asymmetric patterns and population differences (Boyd 1930; Hauser and De Stefano 1989). The most frequently present foramina are the mastoid, the condylar, and the parietal canals (Louis et al. 2009; (Fig. 6.10).

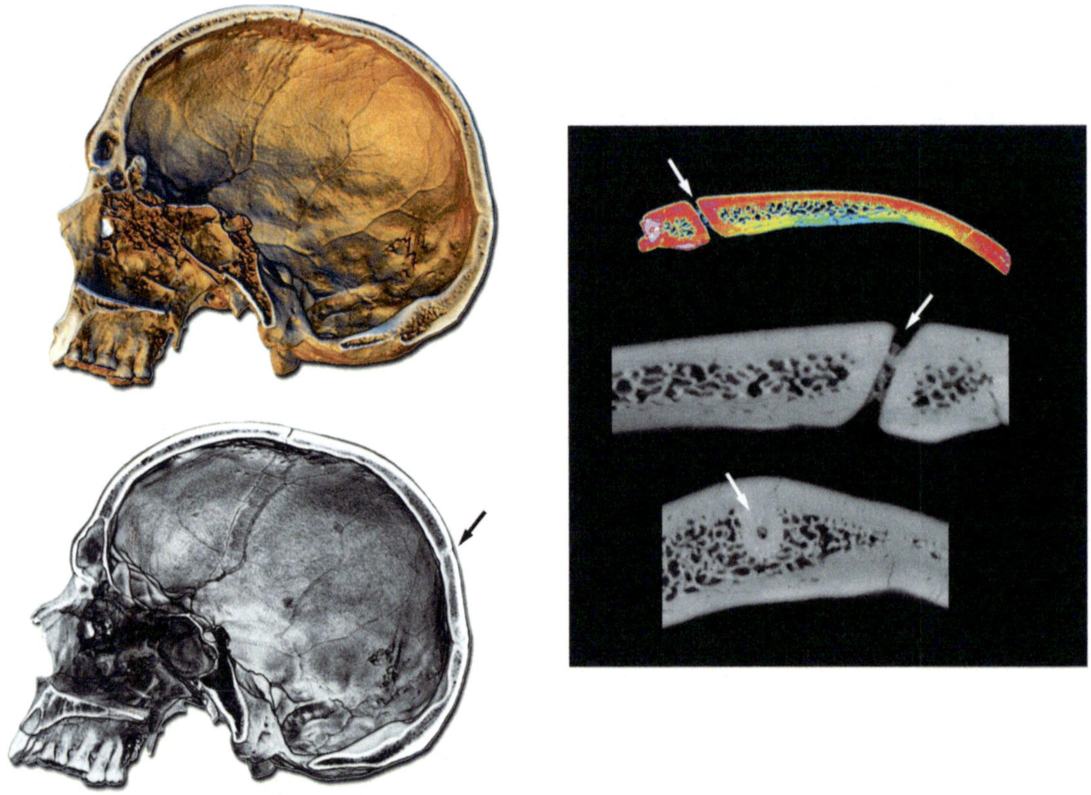

Fig. 6.10 The emissary foramina (*arrows*: the parietal foramina) can be investigated in dry skull or digital samples, taking into consideration their presence, patency, size, and position. Foramina bridge the endocranial and ectocranial surfaces, enveloped by a thick layer of dense bone (Microtomographic sections after Bruner et al. 2016)

6.5 Thermoregulation and Metabolism

The brain is our most expensive organ in terms of energetic resources and management, consuming 20% of the total oxygen and glucose of the body (Aiello and Wheeler 1995; Zenker and Kubik 1996; Bertolizio et al. 2011; Bruner et al. 2012; Mortazavi et al. 2013), compared to 1.4% energy consumption of other mammals (Leonard et al. 2007). The brain requires a constant supply of oxygen and energy-producing substrates to fulfill its functional needs (Touzani and MacKenzie 2007). Accordingly, the vascular network must acknowledge the complexity of these expensive brain functions. The brain undergoes high metabolic rates, but it is extremely sensitive to physical and chemical damages, especially to changes in temperature and oxygen consumption (Karbowski 2009). Selective brain cooling systems (e.g., the carotid rete) are present in many mammal groups, such as carnivores or artiodactyls, to reduce heat accumulation (Jessen 2001). Modern humans have brains that are three times larger than nonhuman primates with similar body size, but we currently have no evidence suggesting the existence of selective cooling systems (Brengelmann 1993; Cabanac 1993; Kuhnen 1995; Karbowski 2009). Through blood flow management, the vascular system probably plays a relevant role in thermoregulation both in physiological (Caputa 2004; Wysocki 2008) and evolutionary terms (Falk and Gage 1997; Isler and van Schaik 2009; Bruner et al. 2011b).

Changes in the frequency of the O/M system and of emissary foramina (mastoid and parietal) were associated with functional adaptations, evolution of bipedal posture, and thermoregulatory function, according to the "radiator theory" (Falk 1986, 1990). This cooling system also depends on numerous minor vascular connections, which are difficult to detect at macroanatomical level (Zenker and Kubik 1996). The thermoregulatory function of emissary veins has been supported by clinical experimental studies. Under certain physical conditions, the blood flow direction in the veins can be detected by a Doppler probe (Cabanac and Brinnel 1985). During hypothermia, the blood flow is minimal or none, but under hyperthermia, the blood runs intracranially in order to cool the cranial cavity. However, the relevance of these flows and their contribution to thermodynamic processes remains unclear. Previously described unidirectional intracranial inflow of the blood within emissary veins was supported by the presence of valves in early ontogenesis, which prevents directional blood flow (Gisel 1964). Yet, the valves are absent in adult stages. Falk's radiator theory is

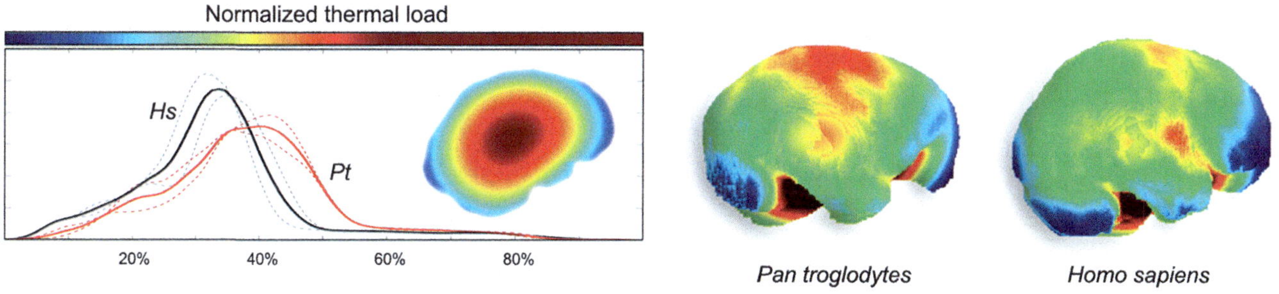

Fig. 6.11 Thermal maps of chimpanzee and human endocasts (*right*). Thermal values are computed following the heat equation and according to the form of the object, simulating the distribution of the thermal loads according to a homogeneous heat production (*red*, warmer areas; *blue*, colder areas). The distribution of the whole values after size normalization (*left*) shows the differences in the heat curves due to brain shape differences. Although brain size is a major factor when dealing with endocranial heat production, changes in brain shape can induce changes in the distribution of the heat loads (Modified after Bruner et al. 2012)

generally accepted in paleoanthropology, although a high prevalence of O/M sinus is also found in anatomically modern humans (Matiegka 1923). Interestingly, artificially deformed skulls show higher prevalence of the enlarged O/M sinus complex, which proves a certain developmental plasticity of the venous system (Dean 1995; O'Loughlin 1996).

Heat dissipation models make it possible to quantify local temperature distribution depending on the cerebral shape (Bruner et al. 2012) (Fig. 6.11). Brain thermoregulation largely depends on physiological processes (Brengelmann 1993) but also on the physical properties of the anatomical elements. The dura mater and the cerebrospinal fluid play an important role in this sense (Zenker and Kubik 1996; Rango et al. 2012). Geometry is one of the factors influencing heat dissipation, and spatial modeling applied to endocranial morphology can show the areas of larger and smaller heat accumulation, as well as changes in heat distribution generated by endocranial morphological variations. The endocranial base is the area with higher heat accumulation, and the frontal and occipital poles represent the areas with lower heat loads (Bruner et al. 2011b). Areas with large loads are not only more sensitive to heat risks but also represent crucial surfaces for heat exchange between the brain and other elements (bones, blood, etc.). In most hominids, the parietal surface displays a noticeable heat load, because of its proximity to the thermal core of the brain mass (Bruner et al. 2014). Conversely, because of parietal bulging, in modern humans, this area shifts to a deeper position, possibly increasing its vulnerability to thermal risks (Bruner and Jacobs 2013).

6.6 Anthropology, Bioarchaeology, and Medicine

Digital anatomy has significantly enhanced the possibilities of studying craniovascular features in cranial samples and fossil specimens (Bruner et al. 2011a, b; Rangel de Lázaro et al. 2016; Eisová et al. 2016). These same traits are of interest in bioarchaeology, which attempts to reconstruct the biological and cultural history of past populations through the study of their skeletal remains (Buikstra and Ubelaker 1994; Larsen 1997). In this case, the application of craniovascular features is generally focused to biodistance studies by using phenotypic aspects of distinct biological characters (metric and nonmetric) to estimate relatedness among populations, groups, and individuals on the basis of expected degree of phenotypic heritability (Hauser and De Stefano 1989; Buikstra et al. 1990; Nikita 2015). The presence of repeated features (such as vascular canals) is often used to evaluate kinship (Berry and Berry 1967; Ossenberg 1976). For example, the Předmostí population in the Czech Republic displays a high frequency of the O/M system, which was interpreted as indicating a close genetic relationship between the individuals (Matiegka 1923). Nonetheless, as mentioned, the same traits have also been hypothesized to be influenced by adaptation (Falk 1986) or by morphological constraints (O'Loughlin 1996).

Further anthropological reports mention the vascular system in relation to specific cultural practice. Trepanation is an invasive surgical procedure which can affect vessels and sinuses attached to the cranial bones and cause epidural bleeding or lethal infections. The position of the trepanation can be even intentionally aimed at damaging the vascular structures, as a part of posttraumatic treatment (Kurin 2013). The uncommon preservation of vascular elements in mummified brain tissues has been described in one case of ancient Egyptian mummies (Isidro et al. 2015).

Craniovascular characters can also serve as markers in forensic anthropology, for individual identification, estimation of cause of death, or to supply information on life history. On the grounds of high individual variation and complexity of the vascular patterns, the craniovascular bony remnants can serve for identification in a similar way to fingerprints (Hershkovitz et al. 1999; Nagesh et al. 2005). Epidural bleeding was discussed as a possible cause of death

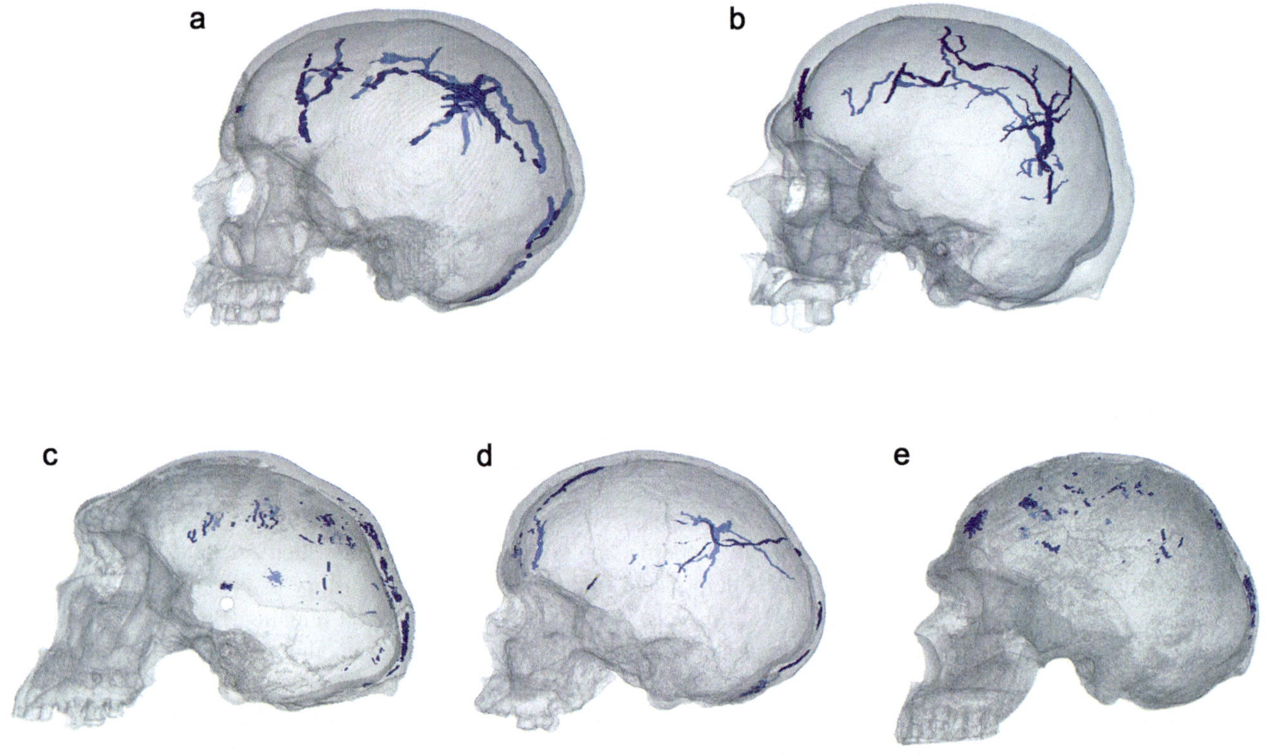

Fig. 6.12 Above, visualization of diploic vessels at low (**a**) (pixel size 0.86; matrix 256 × 256) and high (**b**) (pixel size 0.25 mm; matrix 1024 × 1024) CT scan resolution. In fossils, the occurrence of postdepositional factors can hamper the location and reconstruction of diploic vessels, which can be detected only in a fragmentary and spotted distribution; (**c**) Broken Hill 1 (pixel size 0.46 mm; matrix 512 × 512); (**d**) Mladec 1 (pixel size 0.46 mm; matrix 512 × 512); (**e**) Skhul V (pixel size 0.48 mm; matrix 512 × 512)

in specific historical cases (Vlček et al. 2006). Exceptionally, age-related vascular changes can be used as secondary auxiliary indicators of age at death (Le Gros Clark 1920). However, such method shows low reliability and accuracy and is rarely applied.

It is clear that interest in these characters goes beyond anthropology and evolution, dealing with relevant medical issues. In medicine and human biology, the endocranial and craniovascular system is of considerable interest because of its metabolic and physiological roles and because of its importance in neurosurgery (Ribas et al. 2006). The absence of valves and the multiple connections are crucial aspects in pathological conditions such as pyrexia, strokes, or migraine. The emissary diploic veins can be a gateway of infection within the cranial cavity (Drake et al. 2013). It has also been noted that diploic veins participate in the hematogenous spread of metastases (Kunz and Iliadis 2007; Patel 2009). Diploic veins can avoid clot formation in extradural hematomas, creating fistulous communication with a broken meningeal artery (Ericson et al. 1979). Minimal damage in a blood vessel may provoke a hematoma (blood clot) pressing the brain tissue and potentially leading to death within a few hours (Mitra et al. 2015). Thrombosis of the superior sagittal sinus was described after magnetic resonance imaging (Mancall and Brock 2010). In skull fractures, the meningeal vessels are often injured (Zasler et al. 2006). In most cases, the middle meningeal artery is eliminated during neurosurgical treatments, and we have no information regarding the long-term consequences of this removal (Bruner and Sherkat 2008). The anterior branch of the middle meningeal artery is situated in close proximity to the pteric area (Derezinski 1934) which is the thinnest point of the skull. In traumatized skulls, these vessels are often a source of epidural bleeding resulting from cranial fractures or heavy head shakes that disrupt the vascular tissue (Teegen and Schultz 1994).

The study of the endocranial vascular system with computed tomography is necessarily related to economical and technical factors such as data acquisition, resolution, software, computer performance, and sample size (Fig. 6.12). Digitizing specimens is generally time-consuming and economically demanding. High-resolution data acquisition requires powerful computing machines and storing devices. These new challenges are promoting the development of a specialized expertise based on anatomy, morphometrics, informatics, and computed imaging. The field is undoubtedly multidisciplinary, and it has prompted a vivid historical moment for vascular biology.

Acknowledgments This study is funded by the Wenner-Gren Foundation International Collaborative Research Grant (ICRG) "Cranial anatomy, anthropology, and the vascular system." GRL is funded by the International Erasmus Mundus Doctorate in Quaternary and Prehistory consortium (IDQP). HP is supported by NAKI DF12P01OVV021 and SE by the Leonardo da Vinci programme of the European Commission. EB is funded by the Spanish Government (CGL2012-38434-C03-02). The NESPOS platform provided the modern samples used in this study. The specimens used in the images are kept at the National Museum in Prague (Czech Republic), Museum of Anthropology G. Sergi at the University of La Sapienza in Rome (Italy), and University of Burgos (Spain). We would like to thank Ana Sofia Pereira-Pedro and José Manuel de la Cuétara for their collaboration. Special thanks are also due to Marteyn Van Gasteren for the administrative support.

References

Adachi B (1928) Das Arteriensystem der Japaner. Band 1 Verlag der Kaiserlich-Japanishen Universitat zu Kyoto, Kyoto

Adeeb N, Mortazavi M, Tubbs R, Cohen-Gadol A (2012) The cranial dura mater: a review of its history, embryology, and anatomy. Childs Nerv Syst 28:827–837

Aiello L, Wheeler P (1995) The expensive-tissue hypothesis: the brain and the digestive-system in human and primate evolution. Curr Anthropol 36:199–221

Anderson PJ, Harkness WJ, Taylor W, Jones BM, Hayward RD (1997) Anomalous venous drainage in a cse of non-syndromic craniosynostosis. Childs Nerv Syst 13:97–100

Anzelmo M, Ventrice F, Barbeito-Andrés J, Pucciarelli HM, Sardi ML (2014) Ontogenetic changes in cranial vault thickness in a modern sample of *Homo sapiens*. Am J Hum Biol 27:475–485

Ayanzen RH, Bird CR, Keller PJ, McCully FJ, Theobald MR, Heiserman JE (2000) Cerebral MR venography: normal anatomy and potential diagnostic pitfalls. Am J Neuroradiol 21:74–78

Baab KL, Freidline SE, Wang SL, Hanson T (2010) Relationship of cranial robusticity to cranial form, geography and climate in *Homo sapiens*. Am J Phys Anthropol 141:97–115

Balzeau A (2013) Thickened cranial vault and parasagittal keeling: correlated traits and autapomorphies of *Homo erectus*? J Hum Evol 64:631–644

Barozzino T, Sgro M (2002) Transillumination of the neonatal skull: seeing the light. Can Med Assoc J 167:1271–1272

Bastir M (2008) A systems-model for the morphological analysis of integration and modularity in human craniofacial evolution. J Anthropol Sci 86:37–58

Bastir M, Rosas A, O'Higgins P (2006) Craniofacial levels and the morphological maturation of the human skull. J Anat 209:637–654

Berry AC, Berry RJ (1967) Epigenetic variation in human cranium. J Anat 101:361–380

Bertolizio G, Mason L, Bissonn B (2011) Brain temperature: heat production, elimination and clinical relevance. Pediatr Anesth 21:347–358

Bookstein FL (1991) Morphometric tools for landmark data. Cambridge University Press, Cambridge

Boruah S, Paskoff GR, Shender BS, Subit DL, Salzar RS, Crandall JR (2015) Variation of bone layer thicknesses and trabecular volume fraction in the adult male human calvarium. Bone 77:120–134

Boyd GI (1930) The emissary foramina of the cranium in man and the anthropoids. J Anat 65:108–121

Bremmer J, Berckel BM, Persoon S, Kappelle LJ, Lammertsma A, Kloet R, Luurtsema G, Rijbroek A, Klijn CM, Boellaard R (2011) Day-to-day test–retest variability of CBF, CMRO2, and OEF measurements using dynamic 15O PET studies. Mol. Imaging Biol 13:759–768

Brengelmann G (1993) Specialized brain cooling in humans? FASEB J 7:1148–1153

Breschet M (1829) Recherches anatomiques, physiologiques et pathologiques sur le systeme veineux. Villeret, Paris

Brown P (1994) Cranial vault thickness in Asian *Homo erectus* and *Homo sapiens*. Cour Forsch Inst Senckenberg 171:33–46

Bruner E (2004) Geometric morphometrics and paleoneurology: brain shape evolution in the genus *Homo*. J Hum Evol 47:279–303

Bruner E (2010) The evolution of the parietal cortical areas in the human genus: between structure and cognition. In: Broadfield D, Yuan M, Schick K, Toth N (eds) Human Brain Evolving. The Stone Age Institute, Bloomington, pp 83–96

Bruner E (2015) Functional craniology and brain evolution. In: Bruner E (ed) Human Paleoneurology. Springer, Cham, pp 57–94

Bruner E, Averini M, Manzi G (2003) Endocranial traits. Prevalence and distribution in a recent human population. Eur J Anat 7:23–33

Bruner E, de la Cuétara JM, Masters M, Amano H, Ogihara N (2014) Functional craniology and brain evolution: from paleontology to biomedicine. Front Neuroanat 8:1–15

Bruner E, de la Cuétara JM, Musso F (2012) Quantifying patterns of endocranial heat distribution: brain geometry and thermoregulation. Am J Hum Biol 24:753–762

Bruner E, Holloway R (2010) Bivariate approach to the widening of the frontal lobes in the genus *Homo*. J Hum Evol 58:138–146

Bruner E, Jacobs HIL (2013) Alzheimer's disease: the downside of a highly evolved parietal lobe? J Alzheimer's Dis 35:227–240

Bruner E, de la Cuétara JM, Holloway R (2011a) A bivariate approach to the variation of the parietal curvature in the genus *Homo*. Anat Rec 294:1548–1556

Bruner E, Mantini S, Musso F, de la Cuétara JM, Ripani M, Sherkat S (2011b) The evolution of the meningeal vascular system in the human genus: from brain shape to thermoregulation. Am J Hum Biol 23:35–43

Bruner E, Mantini S, Perna A, Maffei C (2005) Fractal dimension of the middle meningeal vessels: variation and evolution in *Homo erectus*, Neanderthals, and modern humans. Eur J Morphol 42:217–224

Bruner E, Mantini S, Ripani M (2009) Landmark-based analysis of the morphological relationship between endocranial shape and traces of the middle meningeal vessels. Anat Rec 292:518–527

Bruner E, Manzi G (2005) CT-based description and phyletic evaluation of the archaic human calvarium from Ceprano, Italy. Anat Rec 285:643–658

Bruner E, Manzi G (2006) Digital tools for the preservation of the human fossil heritage: Ceprano, Saccopastore, and other case studies. Hum Evol 21:33–44

Bruner E, Píšová H, Martín-Francés L, María Martinón-Torres M, Arsuaga JL, Carbonell E, Bermúdez de Castro JM (2016) A human parietal fragment from the late Early Pleistocene Gran Dolina-TD6 cave site, Sierra de Atapuerca, Spain. Comptes Rendus Palevol (early view)

Bruner E, Sherkat S (2008) The middle meningeal artery: from clinics to fossils. Childs Nerv Syst 24:1289–1298

Buikstra JE, Frankenberg SR, Konigsberg LW (1990) Skeletal biological distance studies in American physical anthropology: recent trends. Am J Phys Anthropol 82:1–7

Buikstra JE, Ubelaker DH (1994) Standards for data collection from humans skeletal remains. Arkansas Archaeological Survey, Fayetteville

Burian R (2004) Molecular epigenesis, molecular pleiotropy, and molecular gene definitions. Hist Philos Life Sci 26:59–80

Butler H (1957) The development of certain human dural venous sinuses. J Anat 91:510–526

Cabanac M (1993) Selective brain cooling in humans: 'fancy' or fact? FASEB J 7:1143–1146

Cabanac M (1995) Human selective brain cooling. Springer Verlag, Heidelberg

Cabanac M, Brinnel H (1985) Blood flow in the emissary veins of the human head during hyperthermia. Eur J Appl Physiol 54:172–176

Caputa M (2004) Selective brain cooling:a multiple regulatory mechanism. J Therm Biol 29:691–702

Chatterjee N, Ansari S, Vakil P, Prabhakaran S, Carroll T, Hurley M (2015) Automated analysis of perfusion weighted MRI using asymmetry in vascular territories. Magn Reson Imaging 33:618–623

Chen F, Deng Y-F, Liu B, Zou L, Wang D, Han H (2011) Arachnoid granulations of middle cranial fossa: a population study between cadaveric dissection and in vivo computed tomography examination. Surg Radiol Anat 33:215–221

Connor S, Jarosz J (2002) Magnetic resonance imaging of cerebral venous sinus thrombosis. Clin Radiol 57:449–461

Copes L (2012) Comparative and experimental investigations of cranial robusticity in mid-pleistocene hominins. Arizona State University

Copes L, Kimbel W (2016) Cranial vault thickness in primates: *Homo erectus* does not have uniquely thick vault bones. J Hum Evol 90:120–134

Curé JK, Van Tassel P, Smith MT (1994) Normal and variant anatomy of the dural venous sinuses. Semin Ultrasound CT 15:499–519

Curnoe D, Green H (2013) Vault thickness in two Pleistocene Australian crania. J Archaeol Sci 40:1310–1318

Das S, Abd LA, Suhaimi FH, Othman FB, Yahaya MF, Ahmad F, Abdul HH (2008) An anatomico-radiological study of the grooves for occipital sinus in the posterior cranial fossa. Bratisl Med J 109:520–524

Dawkins R (1989) The extended phenotype. Oxford University Press, Oxford

Dean V (1995) Sinus and meningeal vessel pattern changes induced by artificial cranial deformation. Int J Osteoarchaeol 5:1–14

Del Papa MC, Perez SI (2007) The influence of artificial cranial vault deformation on the expression of cranial nonmetric traits: its importance in the study of evolutionary relationships. Am J Phys Anthropol 134:251–262

Derezinski C (1934) Variations of the middle meningeal artery in the middle cranial fossa. Dissertation, Loyola University, Chicago

Diamond MK (1991) Homologies of the meningeal-orbital arteries of humans: a reappraisal. J Anat 178:223–241

Dilenge D, Ascherl GF Jr (1980) Variations of the ophthalmic and middle meningeal arteries: relation to the embryonic stapedial artery. Am J Neoradiol 1:45–54

Dora F, Zileli T (1980) Common variations of the lateral and occipital sinuses at the confluens sinuum. Neuroradiology 20:23–27

Drake R, Vogl AW, Mitchell AW (2013) Gray: Anatomía Básica. Elsevier España S.A, Barcelona

Eichmann A, Le Noble F, Autiero M, Carmeliet P (2005) Guidance of vascular and neural network formation. Curr Opin Neurobiol 15:108–115

Eisová S, Rangel de Lázaro G, Píšová H, Pereira-Pedro S, Bruner E (2016) Parietal bone thickness and vascular diameters in adult modern humans: a survey on cranial remains. Anat Rec 299:888–896

Enlow DH (1968) The human face: an account of the postnatal growth and development of the craniofacial skeleton. Hoeber Medical Division, Harper & Row, New York

Enlow DH (1990) Facial growth. WB Saunders Company, Philadelphia

Ericson K, Hakansson S, Lofgren J (1979) Extravasation and arteriovenous shunting after epidural bleeding -a radiological study. Neuroradiology 17:239–244

Falk D (1986) Evolution of cranial blood drainage in hominids: enlarged occipital/marginal sinuses and emissary foramina. Am J Phys Anthropol 70:311–324

Falk D (1990) Brain evolution in *Homo*: the "radiator" theory. Behav Brain Sci 13:333–381. with peer commentary

Falk D (1993) Meningeal arterial patterns in great apes: implications for hominid vascular evolution. Am J Phys Anthropol 92:81–97

Falk D, Gage T (1997) Flushing the radiator? A reply to Braga Boesch. J Hum Evol 33:495–502

Falk D, Nicholls P (1992) Meningeal arteries in rhesus macaques (*Macaca mulatta*): implications for vascular evolution in anthropoids. Am J Phys Anthropol 89:299–308

García Gil O, Cambra-Moo O, Audije Gil J, Nacarino Meneses C, Rodríguez Barbero MA, Rascón Pérez J, González Martín A (2015) Investigating histomorphological variations in human cranial bones through ontogeny. C R Palevol. doi:10.1016/j.crpv.2015.04.006

García-González U, Cavalcanti D, Agrawal A, Gonzalez L, Wallace R, Spetzler RF, Preul MC (2009) The diploic venous system: surgical anatomy and neurosurgical implications. Neurosurg Focus 27(5):1–11

Gauld SC (1996) Allometric patterns of cranial bone thickness in fossil hominins. Am J Phys Anthropol 100:411–426

Georgiou C, Cassell MD (1992) The foramen meningo-orbitale and its relationship to the development of the ophthalmic artery. J Anat 180:119–125

Gisel A (1964) Über die blutströmung im emissarium parietale des neugeborenen. Anat Ant 114:371–373

Giuffrida-Ruggeri V (1904) La capacità del cranio nelle diverse popolazioni italiane antiche e moderne. Atti della Società Romana di Antropologia 10:240–278

Giuffrida-Ruggeri V (1912) Über die endocranischen Furchen der Arteria meningea media beim Menschen. Z Morphol Anthropol 15:401–412

Goebbel J (2013) Computed tomography. In: Czichos H (ed) Handbook of technical diagnostics: fundamentals and application to structures and systems. Springer ScienceBusiness Media, Berlin, pp 249–258

Goldman L (2007) Principles of CT and CT technology. J Nucl Med Tech 35:115–128

Goldsmith RS (1972) Biomechanics of head injury. In: Fung YC, Perrone N, Anliker M (eds) Biomechanics, its foundations and objective. Prentice-Hall, Englewood Cliffs, p 585

Gray H (1913) Anatomy, descriptive and applied. Lea & Febiger, Philadelphia

Gray H, Carter HV (1858) Anatomy: descriptive and surgical. John W. Parker and Son, London

Grimaud-Hervé D (1997) Evolution de l'encéphale chez *Homo erectus* et *Homo sapiens*. Centre National de la Recherche Scientique, Paris

Grimaud-Hervé D (2004) Endocranial vasculature. In: Holloway R, Broadfield DC, Yuan MS (eds) The human fossil record, Brain endocasts: the paleoneurological evidence, vol III. Wiley-Liss, New Jersey, pp 273–284

Gunz P, Mitteroecker P, Neubauer S, Weber G, Bookstein F (2009) Principles for the virtual reconstruction of hominin crania. J Hum Evol 57:48–62

Hajdú J, Marton T, Kozsurek M, Pete B, Csapó Z, Beke A, Papp Z (2008) Prenatal diagnosis of abnormal course of umbilical vein and absent ductus venosus-report of three cases. Fetal Diagn Ther 23:136–139

Hartmann A, Hoyer S (2012) Cerebral blood flow and metabolism measurement. Springer ScienceBusiness Media, New York

Hatipoglu HG, Ozcan HN, Hatipoglu US, Yuksel E (2008) Age, sex and body mass index in relation to calvarial diploe thickness and craniometric data on MRI. Forensic Sci Int 182:46–51

Hauser G, De Stefano GF (1985) Variations in form of the hypoglossal canal. Am J Phys Anthropol 67:7–11

Hauser G, De Stefano G (1989) Epigenetic variants of the human skull. Schweizerbart, Stuttgart

Hershkovitz I, Greenwald C, Rotschild B, Latimer B, Dutour O, Jellema LM, Wish-Baratz S, Pap I, Lenoetti G (1999) The elusive diploic veins: anthropological and anatomical perspective. Am J Phys Anthropol 108:345–358

Horner F (1962) The technique of Transillumination of the skull. Am J Dis Child 103:183–184

Hounsfield GN (1980) Computed medical imaging. Science 210:22–28

Hu HH, Campeau NG, Huston J, Kruger DG, Haider CR, Riederer SJ (2007) High-spatial-resolution contrast-enhanced MR angiography

of the intracranial venous system with fourfold accelerated two-dimensional sensitivity encoding. Radiology 243:853–861
Hwang K, Kim J, Baik S (1999) The thickness of the skull in Korean adults. J Craniofac Surg 10:395–399
Isidro A, Gonzálvez LM, Arbiox A (2015) Brain vessels mummification in an individual of ancient Egypt. Cortex 63:217–219
Isler K, van Schaik CP (2009) The expensive brain: a framework for explaining evolutionary changes in brain size. J Hum Evol 57:392–400
Jefferson G, Stewart D (1928) On the veins of the diploe. Brit J Surg 16:70–88
Jessen C (2001) Selective brain cooling in mammals and birds. Jpn J Physiol 51:291–301
Jivraj K, Bhargava R, Aronyk K, Quateen A, Waljil A (2009) Diploic venous anatomy studied in-vivo by MRI. Clin Anat 22:296–301
Johnston KD, Walji AH, Fox RJ, Pugh JA, Aronyk KE (2007) Access to cerebrospinal fluid absorption sites by infusion into vascular channels of the skull diploë. J Neursurg 107:841–843
Jones FW (1912) On the grooves upon the Ossa Parietalia commonly said to be caused by the arteria Meningea media. J Anat Physiol 46:228–238
Kaplan HA, Browder J, Knightly JJ, Rush BF Jr, Browder A (1972) Variations of the cerebral dural sinuses at the torcular herophili. Importance in radical neck dissection. Am J Surg 124:456–461
Karbowski J (2009) Thermodynamic constraints on neural dimensions, firing rates, brain temperature and size. J Comput Neurosci 27:415–436
Kennedy GE (1991) On the autapomorphic traits of *Homo erectus*. J Hum Evol 20:375–412
Kim LK, Ahn CS, Fernandes AE (2014) Mastoid emissary veins: anatomy and clinical relevance in plastic and reconstructive surgery. J Plast Reconstr Aesthet Surg 67:775–780
Kimball D, Kimball H, Tubbs RS, Loukas M (2015) Variant middle meningeal artery origin from the ophthalmic artery: a case report. Surg Radiol Anat 37:105–180
Kimbel WH (1984) Variation in the pattern of cranial venous sinuses and hominid phylogeny. Am J Phys Anthropol 63:243–263
Knott JF (1881) On the cerebral sinuses and their variation. J Anat Physiol 16:27–42
Kochetkova VI (1978) Paleoneurology. VH Winston, Michigan
Koenig WJ, Donovan JM, Pensler JM (1995) Cranial bone grafting in children. Plast Reconstr Surg 95:1–4
Kopuz C, Aydin ME, Kale A, Demir MT, Corumlu U, Kaya AH (2010) The termination of superior sagittal sinus and drainage patterns of the lateral, occipital at confluens sinuum in newborns: clinical and embryological implications. Surg Radiol Anat 32:827–833
Kresimir LI, Gluncic V, Marusic A (2001) Extracranial branches of the middle meningeal artery. Clin Anat 14:292–294
Kuhnen G (1995) Unilateral selective brain cooling. Pflugers Arch 430:1018–1020
Kunz AR, Iliadis C (2007) Hominid evolution of the arteriovenous system through the cranial base and its relevance for craniosynostosis. Childs Nerv Syst 23:1367–1377
Kurin DS (2013) Trepanation in South-Central Peru during the early late intermediate period (ca. AD 1000-1250). Am J Phys Anthropol 152:484–494
Kyrkanides S, Moore T, Miller JH, Tallents RH (2011) Melvin Moss' function matrix theory revisited. Orthod Waves 70:1–7
Larsen CS (1997) Bioarchaeology: interpreting behavior from the human skeleton. Cambridge University Press, Cambridge
Le Gros Clark WE (1920) On the Pacchionian bodies. J Anat 55:40–48
Lee W, Yoon A, Song M, Wilkinson C (2014) The archaeological contribution of forensic craniofacial reconstruction to a portrait drawing of a Korean historical figure. J Archaeol Sci 49:228–236
Leonard W, Snodgrass J, Robertson M (2007) Effects of brain evolution on human nutrition and metabolism. Annu Rev Nutr 27:311–327
Li Q, Pan SN, Yin YM, Li W, Chen ZA, Liu YH, Wu ZH, Guo QY (2011) Normal cranial bone marrow MR imaging pattern with age-related ADC value distribution. Eur J Radiol 80:471–477
Lieberman DE (1996) How and why humans grow thin skulls: experimental evidence for systemic cortical robusticity. Am J Phys Anthropol 101:217–236
Lieberman D, Ross C, Ravosa M (2000) The primate cranial base: ontogeny, function, and integration. Yearb Phys Anthropol 43:117–169
Lillie E, Urban J, Weaver A, Powers A, Stitzel J (2015) Estimation of skull table thickness with clinical CT and validation with microCT. J Anat 226:73–80
Louis RJ, Loukas M, Wartmann C, Tubbs R, Apaydin N, Gupta AA, Spentzouris G, Ysique JR (2009) Clinical anatomy of the mastoid and occipital emissary veins in a large series. Surg Radiol Anat 31:139–144
Lui Q, Rhoton AL (2001) Middle meningeal origin of the ophthalmic artery. Neurosurgery 49:401–407
Luttenberg J (1959) Arteria meningica media (sulci arteriare meningicae mediae) a její projekce na povrch lebky. Československá morfologie 7(4):335–352
Lynnerup N (2001) Cranial thickness in relation to age, sex and general body build in a Danish forensic sample. Forensic Sci Int 117:45–51
Lynnerup N, Astrup JG, Sejrsen B (2005) Thickness of the human cranial diploe in relation to age, sex, and general body build. Head Face Med 1:1–13
MacDonald ME, Frayne R (2015) Cerebrovascular MRI: a review of state-of-the-art approaches, methods and techniques. NMR Biomed 28:767–791
Mahesh M (2002) Search for isotropic resolution in CT from conventional through multiple-row detector. Radiographics 22:949–962
Mancall EL, Brock DG (2010) Gray's clinical neuroanatomy: the anatomic basis for clinical neuroscience. Elsevier-Saunders, Philadelphia
Manjunath K (2001) Anomalous origin of the middle meningeal artery—a review. J Anat Soc India 4:83–87
Manzi G, Gracia A, Arsuaga JL (2000) Cranial discrete traits in the middle Pleistocene humans from Sima de los Huesos (Sierra de Atapuerca, Spain). Does hypostosis represent any increase in "ontogenetic stress" along the Neandertal lineage? J Hum Evol 38:425–446
Manzi G, Vienna A (1997) Cranial non-metric traits as indicators of hypostosis or hyperostosis. Riv Antropol 75:41–61
Manzi G, Vienna A, Hauser G (1996) Developmental stress and cranial hypostosis by epigenetic trait occurrence and distribution: an exploratory study on the Italian Neandertals. J Hum Evol 30:511–527
Marcozzi V (1942) L'arteria meningea negli nomini recenti, nel Sinantropo, e nelle Scimmie. Riv Antropol 34:407–436
Marsh HE (2013) Beyond thick versus thin: mapping cranial vault thickness patterns in recent *Homo sapiens*. University of Iowa, Iowa
Marsot-Dupuch K, Gayet-Delacroix M, Elmaleh-Breges M, Bonneville F, Lasjuanias P (2001) The petrossquamosal sinus: CT and MRI findings of a rare emissary vein. Am J Neuroradiol 22:1186–1193
Martínez-Maza C, Rosas A, García-Vargas S (2006) Bone paleohistology and human evolution. J Anthropol Sci 84:33–52
Matiegka J (1923) Sulci venosi diluviálních lebek z Předmostí. Antropologie 1:31–38
McLennan JE, Rosenbaum AE, Hauhton VM (1974) Internal carotid origins of the middle meningeal artery. The ophthalmic-middle meningeal and stapedial-middle meningeal arteries. Neuroradiology 29:265–275
Menegaz RA, Sublett SV, Figueroa SD, Hoffman TJ, Ravosa MJ, Aldridge K (2010) Evidence for the influence of diet on cranial form and robusticity. Anat Rec 293:630–641

Mitra I, Duraiswamy M, Benning J, Joy H (2015) Imaging of focal calvarial lesions. Clin Radiol 71:1–10

Moore D (2015) The developing genome: an introduction to behavioral epigenetics. Oxford University Press, New York

Moreira-Gonzalez A, Papay F, Zins J (2006) Calvarial thickness and its relation to cranial bone harvest. Plast Reconstr Surg 117:1964–1971

Mortazavi M, Denning M, Yalcin B, Shoja M, Loukas M, Tubbs R (2013) The intracranial bridging veins: a comprehensive review of their history, anatomy, histology, pathology, and neurosurgical implications. Childs Nerv Syst 29:1073–1078

Moss ML (1968) A theoretical analysis of the functional matrix. Acta Biotheor 18:195–202

Moss ML, Young RW (1960) A functional approach to craniology. Am J Phys Anthropol 18:281–292

Müller U, Baker L, Yeung E (2013) Embodiment and Epigenesis: theoretical and methodological issues in understanding the role of biology within the relational developmental system part B: ontogenetic dimensions. Adv Child Dev Behav 45:39–66

Murlimanju BV, Prabhu LV, Pai MM, Jaffar M, Saralaya VV, Tonse M, Prameela MD (2011) Occipital emissary foramina in human skulls: an anatomical investigation with reference to surgical anatomy of emissary veins. Turk Neurosurg 21:36–38

Nagesh KR, Bastia BK, Menon A, Saralaya KM (2005) Reconstruction of skull by endocranial groove – a case report. J Indian Acad Forensic Med 27:55–56

Nanney D (1958) Epigenetic control systems. Proc Natl Acad Sci U S A 44:712–717

Nawrocki SP (1991) A biomechanical model of cranial vault thickness in archaic *Homo*. Binghamton University, New York

Neubauer S, Gunz P, Hublin JJ (2009) The pattern of endocranial ontogenetic shape changes in humans. J Anat 215:240–255

Neubauer S, Gunz P, Hublin JJ (2010) Endocranial shape changes during growth in chimpanzees and humans: a morphometric analysis of unique and shared aspects. J Hum Evol 59:555–566

Nikita E (2015) A critical review of the mean measure of divergence and Mahalanobis distances using artificial data and new approaches to the estimation of biodistances employing nonmetric traits. Am J Phys Anthropol 157:284–294

O'Loughlin V (1996) Comparative endocranial vascular changes due to Craniosynostosis and artificial cranial deformation. Am J Phys Anthropol 101:369–385

O'Loughlin V (2004) Effects of different kinds of cranial deformation on the incidence of Wormian bones. Am J Phys Anthropol 123:146–155

Ossenberg NS (1976) Within and between race distances in population studies based on discrete traits of the human skull. Am J Phys Anthropol 45:701–715

Patel N (2009) Venous anatomy and imaging of the first centimeter. Semin Ultrasound CT MRI 30:513–524

Patel N, Kirmi O (2009) Anatomy and imaging of the normal meninges. Semin Ultrasound CT MRI 30:559–564

Peña-Melián A, Rosas A, García-Tabernero A, Bastir M, De La Rasilla M (2011) Paleoneurology of two new Neandertal occipitals from El Sidrón (Asturias, Spain) in the context of *Homo* endocranial evolution. Anat Rec 294:1370–1381

Percival CJ, Richtsmeier JT (2013) Angiogenesis and intramembranous osteogenesis. Dev Dyn 242:909–922

Peterson J, Dechow PC (2002) Material properties of the inner and outer cortical tables of the human parietal bone. Anat Rec 268:7–15

Peterson J, Dechow PC (2003) Material properties of the human cranial vault and zygoma. Anat Rec 274A:785–797

Píšová H, Rangel de Lázaro G, Velemínský P, Bruner E (2017) Craniovascular traits in anthropology and evolution: from bones to vessels. J Anthropol Sci. doi:10.4436/JASS.9503

Raciborski A (1841) Histoire des découvertes relatives au système veineux: envisagé sous le rapport anatomique, physiologique, pathologique et thérapeutique, depuis Morgagni jusqu'à nos jours. J-B Baillière, Paris

Rangel de Lázaro G, de la Cuétara JM, Píšová H, Lorenzo C, Bruner E (2016) Diploic vessels and computed tomography: segmentation and comparison in modern humans and fossil hominids. Am J Phys Anthropol 159(2):313–324

Rango M, Arighi A, Bresolin N (2012) Brain temperature: what do we know? Neuroreport 23:483–487

Rea P (2015) Essential clinical anatomy of the nervous system. Academic Press Elsevier, London

Ribas G, Yasuda A, Ribas E, Nishikuni K, Rodrigues A (2006) Surgical anatomy of microneurosurgical sulcal key-points. Neurosurgery 59:177–208

Richtsmeier J, Aldridge K, de Leon V, Panchal J, Kane A, Marsh JL, Yan P, Cole TM (2006) Phenotypic integration of neurocranium and brain. J Exp Zool 306B:360–378

Rilling J (2008) Neuroscientific approaches and applications within anthropology. Am J Phys Anthropol S47:2–32

Rilling JK (2014) Comparative primate neuroimaging: insights into human brain evolution. Trends Cogn Sci 18:45–55

Robert J (2004) Embryology, epigenesis and evolution: taking development seriously. Cambridge University Press, Cambridge

Roche A (1953) Increase in cranial thickness during growth. Hum Biol 25:81–92

Rosas A, Peña-Melián A, Garcia-Tabernero A, Bastir M, De La Rasilla M, Fortea J (2008) Endocranial occipito-temporal anatomy of SD-1219 from the Neandertal El Sidron site (Asturias, Spain). Anat Rec 291:502–512

Ross AH, Jantz RL, McCormick WF (1998) Cranial thickness in American females and males. J Forensic Sci 43:267–272

Rothman D (1937) The endocranial course of the middle meningeal artery in American whites and American negroes. Am J Phys Anthropol 22:425–435

Royle G, Motson R (1973) An anomalous origin of the middle meningeal artery. J Neurol Neurosurg Psychiatry 36:874–876

Saban R (1980) Les empreintes vasculaires endocrâniennes (v.v. méningées moyennes) chez l'Homme de l'Acheuléen en Europe et en Afrique. Anthropogie XVII:133–152

Saban R (1983) Les veines méningées moyennes des australopithèques. Bull Mém Soc d'Anthrop XIII:313–324

Saban R (1995) Image of the human fossil brain: endocranial casts and meningeal vessels in young and adult subjects. In: Changeaux P, Chavaillon J (eds) Origins of the human brain. Clarendon Press, Oxford, pp 11–38

Sabancıoğulları V, Koşar Mİ, Şalk İ, Erdil FH, Öztoprak İ, Çimen M (2012) Diploe thickness and cranial dimensions in males and females in mid-Anatolian population: an MRI study. Forensic Sci Int 219:289e1–289e7

San Millán Ruíz D, Fasel JH, Rufenacht DA, Gailloud P (2004) The sphenoparietal sinus of Breschet: does it exist? An anatomic study. Am J Neuroradiol 25:112–120

San Millán Ruíz D, Gailloud P, Rüfenacht DA, Delavelle J, Henry F, Fasel JHD (2002) The craniocervical venous system in relation to cerebral venous drainage. Am J Neuroradiol 23:1500–1508

Schapira AH (2007) Neurology and clinical neuroscience. Mosby-Elsevier, Philadelphia

Scheuer L, Black S (2000) Developmental Juvenile Osteology. Elservier-Academic Press, London

Schmidt R, Thews G (2013) Human physiology. Springer ScienceBusiness Media, New York

Seeram E (2015) Computed tomography: physical principles, clinical applications, and quality control. Elsevier Health Sciences, St. Louis

Shapiro R, Robinson F (1967) The foramina of the middle fossa: a phylogenetic, anatomic and pathologic study. Am J Roentgenol Radium Ther Nucl Med 101:779–794

Short L, Khambay B, Ayoub A, Erolin C, Rynn C, Wilkinson C (2014) Validation of a computer modelled forensic facial reconstruction technique using CT data from live subjects: a pilot study. Forensic Sci Int 237:147e1–147e8

Silbergleit R, Quint DJ, Mehta BA, Patel SC, Metes JJ, Noujaim SE (2000) The persistent stapedial artery. Am J Neuroradiol 21:572–577

Singh D, Naganawa S, Inoue Y (2004) Anatomical variations of occipital bone impressions for dural venous sinuses arond the torcular Herophili, with special reference to the consideration of clinical significance. Surg Radiol Anat 26:480–487

Skrzat J, Brzegowy P, Walocha J, Wojciechowski W (2004) Age dependent changes of the diploe in the human skull. Folia Morphol (Warsz) 63:67–70

Slice DE (2005) Modern morphometrics in physical anthropology. Springer, New York

Sotero R, Iturria-Medina Y (2011) From blood oxygenation level dependent BOLD signals to brain temperature maps. Bull Math Biol 73:2731–2747

Sperber G (2001) Craniofacial development. BC Decker, Hamilton

Spoor F, Jeffery N, Zonneveld F (2000) Using diagnostic radiology in human evolutionary. J Anat 197:61–76

Sutton MD (2008) Review. Tomographic techniques for fossils. Proc Biol Sci 275:1587–1593

Teegen WR, Schultz M (1994) Epidural hematoma in fetuses, newborns and infants from the early medieval settlements of Elisenhof and Starigard-Oldenburg (Germany). *Homo* 45(suppl):S126

Testut L (1893) Traité d'anatomie humaine. Octave Doin, Paris

Toriumi H, Shimizu T, Shibata M, Unekawa M, Tomita Y, Tomita M, Suzuki N (2011) Developmental and circulatory profile of the diploic veins. Microvasc Res 81:97–102

Torres-Lagares D, Tulasne JF, Pouget C, Llorens A, Saffar JL, Lesclous P (2010) Structure and remodelling of the human parietal bone: an age and gender histomorphometric study. J Craniomaxillofac Surg 38:325–330

Touzani O, MacKenzie ET (2007) Anatomy and physiology of cerebral and spinal cord circulation. In: Schapira AH (ed) Neurology and clinical neuroscience. Elsevier Health Sciences, Philadelphia, pp 540–549

Tsutsumi S, Nakamura M, Tabuchi T, Yasumoto Y, Ito M (2013) Calvarial diploic venous channels: an anatomic study using high-resolution magnetic resonance imaging. Surg Radiol Anat 35:935–941

Tubbs RS, Bosmia AN, Shoja MM, Loukas M, Curé JK, Cohen-Gadol AA (2011) The oblique occipital sinus: a review of anatomy and imaging characteristics. Surg Radiol Anat 33:747–749

Tubbs RS, Walker AM, Demerdash A, Matusz P, Loukas M, Cohen-Gadol AA (2015) Skull base connections between the middle meningeal and internal carotid arteries. Childs Nerv Syst 31:1515–1520

Van Speybroeck L (2002) From epigenesis to epigenetics: the case of C. H. Waddington. Ann N Y Acad Sci 981:61–81

Vlček E, Druga R, Šmahel Z, Bigoni L, Velemínská J (2006) The skull of Wolfgang Amadeus Mozart predicates of his death. Acta Chir Plast 48:133–140

Voie A, Dirnbachera M, Fisher D, Hölsche T (2014) Parametric mapping and quantitative analysis of the human calvarium. Comput Med Imaging Graph 38:675–682

Waddington C (2012) The epigenotype. Int J Epidemiol 41:10–13

Weber GW (2014) Another link between archaeology and anthropology: virtual anthropology. Digit Appl Archaeol Cult Herit 1:3–11

Weber GW (2015) Virtual anthropology. Yearb Phys Anthropol 156:22–42

Weinstein M, Duchesneau P, Weinstein A (1977) Computed angiotomography. Am J Roentgenol 129:699–701

Wind J (1984) Computerized x-ray tomography of fossil hominid skulls. Am J Phys Anthropol 63:265–282

Woolsey TA, Hanaway J, Gado MH (2008) The brain atlas: a visual guide to the human central nervous system. Wiley, New Jersey

Wu XJ, Schepartz LA, Falk D, Liu W (2006) Endocranial cast of Hexian *Homo erectus* from South China. Am J Phys Anthropol 130:445–454

Wua X, Schepartzb L (2009) Application of computed tomography in paleoanthropological research. Prog Nat Sci 19:913–921

Wysocki J (2008) The size of venous foramina and skull capacity in man and selected vertebrate species. Folia Morphol (Warsz) 67:98–103

Wysocki J, Reymond J, Skarzyński H, Wróbel B (2006) The size of selected human skull foramina in relation to skull capacity. Folia Morphol (Warsz) 65:301–308

Yang Z, Guo Z (2015) A three-dimensional digital atlas of the dura mater based on human head MRI. Brain Res 1602:160–167

Zamir M (1999) On fractal properties of arterial trees. J Theor Biol 97:517–526

Zamir M (2001) Fractal dimensions and multifractality in vascular branching. J Theor Biol 212:517–526

Zasler ND, Katz D, Zafonte RD (2006) Brain injury medicine: principles and practice. Demos Medical Publishing, New York

Zelditch M, Swidersky D, Sheets H, Fink W (2004) Geometric morphometrics for biologists. Elsevier, San Diego

Zenker W, Kubik S (1996) Brain cooling in humans—anatomical considerations. Anat Embryol 193:1–13

Zollikofer CPE, Ponce de León MS (2013) Pandora's growing box: inferring the evolution and development of hominin brains from endocasts. Evol Anthropol 22:20–33

Zollikofer CPE, Ponce de León MS (2005) Virtual reconstruction: a primer in computer-assisted paleontology and biomedicine. Wiley-Interscience, Hoboken

Zonoobi D, Kassim A, Shen W (2009) Vasculature segmentation in MRA images using gradient compensated geodesic active contours. J Sign Process Syst 54:171–181

The Brain, the Braincase, and the Morphospace

7

Emiliano Bruner

Abstract

Morphological integration deals with the functional and structural associations, at ontogenetic and evolutionary level, between anatomical traits. Current morphometric tools can be used to analyze anatomical systems in terms of the mutual relationships shared among their components. The brain has no fixed and rigid form, but rather it is largely shaped by a set of mechanical forces involving bones, connectives, and vessels. During morphogenesis, the brain and braincase exert reciprocal influences associated with size and shape changes of soft and hard tissues. The available evidence suggests that such influences are usually based on local interactions, more than on general schemes or long-range effects. The frontal, temporal, and cerebellar lobes have a direct spatial association with the facial block and with the endocranial base, sharing several morphogenetic factors and geometric constraints with these areas. The frontal, parietal, and occipital bones are more directly shaped by the cortical brain surface, but they have constraints associated with bone articulations and reciprocal spatial adjustments. The final phenotype, selected by evolutionary processes, is an admixture of adaptations, secondary consequences, and structural regulations. The set of rules that govern phenotypic variability can be revealed and quantified by using multivariate statistics. The occupation of multivariate morphological space (morphospace) depends on the underlying structural organization and on ecological and phylogenetic constraints. Therefore, the geometric study of morphospace occupation parameters can reveal the rules of variability behind the observed morphological diversity. These intrinsic properties of endocranial variation must be nonetheless interpreted taking into account information from brain, bones, connectives, and vessels, and the data resulting from these quantitative analyses should be used to plan specialized biological surveys.

Keywords

Endocast • Disparity • Morphological diversity • Functional craniology • Paleoneurology • Morphological integration • Multivariate statistics • Principal component analysis

7.1 Introduction

At the beginning of the nineteenth century, paleontologists introduced a key concept in evolutionary biology: the function of anatomical elements can be inferred from their form. Since then, anatomical traits and features have been physically and metrically dissected to evaluate their use, their advantages, and their scopes. Such exploration was

E. Bruner (✉)
Programa de Paleobiología, Centro Nacional de Investigación sobre la Evolución Humana, Paseo Sierra de Atapuerca 3, 09002 Burgos, Spain
e-mail: emiliano.bruner@cenieh.es

E. Bruner et al. (eds.), *Digital Endocasts*, Replacement of Neanderthals by Modern Humans Series,
DOI 10.1007/978-4-431-56582-6_7

inevitably biased by two main factors. First, evolution has no scopes, neither finalities. It is but a self-sustaining process based on the differential reproduction of genotypes, with no finalistic or teleological force behind it. The principle is simple: the more you reproduce, the more your genetic combinations are represented in the next generation. Some traits increase reproduction probability, but many other features simply do not influence this parameter. Selection works easily: it promotes genotypes that increase reproduction and it demotes genotypes that decrease reproduction. Furthermore, it simply ignores any variation which does not influence reproductive success. Therefore, not every trait has a specific function, not every trait has a specific advantage, and most of all not every trait is the final product of a specific and targeted selection.

The second bias concerns the independence of traits. The relationship between genotype and phenotype are far from being linear or simple, and the perspective "one gene – one character" is sometimes more misleading than helpful. Phenotypic traits are often influenced by many different genes (*polygeny*), and each gene influences different characters (*pleiotropy*). Furthermore, genes influence each other, and characters influence each other too (*integration*). Finally, the environment influences and orientates the expression of both genes and characters. Therefore, it is apparent that single genes or characters are not the target of selection: the "whole package" is. Selection will judge the whole complex of genes and characters and not the effect of specific genes or characters. A change in one element will probably involve a change in many others, and selection will evaluate the overall result. Hence, some genetic or phenotypic changes will be the result of a specific selection, but some others will be simply secondary consequences, with no direct effect on the general level of adaptation. Bad (unfavorable) characters, decreasing the fitness of individuals, can even be promoted if associated with good (beneficial) characters which involve a larger advantage in reproduction (this association is called *antagonistic pleiotropy*). Also, neutral secondary consequences may reveal interesting new unexpected functions and become interesting evolutionary investments (*exaptation*).

Because of the finalistic perspective, and because of the attention dedicated to single characters, evolutionary biology has long been devoted to analyzing traits and features as isolated elements of the anatomical system. We must also say that, until recently, quantitative statistics was largely based on univariate or bivariate approaches, limiting the effective capacity of taking into consideration multiple factors. However, a more integrated perspective to the interpretation of the anatomical systems has already been available since the beginning of the past century. D'Arcy Thompson published his first version of "On growth and form" in 1917, and the more complete version in 1942, suggesting that shape variation in living organisms should be analyzed as an integrated spatial system. In his book *Tempo and Mode of Evolution* (1944), George Gaylord Simpson analyzed fossil variation according to a comprehensive and integrated approach based on correlation rules. Moss and Young described the possible patterns of integration between the brain and braincase in 1960, introducing a perspective in functional craniology. Stephen Jay Gould published the book *Ontogeny and Phylogeny* in 1977, establishing a chronological hallmark: following that book, an evolutionary biologist could no longer ignore the existence of those potential mechanisms joining functional and structural processes. All these milestones had however a major limit: they were written in historical periods in which the inadequacy of the analytical tools of the time made it impossible to put most of their enlightening proposals into practice. In fact, D'Arcy Thompson concludes his book evidencing that many aspects of his methods of integrated geometrical analysis of anatomical variation could not be computed at that time, and it had to be left "… to other times, and to other hands" (1942, p. 1087).

At the end of the twentieth century, computers entered any house, any office, and any laboratory. At that time, multivariate statistics was ready to go, and digital tools were providing a further revolution in anatomy and biomedicine. Computed morphometrics made it possible to carry out analyses of the relationships between many variables, of their reciprocal influence, and of the effect of their combinations. Digital anatomy provided the way to work with skulls and brains in living individuals and in large samples, with a new level of resolution and reliability. Anatomy and morphology were resurrected, after a long period of relative stasis. These tools supported new perspectives, new solutions, and new challenges. At the same time, these methods also allowed the recovery of old questions, abandoned decades before because of methodological limits. Now we are rediscovering anatomy, and we are realizing that in many ways, we have more information on molecules and galaxies than on our bones and vessels.

Brain morphology has been a major issue in evolutionary anthropology, largely because it has undergone noticeable changes within the hominid lineages, which match important cognitive differences. Nonetheless, what we have in fossils is not a brain, but rather rough information on its general size and overall shape. The mold of the endocranial cavity (*endocast*) is the main object of study in paleoneurology, and it supplies a three-dimensional item we can see and handle, physically or digitally, in order to consider specific anatomical traits or metric variables. However, an endocast should not be studied as an isolated geometric form, but as the morphogenetic result of the balance between cranial, cerebral, connective, and vascular elements. This is why morphological integration and digital tools have represented essential advances in this field. Moreover, this is why the multivariate space associated with morphological diversity should be explored in search for those hidden rules that mold and constrain the structure of biological forms.

7.2 Functional Craniology and Morphological Integration

In 1958 Olson and Miller published a pioneering book entitled *Morphological Integration*. The principle is pretty intuitive: anatomical elements exert reciprocal influences, and so the aim of the morphometrician is to localize and quantify those patterns underlying such biological relationships. Some relationships may be largely *functional*, associated with metabolic and physiological processes. Others may be more *structural*, namely, based on physical and biomechanical interactions among anatomical elements. During morphogenesis, the anatomical components change in response to both a genetic program and continuous adjustments due to an "exchange" of information with their surrounding elements, influencing directly or indirectly the physiological and anatomical environment. For example, osteoblasts are induced to produce bone in response to biomechanical tensions, while osteoclasts are induced to remove bone in response to biomechanical pressure. The skull is a very complex system in this sense, being the result of intricate interactions between many different elements, tissues, and functions (Cheverud 1982; Enlow 1990; Klingenberg 2013). Similarly, brain biology and evolution must be interpreted according to dynamic networks depending on hard and soft tissues, mechanical strains, vascular and cerebrospinal flows, biochemical balances, ontogenetic forces, genetic programs, and functional limits (Goriely et al. 2015).

A balanced system is essential for normal growth and development and represents the underlying model available for any evolutionary change. Evolutionary variations can be based on changing anatomy according to those underlying rules channeling the direction of change or else by breaking those rules and exploring alternatives. In this sense, the study of the endocranium should not be seen as a simple geometric study of its cast but should be intended as the study of the relationships between the brain and braincase. Accordingly, *paleoneurology* should concern how these relationships have changed during evolution (Bruner 2015).

Size is of course a major factor of biological integration: when changing size, all the elements must be reorganized according to specific scaling rules (Gould 1966; Cheverud 1982; Shea 1992). Brain evolution in hominids underwent both size and shape changes, and a main target of paleoneurological studies is therefore to understand the different factors involved (Neubauer 2015). In general, *isometric* changes (variation of size without variation of shape) are rare, and most anatomical systems rely on *allometric* relationships (i.e., changes of size inducing changes of shape, to maintain functionality or simply as geometric consequences of spatial scaling rules). Evolution can select some changes in size, and shape will change as a secondary consequence. Or, conversely, shape change can represent the result of an adaptation and size variation just a means to an end. Also, the relationships between size and shape can be broken, creating new relationships, to explore different schemes or to avoid limits associated with existing constraints. Of course, in evolution, changes based on underlying rules are more likely, they are simply more probable, and allometric patterns are therefore "highways" that can be used to easily explore morphological alternatives (like *evolutionary lines of least resistance*; Schluter 1996). In contrast, non-allometric changes are more difficult to sustain in terms of evolution, because they must be based on novel scaling rules which must be, anyway, globally functional and viable.

In ontogeny, it can be useful to separate conceptually *growth* (change of size) and *development* (change of shape). There is no agreement regarding whether or not these two components can be really separated in biological terms (Richtsmeier et al. 2002), but their theoretical separation can at least help to differentiate different factors involved in the process. Dealing with cranial morphogenesis, Moss and Young (1960) proposed that size changes are largely due to brain pressure, and shape changes to the redistribution of the growth forces according to tensions exerted mainly by the connective tissues. In particular, the *falx cerebri* and the *tentorium cerebelli* may represent the main tensors: they are anchored to the skull, envelop the brain, and support the vascular channels. Thus, they may act as biomechanical modulators between all the different elements involved in braincase growth and development.

Apart from all the theory on cranial integration, the actual information on this topic is still limited. This is probably due to two main difficulties. First, theory works with sharp concepts and definitions, but nature is generally based on smoother principles and blurred frontiers. For example, we rely on terms like integration and modularity to refer to shared patterns or separated blocks, but most of the real processes deal with different degrees of admixture between them. Integration and modularity are hierarchical concepts, and in most intermediate situations, such dichotomous terminology fails to provide a useful quantitative perspective. The degree of integration of a system says little about the processes involved and about the polarity of causes and consequences among the anatomical elements. A second problem is associated with the heuristic perspective of these kinds of analyses, aimed at localizing and quantifying patterns only through the numerical decomposition of the observed variation. In general we limit the analysis to numerical covariation, which is only a part of the story. Theories and hypotheses on integration should include a further experimental stage which, in general, is very difficult to perform because of ethical or practical reasons.

Nevertheless, despite the fact we lack a clear knowledge of the integration mechanism of the cranial and endocranial systems, there is some information that cannot be neglected when dealing with functional craniology and paleoneurology. We can identify at least two fields of investigation in this sense: we must consider the reciprocal morphogenetic effects among anatomical elements, and at the same time, we must consider their reciprocal spatial position.

Concerning reciprocal influences, it is reasonable to hypothesize that during ontogeny the elements maturing earlier somehow constrain the morphogenesis of the elements maturing later (Bastir and Rosas 2005; Bastir et al. 2006). In general, the braincase matures first, followed by the cranial base and then by the facial block. Reciprocal influences between anatomical elements are also due to the biomechanical properties of the different tissues involved. In this sense, harder tissue (like bone) can have mechanical functions which are more relevant than softer tissues (like brain), and hence may exert a larger influence on the overall structural constraints. It is important to bear in mind that evolutionary and ontogenetic changes may be based on shared processes, but not necessarily match the same scaling rules. The variability of a species can provide information on the underlying mechanisms governing its phenotypic structure and hence on the morphological combinations available for further evolutionary changes. Even so, the morphogenetic constraints and the evolutionary constraints may be based on different principles, shaping the phenotype according to the same raw material but following distinct requirements. For example, it may be hypothesized that maturation sequence can be more important for ontogenetic constraints (depending on the chronological position along the cascade of events), while bone biomechanics can be more important for evolutionary constraints (selection acts on the hard tissue properties, and the soft tissues are adjusted accordingly). In both cases, there is also a connectivity issue, and it can be expected that elements with larger number of structural and functional connections may have a larger influence on the rest of the system (Esteve-Altava et al. 2013).

At intraspecific level, current evidence on modern human cranial variation suggests that integration is largely due to physical interaction and spatial proximity (Fig. 7.1). The midsagittal elements are partially independent from the parasagittal elements (Bastir and Rosas 2006, 2009; Bruner et al. 2017). The ethmoid and the sphenoid bones coordinate the spatial organization of a facial and a neurocranial block, respectively (Esteve-Altava et al. 2013). The three

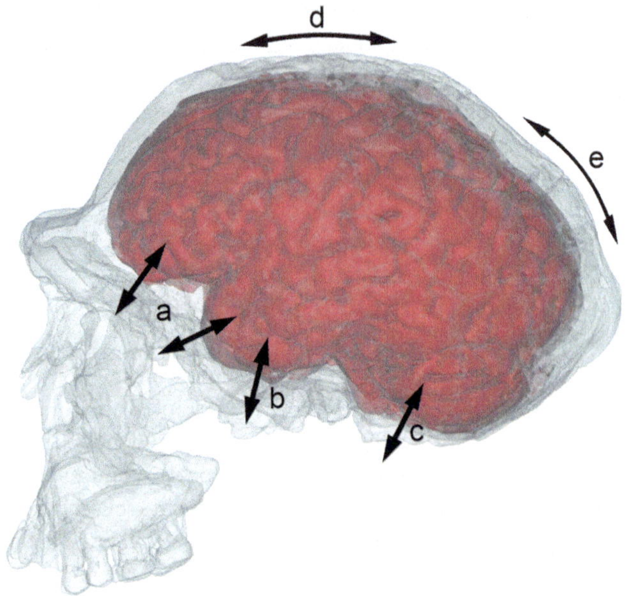

Fig. 7.1 A graphic reconstruction of the brain in a specimen of *Homo ergaster* (KNM-ER3733), after visual superposition with a modern MRI cerebral replica. The morphology of the brain areas positioned at the endocranial base is largely influenced by structural constraints with the underlying cranial elements. The frontal lobes (anterior fossa) and the temporal lobes (middle fossa) interact with the facial block (*a*). In modern humans and Neanderthals, the contact between the frontal lobes and the orbits is particularly pronounced. The temporal lobes further interact with the underlying mandibular elements (*b*). The morphology of the cerebellum (posterior fossa) is influenced by the cranial base, including features associated with cranial base flexion and posture. Conversely, the constraints along the vault are largely due to the longitudinal interaction between frontoparietal (*d*) and parieto-occipital (*e*) adjustments. The frontal bone is the interface between the braincase and the face, with the browridge acting as a structural hinge. The parietal elements (bone and lobe) are constrained between the frontal and occipital spatial changes

endocranial fossae are influenced by different local factors, and apart from local structural adjustment at their respective joining areas, their morphology is therefore not much channeled along shared morphological patterns (Bruner and Ripani 2008). All the endocranial base morphology is sensitive to multiple nonneural factors ranging from the physical influence of the facial block and of the ethmomaxillary complex to issues concerning metabolism and energetics (Bastir and Rosas 2006; Bruner et al. 2017). The cranial base exerts a major influence on the lateral development of the skull (breadths), while longitudinal cranial proportions are less constrained and more independent (Lieberman et al. 2000; Hallgrimsson et al. 2007). The temporal lobes, housed in the middle fossa, are sensitive to the morphology of the cranial base, in particular of the biomechanical relationships with the mandible, positioned right under that area (Bastir et al. 2004; Bastir and Rosas 2005, 2006). It has also been proposed that there is an external biomechanical influence on the braincase exerted by the temporal muscle, even if the hypothesis does not take into account the chronological sequences between frontal expansion and muscular development (Stedman et al. 2004; Mc Collum et al. 2006). The frontal lobes, at least in modern humans and Neanderthals, are positioned right on the orbital roof and are hence constrained by the morphogenetic and spatial relationships with the upper face and with the orbits (Masters 2012; Bruner et al. 2014a; Masters et al. 2015; Beaudet and Bruner 2017; Pereira-Pedro et al. 2017). It is no coincidence that the traces of the sulcal patterns of the brain are most visible on the orbital and temporal surface, two areas of "spatial conflict," in which bone morphogenesis struggles between different and antagonistic growing districts.

The brain itself is modular, and formed by elements and surfaces which undergo different morphogenetic and adaptive processes, with scarce correspondence among their respective spatial variations (Bruner et al. 2010; Gómez-Robles et al. 2014). A degree of integration among brain elements was detected in the rear part of the midsagittal brain section, between the splenium of the corpus callosum and the parietal profile, and it has been tentatively interpreted as the tensor effect of the tentorium cerebelli (Bruner et al. 2010, 2012). There is also a marked integration pattern between parietal and occipital areas (Gunz and Harvati 2007), at least in terms of shape of the bones, although this is not associated with a volumetric correspondence of the underlying cerebral lobes (Allen et al. 2002). In fact, the parietal and occipital bones display a coordinated shape change (the more one bulges, the more the other flattens), but the volumes of the parietal and occipital lobes do not show any correlation, suggesting that there are independent factors that influence their respective size.

The second area of investigation concerns the spatial arrangements between soft and hard tissues. The spatial relationships between skull and brain areas are difficult to establish, and there are a number of specific issues to be taken into consideration depending on the areas investigated and on the species considered. A main methodological limit is the different source of data: brains and bones are generally not detected with the same techniques. Physical dissections preserving both cranial and cerebral anatomy are very difficult to perform, although they can be very informative (Ribas et al. 2006). Currently, the most efficient way to investigate at the same time hard and soft tissues is by integrating physical study, magnetic resonance, and computed tomography (Kobayashi et al. 2014a). In humans, such studies are limited mostly because of the radiation exposure associated with high-resolution imaging. Even in nonhuman taxa, many methodological, economical, logistic, and ethical problems hamper the application of such an integrative approach to large samples. Interestingly, cranial sutures include a minor percentage of collagen that can be detected by MRI signals, allowing the localization of some cranial references in MRI brain stacks (Cotton et al. 2005). Preliminary studies on the brain-skull spatial relationships have shown that at least in macaques, there is a fairly stable spatial relationship between frontal lobes and pteric bones (Kobayashi et al. 2014b). In contrast, in the human vault, there is a good correspondence between bone and brain curvature and surface but a variable spatial correspondence between bone and brain boundaries (Bruner et al. 2015a). For example, larger parietal lobes will increase the bulging of the parietal bones but only to a minor degree the extension of its anatomical boundaries. The larger the parietal lobes, the more they approach the frontal bone (Fig. 7.2). Therefore, there is a good correspondence between the brain and vault shape (curvature) and surface (sulcal patterns) but a less consistent correspondence between the boundaries and position of their respective elements.

Two hypotheses have been provided regarding this partial geometric independence between lobes and bones. The first proposes that this lack of reciprocal spatial correlation can be due to a lack of association between bone and brain anatomical references: the growing brain molds the bones but not their boundaries, which follow dynamics which are associated instead with the relationships with other bones more than with the underlying cortical surface. In some cases, probably for genetic causes, very large pentagonal and symmetric bregmatic bones are added as a fifth bony element of the vault, introducing a new set of sutures, but no changes can be detected in the general neurocranial morphology (Barberini et al. 2008). This suggests a noticeable stability of the vault system and, despite a shape correspondence (surface curvature and local imprints), a spatial independence between bone arrangement and endocranial growth and development. Actually, according to the pattern of spatial connection and articulations, the formation of the vault bones is more passive and less determined by influences of the underlying soft tissues, when compared, for example, with the bones of the facial district (Esteve-Altava and Rasskin-Gutman 2014a).

Alternatively, it can be hypothesized that there may be a spatial correspondence between bones and lobe boundaries during growth and development, but such original correspondence is then lost in later stages, when other parts rearrange the overall anatomical system. In fact, the morphogenesis of the braincase is not linear but formed by different and independent phases (Neubauer et al. 2009). For example, parietal bulging in modern human is associated with early ontogenetic stages, so its spatial effects can be altered by all the successive growth processes. Namely, later changes associated with the frontal areas may alter the spatial organization of the rear vault, achieved during preceding steps.

It must be said that there is an important difference between the functional meaning of the terms "bones" and

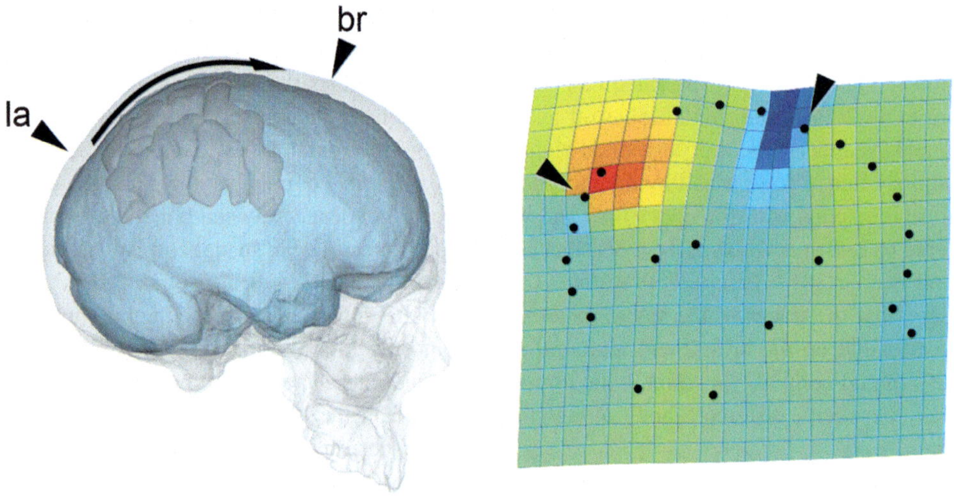

Fig. 7.2 The vault bones are a good proxy to investigate the cortical (brain) sulcal pattern, curvature, and even size. Nonetheless, the spatial correspondence between bones and lobes is less strict. For example, at least when considering the variation among modern adult humans, the size of the parietal lobe influences the longitudinal extension of the parietal bone only to a moderate extent. When the parietal lobe is larger, the bone is more bulging, but the position of bregma (*br*) and lambda (*la*) do not change accordingly. As a result, the anterior part of the lobe "slides" under the bone into a forward position, and the paracentral lobule (the block formed by the central sulcus and the adjacent ascendant circumvolutions) gets closer to the frontal bone (*red* spatial dilation, *blue* spatial compression) (Data from Bruner et al. 2015a)

"lobes." *Bones* are real evolutionary and embryological units. They correspond to specific ossification centers; they are homologous among phylogenetically close taxa and have a stable morphology. *Lobes* are, in contrast, conventional units, with fuzzy and variable boundaries and no stable morphology. Therefore, the geometric organization of bones and lobes is of interest to investigate the dynamics of the braincase morphogenesis, but while the former represent actual structural units, the latter are instead general spatial references of brain districts.

A further note concerns the variation of the parietal bones and lobes in extinct human species. Parietal bulging is a distinctive trait of *Homo sapiens* (Bruner et al. 2003, 2011, 2014a; see Chap. 15). In non-modern humans (i.e., the extinct species of the human genus), brain enlargement was associated instead with a relative reduction of the parietal surface (Bruner 2004). The extreme expression of this pattern can be seen in Neanderthals, which showed a large cranial capacity (comparable with or even larger than modern humans) but a very short and flat parietal outline. Interestingly, this allometric pattern due to the endocranial spatial arrangement is associated, on the ectocranial (bone) counterpart, with supernumerary ossicles (Manzi et al. 1996; Bellary et al. 2013). These supernumerary ossicles, called Wormian bones, are additional ossification centers which are interpreted as *hypostotic traits*, namely, non-pathological characters due to insufficient ossification. Therefore, it has been hypothesized that such an extreme expression of the allometric pattern may have led to a constraint of the morphogenetic model: the relationship between a larger braincase and a shorter parietal length may have reached a limit in Neanderthals, involving a loss of balance between braincase growth (size) and development (shape), with resulting inefficiency in the ossification scheme (Manzi et al. 2000; Bruner 2014). The parietal and occipital bones are integrated in terms of morphology (Gunz and Harvati 2007), and spatial factors may have had a role in generating longitudinal endocranial constraints. In fact, brain size increases with the power of three, but the tensions generated by the endocranial elements (like those associated with biomechanical effects of the connective tensors) increase at a lower rate, forcing the spatial adjustments of the growing cortex. The parietal area is structurally "locked in" between the frontal and occipital ones, having a more limited capacity of spatial adjustment mostly when dealing with its posterior articulation. If this hypothesis is correct, it represents a good example in which allometric schemes associated with brain morphogenesis can meet structural constraints associated with cranial organization. The imbalance associated with supernumerary ossicles is generally non-pathological or sub-pathological. Hence, we ignore whether such structural changes may have been associated or not with some kinds of functional impairment.

The take-home message is: before interpreting a brain form change in terms of neural adaptations and functions, be sure it is not simply a secondary structural constraint exerted by the cranial anatomy or by internal geometric factors. This is why it is advisable to study an endocast with its own skull. If structural factors can be ruled out when dealing with morphological endocranial changes, then functional neural changes are probable. But, if this possibility cannot be tested or has not been tested, any conclusion is speculative and, although interesting, must be confined in the realm of opinions and not put forward as decisive scientific evidence. In some areas (like the vault), geometric changes are more easily interpreted as secondary consequences of brain variation, while in some others (like the cranial base), cranial constraints cannot be, in general, easily ruled out. This does not mean that changes of the endocranial base cannot be the result of brain changes but simply that, in this case, causes and consequences are more difficult to understand, requiring more caution and, possibly, the integration of multiple evidence.

In some cases, sulcal patterns can be more informative than brain geometry, because they are the expression of intrinsic properties of cortical development and less influenced by cranial factors (e.g., Rosas et al. 2014). Nonetheless, sulcal patterns are extremely variable within a single species, their association with specific cytoarchitectonic areas is only approximate, and their functional meaning is not clear. Furthermore, all human species show a very similar sulcal organization, and such homogeneity prevents a reliable use of these traits in human paleoneurology. Hypotheses on possible sulcal differences in hominids (*Australopithecus*, *Paranthropus*, and *Homo*) generally rely on one or few individuals, incomplete specimens, and often on debated sulcal features. The general disagreements on these topics after decades of studies suggest that generalizations and firm inferences are not recommended. For all these reasons, studies on sulcal diversity in human paleoneurology have yet to provide any conclusive result. It is worth nothing that also sulcal morphology is, however, influenced by functional and structural variables. Neurons act as micro-tensors during corticogenesis, redistributing the growth forces according to physical and mechanical laws (e.g., Hilgetag and Barbas 2005; Bayly et al. 2014). A simulation with a gel physical mold showed that passive growth processes can mimic the human sulcal pattern only by virtue of the structural constraints and geometry of a basic fetal brain morphology (Tallinen et al. 2016). This experiment suggests that the sulcal pattern may be but a mechanical consequence of surface/volume adjustments, with no specific functional information. According to this mechanical view of the folding processes, minor changes in the physical composition of neurons or wider cortical areas can induce, on the large scale, sulcal variations (Toro and Burnod 2005). If folding is but a mechanical issue with no functional meaning, sulcal patterns should not be used to supply cognitive or phylogenetic conclusions. Nonetheless, such a mechanical background can, in any case, support different kinds of biological inferences: it can indirectly reveal the underlying scheme of the morphogenetic process (rates, timing, and relative proportions of growth and development in specific cortical areas).

7.3 Stepping into Computed Morphometrics

Studies in morphological integration aim at finding and quantifying the numerical and geometrical schemes behind the organization of the phenotype, by analyzing the structure of the observed variation in order to localize associations and combinations of characters. Such combinations could represent and reveal the architectural aspects of the anatomical organization, which are the ultimate product of the selective processes. Of course, as such these combinations and schemes must have a specific biological meaning, which must be investigated with methods that go beyond morphometrics (histology, physiology, genetics, and so on). The scalpel of the morphometrician is the principle of correlation, used to dissect the variation and to find the rules of variability. Please note the terminological issue: *variation* is the actual range of phenotypic dissimilarity, while *variability* is the ability of a given biological model to vary (Wagner and Altenberg 1996). Such ability to vary depends on those hidden patterns generating integration and relationships among the components. Multivariate approaches are therefore correlation analyses with two main aims: first, to localize those rules (combinations of variables), and second, to quantify such combinations (*vectors*) in order to make individuals or groups comparable within a common scoring framework.

The correlation issue is relevant also because it is independent from its causal interpretations. Excluding methodological biases and incorrect results, a correlation is a fact, independent from the explication of the mechanisms behind the fact. This is why many multivariate tools are descriptive and not inferential: they localize and quantify the numerical rules, but the interpretation of the rule (its validity and its meaning) is something beyond the quantitative fact. Unfortunately, all too often, appealing graphical and numerical outputs are automatically interpreted as conclusions, without a proper passage between the result (a numerical fact, true in that specific analytical frame) and its meaning (an interpretation, according to a wider and independent knowledge).

Two fields have made a major contribution to the renaissance of anatomy in recent decades: digital anatomy and geometric morphometrics (Bookstein 1991; Zelditch et al. 2004; Zollikofer and Ponce de León 2005; Gunz 2015). *Digital anatomy* allowed a comprehensive exploration of anatomical volumes, their virtual dissection, selective isolation, and reconstruction (Spoor et al. 2000). Many techniques can be included in these methods, but the most used are computed tomography for hard tissues and magnetic resonance for soft tissues (Rilling 2008; Hetch and Stout 2015; Weber 2015). The approach is exactly the same as in histology and cytology but with pixels instead of cells: computed sections of the anatomical volume instead of physical slices, algorithms to evidence and separate histological components according to some physical or chemical characteristic (segmentation) instead of biochemical stains (coloration), and visualization through digital platforms and screens instead of microscopes and lens.

Geometric morphometrics is based on the analysis of landmark coordinates (Slice 2007; Mitteroecker and Gunz 2009; Klingenberg 2010; Lawing and Polly 2010). Instead of measuring arcs, chords, and angles, geometric morphometrics captures the spatial relationships among the anatomical elements, by their physical position as recorded through their Euclidean coordinates, in two or three dimensions. Some landmark-based methods like the Euclidean Distance Matrix Analysis (EDMA) analyze the distribution (values and parameters) of all the possible inter-landmark distances, providing a quantitative analysis of the *form* of the object (Richtsmeier et al. 1992). This approach is recommended when analyzing form as a whole and when we want to consider the absolute differences among specimens. These methods evidence the actual differences between groups, giving a comprehensive and quantitative view of all the local and global differences. However, all the many one-to-one associations among elements and proportions cannot be easily handled in terms of statistics, generating huge correlation matrices which are scarcely practical and difficult to interpret.

In most of the other geometric morphometric methods, form is tentatively decomposed in *size* and *shape*, to analyze these two components separately (Bookstein 1991; Rohlf 1999). Size concerns the dimension of the object, whereas shape deals with the reciprocal position and proportions of its elements. Coordinates from all the specimens are normalized so as to be comparable within a similar numerical frame, through processes of superimposition. Such transformation of the coordinate systems standardizes geometry according to a given baseline (*baseline superimposition*) or else minimizes the differences among individuals through translation, scaling, and rotation of the coordinate sets (*Procrustes superimposition*). In both cases, the resulting new coordinates are registered according to a common geometric criterion and can be then used to analyze the correlation among the variation of the anatomical points through multivariate analysis. These methods are effective to localize the existing patterns of covariation within the sample, which is the ultimate scope of morphometrics. However, registration introduces a conventional transformation of the data, which involves the acceptance of assumptions and a priori operational choices. For these reasons, these techniques should be strictly used to compare and analyze patterns of covariation and not absolute differences. Probably we can say that inter-landmark analyses provide less information because of a more restricted analytical capacity, but they supply a "true" result in the sense that they describe actual (real) differences among specimens or groups, independently upon any assumption or criteria. In contrast, superimposition-based methods are far more powerful, but the introduction of operational choices makes the results more dependent on specific methodological decisions. In my opinion, a proper integration of both approaches is helpful when dealing with any specific morphometric issue, mostly when considering that they may provide complementary information.

The use of these recent computed morphometric methods requires some caution, taking into account that the analytical power of these approaches must be properly balanced by an adequate technical control and expertise. Current morphometric programs are definitely user-friendly, and they provide nice outputs even without having a proper competence on the underlying analytical frame. This situation – a complex technical environment, arresting images, and easy-to-use software – can often lead biologists toward a peculiar bimodal distribution of professional careers: those who click buttons without a clear understanding of the underlying procedure and those who dedicate themselves to the exciting methodological challenges forgetting the biological problems. In my opinion, often a midway position is more recommendable and fruitful: a sufficient methodological competence and a robust biological know-how. Needless to say, *multidisciplinarity* (different experts working together) should be the key, instead of *polydisciplinarity* (one person trying to be an expert in everything).

Digital anatomy and geometric morphometrics lead to a main issue: *landmarking*. Because of the smooth and fuzzy morphology of the brain, landmarking in this case requires special attention (see Chaps. 8 and 9). Landmarks are points, but not all points are landmarks. A morphometric landmark must have four characteristics. First, it must be a point and not an area or a surface. Second, it must have a clear and established definition. Third, it must be present in every specimen of the sample. Fourth, it must have a biological meaning. The mammal skull has many such points, but the brain has very few. That is why morphometricians often use *semi-landmarks*, which have some geometrical properties of the landmarks but not necessarily the biological ones (Gunz and Mitterœcker 2013).

When dealing with brain and skull digital data, landmarks can be sampled at least following four imaging approaches (Fig. 7.3). Landmarks can be sampled in two dimensions on specific two-dimensional sections. This approach is rather fast and easy but with a main limit: in the human brain and

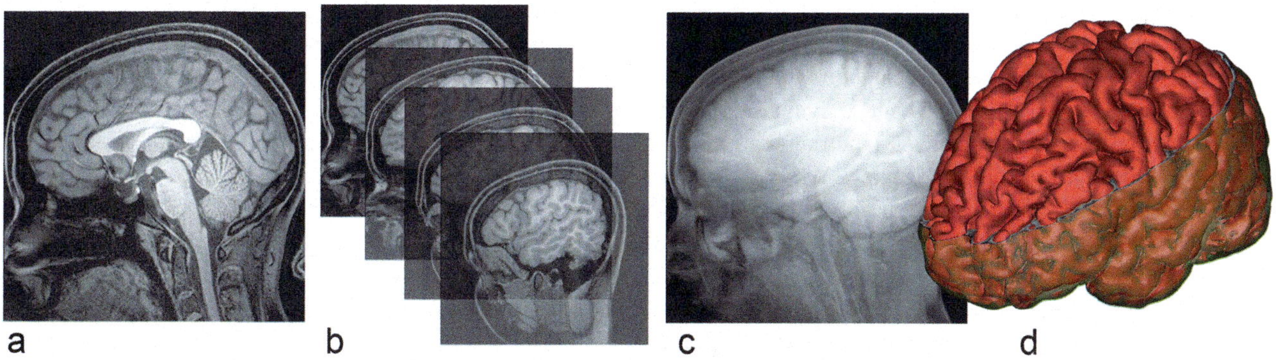

Fig. 7.3 Dealing with skull and brain morphometrics, coordinates can be sampled with four different approaches. Landmarks can be taken on a 2D section (**a**). Unfortunately, the only real homologous section is the midsagittal one or at least its cortical medial approximation. Landmarks can be also captured on different 2D slides, giving 3D coordinates (**b**). Alternatively, 2D landmarks can be sampled on 2D projection of the whole 3D volume, to integrate 3D morphology and 2D analyses (**c**). Virtual 3D reconstruction (**d**) can be used when landmarks must be sampled on the anatomical surface. All these approaches have advantages and disadvantages

skull, the only plane with a real homologous and functional meaning is the midsagittal one. Other planes are but conventional planes, with no homologous value. Alternatively, two-dimensional landmarks can be sampled all through a stack of images, obtaining three-dimensional coordinates from two-dimensional slices. This may be time consuming but very complete and effective. The main problem may be a difficulty in sampling some structures which need a three-dimensional rendering to be properly localized. A third way is a hybrid between two and three dimensions: two-dimensional landmarks can be sampled on two-dimensional images resulting from the projection of three-dimensional objects. Such projections in digital anatomy are called *scout views* and can be computed according to different algorithms evidencing different aspects of the anatomical volumes. Scout views are very informative to give a quick orthogonal perspective of the 3D anatomy on a 2D plane. Finally, landmarks can be sampled in three dimensions directly on the 3D reconstruction of the object. This may be rather time consuming because it requires the previous reconstruction of the virtual replicas, and it is useful mostly when analyzing the external surface of an object.

Once we have metrics, the next step is to investigate the observed phenotypic variability of the numerical model of this anatomical system. As mentioned above, most multivariate tools are based on a correlation approach: all the correlations among the variables are considered, so as to localize and quantify those combinations and associations of characters which are "hidden" behind the observed variation. The technique most frequently used to evaluate the morphological structure of a given sample is principal component analysis or PCA (see Jolliffe 2002 for a complete reference on this method). PCA is based on the actual covariation/correlation among the variables. In this sense, as previously mentioned, it is not an inferential tool but a descriptive one: its results are "always true" in the sense that, through a numerical quantification, they simply describe the structure of the observed variation. The interpretation of the results may be adequate or not, and the selection of variables or specimens may be adequate or not, but the results themselves are nothing more than an objective description of the numerical structure, according to a given algebraic criterion. PCA is the main door of the morphometrician to dissect numerically the model hidden behind the phenotypic organization. Working with morphological variables, PCA protocol reveals a continuous high-dimensional space in which, according to shared anatomical rules, each point represents a specific combination of values (*morphospace*).

7.4 Exploring the Morphospace

7.4.1 The Structure of the Morphospace

The algebraic background of PCA is matrix analysis and, for an evolutionary biologist, it may be sufficient to know that the computation finds those special combinations of variables (*eigenvectors*) responsible for the main variation of the sample, by virtue of some underlying correlations (see Polly et al. 2013). In bivariate analysis, regression methods are often based on asymmetric approaches like *minimum least square*, which considers the relationships of one variable according to another. Other regression methods (like *major axis* or *reduced major axis*) are symmetrical: they localize and quantify the best linear equation correlating the two variables to each other by minimizing the orthogonal distances between the data and the line fit (see Martin and Barbour 1989). PCA is simply an iterative major axis analysis: it localizes and quantifies the first axis of covariation, then a second axis orthogonal to the first one, then a third, and so on. Each axis is a vector, in the sense that it is numerically a specific linear combination of the original variables. Each axis will be sufficient to synthesize a part of the observed variation (*eigenvalue*), putting the information from different variables together. When using the *covariance matrix* for this analysis, each variable will have a weight (its influence on the results) proportional to its range of variation or to its absolute dimensions. Therefore, if one wants to normalize the effect of the variables and to give them a similar importance, the *correlation matrix* should be used instead, or else the normalized values of all the variables within the sample (z-scores). In geometric morphometrics, coordinates are already normalized through the superimposition passages (*registration*), and this distinction is not necessary.

PCA is often said to be used to "reduce information," by reducing dimensionality. Few components (vectors) can be used instead of many original variables, and this is surely useful for many theoretical and practical reasons. Nonetheless, this may be a meager description of its methodological value. Apart from synthesizing information (which is, anyway, an excellent property), PCA filters and extracts the relevant associations, localizes hidden relationships, and quantifies such relationships, revealing the underlying rules (combinations of characters) behind the observed phenotypic variation. This is more than reducing information. If there is some strong correlation pattern behind the observed variation, there must be a reason why. Whatever reason, it deals with the biological scheme behind the phenotype and probably also with evolution and selection. I will call *structure of the morphospace* the hierarchical organization of factors channeling the phenotypic variability and *form of the morphospace* the geometric pattern of distribution of the sample within this space. Here, *structure* refers to the arrangement of the covariation schemes, while *form* refers to the geometry of the sample-specific occupation. This distribution can be quantified through geometrical parameters describing the occupation of this space for a specific group or subsample. The analysis of these two intrinsic features of the variability is often neglected, PCA being more frequently used only to describe the specific patterns of changes associated with each vector and not the schemes of distribution of the individuals.

The first output of a PCA is often totally ignored: the *scree plot*. A scree plot shows the sequence of variation explained by each vector, generally as a percentage of the total variance. A given biological model (the scheme behind a phenotype) will produce a given scree plot, so the information contained there may be really relevant. Unfortunately, most studies skip this part of the analysis, simply limiting the following considerations to the first two principal components, only because they fit nicely into a bidimensional page or screen. A detailed analysis of the scree plot is necessary for two main reasons. First, it is mandatory to evaluate whether those vectors may represent real biological axes of variability or else random noise due to stochastic factors. Second, it discloses information on the underlying phenotypic organization, revealing the hierarchical sequence of integrative factors (independent components). Namely, the scree plot supplies a quantitative description of the structure of the multivariate space.

There are different methods to evaluate the stability and meaningfulness of the principal components (e.g., Jackson 1993). Remember that this is a descriptive technique, so there is no established method or fixed thresholds, and most of them are "rules of thumb." The analysis shows things as they are, and the researcher must decide how to interpret the output according to the available information, including additional information independent from the statistical analysis itself. A very basic rule is to discard components that explain less than a given amount of variation (generally 5%) or conversely accept vectors until they reach a given amount of variance (often 90–95%). A practical and efficient "visual way" is to note when there is a patent fall of the variance explained, creating a discontinuity between a structured set of vectors and a "fading tail" of vectors which can represent noise. A very good approach is also to simulate a random variation and consider only those vectors that explain more than a random result. Resampling methods can be also used to evaluate a range of variance explained by each axis, to add a range of uncertainty. In this case, vectors whose variance does not overlap with the following axes can be interpreted as stable.

These approaches are all extremely important to provide reliable information on which axes may be the result of a real biological property. Within the observed variation (i.e., with that sample and according to those variables), the scree plot can actually quantify the degree of integration of the morphological system and its underlying structure (Wagner 1984). In fact, if a vector is a true biological signal, it is revealing (and quantifying within that sample) a combination of characters, that is an integrated system of trait variation (Fig. 7.4). These combinations or correlations are due to the structural and functional association among the anatomical elements, channeling the variation according to specific rules of variability. Such channeling can be stronger or weaker, depending on the strength of the associations and the strength of the constraints. A morphological multivariate space characterized by a single component may be showing the existence of a single pattern of integration. Otherwise, there may be two or three covariation rules, defining two- or three-dimensional morphospaces. The rest of the components ("scree") can be probably associated with noise, errors, or with variation not structured according to any apparent rules. In fact, when there are no "dominant" axes, but just a smooth set of decreasing components, this may suggest that the system is not integrated at all. Vectors not

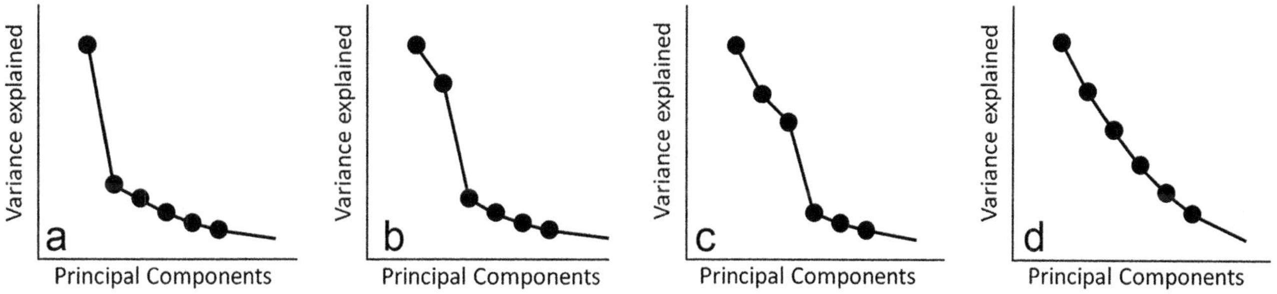

Fig. 7.4 The "scree plot" associated with a principal component analysis is a quantitative description of the structure of the morphospace. When there are stable and isolated components (**a–c**), these can represent biological covariation vectors, localizing and quantifying association between characters due to a given degree of integration. Otherwise, the absence of dominant components (**d**) may be the consequence of a lack of marked integration. Different criteria can be used to evaluate whether or not a component represents a real biological signal. The structure of the morphospace must not be intended as an absolute value, depending on the number of variables and the sample. It must be also interpreted through distinct perspectives whether it refers to intraspecific or intraspecific variation

associated with consistent covariation patterns may be still useful for the aim of reducing information, but their composition in terms of variables and effects should not be intended as a real biological character. As always in statistics, the fact that a component belongs to the minor "scree" does not mean that it does not represent a real biological character, but just that the current analysis, for whatever reason, is not able to reveal this possibility with a sufficient consistency.

As mentioned, strong patterns of covariance can generate evolutionary lines of least resistance (Schluter 1996), which can be crucial in channeling evolutionary changes. Such "preferred" trajectories may facilitate evolutionary variations in specific directions or else be themselves the consequence of some evolutionary pressure. In both cases, they reveal an underlying structure of the sample, which has an intimate relationship with the adaptive landscape.

Once more, it is important to understand that these are exploratory heuristic tools and not inferential tests. Hence, these kinds of analyses can suggest a given degree of integration, but they must not be used as a definite verification. Many cases will show intermediate distributions, which are difficult to interpret and that will require future analyses. In some other situations, the structure of the morphospace can instead reveal patent integration schemes. In both cases, a proper evaluation of the morphospace distribution is mandatory, and the number of components that must be considered (even when using this approach only to reduce dimensionality) must depend on these kinds of considerations and not on the bidimensional limits of the graphic supports.

Taking into account the pros and cons of these approaches, it is apparent that we can use these methods with three different purposes:

1. As proper heuristic tools to "have a look" within the variation of a sample, so as to collect information to generate future hypotheses and to direct future investigations.
2. As an evaluation tool, to assess if and how much a given structure or distribution of the variation supports or does not support a specific (a priori) hypothesis.
3. Independently from the existence of any hypothesis, these methods can be used to localize specific morphological areas or specific morphological traits that need to be subsequently investigated in biological terms (e.g., histologically, genetically, physiologically).

In all cases, the multivariate analysis is often not the aim of the study, but a preliminary step to direct future research. A multivariate approach provides a powerful environment, but also a very reductive one, being limited to a specific sample and to specific variables. Its interpretation requires additional external (and possibly experimental) information.

I would also like to mention *cluster analysis*, another descriptive tool in multivariate statistics. In this case, the ordination is not necessarily based on correlation but more on similarities or differences. Given a specific sample, a specific set of variables, a specific measure of affinity or distance, and a given criterion of coupling, the analysis shows what specimens are more similar and to what extent. Clustering depends on two main operational choices: the choice of the metrics used to calculate the distance between objects (for example the cumulative or mean difference when considering all the variables) and the choice of an algorithm of grouping. Different operational choices may lead to different clustering results. In this case, a general index of the "goodness" of the branching scheme is given by the *cophenetic correlation coefficient*, which is the correlation between the absolute distance between each pair of elements and their distance along the resulting branching scheme. As in PCA, the cluster result is always "true," in the sense that it shows the cluster structure according to those operational criteria. But, as always, the interpretation will depend on additional information. Unfortunately, all too often the output of a clustering procedure is directly interpreted as the true relationships among the elements. In zoology, for example, phenograms and genograms are too often misinterpreted as "trees," interpreting the numerical output (the dendrogram) directly as the conclusion (phylogeny), with no intermediate steps. It must be remembered, instead, that clusters are but fixed, unique, and true graphical outputs of an algorithm, while phylogenetic trees are inductive and deductive speculations of the researcher, based on personal knowledge and on the ability to integrate additional independent data. A dendrogram is a fact, while phylogeny is a hypothesis. Numerical outputs, in morphology, genetics, ecology, or whatever fields, can support or not support a given hypothesis (functional, ecological, phylogenetic), but they do not represent the hypothesis itself.

7.4.2 The Form of the Morphospace

Also the distribution of the individuals within the morphospace can provide further essential information which is often neglected (Fig. 7.5). The "occupation properties" of different groups within the morphospace can be analyzed and quantified, to evaluate the mean position, the degree of variation, and the pattern of variation. Within a given morphospace, the pattern of occupation and distribution of a group will depend, beyond shape differences, on structural, ecological, and phylogenetic constraints, and the geometry of this occupation directly provides some information on the underlying evolutionary forces allowing or not allowing specific forms (Roy and Foote 1997). That is, the quantification of the occupation parameters of a group within a given morphospace can provide an estimate of its *morphological disparity* (see Wills 2001 for a detailed review on this topic).

These concepts and methods have been traditionally developed in invertebrate paleontology (a field that frequently deals with high-rank taxonomic variations) and scarcely applied in anthropology. There is only a vague agreement on concepts and definitions, but the term *disparity* can be generally used to describe the "morphological variety" within a given group (Hammer et al. 2001; p. 136). Nonetheless, in some specific contexts, it may be useful to distinguish between *richness* (number of different taxa within a group), *morphological variety* (the form variance, as the amount of morphological space occupied), and *diversity* (a general term of diversification) (Foote 1992). To evaluate the morphological distributions within a sample, *theoretical morphospaces* can be generated with axes which follow some specific criterion of variation (McGhee 2007). In this case, the space will not depend on the sample composition, being determined a priori. Thus, the spatial model is predetermined according to some rules, to evaluate the behavior of the sample within that space (e.g., Esteve-Altava and Rasskin-Gutman 2014b). However, in most cases the space will be the result of a sample-dependent procedure (like in PCA), representing an *empirical morphospace* obtained after ordination of the observed variation. The occupation of the sample within this space can follow phylogenetic or ecological principles, and this is why many studies integrate this approach with cladistic or environmental adjustments and perspectives. Nonetheless, the occupation of the space can be investigated with the only aim to analyze the structural properties and possibilities of the phenotype, independently from any systematic hypothesis and with a clear phenetic perspective.

The most intuitive parameter to investigate when dealing with the occupation of a morphospace is the amount of the occupation, which is a proxy of the degree of diversity. Within a shared morphospace, a group which occupies a larger region of the space displays theoretically more phenotypic disparity. At univariate level, variance, standard deviation, or range can be used as statistical parameters to estimate and compare diversity among different groups, depending on the nature of the variables investigated. The range is more sensitive to outliers and sample effects than variance, and therefore these two parameters should be employed depending on whether extreme or normal properties are the target of the study. At multivariate level, the same parameters can be summed or multiplied for a set of principal components, to provide a general index of the amount of occupation of the multivariate space for each group. Other methods include the parcellation of the morphospace into conventional units to assess the rough hypervolume occupied or the average distance of the specimens from its group centroid. All these parameters can give different results, and hence different choices must be properly evaluated case by case. For example, metrics based on summing the values from different axes (ranges or variances) are more useful when groups show a similar "shape" of distribution. Otherwise, if the shape of their occupation is distinct, the hypervolume may not be necessarily correlated with its PC projections.

It is clear that all these estimates of diversity are proxies with a strict comparative value, being associated with a given sample and with a given set of variables. That is, individuals and groups within a sample must necessarily share the same morphospace to be compared, and those values have no absolute meaning outside that specific analysis. Also, because of the many different ways to quantify the

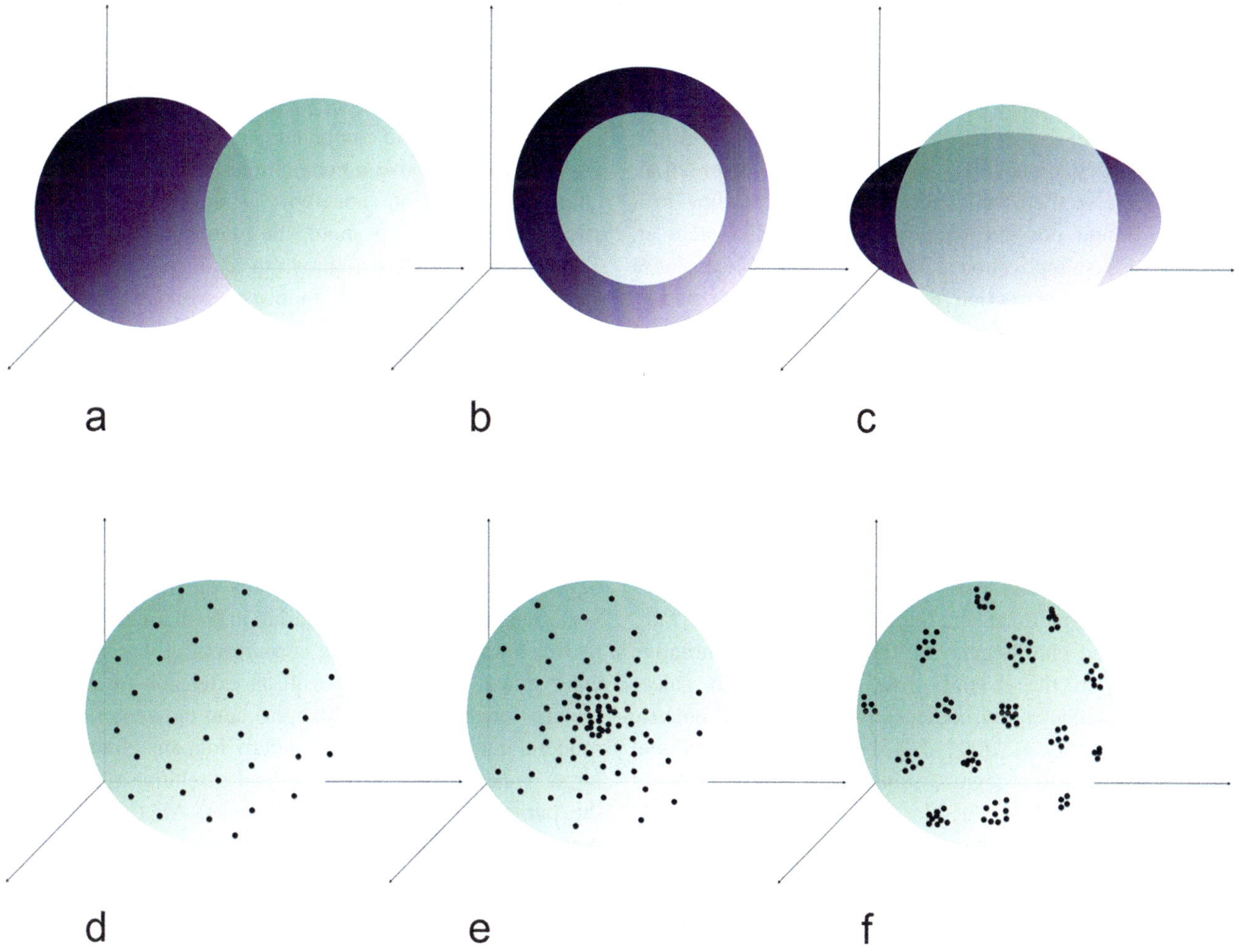

Fig. 7.5 The "form" of the morphospace itself is a quantitative description of the underlying biological model, with its evolutionary meanings and constraints. Two groups can have (**a**) different means but the same degree and structure of variation, (**b**) the same means and structure of variation but a different degree of variation, and (**c**) the same mean and degree of variation but a different structure of variation. Also, within a group, individuals may be differently scattered: (**d**) randomly, (**e**) normally or with gradual density, or (**f**) clustered. These different patterns denote different evolutionary (ecological, anatomical, or genetic) constraints, in terms of morphological structure

occupation of the morphospace, and taking into account that different metrics can give different results, it is necessary to specify and define in each study what kind of diversity is been investigated and through what specific distribution parameters.

The size of the morphospace occupation represents hence the main proxy to quantify diversity. Nonetheless, other distribution parameters can be evaluated within a multivariate context. Two different groups can display, for example, different average values but the same amount and same pattern of variation; or similar average values and pattern of variation but different amount of variation; or similar mean value and amount of variation but different pattern of variation. Further information can be also obtained by the pattern of spatial distribution of the individuals within their group space, which can be, for example, homogeneous, random, bell shaped, or clustered. Metrics associated with the distribution of the nearest neighbor values can give information on the pattern of continuity of the morphospace occupation. Also in this case, the fact that a group displays one kind of spatial distribution or another provides information on the hidden structure of the phenotype and on its possible constraints. If there are sufficient data to supply a chronological perspective, the patterns of morphospace occupation can be studied through time, providing essential information on species dynamics (Erwin 2007).

It is worth noting that a multivariate space generated by the covariance patterns of a sample is a continuous field of variation: each point corresponds to a given shape or phenotype. Hence, "empty spaces" can reveal combinations of characters that are not described in that sample. Apart from limits of the sample or random fluctuations, this perspective is interesting to discover and investigate phenotypes that are more or less probable, or even not viable, due to evolutionary, anatomical, or ecological limits (McGhee 2007). Finally, it must be also mentioned that these same approaches can be also used to investigate multivariate aspects of single specimens within the frame of the observed variation, that is, to evaluate their individual reliability. In paleontology, the quantification of their overall morphology within the morphospace is useful, for example, to consider the consistency of a fossil reconstruction (Bruner et al. 2015b), or else their multivariate position according to a set of different possible reconstructions (Neubauer et al. 2012). This last application is particularly valuable, because it takes fossil reconstruction from a more subjective perspective toward a reliable and probabilistic quantitative framework.

7.4.3 Endocranial Diversity in Apes and Humans

As a simple case study, we can use the hominoid endocranial variation (humans and apes), by analyzing their endocasts through traditional diameters (Holloway et al. 2004). Traditional arcs and chords generally provide only quantitative information on the overall gross proportions in paleoneurology, being largely influenced by size and allometric factors (Bruner et al. 2006, 2015b). Although traditional metrics may fail to recognize subtle cortical differences, it is useful to provide a general estimation of the degree of form variability. Form is here intended in its geometrical meaning, that is, both shape and size components with no distinction. In this case study, the human sample includes three groups, namely, *H. sapiens* ($N = 20$), *H. ergaster/erectus* ($N = 19$), and *H. neanderthalensis* ($N = 11$), and the ape sample includes four species, namely *Pan paniscus* ($N = 43$), *Pan troglodytes* ($N = 32$), *Gorilla gorilla* ($N = 39$), and *Pongo pygmaeus* ($N = 22$). The metric set includes frontal, biasterionic, and maximum endocranial width, hemispheric length, bregma-asterion length, maximum (parietal) arc and parietal bone arc length, and vertex-temporal and basion-bregma height. Please note that in this case most metrics involve cranial (bone) and not cerebral (lobe) references. Samples and metrics come from the large database of Ralph Holloway, based on his extensive and comprehensive collection of physical endocranial casts (see Chap. 1). The analysis was computed with PAST 2.17c (Hammer et al. 2001), by using a correlation matrix to eliminate the effect of different metric ranges among variables. Figure 7.6 shows the principal component analysis for the ape and human samples and their 95% confidence ellipses.

The structure of the morphospace is very similar in the two cases. It roughly corresponds to the case shown in Fig. 7.4a, in which we have only one consistent vector of variation and a successive set of minor components that cannot be reliably distinguished from random noise or individual (idiosyncratic) variations. The first principal component strongly polarizes the multivariate distribution, explaining about 70% of the variance. This vector is an allometric component associated with an increase of all the variables. The second component explains about 10% of the variance, being above a conventional eigenvalue threshold (Jolliffe's cutoff) but not above a random effect (broken stick). Accordingly, it should be interpreted with caution. The second component contrasts the endocranial widths with the parietal size: broader endocasts with smaller parietal areas vs. narrower endocasts with larger parietal areas. Therefore, in both apes and humans, a relative enlargement of the parietal dimensions is associated with a relative decrease of the endocranial widths. In the case of apes, the size vector separates gorillas from the rest of the sample, but the second component, associated with parietal bone length, does not display any noticeable differences among groups (although bonobos show a slightly pronounced parietal development). In the case of humans, the allometric component generates a continuous variation from archaic hominids to Neanderthals. Along the second axis, the parietal bone enlargement is associated with an extension of all its diameters (length, width, and height), and the vector separates modern humans from non-modern humans. Interestingly, larger Neanderthals display instead a relative reduction of the parietal size and wider endocasts. All the successive components are below any threshold of stability and should not be considered as reliable multivariate vectors.

In the case of apes, this study with traditional metrics fails to reveal a main endocranial shape difference among apes, namely, a pattern from more rounded (orangs) to more elongated (gorillas) brains, possibly a structural consequence of opposite spatial relationships between the braincase and face (Shea 1989; Bienvenu et al. 2011). Nonetheless, if the analysis is computed on group mean values (between-group PCA – see below), the first component is still allometric, but the second separates orangs because of higher and shorter endocasts (data not shown). In the case of humans, this analysis replicates the same results found in other studies with different samples and different variables (Bruner et al. 2003; Wu and Bruner 2016; see also Chap. 15). It is interesting to note that in both cases, the variation is characterized by a size-related component and a second minor vector associated with parietal changes. Therefore, the parietal areas are once more confirmed to be a relevant source of variation (Holloway 1981; Bruner et al. 2014b). Nonetheless, this study also provides a quantitative figure concerning the intragroup degree of variation. Taking into account the first two principal components, an index of diversity can be the area of the group-specific 95% ellipses of variation. The area of the ellipses supplies a direct quantification of the amount of occupation of this shared morphospace. In the case of apes,

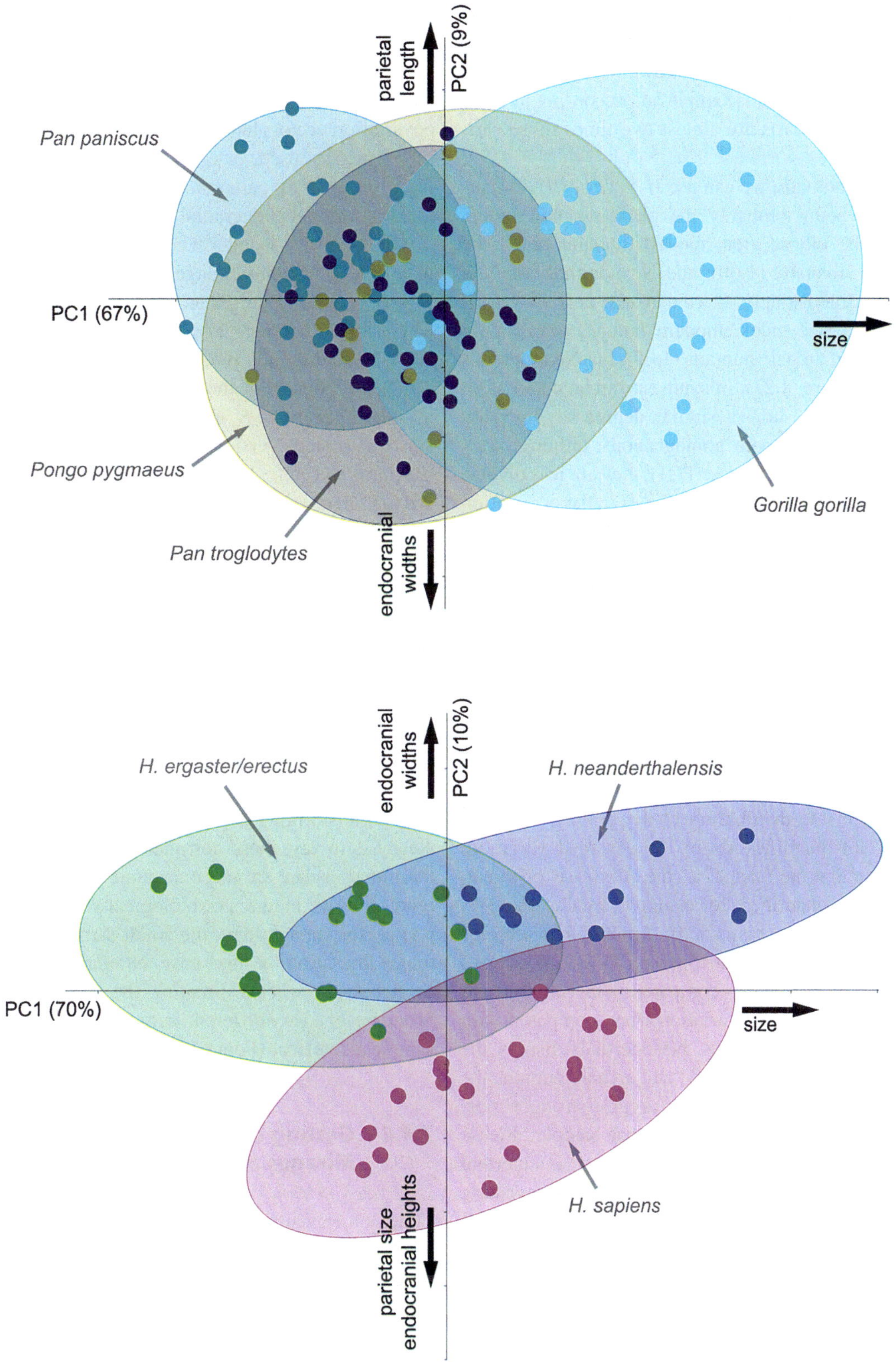

Fig. 7.6 Principal component analysis for the ape (*above*) and human (*below*) endocast samples (first and second components). The occupation of the morphospace by each group (*diversity*, *variety*, or *disparity*) can be quantified and compared following ranges or variances for each axis (cumulative or product) or else through the areas and axes of the confidence ellipses. The shape and size of the spatial occupation supply information on the variability of the morphotype. *Homo ergaster*/*Homo erectus*, Buia; Daka; KNM 3733 and 3883; OH9; Salé; Sambungmacan 3; Sangiran 2, 4, 12, and 17; Sinanthropus LE, II, IID, and IIIL; Solo V, VI, and XI; and Trinil 2; *Homo neanderthalensis*, Amud, Feldhofer, Forbes Quarry, Guattari, Krapina 3 and 6, La Chapelle-aux-Saints, La Ferrassie, Reilingen, and Spy 1 and 2

bonobos are the less variable group, and we can measure the other groups in terms of ratio with the bonobo value. Chimps are slightly more variable (1.19), followed by orangs (1.92) and gorillas (2.24). So, in this sample and according to these metrics, orangs' endocasts are almost two times more variable than bonobos.

In the case of humans, we can use *H. ergaster*/*H. erectus* as a reference, being probably the plesiomorphic group. According to the ellipse area, modern humans show the same degree of diversity (1.00), and Neanderthals are less variable (0.86). Interestingly, if we consider only PC1 variance, which is a size index, modern humans are slightly more variable than archaic humans (1.05), but Neanderthals display a larger figure (1.27). In synthesis, archaic and modern human endocasts show a similar degree of diversity, while Neanderthals are more homogeneous: although they are more variable in terms of size (PC1), they are less diverse in terms of shape (PC2). These differences in the occupation parameters of the morphospace suggest caution when using summed diversity values based on PC vectors. For example, in this case the different patterns of distribution lead to different results if trying to quantify species diversity with a sum of variance (in this case, Neanderthals are more variable than the other groups by virtue of the distribution along PC1) or the sum of standard deviations (in this case, all three groups show roughly similar values of diversity, which are only slightly larger in modern humans). Therefore, a numerical proxy for diversity must take into account possible differences in the distribution of the groups, trying to give a consistent quantification of the observed variation. Please note also that the "shape" of the ellipses is different, as it can be easily quantified, for example, by the ratio (%) between their minor and major axis. Among apes, chimps have the most "rounded" distribution (75%), showing a smaller size variation when compared with orangs and gorillas but a larger parietal variation when compared with bonobos. In the case of humans, Neanderthals display the most "stretched" distribution (0.16) and archaic humans the most scattered one (0.31). This may suggest stronger morphological channeling in the Neanderthal sample. Please note also that both modern humans and the Neanderthal clade show a similar inclination of their ellipses, suggesting similar parietal relative reduction as brain size gets bigger, as already hypothesized on the basis of shape analysis (Bruner 2004). However, the sample size is too small and too scattered to test this topic with a proper quantitative approach, at least in fossil species. Using the same metrics and comparing apes together with a sample of 35 modern humans (results not shown), the endocranial morphology in *H. sapiens* is two times more variable than gorillas and almost three times more variable than chimps.

The degree of morphological variation is a consequence of biological, ecological, and phylogenetic factors and constraints. In the case of apes, sexual dimorphism has a major role as a source of variability. In the case of human species, encephalization is the most relevant factor. The fact that apart from size and parietal dimensions, no noticeable covariation vectors have been identified may be due to limits in the metrics used, or else to an actual absence of further clear integration patterns channeling the morphotype, at least at a global endocranial level.

One may also wonder whether such a degree of variation may reflect taxonomic ranges. In primates, cranial variation is not necessarily correlated with the taxonomic rank (Tattersall 1986; Albrecht and Miller 1993; Plavcan and Cope 2001; Collard and Wood 2000; Holliday 2003; Bruner 2013). In the case of the human genus, we currently know that *H. ergaster* and *H. erectus* may be the same species or at least a scaled version of the same biological model (Rightmire 1998, 2013; Wood and Collard 1999; Baab 2016). There is also the possibility that *H. heidelbergensis* (or at least a part of it) and Neanderthals may be a single paleospecies, making any taxonomic boundary blurred and, accordingly, notably extending the range of variation of this lineage (Stringer 2012; Arsuaga et al. 2014). In terms of endocasts, differences between *H. ergaster*/*erectus* and *H. heidelbergensis* are largely a matter of size, while differences between *H. heidelbergensis* and Neanderthals also involve changes in proportions (Bruner et al. 2003, 2015b). All this information is external to the quantitative output presented here, but it must be carefully considered when transforming results into evolutionary hypotheses. Needless to say, these considerations on the occupation of the morphospace are more accurate the more the sample is representative in terms of biological variation and statistical power. In paleontology, the small sample size represents a major limitation, in this sense. Numerical adjustments may be necessary when estimating the degree of variation in groups with a very different sample size, to evaluate possible sample size effect (Foote 1992).

7.4.4 Getting Lost in Multivariate Spaces: Mirages and Houses of Mirrors

The geometrical and spatial properties of the morphospace can be studied like any other distribution or form. Means and ranges of the group occupation through the multivariate vectors, as well as the structure of the morphospace as revealed by the scree plot, represent a numerical quantification of the phenotypic constraints and potentialities, which are after all the actual prime matter available to selection and evolution. Of course, all these parameters are strictly associated with the variables and sample of a specific analysis and cannot be used as absolute values. That is, groups must be necessarily compared within the same multivariate space, and the analysis of the structure and form of the variation is hence study-dependent. It is also worth noting that PCA is an exploratory tool: it reveals and quantifies patterns and rules, but it cannot explain the reason behind those rules. A robust interpretation of those patterns can be only achieved through a second step: analyzing the real anatomical components, namely, cells, tissues, and organs, and looking for the direct biological evidence behind those quantitative results. Multivariate morphometrics can give a numerical ordination of the variability, but also orientate further biological research, focusing attention on specific components, areas, relationships, or processes, of the anatomical system.

Between a pure numerical approach and successive biological analyses, it may be worth considering an intermediate step: a study of the real specimens according to the multivariate evidence. For example, after having localized and quantified a vector associated with a main pattern of integration thanks to the multivariate analysis, the (real) specimens at the extremes of that vector can be further analyzed, being representative of the maximum expression of that crucial pattern (Fig. 7.7). This may have at least three main scopes. First, it is a physical validation of the multivariate results, showing the actual phenotype of those specimens. Namely, the observed real morphology must match the multivariate shape characterization. Second, it shows the actual extension (range) of the phenotypic modification. This helps to have a more direct idea of the degree of variation involved. Third, individuals belonging to the extremes of the normal axes of variation can show additional information on characters which are not included in the analysis, but that are associated with those specimens and revealed through case-wise inspection. That is, comparing the extreme phenotypes, other traits associated with that pattern can be found, including morphological or non-morphological features.

Nonetheless, the abstract nature of the multivariate analysis requires some caution because all too often the quantitative model (a numerical representation) is confounded with its real anatomical system (an existing biological structure). These metric approaches are based on a spatial distribution of the values within the multivariate dimensions after rotation and projection of these multidimensional spaces: the individuals scattered along many axes are rotated and projected according to the correlation among the variables, so as to localize numerical rules of variation. As evidenced, excluding minor technical issues, the final results are a description of the actual structure of that scattered sample. The strength of such information will depend on the complexity (or the absence) of those rules. If there are contrasting factors or heterogeneous patterns, the projection will be an admixture of different components. It will be true mathematically but, in terms of biology, it will be a

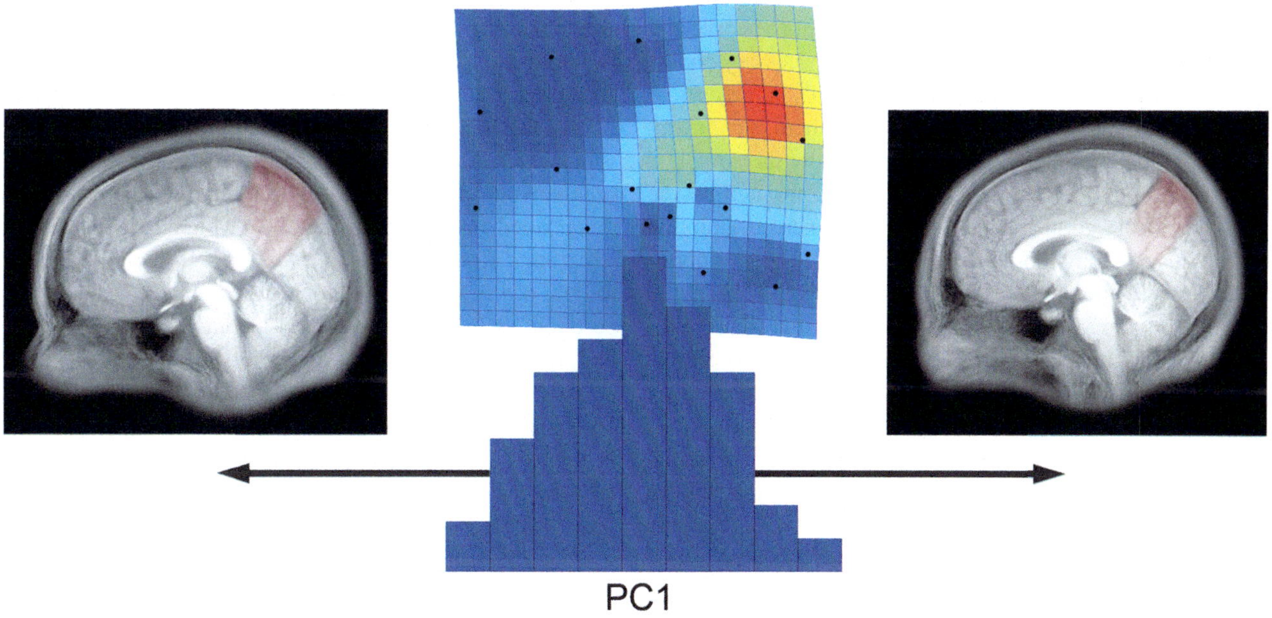

Fig. 7.7 In a study on the normal adult midsagittal brain section, it was found that precuneus dilation was the first source of variability (deformation grid: *red* expansion). Superimposing the ten specimens with larger PC1 values (right MRI superimposition) and the ten specimens with smaller PC1 values, the actual variation of the precuneus can be easily checked and evaluated (in *red*). The distribution of the specimens along this component (histogram) shows that the variation is not symmetric but skewed: there are more individuals with a larger precuneus than individuals with a smaller precuneus. Note that this morphological change also influences the morphology of the space interposed between the splenium, the midbrain colliculi, and the cerebellum. This specific anatomical variation was not implicit in the data, but it can be revealed when analyzing directly these extreme phenotypes

chimerical inexistent scheme. In this sense, the most relevant note concerns the interpretation of the *intraspecific* and *interspecific* variation (Martin and Barbour 1989). Intraspecific and interspecific differences are based on different mechanisms and principles and must be analyzed and evaluated through different perspectives and different approaches. Intraspecific (individual) variation is often based on random individual differences along correlation patterns associated with the underlying biological structure. Thus the correlation among variables is largely the result of some specific biological factor, and the individual discrepancies are randomly/normally distributed along such a biological factor. In contrast, intraspecific (evolutionary and phylogenetic) differences may not necessarily be distributed along a real biological vector, but along general "tendencies," which may or may not be structured on anatomical relationships. A given correlation may be due to a specific set of relationships channeling the variation along preferred directions or else may be simply a mathematical hybrid projection of different independent lineages. Furthermore, the discrepancies of each species from the general trend are not necessarily randomly distributed, because that discrepancy may represent a specific (directional) adaptation. Apart from different interpretations of the results, these differences between intra- and interspecific variability involve an important recommendation: mixing these two sources of variation can seriously bias the multivariate output, creating hybrid axes which are but meaningless numerical vectors. Unfortunately, mixing species and individuals is a common practice, also as a necessary consequence of the small sample sizes available in fields like paleontology or neuroanatomy. When there are many groups, a good compromise is represented by the so-called *between-group PCA*: the morphospace is computed on the averages of the groups, and then the specimens are plotted within this resulting morphospace according to their individual values (Mitteroecker and Bookstein 2011). Therefore, the vectors are computed according to the intergroup patterns, but the position of all the specimens is shown within the resulting morphospace, giving a comparative perspective of their intragroup diversity and distribution.

In all cases, we must always take into account the actual elements we are working with: numbers. We must always bear in mind that we are working with models of something. Computed imaging provides a model of the anatomical system, accounting for spatial (pixel position) and physical (densities) properties of the anatomical elements. Landmarks provide a numerical model of the spatial relationships among the anatomical elements. The endocast is a model too, as a proxy for the brain. The morphospace itself is a model, in the sense that it provides a numerical ordination of the observed covariance among coordinates. All these tools provide "models" because they reproduce or simulate some specific features of the original anatomical entities and some of their relationships. Like any model, they are not complete and, most of all, they are not the object they represent. Any model is based on the reproduction of some features, and can only give information on those features, within a given range of expression and within given conditions. It may be hence useful to take into account that we are not working with brains and skulls but with numerical models of some of their properties.

A final remark concerns the bridge between theory and method: technique. These fields have a very relevant technical component. On the one hand, this means new opportunities and a very specialized expertise, mixing biology and evolution with informatics, statistics, and graphics. Nowadays "the morphometrician" is a specialized professional figure constantly updating on programs and tools, with a proper methodological training. A morphometrician is no more "the guy computing means and t-tests," so it is highly recommendable to avoid improvised approaches simply because programs are "user-friendly" and allow a quick visualization of a nice graphic output. Thus, if you are not a morphometrician and sometimes need a morphometric analysis, it would be better to contact somebody dedicated to this field. This strong technical component exerts an influential fascination on biologists, who are constantly tempted to dedicate more and more time and energy to these methodological aspects. Exceptions apart, in general, biologists will be good at biology, mathematicians will be good at mathematics, and engineers will be good at engineering. It may sound obvious, but the temptation to get involved in exotic things is always attractive, and the result may be odd: biologists programming algebra, mathematicians measuring genes, and engineers proposing evolutionary theories. In some cases, it may work; in others it doesn't. To avoid problems and to optimize time and energy, there is something called "multidisciplinarity," in which different specialists from different fields work together. If the biologist gets too trapped in the technical charm of numbers, the risk is to be transformed into a "glorified technician," which can be fine, or not. Depending on the priorities of the researcher, what is important to remember is that we all have a determined "life budget" to invest in our research, and therefore it is important to have a clear idea about whether our objective is biological or methodological, trying to find a stable (and personal) balance between these two interests.

7.5 Brains and Braincases

Differently from bones and other rigid anatomical elements, the brain lacks its own shape. Its form largely depends on forces associated with extrinsic anatomical components. The meninges anchor the brain to the skull, exerting a negative pressure on its surfaces. Blood fills its volume, exerting a positive internal pressure like a hydrostatic skeleton. So what we call "shape of the brain," apart from its intrinsic sulcal pattern, is nothing more than the result of a dynamic biomechanical system which involves bones, vessels, and connectives. Ontogeny and phylogeny must necessarily acknowledge such functional and structural environments, and morphogenesis is a key feature to understand balanced and unbalanced variations. This issue necessarily requires a multidisciplinary approach, in which anthropologists, neurobiologists, histologists, embryologists, geneticists, and engineers can share expertise, methods, and databases. The collaboration between evolutionary biologists and medical doctors can be particularly fruitful. Many endocranial and cerebral characters and processes are still not known for our species. Therefore, a proper study in large samples and living individuals is mandatory before investigating the same issue in few fragmentary bony remains. More importantly, many of those traits can have a medical importance. These features deal with the integration among cranial elements, with the integration between brain and bones, with vascular anatomy and functions, with geographic, sexual, and ontogenetic variations, and with the influence of environmental factors. Pathologies associated with the reciprocal relationships between the brain and braincase range from bone ossification processes to craniosynostosis (Aldridge et al. 2002; Carter and Anslow 2009). Pathological studies should go beyond the analysis of the single elements and investigate the relationships among the anatomical components, the separation between causes and consequences, the relationships between function and structure, and the existence of limits and constraints in the surrounding anatomical environment. During human evolution, encephalization involved important adjustments among cerebral and cranial elements, among the braincase and the facial block, and a marked increase in thermal loads, and such arrangements may have introduced new limits and drawbacks (Bruner et al. 2014a). An evolutionary perspective can hardly find a solution for specific medical issues, but it can certainly orientate biomedical research by offering a different point of view, which is more integrated at the anatomical level and more comprehensive in terms of biological variability. After all, evolutionary biologists and medical doctors have the same questions for different scopes, and they use the same tools for different applications. The former are interested in long-range causes, working with a comparative perspective, good theoretical expertise, and a skilled analytical capacity. The latter are interested in short-range causes, working with a perspective toward applications, a good practical expertise, and a skilled diagnostic capacity. Joined collaborations between these two professional areas can seriously improve research potential in both fields.

Morphometrics is about relationships, being able to capture hidden rules behind phenotypes. Nonetheless, numerical rules are only quantitative representations of those relationships which are made of bones, neurons, and other tissues. Evolutionary hypotheses must be based on diverse biological information and not only on numbers. In this sense, I am glad to conclude this article with an outstanding citation of D'Arcy Thompson, which is particularly illuminating considering its historical period:

> Mr. Heimann tells me that he has tried, but without success, to obtain a transitional series between the human skull and some prehuman, anthropoid type, which series (as in the case of Equide) should be found to contain other known types in direct linear sequence. It appears impossible, however, to obtain such a series, or to pass by successive and continuous gradations through such forms as Mesopithecus, Pithecanthropus, *Homo neanderthalensis*, and the lower or higher races of modern man. The failure is not the fault of our method. It merely indicates that no one straight line of descent, or of consecutive transformation, exists; but on the contrary, that among human and anthropoid types, recent and extinct, we have to do with a complex problem of divergent, rather than of continuous, variation. (On Growth and Form, 1942; p. 1095)

Acknowledgments This paper is funded by the Spanish Government (CGL2015-65387-C3-3-P). I am grateful to Naomichi Ogihara, Hideki Amano, Aida Gómez-Robles, David Costantini, David Polly, José Manuel de la Cuétara, Sofia Pereira-Pedro, Gizéh Rangel de Lázaro, Hana Píšová, Markus Bastir, Michael Masters, Sheela Athreya, Diego Rasskin-Gutman, Borja Esteve-Altava, and Simon Neubauer for their collaboration and suggestions on the topics presented in this manuscript. A special acknowledgment goes to Ralph Holloway, for providing his invaluable database, for his constant support, and for his friendship.

References

Albrecht GH, Miller JM (1993) Geographic variation in primates: a review with implications for interpreting fossils. In: Kimbel WH, Martin L (eds) Species, species concepts, and primate evolution. Plenum Press, New York, pp 123–161

Aldridge K, Marsh JL, Govier D, Richtsmeier JT (2002) Central nervous system phenotypes in craniosynostosis. J Anat 201:31–39

Allen JS, Damasio H, Grabowski TJ (2002) Normal neuroanatomical variation in the human brain: an MRI-volumetric study. Am J Phys Anthropol 118:341–358

Arsuaga JL, Martinez I, Arnold LJ, Aranburu A, Gracia-Tellez A, Sharp WD, Quam RM, Falgueres C, Pantoja-Perez A, Bischoff J, Poza-Rey E, Pares JM, Carretero JM, Demuro M, Lorenzo C, Sala N, Martinon-Torres M, Garcia N, de Velasco A (2014) Neandertal roots: cranial and chronological evidence from Sima de los Huesos. Science 344:1358–1363

Baab KL (2016) The role of neurocranial shape in defining the boundaries of an expanded *Homo erectus* hypodigm. J Hum Evol 92: 1–21

Barberini F, Bruner E, Cartolari R, Franchitto G, Heyn R, Ricci F, Manzi G (2008) An unusually-wide human bregmatic Wormian bone:

anatomy, tomographic description, and possible significance. Surg Radiol Anat 30:683–687
Bastir M, Rosas A (2005) Hierarchical nature of morphological integration and modularity in the human posterior face. Am J Phys Anthropol 128:26–34
Bastir M, Rosas A (2006) Correlated variation between the lateral basicranium and the face: a geometric morphometric study in different human groups. Arch Oral Biol 51:814–824
Bastir M, Rosas A (2009) Mosaic evolution of the basicranium in *Homo* and its relation to modular development. Evol Biol 36:57–70
Bastir M, Rosas A, Kuroe K (2004) Petrosal orientation and mandibular ramus breadth: evidence for an integrated petroso-mandibular developmental unit. Am J Phys Anthropol 123:340–350
Bastir M, Rosas A, O'Higgins P (2006) Craniofacial levels and the morphological maturation of the human skull: spatiotemporal pattern of cranial ontogeny. J Anat 209:637–654
Bayly PV, Taber LA, Kroenke CD (2014) Mechanical forces in cerebral cortical folding: a review of measurements and models. J Mech Behav Biomed Mater 29:568–581
Beaudet A, Bruner E (2017) A frontal lobe surface analysis in three archaic African human fossils: OH 9, Buia, and Bodo. Comptes Redus Palevol 16:499–507
Bellary SS, Steinberg A, Mirzayan N, Shirak M, Tubbs RS, Cohen-Gadol AA, Loukas M (2013) Wormian bones: a review. Clin Anat 26:922–927
Bienvenu T, Guy F, Coudyzer W et al (2011) Assessing endocranial variations in great apes and humans using 3D data from virtual endocasts. Am J Phys Anthropol 145:231–246
Bookstein FL (1991) Morphometric tools for landmark data. Cambridge University Press, New York
Bruner E (2004) Geometric morphometrics and paleoneurology: brain shape evolution in the genus *Homo*. J Hum Evol 47:279–303
Bruner E (2013) The species concept as a cognitive tool for biological anthropology. Am J Primatol 75:10–15
Bruner E (2014) Functional craniology, human evolution, and anatomical constraints in the Neanderthal braincase. In: Akazawa T, Ogihara N, Tanabe HC, Terashima H (eds) Dynamics of learning in Neanderthals and modern humans, vol 2. Springer, Japan, pp 121–129
Bruner E (2015) Human paleoneurology. Springer, Switzerland
Bruner E, Ripani M (2008) A quantitative and descriptive approach to morphological variation of the endocranial base in modern humans. Am J Phys Anthropol 137:30–40
Bruner E, Manzi G, Arsuaga JL (2003) Encephalization and allometric trajectories in the genus *Homo*: evidence from the Neandertal and modern lineages. Proc Natl Acad Sci U S A 100:15335–15340
Bruner E, Manzi G, Holloway R (2006) Krapina and Saccopastore: endocranial morphology in the pre-Wurmian Europeans. Period Biol 108:433–441
Bruner E, Martin-Loeches M, Colom R (2010) Human midsagittal brain shape variation: patterns, allometry and integration. J Anat 216:589–599
Bruner E, De La Cuétara JM, Holloway R (2011) A bivariate approach to the variation of the parietal curvature in the genus *Homo*. Anat Rec 294:1548–1556
Bruner E, de la Cuétara JM, Colom R, Martin-Loeches M (2012) Gender-based differences in the shape of the human corpus callosum are associated with allometric variations. J Anat 220:417–421
Bruner E, De la Cuétara JM, Masters M, Amano H, Ogihara N (2014a) Functional craniology and brain evolution: from paleontology to biomedicine. Front Neuroanat 8:19
Bruner E, Rangel de Lázaro G, de la Cuétara JM, Martín-Loeches M, Colom R, Jacobs HIL (2014b) Midsagittal brain variation and MRI shape analysis of the precuneus in adult individuals. J Anat 224:367–376
Bruner E, Amano H, de la Cuétara JM, Ogihara N (2015a) The brain and the braincase: a spatial analysis on the midsagittal profile in adult humans. J Anat 227:268–276
Bruner E, Grimaud-Hervé D, Wu X, De la Cuétara JM, Holloway R (2015b) A paleoneurological survey of *Homo erectus* endocranial metrics. Quat Int 368:80–87
Bruner E, Pereira-Pedro AS, Bastir M (2017) Patterns of morphological integration between parietal and temporal areas in the human skull. J Morphol 278:1312–1320
Carter RMS, Anslow P (2009) Imaging of the calvarium. Semin Ultrasound CT MRI 30:465–491
Cheverud JM (1982) Relationships among ontogenetic, static, and evolutionary allometry. Am J Phys Anthropol 59:139–149
Collard M, Wood B (2000) How reliable are human phylogenetic hypotheses? Proc Natl Acad Sci U S A 97:5003–5006
Cotton F, Ramirez Rozzi FR, Vallee B, Pachai C, Hermier M, Guihard-Costa AM, Froment JC (2005) Cranial sutures and craniometric points detected on MRI. Surg Radiol Anat 27:64–70
Enlow DH (1990) Facial growth. Sounders, Philadelphia
Erwin DH (2007) Disparity: morphological pattern and developmental context. Palaeontology 50:57–73
Esteve-Altava B, Rasskin-Gutman D (2014a) Beyond the functional matrix hypothesis: a network null model of human skull growth for the formation of bone articulations. J Anat 225:306–316
Esteve-Altava B, Rasskin-Gutman D (2014b) Theoretical morphology of tetrapod skull networks. C R Palevol 13:41–50
Esteve-Altava B, Marugán-Lobón J, Botella H, Bastir M, Rasskin-Gutman D (2013) Grist for Riedl's Mill: a network model perspective on the integration and modularity of the human skull. J Exp Zool 320:489–500
Foote M (1992) Rarefaction analysis of morphological and taxonomic diversity. Paleobiology 18:1–16
Gómez-Robles A, Hopkins WD, Sherwood CC (2014) Modular structure facilitates mosaic evolution of the brain in chimpanzees and humans. Nat Commun 5:4439
Goriely A, Geers MGD, Holzapfel GA, Jayamohan J, Jérusalem A, Sivaloganathan S, Squier W, Van Dommelen JAW, Waters S, Kuhl E (2015) Mechanics of the brains: perspectives, challenges, and opportunities. Biomech Model Mechanobiol 14:931–965
Gould SJ (1966) Allometry and size in ontogeny and phylogeny. Biol Rev Camb Philos Soc 41:587–638
Gould SJ (1977) Ontogeny and phylogeny. Belknap Press, Harvard University Press, Cambridge, MA
Gunz P (2015) Computed tools for paleoneurology. In: Bruner E (ed) Human paleoneurology. Springer, Cham, pp 39–55
Gunz P, Harvati K (2007) The Neanderthal "chignon": variation, integration, and homology. J Hum Evol 52:262–274
Gunz P, Mitterœcker P (2013) Semilandmarks: a method for quantifying curves and surfaces. Hystrix 24:103–109
Hallgrimsson B, Lieberman DE, Liu W, Ford-Hutchinson AF, Jirik FR (2007) Epigenetic interactions and the structure of phenotypic variation in the cranium. Evol Dev 9:76–91
Hammer Ø, Ryan P, Harper D (2001) PAST: paleontological statistics software package for education and data analysis. Palaeontol Electron 4:9
Hetch E, Stout D (2015) Techniques for studying brain structure and function. In: Bruner E (ed) Human paleoneurology. Springer, Switzerland, pp 209–224
Hilgetag CC, Barbas H (2005) Developmental mechanics of the primate cerebral cortex. Anat Embryol 210:411–417
Holliday TW (2003) Species concepts, reticulation, and human evolution. Curr Anthropol 44:653–673
Holloway RL (1981) Exploring the dorsal surface of hominoid brain endocasts by stereoplotter and discriminant analysis. Philos Trans R Soc Lond B 292:155–166

Holloway RL, Broadfield DC, Yuan MS (2004) Brain endocasts: the paleoneurological evidence. Wiley, Hoboken

Jackson DA (1993) Stopping rules in principal components analysis: a comparison of heuristical and statistical approaches. Ecology 74: 2204–2214

Jolliffe IT (2002) Principal component analysis. Springer, New York

Klingenberg CP (2010) Evolution and development of shape: integrating quantitative approaches. Nat Rev Genet 11:623–635

Klingenberg CP (2013) Cranial integration and modularity: insights into evolution and development from morphometric data. Hystrix 24:43–58

Kobayashi Y, Matsui T, Haizuka Y, Hirai N, Matsumura G (2014a) Cerebral sulci and gyri observed on macaque endocasts. In: Akazawa T, Ogihara N, Tanabe HC, Terashima H (eds) Dynamics of learning in Neanderthals and modern humans, vol 2. Springer, Japan, pp 131–137

Kobayashi Y, Matsui T, Haizuka Y, Hirai N, Matsumura G (2014b) The coronal suture as an indicator of the caudal border of the macaque monkey prefrontal cortex. In: Akazawa T, Ogihara N, Tanabe HC, Terashima H (eds) Dynamics of learning in Neanderthals and modern humans, vol 2. Springer, Japan, pp 139–143

Lawing AM, Polly PD (2010) Geometric morphometrics: recent applications to the study of evolution and development. J Zool 280:1–7

Lieberman DE, Pearson OM, Mowbray KM (2000) Basicranial influence on overall cranial shape. J Hum Evol 38:291–315

Manzi G, Vienna A, Hauser G (1996) Developmental stress and cranial hypostosis by epigenetic trait occurrence and distribution: an exploratory study on the Italian Neandertals. J Hum Evol 30: 511–527

Manzi G, Gracia A, Arsuaga JL (2000) Cranial discrete traits in the Middle Pleistocene humans from Sima de los Huesos (Sierra de Atapuerca, Spain). Does hypostosis represent any increase in "ontogenetic stress" along the Neanderthal lineage? J Hum Evol 38: 425–446

Martin RD, Barbour AD (1989) Aspects of line-fitting in bivariate allometric analyses. Folia Primatol 53:65–81

Masters M (2012) Relative size of the eye and orbit: an evolutionary and craniofacial constraint model for examining the etiology and disparate incidence of juvenile-onset myopia in humans. Med Hypotheses 78:649–656

Masters M, Bruner E, Queer S et al (2015) Analysis of the volumetric relationship among human ocular, orbital and fronto-occipital cortical morphology. J Anat 227:460–473

McCollum MA, Sherwood CC, Vinyard CJ et al (2006) Of muscle-bound crania and human brain evolution: the story behind the MYH16 headlines. J Hum Evol 50:232–236

McGhee GR (2007) The geometry of evolution: adaptive landscapes and theoretical morphospaces. Cambridge University Press, Cambridge

Mitteroecker P, Bookstein F (2011) Linear discrimination, ordination, and the visualization of selection gradients in modern morphometrics. Evol Biol 38:100–114

Mitteroecker P, Gunz P (2009) Advances in geometric morphometrics. Evol Biol 36:235–247

Moss ML, Young RW (1960) A functional approach to craniology. Am J Phys Anthropol 18:281–292

Neubauer S (2015) Human brain evolution: ontogeny and phylogeny. In: Bruner E (ed) Human paleoneurology. Springer, Cham, pp 95–120

Neubauer S, Gunz P, Hublin J-J (2009) The pattern of endocranial ontogenetic shape changes in humans. J Anat 215: 240–255

Neubauer S, Gunz P, Weber GW, Hublin JJ (2012) Endocranial volume of Australopithecus africanus: new CT-based estimates and the effects of missing data and small sample size. J Hum Evol 62: 498–510

Olson EC, Miller RL (1958) Morphological integration. University Chicago Press, Chicago

Pereira-Pedro S, Masters M, Bruner E (2017) Shape analysis of spatial relationships between orbitoocular and endocranial structures in modern humans and fossil hominids. J Anat. https://doi.org/10.1111/joa.12693

Plavcan JM, Cope DA (2001) Metric variation and species recognition in the fossil record. Evol Anthropol 10:204–222

Polly PD, Lawing AM, Fabre AC, Goswami A (2013) Phylogenetic principal components analysis and geometric morphometrics. Hystrix 24:33–41

Ribas GC, Yasuda A, Ribas EC, Nishikuni K, Rodrigues AJ Jr (2006) Surgical anatomy of microneurosurgical sulcal key points. Neurosurgery 59:177–210

Richtsmeier JT, Cheverud JM, Lele S (1992) Advances in anthropological morphometrics. Annu Rev Anthropol 21:283–305

Richtsmeier JT, DeLeon VB, Lele SR (2002) The promise of geometric morphometrics. Am J Phys Anthropol S35:63–91

Rightmire GP (1998) Evidence from facial morphology for similarity of Asian and African representatives of *Homo erectus*. Am J Phys Anthropol 106:61–85

Rightmire GP (2013) *Homo erectus* and Middle Pleistocene hominins: brain size, skull form, and species recognition. J Hum Evol 65:223–252

Rilling JK (2008) Neuroscientific approaches and applications within anthropology. Am J Phys Anthropol 137:2–32

Rohlf FJ (1999) Shape statistics: procrustes superimpositions and tangent spaces. J Classif 16:197–223

Rosas A, Peña-Melián A, García-Tabernero A, Bastir M, De La Rasilla M (2014) Temporal lobe sulcal pattern and the bony impressions in the middle cranial fossa: the case of the El Sidrón (Spain) Neandertal sample. Anat Rec 297:2331–2341

Roy K, Foote M (1997) Morphological approaches to measuring biodiversity. Trends Ecol Evol 12:277–281

Schluter D (1996) Adaptive radiation along genetic lines of least resistance. Evolution 50:1766–1774

Shea BT (1989) Heterochrony in human evolution: the case for neoteny reconsidered. Am J Phys Anthropol 32:69–101

Shea BT (1992) Developmental perspective on size change and allometry in evolution. Evol Anthropol 1:125–134

Simpson GG (1944) Tempo and mode in evolution. Columbia University Press, New York

Slice DE (2007) Geometric morphometrics. Annu Rev Anthropol 36: 261–281

Spoor F, Jeffrey N, Zonneveld F (2000) Using diagnostic radiology in human evolutionary studies. J Anat 197:61–76

Stedman HH, Kozyak BW, Nelson A, Thesier DM, Su LT, Low DW, Bridges CR, Shrager JB, Minugh-Purvis N, Mitchell MA (2004) Myosin gene mutation correlates with anatomical changes in the human lineage. Nature 428:415–418

Stringer C (2012) The status of *Homo heidelbergensis* (Schoetensack 1908). Evol Anthropol 21:101–107

Tallinen T, Chung JY, Rousseau F, Girard N, Lefèvre J, Mahadevan L (2016) On the growth and form of cortical convolutions. Nat Phys 12:588–593

Tattersall I (1986) Species recognition in human paleontology. J Hum Evol 15:165–175

Thompson D'A (1942) On growth and form. Cambridge University Press, Cambridge

Toro R, Burnod Y (2005) A morphogenetic model for the development of cortical convolutions. Cereb Cortex 15:1900–1913

Wagner GP (1984) On the eigenvalue distribution of genetic and phenotypic dispersion matrices: evidence for a nonrandom organization of quantitative character variation. J Math Biol 21:77–95

Wagner GP, Altenberg L (1996) Complex adaptations and the evolution of evolvability. Evolution 50:967–976

Weber GW (2015) Virtual anthropology. Am J Phys Anthropol 156: 22–42
Wills MA (2001) Morphological disparity: a primer. In: Adrain JM, Edgecombe GD, Lieberman BS (eds) Fossils, phylogeny, and form. Kluwer Academic, New York, pp 55–144
Wood B, Collard M (1999) The changing face of genus *Homo*. Evol Anthropol 8:195–207
Wu X, Bruner E (2016) The endocranial anatomy of Maba 1. Am J Phys Anthropol 160:633–643
Zelditch ML, Swidersky DL, Sheets HD, Fink WL (2004) Geometric morphometrics for biologists. Elsevier, San Diego
Zollikofer CPE, Ponce de León MS (2005) Virtual reconstruction: a primer in computer-assisted paleontology and biomedicine. Wiley-Liss, New York

Landmarking Brains

8

Aida Gómez-Robles, Laura D. Reyes, and Chet C. Sherwood

Abstract

Geometric morphometric techniques are extensively used in paleontology and evolutionary biology to describe and quantify shape variation. These approaches, however, are rarely used in neuroscience. Approaches emphasizing qualitative anatomical description and volumetric measurements of brain structures dominate evolutionary neuroscience, whereas automated computing-intensive approaches are the norm in the field of human neuroimaging. Such approaches often are not compatible with formal quantitative assessments of brain evolution (anatomical descriptions), overlook fundamental aspects of shape variation (volumetric measurements), or involve intensive processing of neuroimaging scans, which can complicate straightforward neurobiological interpretation of results. Here we review how geometric morphometrics can provide a useful toolkit to analyze brain variation in a comparative and evolutionary context. We suggest different methodological alternatives within geometric morphometrics, highlighting their advantages and disadvantages. We also discuss how strengths of automated neuroimaging techniques can be combined with geometric morphometric analytical tools.

Keywords

Comparative neuroanatomy • Geometric morphometrics • Human brain evolution • Neocortex

8.1 Introduction

The study of human brain evolution requires a detailed evaluation of anatomical variation across extant and extinct species. Paleoneurology focuses on endocranial size and shape changes across hominins, but endocasts retain very limited information regarding brain organization, which only can be inferred from endocranial morphology and from the imprints of sulcal anatomy. For this reason, studies of the hominin fossil record can gain greater insight when supplemented with studies of brain variation among modern humans and other living primates. In this regard, two approaches are possible and complementary. The first one, which is commonly used in evolutionary neuroscience, involves macroevolutionary comparisons of different species, ranging from smaller-scale taxonomic groups (great

A. Gómez-Robles (✉)
Department of Genetics, Evolution and Environment, University College London, Gower St., London, WC1E 6BT, UK

Center for the Advanced Study of Human Paleobiology, Department of Anthropology, The George Washington University, 800 22nd St NW, Washington, DC 20052, USA
e-mail: a.gomez-robles@ucl.ac.uk

L.D. Reyes • C.C. Sherwood
Center for the Advanced Study of Human Paleobiology, Department of Anthropology, The George Washington University, 800 22nd St NW, Washington, DC 20052, USA
e-mail: laura.d.reyes08@gmail.com; sherwood@gwu.edu

E. Bruner et al. (eds.), *Digital Endocasts*, Replacement of Neanderthals by Modern Humans Series,
DOI 10.1007/978-4-431-56582-6_8

apes) to larger-scale groups (primates and mammals) (Finlay and Darlington 1995; Barton and Harvey 2000; de Winter and Oxnard 2001; Oxnard 2004; Smaers and Soligo 2013). This approach can help us understand the evolution of species-specific brain structural organization, functional and developmental constraints, as well as the evolutionary patterns underlying them. The second approach, which is far less common in evolutionary neuroscience, involves micro-evolutionary comparisons based on large intraspecific samples (Gómez-Robles et al. 2013, 2014, 2015, 2016). This second approach is necessary to define evolutionary mechanisms that operate within populations and that may influence evolutionary diversification. An integration of both approaches would entail an in-depth comparison of intraspecific patterns of variation across different species and an assessment of how these species-specific patterns and mechanisms have evolved through time (Steppan et al. 2002). This research avenue, however, remains largely unexplored due to the difficulties of obtaining representative sample sizes of a reasonably large number of species. Whichever the approach, a challenge remains regarding the best way to quantify and analyze cerebral anatomical variation. In this contribution, we review different methodologies to measure brain anatomical variation, ranging from classic descriptive approaches to fully automated and computationally intensive modern neuroimaging techniques. After reviewing the advantages and disadvantages of these options, we focus on geometric morphometric techniques as a promising and relatively underused approach to link micro- and macroevolution of the brain, emphasizing some practical issues.

8.2 Classic Approaches to Studying Brain Variation

The most basic way to assess brain variation involves simple anatomical descriptions and scoring systems, which can be based on direct observations of postmortem brains, on medical scans, such as magnetic resonance imaging (MRI) scans, or on virtual and physical models obtained from them. Descriptive scoring systems are often used to characterize patterns of sulcal variation (Ono et al. 1990; Chiavaras and Petrides 2000; Zlatkina and Petrides 2010), and this descriptive approach can be used to compare species or individuals within the same species. Although this is simple and straightforward, it entails serious drawbacks, including the unavoidable subjectivity of qualitative descriptions and the difficulty in quantifying variation, which cannot be accurately measured even when scored on a scale. Not being able to quantify anatomical similarities and differences hampers further evaluations beyond the descriptions themselves, thus precluding additional analyses of evolutionary mechanisms and scenarios.

Another approach that is commonly used in evolutionary neuroscience consists of the comparison of volumetric measurement of different brain structures, either based on neuroimaging scans or histological sections (Stephan et al. 1981). The use of these quantitative variables in a multivariate context opens numerous downstream options to explore scenarios of brain evolution, including possible developmental constraints on evolutionary change (Finlay and Darlington 1995), lineage-specific patterns of variation (Smaers and Soligo 2013), and covariation networks (Barton and Harvey 2000; Whiting and Barton 2003; Smaers et al. 2011). However, size—as inferred from volume—is not the only relevant parameter of variation, and important aspects of anatomical variation can be contained in the shape of different brain structures. The shape of these structures (including their position and orientation within the whole brain) is not reflected in volumetric measures, but it can have important implications in enhancing or hindering connectivity between different brain regions and thus have fundamental functional implications.

8.3 Neuroimaging Techniques

The development of modern neuroimaging techniques is responsible for the explosion in the number of studies concerning the structure of the brain, mostly focused on humans. At the structural level, these studies rely on MRI scans. The derivations of the study of MRI scans are plentiful, and they include the study of different parameters including regional volumes (Allen et al. 2002), cortical thickness (Fischl and Dale 2000), surface areas (Van Essen et al. 2012), sulcal anatomy (Keller et al. 2007), and patterns of gray and white matter distribution (Sowell et al. 2002), to name a few. In addition, some methods have been developed to measure connectivity between different brain areas, such as diffusion tensor imaging (DTI) and diffusion spectrum imaging (DSI). Although extremely useful to explore human brain variation, these approaches are not free of limitations and biases, which may remain unnoticed due to the automated processing associated with many of these techniques and the lack of control over parts of the process. Furthermore, commonly used neuroscience software can yield significant differences when processing the same brain samples with different versions, workstations, or operating systems (Gronenschild et al. 2012). These observations demonstrate that neuroimage processing can have a significant effect on the evaluated patterns of variation, which is particularly problematic because many result in effects that are significant, although quantitatively small.

Because of the enormous anatomical variation of the human neocortex, the quantification of its morphology tends to rely on registration of individual brains to a common template coordinate space such that different regions can be studied in spatial correspondence across different individuals (see section "Beyond Landmarks" below). This method is used, for example, when parcellating the neocortex into different cortical regions according to automated algorithms that allow for comparisons in terms of surface area, cortical thickness, and volume (Desikan et al. 2006). One undesirable implication of this step is that variation in the sample is disregarded and its presence can affect the validity of the results obtained using these automated tools. For example, if sulcal patterns in a given individual are extremely different from the template, the resulting parcellation following registration will not observe individual sulcal patterns but instead will be forced to match those of the template (Fig. 8.1). This may result in incorrect parcellations, since many software packages cannot effectively handle variation in sulcal morphology. Variation, which is the most important parameter when studying evolution, is thus minimized or removed when using such automated tools and protocols to ensure comparable results across individuals.

Problems can be even more serious when extending analyses to nonhuman species. Because most available software and protocols are optimized to analyze human brains, the study of nonhuman primates often requires case-specific adjustments. The greater the variation across different species to be analyzed, the greater the need for species-specific protocols that cannot be standardized across the whole study. This problem is even further exaggerated because studies of different species often pool scans obtained at different institutions with different types of equipment and protocols. In practical terms, this means that interspecific studies incorporate not only biological variation across species but also the error associated with significant methodological differences in scan acquisition and processing both across and within species. It can be argued that methodological differences are likely to be negligible in comparison with biological differences across species, but it is important to remember that automated neuroimaging approaches are extremely sensitive to nonbiological variation and that extensive scan acquisition and processing differences can have substantial effects on the assessment of patterns of variation.

8.4 Geometric Morphometrics

Geometric morphometrics can be defined as the statistical analysis of shape based on the coordinates of homologous points or landmarks (Adams et al. 2013). This approach is extensively used in evolutionary biology (including

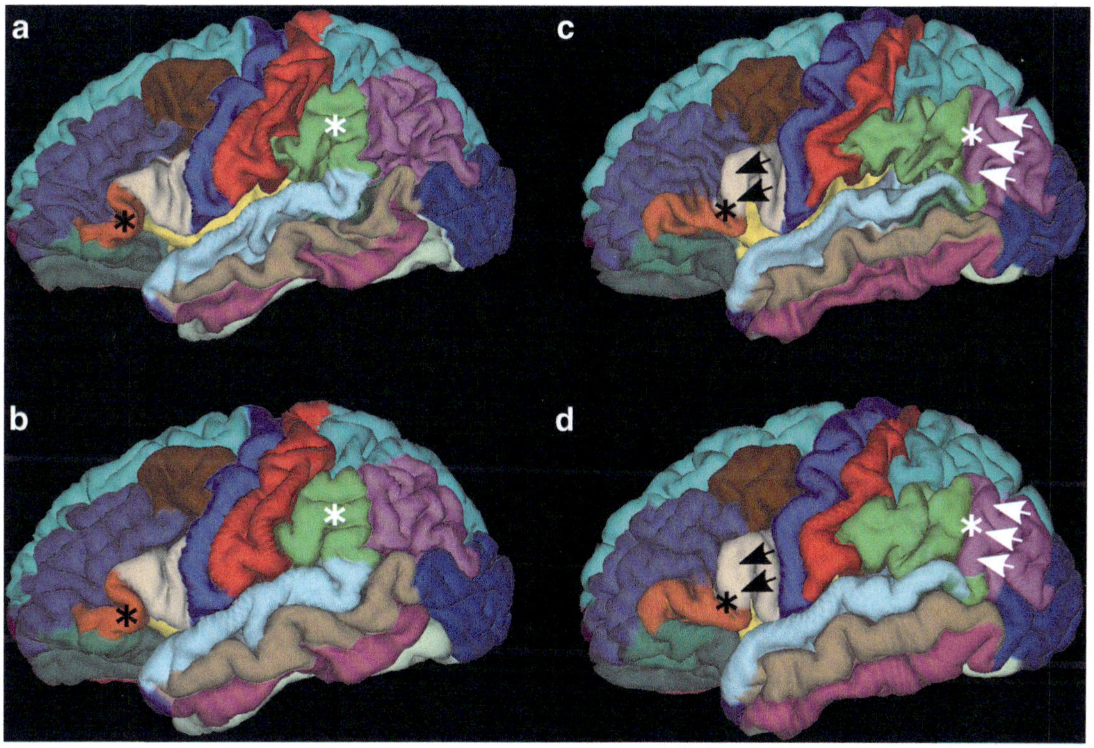

Fig. 8.1 FreeSurfer parcellation errors in highly variable human brains. (**a**, **b**) parcellated cortex in individuals with a common brain anatomy in a mid-thickness cortical reconstruction (**a**) and at the pial surface (**b**). *Black asterisks* indicate the *pars triangularis* (in *orange*) and *white asterisks* indicate the supramarginal gyrus (in *bright green*). (**c**, **d**) parcellated cortex in an individual with a less common cortical anatomy, with a large *pars triangularis* and a very long superior temporal gyrus. As a result, a substantial proportion of the *pars triangularis* is misclassified as part of the *pars opercularis* (in *beige* and indicated by the *black arrow heads*), and a substantial proportion of the supramarginal gyrus is misclassified as part of the angular gyrus (in *pink* and indicated by the *white arrow heads*)

paleontology, biological anthropology, and comparative anatomy) to assess patterns of anatomical variation across individuals and species, but much more seldom used in comparative neuroanatomy or in the study of normal brain variation and development (but see Bookstein et al. 2001; Bruner et al. 2010). We believe that the broad availability of fully automated techniques to neuroscientists has hampered the dissemination of geometric morphometric approaches to human neuroscience, whereas evolutionary neuroscience has mostly relied on volumetric measures due to the perception that the volume of different brain structures serves a close proxy of neuron numbers.

Geometric morphometrics, however, can offer a number of benefits to evolutionary neuroscientists. A geometric morphometric approach allows researchers to evaluate not only the size of different brain regions but also their shape and spatial relations to each other. As stated previously, some anatomical details can have important functional implications, as they can be macromorphological manifestations of microstructural changes that determine brain function, such as gray matter neuronal and glial numbers and distributions, and short- and long-distance connectivity in the underlying white matter. The quantitative descriptors of shape variation that derive from these analyses are readily available to more complex downstream analyses, such as the analysis of covariation networks (Gómez-Robles et al. 2014), quantitative genetics, which can shed light on heritability (Gómez-Robles et al. 2015, 2016), or phylogenetic analyses that can be used to describe large-scale evolutionary patterns (see Aristide et al. 2016 for an example of this methodological approach applied to endocranial variation in New World monkeys).

Geometric morphometric approaches rely on the identification of homologous landmarks, which are based on anatomical criteria and definitions. This poses some advantages and disadvantages. It can be argued that automated techniques do not have to deal with inter- or intraobserver error. However, this does not mean that these techniques are unbiased. This problem has been recognized in the geometric morphometrics literature as a trade-off between the use of homology-free techniques, in which a certain number of semilandmarks are placed according to some algorithms, and homology-based approaches, in which landmarks are placed according to anatomical criteria (Klingenberg 2008; Polly 2008). At a theoretical level, different preferences may exist concerning whether human observer biases are more or less desirable than automated biases. At a more practical level, it has been demonstrated that divergent results are obtained when analyzing the same dataset with these different approaches (Gómez-Robles et al. 2011; Gonzalez et al. 2016). We therefore suggest that different methodological approaches should be tested and consistencies sought whenever possible, ideally using both automated processing tools and homology-based methods to test the same hypotheses (Gómez-Robles et al. 2014).

A geometric morphometric approach is limited by the ability to identify anatomically homologous landmarks on the brain, especially on the surface of the human neocortex, which is characterized by an extreme degree of gyrification and interindividual variation (Zilles and Amunts 2013). Although some regions that can be identified through gross anatomy and/or cytoarchitecture have well established homology across different brains (e.g., Brodmann areas), identifying homologous landmarks is a challenging task. The result of this is that some areas of the brain can remain relatively underrepresented when studies are based on homologous landmarks. One possible way to overcome this limitation is based on the use of homologous curves and surfaces, which is usually accomplished through the use of sliding semilandmarks.

8.5 Landmarks on the Brain

Unlike the skull, the brain consists of soft tissue, which poses additional challenges within a geometric morphometric context. Although brain size and structure change over the course of development (Kochunov et al. 2005), the general sulcal configuration of the cerebral cortex remains fairly stable once established, and age-related changes in the anatomy of healthy brains are not greater in magnitude than age- or activity-dependent bone remodeling. Therefore, even if the brain is a soft tissue, its anatomy is readily subjected to geometric morphometric analyses. Indeed, other soft tissues have been studied using the same methodological approach (Klingenberg et al. 2010). However, conservation and handling of soft tissue requires particular chemicals and laboratory facilities, which limits the availability of postmortem brain tissue. Direct landmarking of postmortem brains (e.g., using a Microscribe digitizer) is highly impractical and potentially damaging for delicate brain tissue. Additionally, postmortem brains tend to suffer from different degrees of anatomical distortion when they are extracted from the skull and undergo chemical fixation, which can strongly interfere with the evaluation of biological patterns of shape variation.

For these reasons, the most convenient way to place landmarks on brains is through the use of in vivo collected MRI scans (Aldridge 2011; Gómez-Robles et al. 2013). The scans can be used to reconstruct three-dimensional models of the cortical surface and subcortical structures, or landmarks can be placed directly on MRI sections (Bookstein et al. 2001; Gómez-Robles et al. 2014). Obtaining landmarks from MRI sections is generally straightforward and can be performed with different software packages (e.g., MIPAV, McAuliffe et al. 2001) as long as anatomical locations of interest can be identified on 2D sections. Although some geometric morphometric studies have focused on particular aspects of 2D organization such as the midsagittal profile of the cerebral cortex (Bruner et al. 2010) or the shape of the corpus callosum (Bookstein et al. 2001), most studies tend to analyze brain anatomy using 3D configurations of landmarks, even if they are obtained from 2D sections. Reconstruction of the cortical surface benefits from the use of specialized neuroimaging software packages. Among them, BrainVisa (Cointepas et al. 2001) and FreeSurfer (Fischl 2012) offer extraordinary results for cortical surface reconstruction, and FSL (Jenkinson et al. 2012) is particularly useful in reconstructing subcortical structures. Because these packages are optimized to work with human brains, adjustments are usually necessary when analyzing other species. Once 3D models are obtained, it is possible to work with them following the classic 3D geometric morphometric workflow.

Because the use of geometric morphometrics to study the brain is relatively rare, landmark definition remains an issue that requires optimization (Fig. 8.2). The resolution of MRI scans does not allow for observation of detailed cytoarchitectural subdivisions, so landmarks are often defined based on estimates of repeatable anatomical locations (e.g., the centroid of the amygdala, the most anterior point of the caudate nucleus, etc.). These definitions suffer from the well-known problem of type 3 landmarks, which are those anatomical locations that do not have a precise definition and, therefore, cannot be unmistakably identified (Bookstein 1997). This problem, however, is very common in geometric morphometric studies, and it is not exclusive to brain landmarks. Sulcal landmarks often have more accurate definitions (e.g., the intersection between the precentral sulcus and the inferior frontal sulcus), but locating them is usually far from straightforward due to the anatomical variability of the brain. For example, it is very common for some sulci to be interrupted, branched, and/or formed by different segments, which may make landmark identification challenging.

These difficulties in identifying brain landmarks result in higher error rates in landmark placement than those observed in cranial studies. It has been recently suggested, however, that a refinement of protocols and landmark definitions can help decrease those error rates in spite of the inherent variability and intrinsic difficulty in landmarking brains (Chollet et al. 2014). It has also been noted that certain landmarks are subjected to higher error rates than others. Cortical landmarks (especially those located on sulci within neocortical association regions) are the most variable ones due to biological factors (Fig. 8.3), but they also have the highest error rates. In any case, preliminary inter- and intra-rater error studies can be a useful tool to determine those landmarks whose amount of error can obscure biological signals of interest and therefore should be removed from final analyses.

8.6 Selecting Landmarks

As in other geometric morphometric studies, the selection of a landmark configuration in neuroanatomical studies will depend on the question at hand. While some studies will focus on the anatomy of particular brain regions, others will aim at describing the general shape of the whole brain. Landmark configurations on the neocortex aim at describing sulcal variation, which is the result of an interaction of developmental and physical processes that are not completely understood. For example, recent studies have shown that physical forces can have a predominant role in driving gyrification (Mota and Herculano-Houzel 2015; Tallinen et al. 2016), which might make us expect certain randomness in folding patterns. However, gyrification patterns are in general quite consistent in many mammalian species with relatively simple folding patterns. Even in species with complex sulcal patterns, such as humans, certain sulci—those that develop early during ontogenetic development, surround primary sensory and motor areas, and are shared with other species—are relatively consistent (Zilles and Amunts 2013). Therefore, a sound strategy to select landmarks in human brains is to focus on those evolutionarily and developmentally primary sulci.

Our previous studies have focused on those sulci, including, among others, the central sulcus and Sylvian fissure, the precentral sulcus, the latero-orbital and inferior frontal sulci, and the superior temporal and the parieto-occipital sulcus (Gómez-Robles et al. 2013, 2014, 2015, 2016). Some of these sulci, such as the central sulcus, the parieto-occipital sulcus, and the latero-orbital sulcus (or the central, lunate, and fronto-orbital sulci in chimpanzees), have fairly uninterrupted courses and easy-to-identify landmarks. Figure 8.3 shows that dispersal around these landmark positions is the lowest across all cortical landmarks. Other sulci, however, show complex anatomies and frequent ramifications and subdivisions that greatly increase variation. Among them, the Sylvian fissure and the superior temporal sulcus show the greatest amounts of variation (Fig. 8.3). In these cases, some criteria are required to standardize landmark placement. These criteria can be anatomical (i.e., when sulci show ramifications, the most anterior one is used), but anatomical criteria can give the same importance to developmentally primary sulci—deep sulci appearing early during development—and to secondary and tertiary sulci—swallow dimples appearing later during development—simply because they happen to be located in the same position. For this reason, we recommend placing landmarks on the main course of sulci when they show ramifications, which

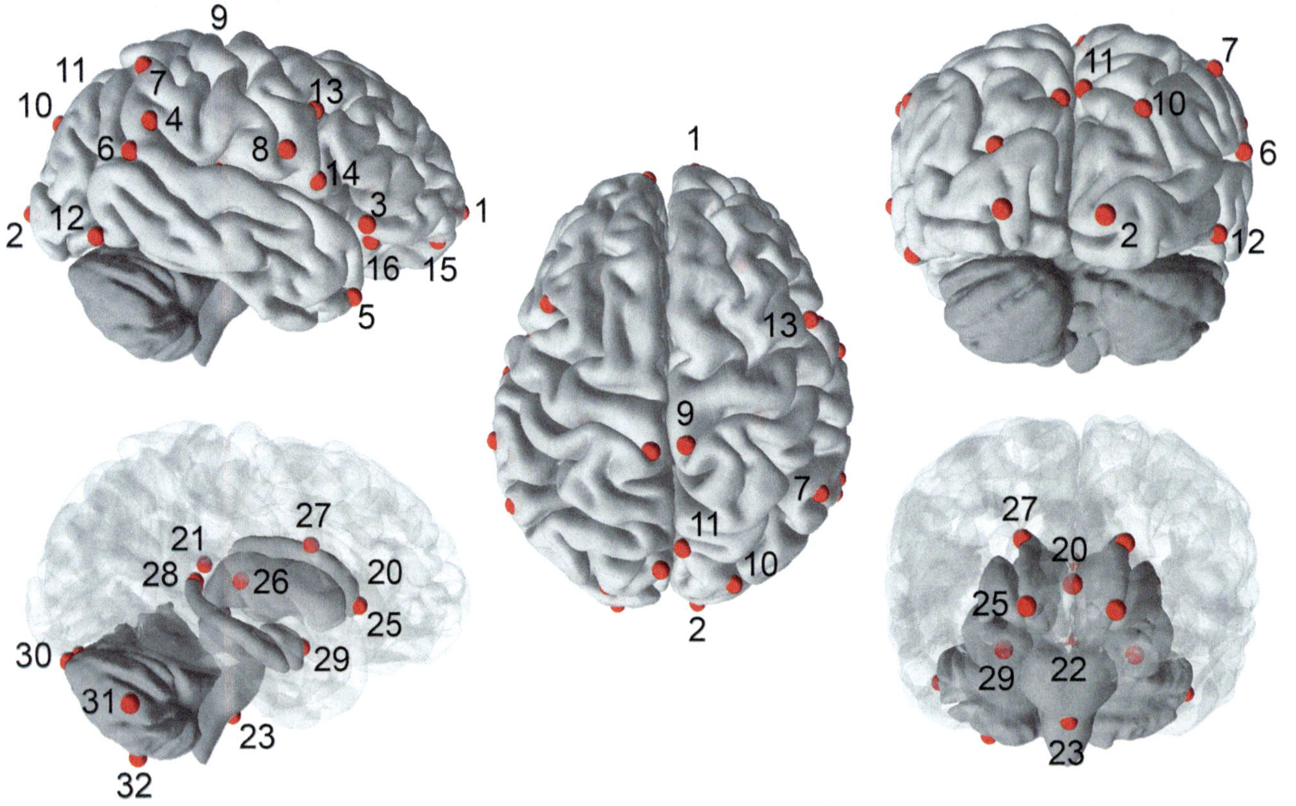

Fig. 8.2 Examples of cortical and subcortical brain landmarks shown in *left* lateral view (with opaque and transparent cortical surface, *left*), dorsal view (*opaque*, *center*), caudal view (*opaque*, *right top*), and rostral view (*transparent*, *right bottom*). *1* Frontal pole, *2* occipital pole, *3* anterior end of the Sylvian fissure, *4* posterior end of the Sylvian fissure, *5* anterior end of the superior temporal sulcus, *6* inflection point between the horizontal segment and the ascending segment of the superior temporal sulcus, *7* most posterior point of the superior temporal sulcus, *8* inferior termination of the central sulcus, *9* superior termination of the central sulcus, *10* intersection between the intraparietal sulcus and the transverse occipital sulcus, *11* intersection of the parieto-occipital sulcus with the midline, *12* occipital notch, *13* intersection of the inferior frontal sulcus with the precentral sulcus, *14* inferior end of the precentral sulcus, *15* anterior end of the latero-orbital sulcus, *16* posterior end of the latero-orbital sulcus, *17*, *18*, *19* insular landmarks (not shown), *20* centroid of the genu of the corpus callosum, *21* centroid of the splenium of the corpus callosum, *22* superior aspect of the pons, *23* inferior aspect of the pons, *24* point where superior cerebellar peduncles meet (not shown), *25* most anterior point of the caudate nucleus, *26* most posterior point of the putamen nucleus, *27* most superior and central point of the caudate nucleus, *28* most superior point of the hippocampus, *29* centroid of the anterior aspect of the amygdala, *30* most posterior point of the cerebellum, *31* most lateral point of the cerebellum, *32* most inferior point of the cerebellum

can be identified by comparing pial surface models with inflated models that can be obtained and visualized using different software packages, including BrainVisa (Cointepas et al. 2001), FreeSurfer (Fischl 2012), Caret (Van Essen et al. 2001), etc. (see Fig. 8.4). Comparisons with automatically parcellated models can be also useful, although these comparisons should be employed critically, considering the potential for parcellation errors (see Fig. 8.1).

Even after carefully considering these caveats and observing these recommendations, some sulci can be too variable to be accurately landmarked, particularly when describing human brains. In our experience, the superior frontal region remains particularly difficult, as the superior frontal sulcus shows an extreme degree of variation with respect to other sulci. Variation within the temporal lobe is also particularly hard to describe through classic landmarks, since the superior and inferior temporal sulci show extensive variation and lack clear homologous points across their course through the temporal lobe. Variation in certain regions within the medial surface of the cerebral cortex, such as the cuneus and precuneus, can be accurately described with landmarks, whereas the anterior cingulate cortex is more challenging due to the variability of the anterior aspect of the cingulate sulcus. Subcortical landmarks, which can be digitized on 3D models of subcortical structures or on MRI sections, tend to show substantially less variation than cortical landmarks (Fig. 8.3).

Apart from describing the variation of the whole brain, certain studies may need to focus on the anatomy of certain regions. Indeed, some studies have isolated certain sulci to measure variation across different species [e.g., central sulcus variation across primates (Hopkins et al. 2014)], which

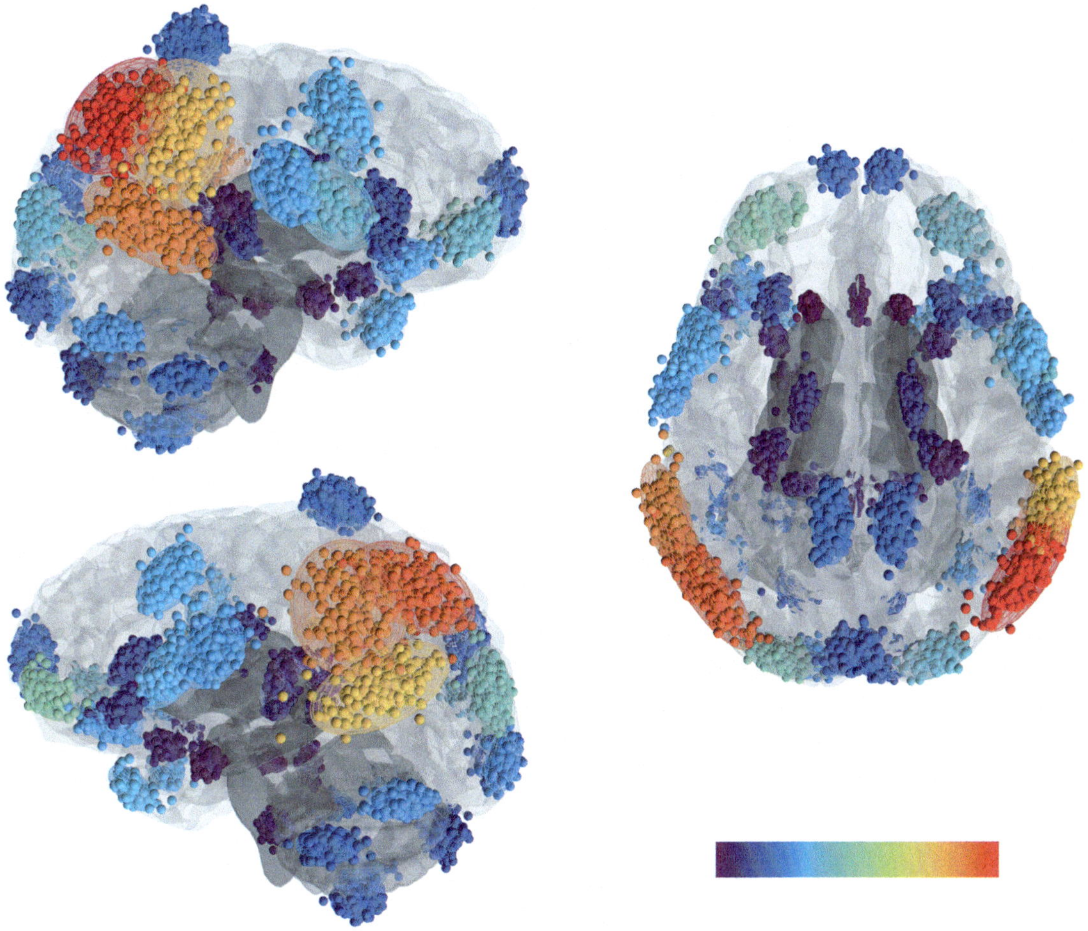

Fig. 8.3 Procrustes-superimposed landmarks in a sample of 189 human brains overlaid on a human brain model. Landmarks are color coded to represent dispersion around their location in the mean or consensus shape of the complete sample (*red* represents high dispersion and *purple* represents low dispersion) (See text for discussion on different landmarks)

can be accomplished through geometric morphometrics or through other quantitative methods of analysis.

8.7 Beyond Landmarks

The study of brain anatomical variation through geometric morphometric does not need to rely solely on landmarks but can alternatively be based on the analyses of curves and surfaces through the use of semilandmarks. This approach has been extensively and successfully applied to the study of hominin and hominoid endocasts (Neubauer et al. 2010; Gunz et al. 2010), but it remains unexplored in the study of actual brain morphology. In practical terms, using semilandmarks to capture anatomical variation of the brain can be based on the selection of homologous regions, although the sulcal and gyral complexity of the brain poses certain challenges, such as the demarcation of areas from surface morphology that share underlying microstructural features (Sherwood et al. 2003; Keller et al. 2012) and the use of landmark patches that will fully describe this sulcal and gyral complexity.

A relatively direct methodological approach is defining the anatomy of sulci, which can be considered homologous curves, based on semilandmarks. This approach can be successful for those sulci that are highly conserved in different species (or, at least, in different species within a given group). The most problematic derivative of this approach is again the study of human brains, where interindividual variation is pronounced and includes ramification, subdivision, and anastomosis of sulci, which interfere with recognizing the basic course of each sulcus. As with using homologous landmarks, it is important to define clear criteria to identify sulcal semilandmarks, especially when such a high degree of anatomical variability is to be found. The use of surface semilandmarks can be problematic due to the intrinsic difficulty in defining anatomically homologous regions and to accurately place semilandmarks along sulci, which can

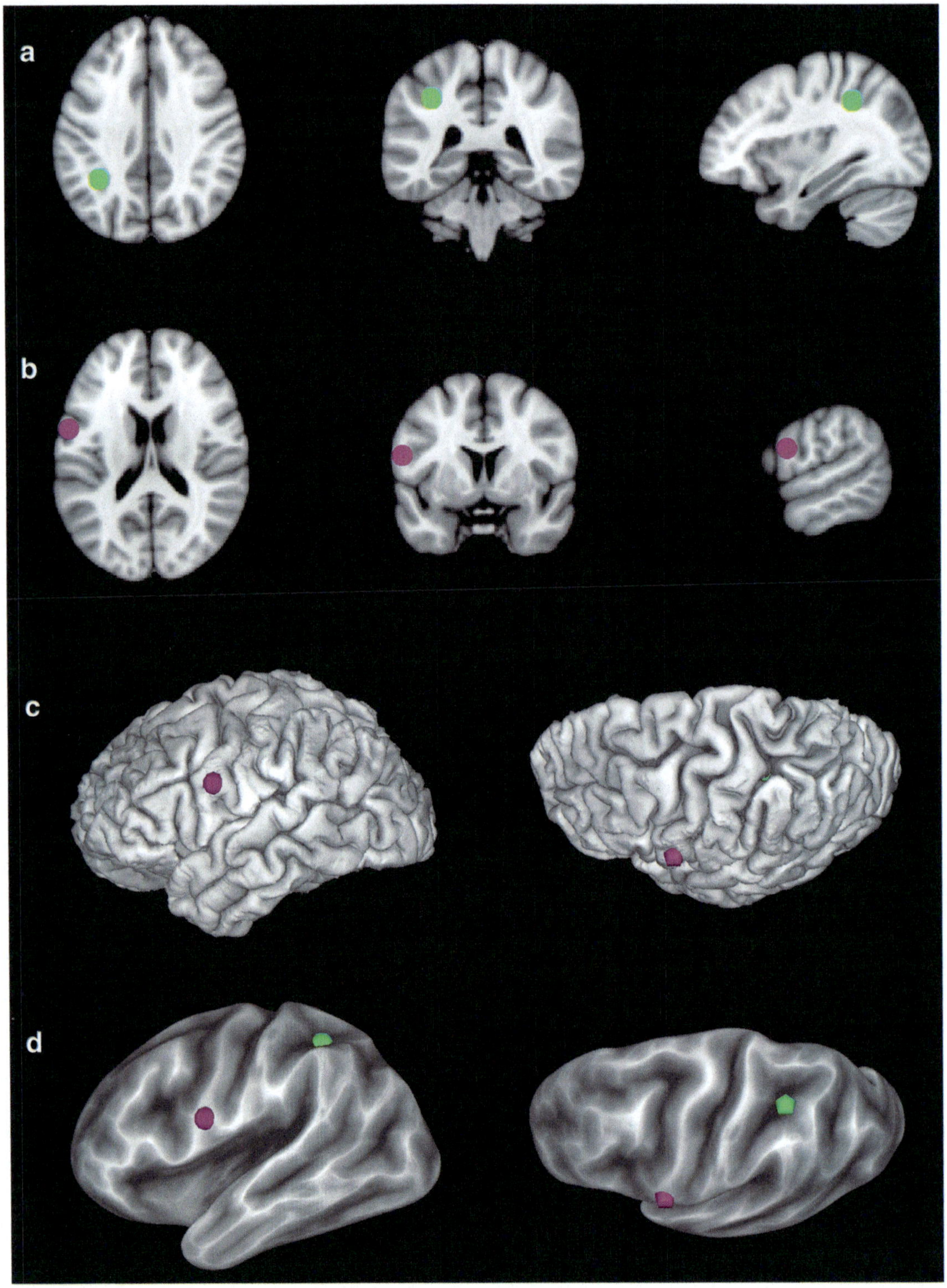

Fig. 8.4 A volume-based versus surface-based approach to visualizing points on the cerebral cortex in the human brain. A volume-based approach maintains a 2D representation of the cortex, while a surface-based approach maintains the topology of the cortical surface and portrays the cortex as a 3D sheet that shows parts of the cortex usually obscured by cortical folding. This concept is shown here by two points of interest on the cortex, one in the intraparietal sulcus (**a**, *green*) and one in the motor cortex (**b**, *pink*). A volume-based approach requires three separate views to demonstrate the location of the point (**a**, **b**). A reconstruction of the surface of the brain (**c**) allows for the points to be visualized in 3D, but the point within the depth of the intraparietal sulcus is obscured by cortical folding. A surface-based approach inflates the gray matter of the brain (**d**) so that it becomes a sheet and allows for both points to be visualized. The inflated brain also maintains information about the location of sulci and gyri with the color of the surface: *darker* colors indicate the sulci, while *lighter* colors indicate gyri (Images from Human Connectome Project data)

themselves vary in the depth of their inward folds. An intermediate approach relying on automated processing tools is, however, possible.

The development of surface-based registration (SBR) in neuroimaging allows for the use of anatomical features to align brain surfaces across subjects to a standard space (Anticevic et al. 2012). SBR is based on the production of cortical surfaces from structural MRI scans and aligns surface vertices across scans. As mentioned previously, cortical surfaces can be produced using various neuroimaging software packages. For example, FreeSurfer software creates cortical surfaces by first detecting the boundary between white and gray matter, and then this outline is moved outward until it reaches the pial surface separating gray matter from the area containing cerebral spinal fluid (Dale et al. 1999; Fischl et al. 1999a; Fischl and Dale 2000; Fischl 2012). The cortical surface is then produced by inflating the pial surface so that the entire cortical surface can be represented by a mesh made up of vertices. The mesh is then used to produce an image of the surface, and this surface image retains information about sulcal depth to show the features of the original brain shape.

The major advantage of SBR compared to a volume-based approach is that SBR methods allow for the visualization of the entire cortex as a sheetlike structure without being affected by cortical folding. The effect of a volume-based versus surface-based approach to visualizing points on the cortex is shown in Fig. 8.4. One possible drawback of using SBR is that it cannot be used to directly investigate subcortical structures, which therefore require the use of volume-based methods (Glasser et al. 2013). However, subcortical structures do not suffer from the same issues as the cortical surface since they do not experience the same type of folding, and volume-based methods are generally sufficient. Another possible drawback is that the initial segmentation of white and gray matter prior to creating the surfaces relies on detecting differences in intensities between the two tissue types (Dale et al. 1999; Fischl et al. 1999a; Fischl and Dale 2000; Fischl 2012). In practice, the use of intensities results in errors if there is not enough contrast between the gray and white matter, which may require extensive manual editing. Acquiring images with good contrast can reduce the potential for this effect and maintain the automated nature of SBR.

There are two methods of SBR commonly in use: the automated landmark-based approach (Van Essen 2005; Anticevic et al. 2012) and the FreeSurfer approach (Fischl et al. 1999b; Fischl 2012). The automated landmark-based approach uses six core landmarks to register individual surfaces to a template in standard space (note that the meaning of landmark here is different from the meaning of landmark in a geometric morphometric context): the central sulcus, Sylvian fissure, superior temporal gyrus, dorsal medial wall, ventral medial wall, and calcarine sulcus (Van Essen 2005). In the FreeSurfer approach, each individual white matter surface is registered to a template based on the pattern of cortical folding at each vertex; cortical folding and curvature patterns for each individual are aligned with the template (Fischl et al. 1999b). In both methods, the surfaces obtained for each individual are inflated into a sphere, and either the six core landmarks or pattern of cortical folding is registered to the same landmarks on the template sphere (Fischl et al. 1999b; Van Essen 2005; Anticevic et al. 2012).

Recently, SBR methods have been adapted for use with the Human Connectome Project (HCP). The HCP is a large-scale neuroimaging data collection effort that has revolutionized the way that this type of data is collected, managed, and analyzed (Marcus et al. 2011; Van Essen et al. 2013). The HCP has currently obtained high-resolution scans from 1200 individuals that include a number of different imaging modalities including structural MRI, functional MRI (fMRI), and diffusion tensor imaging (DTI). All HCP data was collected and processed in the same manner allowing for a large, consistent database available for study. The HCP pipeline utilizes aspects of the FreeSurfer protocols discussed previously but provides updated methods for improving cortical surfaces and SBR (Glasser et al. 2013).

SBR improves spatial localization in the cerebral cortex and accounts for individual variability because it maintains the topology of the cortical surface (Van Essen et al. 1998; Fischl et al. 2008; Anticevic et al. 2008). In SBR, the neural anatomy is used to align vertices across cortical surfaces, resulting in the alignment of homologous features between hemispheres and across individuals. Therefore, any selected vertex will be homologous across all individuals. A certain number of randomly chosen vertices can be selected to reduce landmark number to a figure that will accurately represent brain anatomy (or the anatomy of certain region of the brain) and that will still be appropriate for subsequent statistical analyses in a geometric morphometric context (Fig. 8.5). It is important to note, however, that homology in this case is achieved a priori through surface-based registration. This approach is conceptually similar to the automated shape aligning and comparing approach described by Boyer and colleagues (Boyer et al. 2015) and to other similar automated methods proposed recently (Pomidor et al. 2016), but it makes use of highly optimized surface registration techniques commonly used in neuroscience. We realize, however, that this may look like a heterodox twist of geometric morphometrics for some theoreticians and debate on the applicability of this landmarking approach to classic geometric morphometric analyses is warranted.

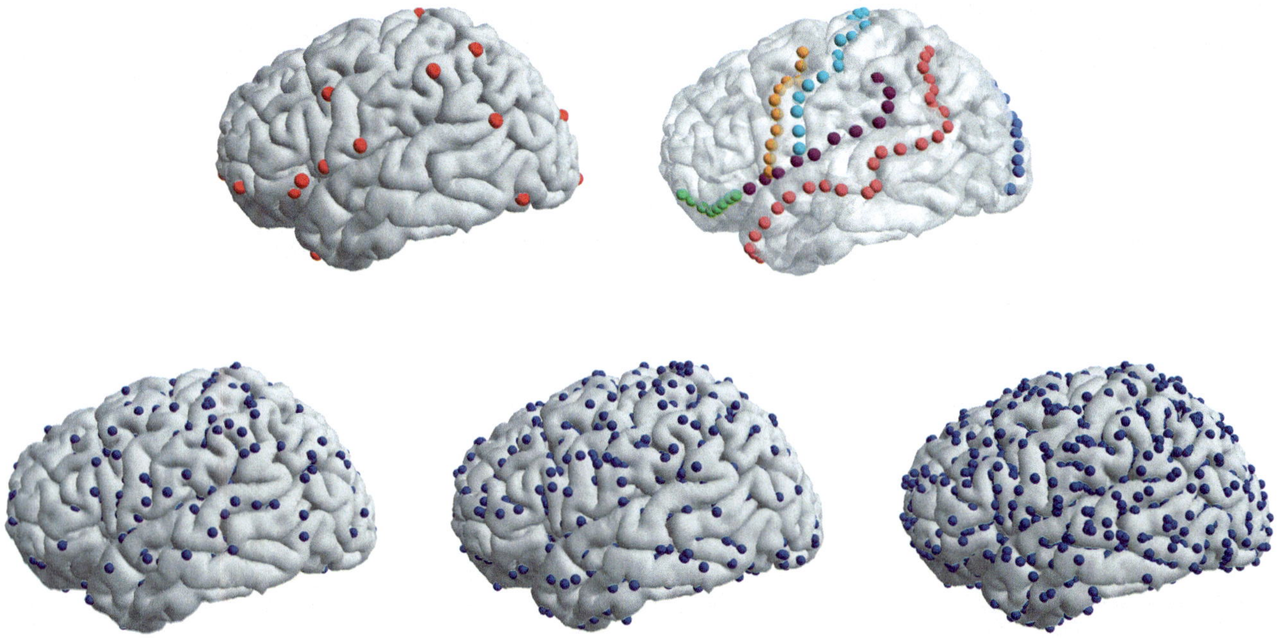

Fig. 8.5 Sampling strategies for geometric morphometric analyses of the human neocortex. *Top left*: anatomically homologous landmarks are represented in *red* (definitions follow those in Fig. 8.2). *Top right*: sulcal semilandmarks. Different sulci are represented in different colors (*green* latero-orbital sulcus, *orange* precentral sulcus, *teal* central sulcus, *purple* Sylvian fissure, *pink* superior temporal sulcus, *light blue* parieto-occipital sulcus. *Bottom*: random vertex sampling after surface-based registration showing a random selection of 500 (*left*), 1000 (*center*), and 2000 (*right*) surface cortical vertices (vertices located on the *right* hemisphere and on the medial wall and buried within sulcal depths are not shown)

The combination of these automated techniques with geometric morphometrics and with other methodological toolkits—including phylogenetic comparative methods, quantitative genetics, network analysis, etc.—promises great advances in the study of evolutionary neuroanatomy, the main limiting factor being obtaining high-quality scans for nonhuman species. In this regard, the major challenge remains in the comparison of very different brains belonging to diverse species because automated tools are particularly problematic when there is extensive variation. Geometric morphometrics can prove useful for such comparisons if a sufficient number of anatomically homologous locations can be identified across species, which can be aided by comparisons of external morphology with microstructural traits.

8.8 Concluding Remarks

Geometric morphometric techniques allow researchers to quantify brain anatomical variation and to explore scenarios of brain evolution and developmental mechanisms underlying them. Methodological alternatives are plentiful, and they range from the classic use of homologous landmarks to the use of curve and surface semilandmarks to the combination of automated approaches routinely used in human neuroscience with geometric morphometric analytical tools. Each of these alternatives poses different challenges and limitations, but they also offer promising opportunities to quantify brain shape variation and, consequently, to better understand brain evolution.

Acknowledgments We are grateful to Emiliano Bruner, Naomichi Ogihara, and Hiroki Tanabe for their invitation to contribute to this volume.

References

Adams DC, Rohlf FJ, Slice DE (2013) A field comes of age: geometric morphometrics in the 21st century. Hystrix 24:7–14

Aldridge K (2011) Patterns of differences in brain morphology in humans as compared to extant apes. J Hum Evol 60:94–105

Allen JS, Damasio H, Grabowski TJ (2002) Normal neuroanatomical variation in the human brain: an MRI-volumetric study. Am J Phys Anthr 118:341–358

Anticevic A, Dierker DL, Gillespie SK et al (2008) Comparing surface-based and volume-based analyses of functional neuroimaging data in patients with schizophrenia. NeuroImage 41:835–848

Anticevic A, Repovs G, Dierker DL et al (2012) Automated landmark identification for human cortical surface-based registration. NeuroImage 59:2539–2547

Aristide L, dos Reis SF, Machado AC et al (2016) Brain shape convergence in the adaptive radiation of new world monkeys. Proc Natl Acad Sci U S A 113:2158–2163

Barton RA, Harvey PH (2000) Mosaic evolution of brain structure in mammals. Nature 405:1055–1058

Bookstein FL (1997) Morphometric tools for landmark data: geometry and biology. Cambridge University Press, Cambridge
Bookstein FL, Sampson PD, Streissguth AP, Connor PD (2001) Geometric morphometrics of corpus callosum and subcortical structures in the fetal-alcohol-affected brain. Teratology 64:4–32
Boyer DM, Puente J, Gladman JT et al (2015) A new fully automated approach for aligning and comparing shapes. Anat Rec 298:249–276
Bruner E, Martin-Loeches M, Colom R (2010) Human midsagittal brain shape variation: patterns, allometry and integration. J Anat 216:589–599
Chiavaras MM, Petrides M (2000) Orbitofrontal sulci of the human and macaque monkey brain. J Comp Neurol 422:35–54
Chollet MB, Aldridge K, Pangborn N et al (2014) Landmarking the brain for geometric morphometric analysis: an error study. PLoS One 9:e86005
Cointepas Y, Mangin J-F, Garnero L et al (2001) BrainVISA: software platform for visualization and analysis of multi-modality brain data. NeuroImage 13:98
Dale AM, Fischl B, Sereno MI (1999) Cortical surface-based analysis: I. Segmentation and surface reconstruction. NeuroImage 9:179–194
Desikan RS, Ségonne F, Fischl B et al (2006) An automated labeling system for subdividing the human cerebral cortex on MRI scans into gyral based regions of interest. NeuroImage 31:968–980
de Winter W, Oxnard CE (2001) Evolutionary radiations and convergences in the structural organization of mammalian brains. Nature 409:710–714
Finlay B, Darlington R (1995) Linked regularities in the development and evolution of mammalian brains. Science 268:1578–1584
Fischl B (2012) FreeSurfer. NeuroImage 62:774–781
Fischl B, Dale AM (2000) Measuring the thickness of the human cerebral cortex from magnetic resonance images. Proc Natl Acad Sci 97:11050–11055
Fischl B, Sereno MI, Dale AM (1999a) Cortical surface-based analysis: II: inflation, flattening, and a surface-based coordinate system. NeuroImage 9:195–207
Fischl B, Sereno MI, Tootell RBH, Dale AM (1999b) High-resolution intersubject averaging and a coordinate system for the cortical surface. Hum Brain Mapp 8:272–284
Fischl B, Rajendran N, Busa E et al (2008) Cortical folding patterns and predicting cytoarchitecture. Cereb Cortex 18:1973–1980
Glasser MF, Sotiropoulos SN, Wilson JA et al (2013) The minimal preprocessing pipelines for the human connectome project. NeuroImage 80:105–124
Gómez-Robles A, Olejniczak AJ, Martinón-Torres M et al (2011) Evolutionary novelties and losses in geometric morphometrics: a practical approach through hominin molar morphology. Evolution 65:1772–1790
Gómez-Robles A, Hopkins WD, Sherwood CC (2013) Increased morphological asymmetry, evolvability and plasticity in human brain evolution. Proc R Soc B Biol Sci 280:20130575
Gómez-Robles A, Hopkins WD, Sherwood CC (2014) Modular structure facilitates mosaic evolution of the brain in chimpanzees and humans. Nat Commun 5:4469
Gómez-Robles A, Hopkins WD, Schapiro SJ, Sherwood CC (2015) Relaxed genetic control of cortical organization in human brains compared with chimpanzees. Proc Natl Acad Sci U S A 112:14799–14804
Gómez-Robles A, Hopkins WD, Schapiro SJ, Sherwood CC (2016) The heritability of chimpanzee and human brain asymmetry. Proc R Soc B Biol Sci 283:20161319
Gonzalez PN, Barbeito-Andrés J, D'Addona LA et al (2016) Technical note: performance of semi and fully automated approaches for registration of 3D surface coordinates in geometric morphometric studies. Am J Phys Anthropol 160:169–178
Gronenschild EHBM, Habets P, Jacobs HIL et al (2012) The effects of FreeSurfer version, workstation type, and Macintosh operating system version on anatomical volume and cortical thickness measurements. PLoS One 7:e38234
Gunz P, Neubauer S, Maureille B, Hublin J-J (2010) Brain development after birth differs between Neanderthals and modern humans. Curr Biol 20:R921–R922
Hopkins WD, Meguerditchian A, Coulon O et al (2014) Evolution of the central sulcus morphology in primates. Brain Behav Evol 84:19–30
Jenkinson M, Beckmann CF, Behrens TEJ et al (2012) FSL. NeuroImage 62:782–790
Keller SS, Highley JR, Garcia-Finana M et al (2007) Sulcal variability, stereological measurement and asymmetry of Broca's area on MR images. J Anat 211:534–555
Keller SS, Deppe M, Herbin M, Gilissen E (2012) Variability and asymmetry of the sulcal contours defining Broca's area homologue in the chimpanzee brain. J Comp Neurol 520:1165–1180
Klingenberg CP (2008) Novelty and "homology-free" morphometrics: what's in a name? Evol Biol 35:186–190
Klingenberg CP, Wetherill L, Rogers J et al (2010) Prenatal alcohol exposure alters the patterns of facial asymmetry. Alcohol 44:649–657
Kochunov P, Mangin JF, Coyle T et al (2005) Age-related morphology trends of cortical sulci. Hum Brain Mapp 26:210–220
Marcus DS, Harwell J, Olsen T et al (2011) Informatics and data mining tools and strategies for the human connectome project. Front Neuroinform. doi:10.3389/fninf.2011.00004
McAuliffe MJ, Lalonde FM, McGarry D, et al (2001) Medical image processing, analysis and visualization in clinical research. In: 14th IEEE Symposium on Computer-Based Medical Systems, 2001. CBMS 2001. Proceedings, pp 381–386
Mota B, Herculano-Houzel S (2015) Cortical folding scales universally with surface area and thickness, not number of neurons. Science 349:74–77
Neubauer S, Gunz P, Hublin J-J (2010) Endocranial shape changes during growth in chimpanzees and humans: a morphometric analysis of unique and shared aspects. J Hum Evol 59:555–566
Ono M, Kubick S, Abernathey C (1990) Atlas of the cerebral sulci. Thieme, Stuttgart
Oxnard CE (2004) Brain evolution: mammals, primates, chimpanzees, and humans. Int J Primatol 25:1127–1158
Polly PD (2008) Developmental dynamics and G-matrices: can morphometric spaces be used to model phenotypic evolution? Evol Biol 35:83–96
Pomidor BJ, Makedonska J, Slice DE (2016) A landmark-free method for three-dimensional shape analysis. PLoS One 11:e0150368
Sherwood CC, Broadfield DC, Holloway RL et al (2003) Variability of Broca's area homologue in African great apes: implications for language evolution. Anat Rec 271A:276–285
Smaers JB, Soligo C (2013) Brain reorganization, not relative brain size, primarily characterizes anthropoid brain evolution. Proc R Soc B Biol Sci 280:20130269
Smaers JB, Steele J, Zilles K (2011) Modeling the evolution of cortico-cerebellar systems in primates. Ann N Y Acad Sci 1225:176–190
Sowell ER, Thompson PM, Rex D et al (2002) Mapping sulcal pattern asymmetry and local cortical surface gray matter distribution in vivo: maturation in perisylvian cortices. Cereb Cortex 12:17–26
Stephan H, Frahm H, Baron G (1981) New and revised data on volumes of brain structures in insectivores and primates. Folia Primatol (Basel) 35:1–29

Steppan SJ, Phillips PC, Houle D (2002) Comparative quantitative genetics: evolution of the G matrix. Trends Ecol Evol 17:320–327

Tallinen T, Chung JY, Rousseau F et al (2016) On the growth and form of cortical convolutions. Nat Phys 12:588–593

Van Essen DC (2005) A population-average, landmark- and surface-based (PALS) atlas of human cerebral cortex. NeuroImage 28:635–662

Van Essen DC, Drury HA, Joshi S, Miller MI (1998) Functional and structural mapping of human cerebral cortex: solutions are in the surfaces. Proc Natl Acad Sci 95:788–795

Van Essen DC, Drury HA, Dickson J et al (2001) An integrated software suite for surface-based analyses of cerebral cortex. J Am Med Inform Assoc 8:443–459

Van Essen DC, Glasser MF, Dierker DL et al (2012) Parcellations and hemispheric asymmetries of human cerebral cortex analyzed on surface-based atlases. Cereb Cortex 22:2241–2262

Van Essen DC, Smith SM, Barch DM et al (2013) The WU-Minn human connectome project: an overview. NeuroImage 80:62–79

Whiting B, Barton R (2003) The evolution of the cortico-cerebellar complex in primates: anatomical connections predict patterns of correlated evolution. J Hum Evol 44:3–10

Zilles K, Amunts K (2013) Individual variability is not noise. Trends Cogn Sci 17:153–155

Zlatkina V, Petrides M (2010) Morphological patterns of the postcentral sulcus in the human brain. J Comp Neurol 518:3701–3724

Landmarking Endocasts

9

Ana Sofia Pereira-Pedro and Emiliano Bruner

Abstract

The brain and the braincase are tightly integrated during growth and development, and the cerebral cortex leaves morphological traces on the internal cranial surface. Paleoneurology studies the endocranial moulds – *endocasts* – in order to make inferences about brain evolution. The use of shape analysis based on landmark data is a current standard in morphometrics, but placing landmarks on endocasts involves inherent challenges. Here, we address endocast shape analysis focusing on the problems associated with landmarking their anatomical areas. First, we review the application of shape analysis in paleoneurology, from stereoplotting to sliding semilandmarks. Then, we address the problematics of landmarking endocasts' smooth surfaces. Finally, we present a case study comparing the uncertainty in landmarking physical and digital endocasts. The mean absolute intra-observer error calculated on a digital sample is 0.86 mm, which is comparable with the error obtained by other authors on cranial, cerebral and endocast analyses. Landmarking on digital replicas, overall, displays larger uncertainty when compared with the physical casts, but the differences are not significant, and both methods give similar results when dealing with Hominoids' interspecific variation. Although patterns of uncertainty seem to be largely idiosyncratic, larger errors are usually found on the parietal surface. This study concerns only intra-observer error. Anatomists can have a different perception or interpretation of the cortical references, and landmarking can be improved by performing joint sampling based on the agreement of different experts.

Keywords

Paleoneurology • Endocasts • Shape analysis • Geometric morphometrics • Hominoids • Digital anatomy • Intra-observer error

9.1 Introduction

The size and shape of the braincase are the result of a functional and structural interaction with the brain during ontogeny and evolution (Moss and Young 1960; Richtsmeier et al. 2006; Bruner et al. 2014a). During morphogenesis, the skull bones are separated by the growing brain, while the expansion of the neuronal tissue is influenced by the connectives and cranial elements. At the cranial base, the final phenotype is influenced by multiple factors and complex morphological dynamics (Lieberman et al. 2000; Bruner and Ripani 2008), whereas the vault presents simpler relationships, in which brain changes are the main source of cranial shape variations.

Moss and Young (1960) introduced the concept of functional craniology to emphasize the importance of analysing endocranial interactions in order to understand the cranial

A.S. Pereira-Pedro (✉) • E. Bruner
Programa de Paleobiología, Centro Nacional de Investigación sobre la Evolución Humana, Paseo Sierra de Atapuerca 3, 09002 Burgos, Spain
e-mail: sofia.aspp@gmail.com; emiliano.bruner@cenieh.es

E. Bruner et al. (eds.), *Digital Endocasts*, Replacement of Neanderthals by Modern Humans Series,
DOI 10.1007/978-4-431-56582-6_9

form. The skull constitutes a functional matrix composed of sutures, bones, brain, muscles and connectives, orienting and channelling the morphogenetic variations (Bruner 2015; Goriely et al. 2015). In general, integration concerns the degree of covariation between different structures (Cheverud 1996), which instead are referred to as *modules* when their internal cohesion is stronger than the association with the rest of the anatomical system (Klingenberg 2010). Integration and modularity can be studied at individual or population level (Cheverud 1996; Klingenberg 2010). The first deals with the interaction of structures during development, reflecting their functional association, while the second deals with the joint inheritance of these closely integrated complexes, which evolve together (Cheverud 1996).

Integration can be inferred through the study of correlation between morphological characters (Klingenberg 2013). Size is a major factor constraining the covariation schemes and hence generating allometric integrative relationships (Klingenberg 2010). These kinds of studies, focusing on the variation and correlation between different anatomical regions in terms of geometry and morphology, require the use of heuristic and comprehensive statistical tools, such as those offered by geometric morphometrics (Bookstein 1991; Mitteroecker and Gunz 2009).

Although comparative primate studies (especially those including anthropoids) can give some insight into human brain evolution, direct evidence can only be provided by the human fossil record. *Paleoneurology* studies brain anatomy through the morphology of the endocasts, which are moulds of the endocranial cavity (Holloway et al. 2004). Endocasts can sometimes be formed naturally by matrix deposition in the endocranial cavity, but traditionally they were prepared artificially with moulding materials (Holloway 1975). Nowadays, they are digitally reconstructed from three-dimensional (3D) imaging techniques (e.g. Gunz et al. 2009). Due to the close spatial proximity and integration between bones and the brain during growth and development, the traces of the cerebral gyri and sulci are partially left on the endocranial surface (Holloway et al. 2004; Kobayashi et al. 2014). The endocast supplies two kinds of information. First, it shows the overall geometry of the brain, including its shape and size components. Second, it can show the main circumvolutions, that is, the sulcal pattern. These features are relevant because they can reveal differences in the sulcal organization, but most of all, they are essential to show the relative proportions of the brain areas, when used to outline cortical districts or functional surfaces. Although there is an important loss of information from the brain to endocasts, these features represent the only direct evidence of brain anatomical variations in fossil species. Caution is nonetheless recommended in this sense, at least when dealing with three major limits (Bruner 2015). First, an endocast is not a brain but simply its external mould. Second, the endocast is the result of a morphogenetic system, and its geometry must be interpreted accordingly, in terms of functional and structural relationships among soft and hard tissues. Third, because of the reciprocal influence between the brain and braincase, an endocast should not be evaluated alone but instead together with its cranial counterparts.

A main issue regarding endocast analysis is the fact that imprints of the brain areas are blurred and smoothed by the meningeal layers that surround the brain. As a consequence, some features may be absent or poorly shaped, hampering a complete and objective assessment of the cortical morphology. A consequence of this is that the evaluation of the uncertainty associated with the localization of these elements is, in this field, more important than for other morphological districts.

9.2 Landmarking in Paleoneurology

Traditional analyses of endocasts focused on estimations of brain volume or cranial capacity of the fossils or on linear morphometric measurements such as arcs, chords or angles (see Holloway et al. 2004; Holloway 1981). A pioneering geometric analysis of endocast morphology was published by Ralph Holloway in 1981 by using a stereoplotter craniostat, a technique adapted from Oyen and Walker (1977) in geology to obtain 3D coordinates (Fig. 9.1). It consisted of a rotating circle and a perpendicular arc with a needle attached which indicated, for any point, the azimuth, the elevation and the radial distance, respectively. Because azimuth and elevation are equivalent to the geographical latitude and longitude, the coordinates could be projected onto a map, or stereographic net, which represented the topography of the object. In his experimental study, Holloway used as a reference the central point between the frontal and occipital poles. A system of vertical and horizontal transects (analogous to parallels and meridians) was projected over the endocast surface, with intervals of 10° or 20°. For each intersection point between the vertical and horizontal lines, he measured the radial distance to the central homologous reference point. In order to correct for allometry, he constructed regression lines between each log radial distance and log volume and then calculated the expected values for each coordinate using the correspondent regression equation. He used as variables the residual values from the subtraction between the measured and expected values of the radial distance. Among other results, this study showed a remarkable degree of variation at the parietal areas.

Spatial analyses based on coordinates have been used ever since in anthropology, primatology and paleontology, although generally limited to 2D form description after baseline normalization (e.g. Huxley 1863; Weidenreich

Fig. 9.1 A pioneering Ralph Holloway with the stereoplotter craniostat in 1981, sampling coordinates from hominoid endocasts (Courtesy of Ralph Holloway)

1941; Verheyen 1957). D'Arcy Thompson (1942) proposed the most complete and comprehensive perspective in this sense, integrating theories in spatial analysis and shape deformation. However, landmark-based morphometrics was applied only after integration with multivariate statistics (Adams et al. 2004; Zelditch et al. 2004). In geometric morphometrics, a landmark coordinate set represents a single specimen. All the specimens are normalized according to a given criterion, so as to superimpose all the sets onto a shared reference space through translation, scaling and rotation. *Baseline registration* superimposes all the specimens along a shared chord, which usually represents a major length or a functional distance. *Procrustes superimposition* minimizes shape differences by translation of all the sets on a same centroid, scaling to unitary size and rotating in order to minimize the distance between corresponding landmarks. After registration, residuals between each landmark and the mean value can be used for multivariate statistics, namely, to investigate overall geometrical differences between individuals or groups or to investigate the correlation patterns underlying the observed variation. Apart from ordering the specimens and reducing the information to a smaller number of variables, such quantitative approaches can reveal the schemes of correlations behind the structure of the phenotype. Another way to perform shape analysis from landmarks is through the *thin-plate spline* interpolant functions, which generate vectors of shape changes (principal warps) that can be scored for each specimen (partial warps) and used for multivariate statistics (Bookstein 1991). Thin-plate spline is also useful to display any landmark-based multivariate vector through deformation grids, warping a reference space into a target space to show a geometric variation through an efficient and synthetic visualization method.

Registrations based on the whole configuration (like the Procrustes approach) have a major downside: the overall variation is distributed all through the landmark set. Because of this 'Pinocchio effect', if the variation is due to a single element or to a specific area, it will be distributed on average also to the other ones, generating a biased perspective of the actual changes. In reality, this is a problem which occurs mostly when using superimposition to compare groups according to their absolute differences, because their registration may be influenced by an uneven distribution of the variation. On the other hand, this limit is less relevant when considering the patterns of covariance, which are based on the correlations between traits and represent the ultimate scope of the multivariate survey.

All these quantitative perspectives are hence strongly rooted in the geometrical models of the anatomical system represented by landmarks. Landmarks are anatomical points, defined by Cartesian coordinates in two or three dimensions, chosen according to a criterion of *biological* or *geometrical* correspondence (Bookstein 1991; Zelditch et al. 2004). Depending on the type of study, correspondence may be phylogenetic, developmental, structural or functional (Richtsmeier et al. 2002; Oxnard and O'Higgins 2015). Given that landmarks are the basic unit of geometrical modelling, their proper definition is essential to guarantee stability and reproducibility of results (Klingenberg 2008). Correspondence is fundamental, as landmarks will be mathematically 'matched one to one' (Zelditch et al. 2004), and thus they must label equivalent or homologous loci on all specimens in the study dataset. Besides the correspondence between landmarks, their ability to represent the geometry of the object and a given biological meaning are also fundamental criteria for choosing the landmark configuration (Richtsmeier et al. 2002; Zelditch et al. 2004). Bookstein (1991) distinguished three types of landmarks based on their biological significance. Type I landmarks are well defined by three surrounding structures, for instance, the cranial sutures, and can be associated with a direct biological meaning. Type II landmarks are usually identified as points of maximum curvature, like tips or valleys, and are often used as general morphological references of areas influenced by biomechanical or morphogenetic forces (e.g. like the tip of a tooth). Type III landmarks are geometrically defined relative to other structures, for instance, a centroid point or the furthermost point in relation to another element. These points are strictly geometrical references, with a lesser degree of freedom because of their dependence on other elements, and their displacement is only meaningful in one direction. Because of their dependence on other information and their minor degrees of freedom, these landmarks are often called *semilandmarks*. Nomenclature apart, it is worth noting that the boundaries among these classes are not sharp, and there may be cases in which landmarks lie in intermediate positions. Type I landmarks are the most informative and reliable (Zollikofer et al. 1998; Free et al. 2001), but generally all three types are necessary to generate a geometrical model sufficiently complete to represent most of the anatomical elements involved. Missing important areas of variation in a spatial configuration can give a biased covariance matrix, but at the same time, an excess of landmarks will introduce redundancy or unequal weights in the analysis. Therefore, apart from guaranteeing reliable definitions for landmarks, the configuration must provide a properly balanced geometrical model of the anatomical system under scrutiny.

Geometric morphometrics was introduced in paleoanthropology at the end of the 1990s (Bookstein et al. 1999; Ponce de León and Zollikofer 2001; Lieberman et al. 2002) and in paleoneurology a few years after (Bruner et al. 2003; Bruner 2004). Since the beginning, it was clear that endocasts generally lack type I landmarks, most of the points representing surfaces, bosses and curves. Despite the limits associated with endocast geometric modelling, the interspecific differences within the human genus were sufficient to reveal a general allometric pattern and a specific trait associated with modern humans: an expansion of the parietal surface.

A major improvement in surface analysis was represented by the introduction of *sliding landmarks*, namely, semilandmarks placed along a geometric curve which are allowed to move along a plane according to a specific normalization algorithm (Bookstein 1997; Gunz et al. 2005; Gunz and Mitteroecker 2013). Landmarks are originally placed uniformly in two or three dimensions throughout a curve or surface, and then their position is iteratively recalculated according to the position of the neighbouring landmarks, as to fit the coordinate system within a general geometric rule, as to make all the specimens comparable according to a shared form template. To achieve such geometric correspondence (a sort of 'geometric homology') between surface landmarks, points are repositioned following a minimization criterion, generally a minimization of the deformation from an average consensus (calculated with the thin-plate spline interpolant functions in terms of bending energy) or else a minimization of the shape difference after Procrustes superimposition (calculated as Procrustes distance). With sliding semilandmarks, the homology is not point-to-point, but curve-to-curve or surface-to-surface, in two or three dimensions, respectively. There was a considerable improvement with the introduction of sliding landmarks. Nonetheless, some caution is required when introducing algorithms and minimization criteria into shape registration and comparative anatomy. Any numerical transformation can introduce artificial constraints in the shape comparison due to incorrect assumptions or biased operational choices (numerical transformations and adjustments for normalization and registration of the values, importance and weights of the variables, different regression models and so on). This can influence the statistical ordination and the biological interpretation of the outputs. It must be also taken into account that sliding landmarks are not associated with specific anatomical boundaries. This leads to the main drawback: variations specific to a given anatomical element will be distributed all across the surface. For example, if two areas, A and B, are covered by a continuous sliding set, an increase of A alone will be interpreted as an increase of both A and B, which contribute equally to the final difference. Therefore, within an area covered with sliding landmarks, we can observe the general form changes, but we cannot know what elements are involved. In this case, the object is analysed as a general geometric form and not as a system of anatomical elements. This is an advantage when dealing with objects that have uncertain boundaries, but it must be seriously taken into account when interpreting the results. In paleoneurology, this limit can be particularly crucial, because the continuous endocranial surface is in fact formed by distinct cortical territories. Analysing this surface as a homogeneous and undifferentiated geometrical object can lead to misinterpretations of the morphometric output, if such limitation is not considered properly.

Finally, it must be also stressed that different sliding criteria can produce different results, and methodological considerations are hence necessary before a specific operational choice is put forward to design and interpret a morphometric analysis based on additional sliding algorithms (Bruner and Bastir 2009).

Sliding landmarks have been applied to quantify and evaluate endocranial variation in modern humans (Neubauer et al. 2009), chimpanzees (Neubauer et al. 2010), apes (Scott et al. 2014) and Neandertals (Gunz et al. 2010). Sliding landmarks are also extremely helpful for the virtual reconstruction of damaged or incomplete fossils (Gunz et al. 2009; Amano et al. 2015).

9.3 Challenges of Landmarks Analysis

As Rohlf and Marcus (1993) acknowledged, the greatest advantage of geometric morphometrics is the ability of the landmark coordinates to capture information on the geometry of the specimens and the highly visual graphic outcomes. Indeed, these methods emphasize the graphical display of patterns and variations, allowing an easier and more intuitive interpretation of the results. However, this same advantage can be counterproductive, because without proper experience, the interpretation of the biological meaning can be influenced by a simplified reading of the visual scheme. The graphical results are nothing but numerical representations, and the biological significance is then provided by the researcher on the basis of previous information on the anatomical elements involved (O'Higgins 2000). Biological inferences must be proposed taking into consideration a specific research question, the landmarks used in the geometrical models and the numerical procedure used to normalize and compare the specimen (Adams et al. 2004; Klingenberg 2008). Hence, a deep knowledge of the assumptions and limitations of the landmark data and analysis are essential to obtain meaningful conclusions. When applying landmark data in paleoneurology, one must take into account both the limitations underlying the use of landmarks and those inherent to the endocasts.

Several issues contribute to the difficulty in landmarking endocasts (Fig. 9.2). The first issue concerns the preservation of the specimen. Obviously, the cortical surface can be more easily observed in well-preserved endocasts. Apart from the conditions of preservation of the bone surface, in paleontology, many endocasts have been reconstructed from incomplete specimens, introducing a degree of subjectivity in the cast, depending on the experience of the anatomists and on the percentage of missing information. Additionally, landmarks from endocasts must rely only on the superficial traits and particularly only on those that have left imprints. Interestingly, smaller brains usually leave clearer imprints (Kobayashi et al. 2014; Zollikofer and Ponce de León 2013), probably due to allometric effects influencing the relative thickness of the meninges or the endocranial pressure. Because of the spatial constraints with their anatomical counterparts, the basal areas (orbital and temporal) often display more visible traces of the circumvolutions, when compared with the vault areas (frontal, parietal and occipital). Unfortunately, the basal areas are far less represented in the fossil record than the upper cranial ones, because of their extreme fragility. Therefore, the most frequently available elements in paleoanthropology (the frontal, parietal and occipital bones) are also those with less sharp sulcal details.

A second problem is directly associated with the nature of the brain itself (see Chap. 8). Brain lobes and gyri are conventional anatomical entities, with unclear and blurred boundaries. Most regions cannot be represented strictly in terms of landmark points, as they can only be identified as 'areas'. In this sense, landmarks are intended to show the approximate spatial position of those areas. Moreover, the homology between some cortical regions in different species is not known. Hence, more than in other fields, in this case, the expertise of the observer may influence both the identification of the brain region and the repeatability in locating the landmarks. Because of the limits associated with brain and endocast landmarking, in paleoneurology, it is even more difficult to design balanced geometrical models. Uneven distributions of landmarks can miss some important areas of variation or leaving areas underrepresented. With these limits in mind, methodological decisions are of utmost importance for a consistent study.

At present, most of the studies in morphometrics rely on digital data, since the introduction of biomedical imaging in paleoanthropology (Zollikofer et al. 1998; Spoor et al. 2000). The availability of software for landmarking virtual specimens has increased. However, analysing digital data can add further methodological questions. Working with virtual reconstructions hampers the possibility to touch and handle the object, which are complementary and useful procedures that assist the exploration of the surface. Furthermore, the appearance of the object may in part depend on perspective and shading factors, introducing a biased perception of forms and boundaries. Landmark repeatability is similar for all coordinates when sampling dry skulls (Corner et al. 1992), but the same is not true when landmarking images from computed tomography (CT). A further methodological issue concerns directly the quality and homogeneity of the data. The mean error will depend on the resolution of the scans. Within the same CT sample, Richtsmeier et al. (1995) observed lower repeatability of the z coordinate, because in that case, x and y displayed higher resolution (pixel size of 0.39 mm) than the z coordinate (slice thickness of 1.50 mm). It is worth noting that, because of the Pinocchio effect, if there is a larger error in one or few landmarks, it will be distributed through the remaining landmarks as well (von Cramon-Taubadel et al. 2007).

9.4 Landmark Uncertainty

The repeatability and precision of landmarks are of the greatest importance when using a given geometrical model to represent a biological structure. Several studies have addressed the observer error in landmarking, in different kinds of data and with different computation procedures. Different parameters can be used to quantify landmark

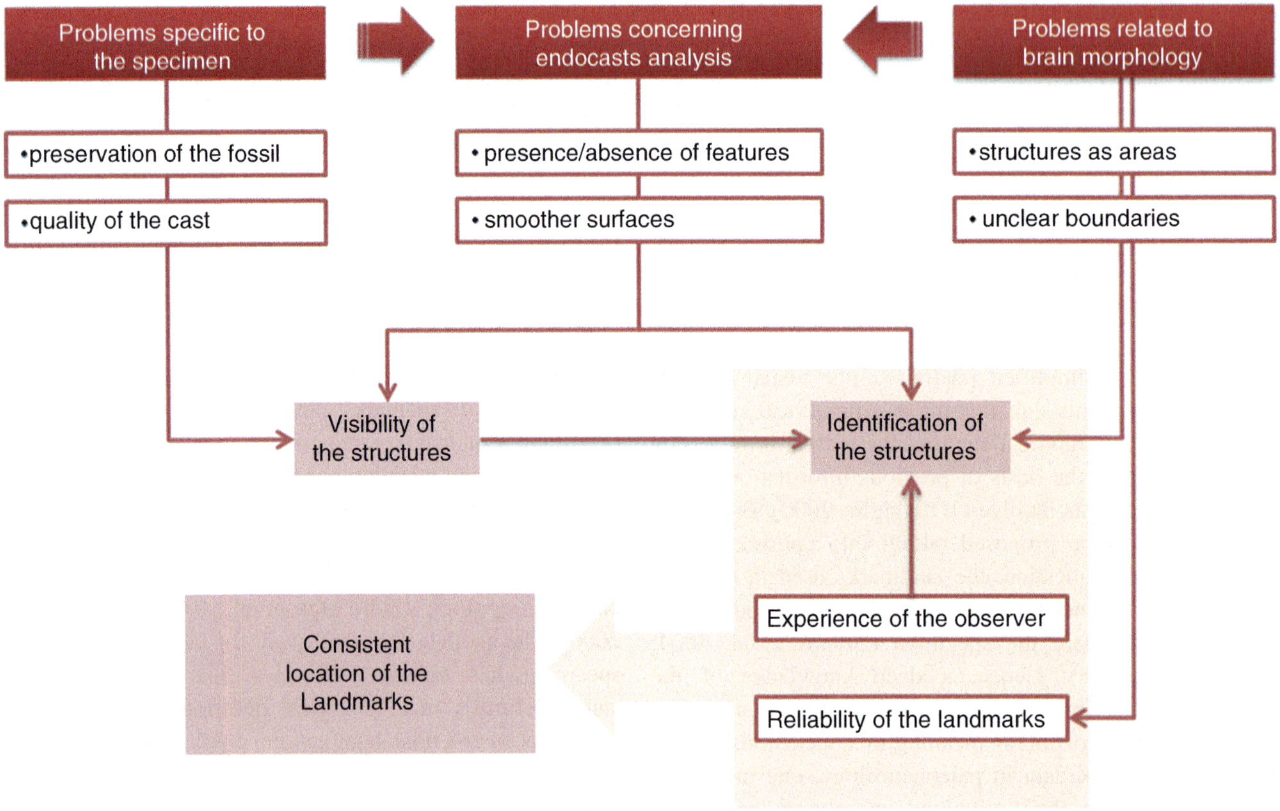

Fig. 9.2 Landmarking endocasts presents many challenges. First, the features displayed by the endocast depend on the quality and preservation of the specimen as well as on the clarity of the cortical imprints. Resolution of the specimen and its endocast will influence the visualization of the structures. The identification of the structures is then influenced by the variability of brain morphology, which structures are present and how visible they are, and by the experience of the observer. The poor visibility of boundaries related to brain morphology limits the use of reliable landmarks. Hence, the consistent location of landmarks depends on the experience of the observer, the quality of the structures imprinted and the type of landmarks that can be used for the analysis

error, like the variance or standard deviation of the coordinates' means, the linear distance or absolute difference between the location of a landmark in a given configuration and its position in the mean configuration of the sample, and the linear distance or absolute difference between the location of the same landmark in two different sessions (see Table 9.1).

Landmark precision has been frequently addressed in cranial material. Corner et al. (1992) digitized craniofacial landmarks on the same dry skull of *Macaca fascicularis* for 20 sessions. The skull and the digitizer were never moved between sessions, so that the coordinate system was constant and the landmarks could be directly compared without the need for superimposition by calculating the standard deviation for each axis (x, y and z). They obtained intra-observer error values of 0.27 and 0.26 mm, averaging all trials and the three axes. Von Cramon-Taubadel et al. (2007) measured the intra-observer error for only three landmarks, with the same computation as Corner et al., obtaining similar values for two observers: 0.25 and 0.27 mm. Richtsmeier et al. (1995) calculated landmark precision as the linear distance between the coordinates of the same landmark in repeated measures of the same CT image, obtaining an error of 0.27 mm, similar to that obtained for dry skulls (Table 9.1). Valeri et al. (1998) compared data from dry crania and the 3D reconstruction of CT scans in landmarking the main cranial bosses, named 'fuzzy landmarks'. They calculated intra-observer error through the variance-covariance matrices as estimated separately for each specimen, obtaining an error of 0.96 mm for the dry skull, a value larger than those from the previous studies which used well defined landmarks. For the 3D reconstruction, they digitized the landmarks 30 times, with an error of 1.15 mm. However, when compared to the last ten trials, they observed that there was a learning curve, as the intra-observer error, 0.75 mm, was comparable to those they had obtained for dry skulls.

The discrepancy in landmarking the brain is generally accomplished in 3D reconstructions of magnetic resonance images. Maudgil et al. (1998) were the first to address this issue in a study divided in two parts: first they addressed the homology of the brain sulci and then defined and tested landmarks in the intersections of those sulci. They obtained

Table 9.1 Intra-observer error in landmarking compiled from literature

			Error (mm)	
Data	Source	Obs	Mean (range)	Computation
Skull	Corner et al. (1992)	1	0.27 (0.12–0.36)	Standard deviation
		2	0.26 (0.21–0.32)	
	von Cramon-Taubadel et al. (2007)	1	0.25	
		2	0.27	
	Valeri et al. (1998)	1	0.96 (0.42–1.77)	
CT	Richtsmeier et al. (1995)	1	0.27 (0.12–0.48)	Euclidean distance
	Valeri et al. (1998)	1	1.15 (0.22–1.43)	Standard deviation
		1[a]	0.75 (0.19–0.90)	
3D MRI	Maudgil et al. (1998)	1	3.72 (2.8–5.0)	Euclidean distance
	Aldridge (2011)	1	1.97	Standard deviation
	Chollet et al. (2014)	1	1.9 (1.0–5.6)	Euclidean distance
Endocasts	Bienvenu et al. (2011)	1	1.37 (0.41–2.71)	Absolute difference

[a]These results regard the last ten trials only

an intra-observer error of 3.27 mm, calculated as the Euclidean distance between the corresponding landmarks. Analysing brain reconstruction, but with different computation, Aldridge (2011) and Chollet et al. (2014) obtained similar average values (1.97 and 1.9 mm, respectively), though lower than those of Maudgil (see Table 9.1). In general, brain landmarks display larger error than cranial landmarks, regardless of the parameter considered. However, landmark error on brains and endocasts was calculated on 3D digital samples, while cranial landmarks were sampled both on dry skulls and CT data. According to Aldridge (2011), precision is usually lower for cortical landmarks, in particular those located on the intersections of sulci, as some displayed errors larger than 2 mm. In fact, both Maudgil et al. (1998) and Chollet et al. (2014) consider the possibility of misidentification of some sulci, namely, the precentral and central sulci in the first study and the superior temporal sulcus in the second. It must be stressed that brain individual anatomical variation is outstanding, and the irregularity of the sulcal patterns may further complicate a univocal identification of conventional anatomical elements. Accordingly, anatomists need to partially rely on their personal experience and subjectivity to label cortical areas and cerebral folds, and a sulcal configuration can be interpreted differently by different anatomists. Although such inconsistency is usually minor, it must be acknowledged when dealing with brain cortical assessments.

The only analysis of landmark error in endocasts was performed by Bienvenu et al. (2011) on a sample of digital casts from five species of extant hominoids (orangutan, gorilla, chimpanzee, bonobo, and human). To address the intra-observer error, they sampled one specimen of each species twice and calculated the error as the difference between the coordinates of the homologous landmarks obtained in the two trials. The resulting intra-observer error, 1.37 mm, is comparable to those of brain and cranial bosses (Table 9.1) and was considered acceptable by the authors.

All of these error computations regard the precision of individual landmarks. Various authors advise against using Procrustes superimposition to calculate specific landmark error because the registration distributes the error of some landmarks throughout the configuration (Richtsmeier et al. 2002; von Cramon-Taubadel et al. 2007). The unit of the analysis is the specimen, and not the landmark, and 'the configuration is the datum' (Zelditch et al. 2004: p. 26). Accordingly, it may be useful to consider the uncertainty of the whole configuration more than the uncertainty of a single point. Von Cramon-Taubadel et al. (2007) showed the difference by comparing a full superimposition to a partial superimposition using three landmarks as a reference. The resulting error calculated after the partial superimposition (intra, 0.41 mm; inter, 0.76 mm) was larger than that calculated after the full superimposition (intra, 0.37 mm; inter, 0.48 mm). For this reason, it may be also of interest to analyse how observer error affects the landmarks as a group. Thus, assessing landmark error in the shape space, after registration, is fundamental to see how the landmark set is affected and how it influences the results. For instance, Hale et al. (2014) compared data taken from skulls and their correspondent 3D reconstructions to evaluate the differences in using the two types of samples. Although they found some separation between the two types of samples along the first principal component, these were still clustered together in the PCA. In addition, they obtained a misclassification rate of 48% in a cross-validation comparing the skull and CT samples. Therefore, they concluded that the influence of the type of data in the results was minimal, though the population variation must not be exceeded by the variation introduced by using different types of data.

9.5 Case Study: Landmark Uncertainty in Physical and Laser Scanned Endocasts

As we have stressed, endocasts are an important source of information in paleoneurology. However, information on the uncertainty and error associated with endocast landmarking is still scarce, probably due to their relatively recent inclusion in morphometric analyses. Moreover, the influence of the kind of data, i.e. physical or digital, has yet to be addressed. Therefore, because of the limits found when landmarking endocasts in human evolution, it is essential to evaluate the effect of observer error, individual and specific variation and type of data. A simple approach is to compare a sample of high-quality hominoid endocasts to their physical and digital versions (see Fig. 9.3), using cortical landmarks that have already been used in paleoneurology (Table 9.2; Fig. 9.4). In this case, we have used eight endocasts: human, gibbon, siamang, orangutan, gorilla, bonobo, male chimpanzee and female chimpanzee.

As in Corner et al. (1992), when the coordinate system is fixed, as in digital specimens, it is possible to calculate the absolute error associated with the placement of each landmark. In this case, absolute error can be addressed using only the digital sample as the coordinate system is fixed within the 3D images, and thus the variation in landmark location is only due to observer error. When considering the digital sample, we can observe that error values range from 0.15 to 3.44 mm, with a mean of 0.86 $\pm$ 0.51 mm (Fig. 9.5). Nonetheless, most of the values fall below 1.2 mm. These values are comparable to those obtained by Bienvenu et al. (2011) for digital endocasts, even though they used Euclidean distances. Interestingly, error values obtained for endocasts are generally larger than those obtained for cranial material (skull and CT) and smaller than those obtained for 3D MRI (see Table 9.1). Maudgil et al. (1998) and Chollet et al. (2014) obtained larger errors, probably due to difficulties in identifying some of the sulci, namely, the precental and central (Maudgil et al. 1998) and the temporal sulcus (Chollet et al. 2014). Some of the landmarks associated with larger error in our case study are actually associated with the same sulci. The lateral sulcus was also a problematic landmark in the study by Bienvenu et al. (2011). Nonetheless, the parietal landmarks, namely, the supramarginal gyrus (SG) and the angular gyrus (AG), seem to be the most challenging. These landmarks are not associated with grooves but with bulging and poorly defined areas, and they are placed on the endocast surface taking into account their reciprocal position, and with reference to the location of the lateral sulcus. However, Valeri et al. (1998) found that the parietal bosses displayed the lowest error among all the cranial bosses. It seems the cranial bulging areas are better defined on the external cranial surface than on the internal surface. Interestingly, although there is a general pattern of larger error associated with the parietal landmarks, some of the specimens have specific landmarks which display large deviations (e.g. the frontal poles in the human). Differences between hemispheres can also be observed. In fact, in some specimens, specific landmarks may display larger error on one side than on the other, although differences do not reach a statistical significance. In this sample, the largest amount

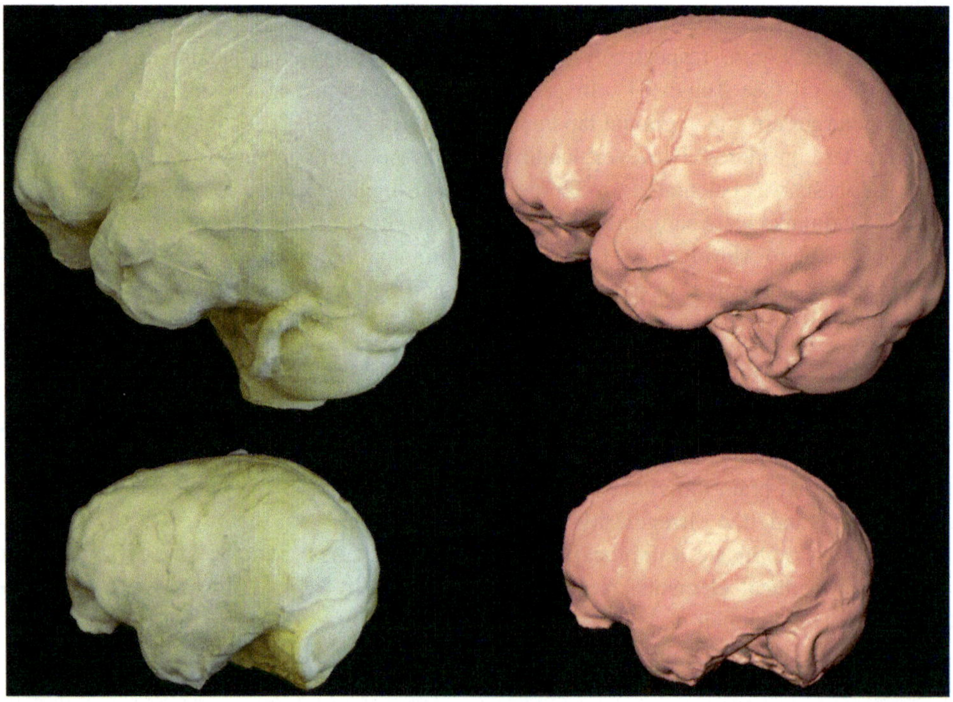

Fig. 9.3 Examples of the physical (*left*) and digital (*right*) endocasts used for the analysis. Physical casts (Bone Clones Inc.) are reproduced from original specimens, and the digital casts are replicas of the physical ones after laser scanning (NextEngine Inc.; resolution: 0.13 mm). The endocasts illustrated here are from a human (*Homo sapiens*), above, and a male chimpanzee (*Pan troglodytes*), below. The remaining endocasts used in the sample are from a male bonobo (*Pan paniscus*), a female chimpanzee, a male gibbon (*Hylobates moloch*), a male gorilla (*Gorilla gorilla*), a male orangutan (*Pongo pygmaeus*) and a male siamang (*Symphalangus syndactylus*)

Table 9.2 Description of the landmarks

Landmark		Acronym	Description
Bilateral	Frontal poles	FP	Anteriormost point, following maximum length
	Occipital poles	OP	Posteriormost point, following maximum length
	Temporal poles	TP	Tip point
	Cerebellar poles	CP	Lowermost point
	Broca's cap	BC	Maximum prefrontal width, namely, the posterior bulging area of the third frontal gyrus, corresponding to the Broca's cap in humans
	Lateral sulcus	LS	Posterior limit of the LS
	Supramarginal gyrus	SG	Maximum curvature point, above the LS
	Angular gyrus	AG	Maximum curvature point, behind the SG
Sagittal	Central sulcus	CS	Intersection of the CS with inter-hemispheric fissure
	Perpendicular sulcus	PS	Parieto-occipital boundary, central point
	Internal occipital protuberance	IOP	Inter-hemispheric and cerebro-cerebellar separation, central point

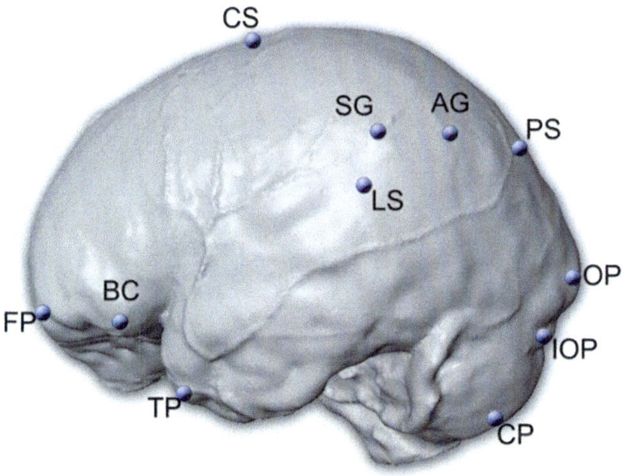

Fig. 9.4 Landmarking uncertainty was evaluated on the following cortical landmarks: frontal poles (*FP*), occipital poles (*OP*), temporal poles (*TP*), cerebellar poles (*CP*), Broca's cap (*BC*), lateral sulcus (*LS*), supramarginal gyrus (*SG*), angular gyrus (*AG*), central sulcus (*CS*), perpendicular sulcus (*PS*) and internal occipital protuberance (*IOP*). Details on landmarks can be seen in Table 9.2. The landmarks were collected from each endocast with a Microscribe G2X digitizer (Immersion Corporation: resolution, 0.23 mm) for the physical moulds and with Landmark Editor (Wiley et al. 2005) for the digital ones. Each specimen was sampled ten times (ten times the physical replica and ten times the digital replica) by the same observer (ASPP), in a random order to minimize systematic errors caused by improvement through time, memorization (Valeri et al. 1998) or fatigue-induced quality deterioration (Hammer and Harper 2006)

of error is displayed by the human endocast, with the maximum values on the left angular gyrus (3.44 mm) and the left supramarginal gyrus (2.67 mm), while the gibbon and the siamang show the smallest figures. It is worth noting that smaller endocasts seem to display the endocranial features more clearly (Kobayashi et al. 2014). Even so, the relative error is very small and comparable between all specimens (Table 9.3).

In paleoanthropology, the use of virtual replicas is increasing, and it is important to investigate whether there is a substantial difference in landmark localization when using digital or physical samples. To address this question, we compared the physical and digital data in the same analysis after Procrustes superimposition. Although comparing error after superimposition can introduce a bias due to the redistribution of the difference throughout the configuration (Richtsmeier et al. 2002; von Cramon-Taubadel et al. 2007), the distribution of the variation at each landmark can be informative when the analysis aims to provide within-sample differences (Gómez-Robles et al. 2014). In this case, we are not interested in the actual biological meaning of the landmark variation, but only in their response to the methodological procedure, i.e. to the different analytical behaviours of the digital and physical samples. Similarly to Hale et al. (2014) for skulls and CT scans, the effects of mixing silicon and laser scan endocasts in the same analysis are minimal. The specimens (in this case representing species) are clearly separated, and the physical and digital replicas from the same specimen group together (Fig. 9.6). This means that the error is low when compared with the species-specific differences, and thus these landmarks can be reliably used in evolutionary comparisons. Evidently, in this example, we have used one specimen per species (except for male and female chimps), so we have not considered intraspecific variation. A large intraspecific variation could add a further source of noise, and we ignore the degree of possible spatial overlapping between species. Nonetheless, it must be remarked that intraspecific and interspecific variations are the result of different processes and have different numerical properties (Martin and Barbour 1989) and hence should not be mixed into a same multivariate space. A multivariate analysis on individual values should be computed only when dealing with intraspecific variation, while interspecific analyses should be performed on species means (see Chap. 7). A very interesting integration between these two alternatives is between-group PCA, computing the vectors on the group mean values and then projecting individual values within the resulting morphospace.

Apart from the separation among species, in our survey, it is worth noting that physical and digital replicas are slightly different, indicating that there are systemic (endocast-specific) factors related to the different data source which can influence the spatial perception of the landmarks. For the

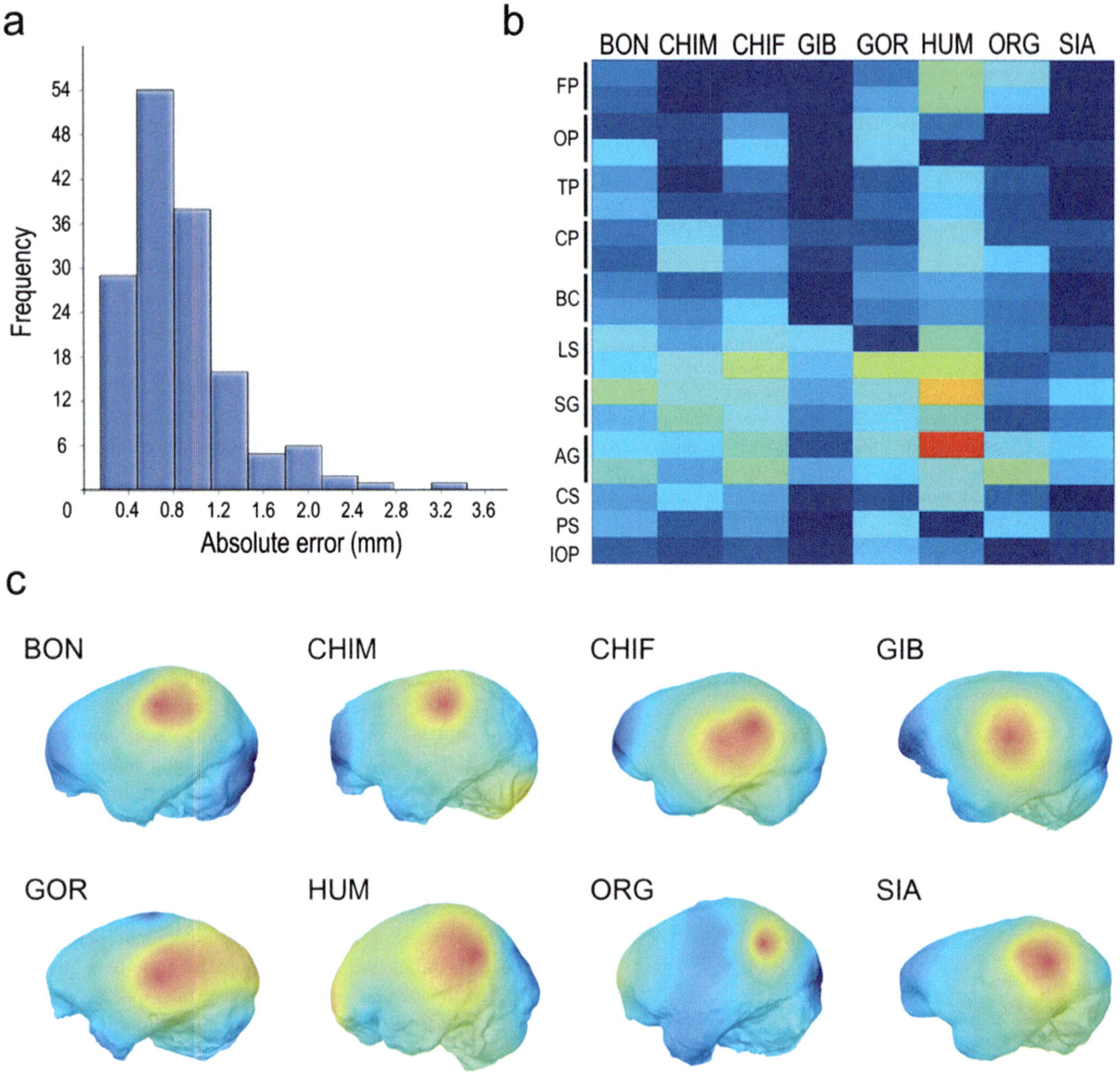

Fig. 9.5 Absolute intra-observer error was calculated on the digital sample as the standard deviation for each coordinate (x, y, z) and then the mean of these values for each landmark point. The statistics were performed with PAST 2.17c (Hammer et al. 2001). The majority of values are lower than 1.2 mm, as can be seen on the histogram (**a**). In the matrix of error distribution (**b**) per landmark and per endocast (for bilateral landmarks left is represented on the upper cell), values increase from *blue* to *red*, showing that the highest values are located on the landmarks LS, SG and AG on both hemispheres. The same patterns are shown on the maps of error pattern (**c**), graphically computed by plotting the landmark-wise values (averaged for bilateral landmarks) and the values interpolated by a gridding function with a multiquadratic algorithm. Note that the only aim of these maps is to provide an immediate visual output for the differences between specimens regarding the distribution of discrepancy on each endocast and are not intended to be an interpolation analysis. Specimens are *BON* bonobo, *CHIM* male chimpanzee, *CHIF* female chimpanzee, *GIB* gibbon, *GOR* gorilla, *HUM* human, *ORG* orangutan, *SIA* siamang

human cast, digital and physical separation is loaded onto PC2, in female chimp onto PC1, while in other specimens, there are intermediate situations. Because the pattern is not the same for all the specimens, these factors might be specimen-specific rather than due to a general bias introduced with digital casting and virtual perception of the anatomical surface. The digital sample displays larger discrepancies than the physical sample (Fig. 9.7a, b), but the difference is not significant. In fact, some landmarks show larger variation on the physical sample and others on the digital sample (Fig. 9.7c). The same can be evidenced for the hemispheric differences: patterns of error are not constant

Table 9.3 Absolute and relative errors

	AE	EM	RE
GIB	0.49	59.7	0.8%
SIA	0.52	63.8	0.8%
ORG	0.79	92.5	0.9%
CHIF	0.98	94.8	1.0%
BON	0.92	96.0	1.0%
CHIM	0.81	99.2	0.8%
GOR	0.94	100.8	0.9%
HUM	1.40	145.5	1.0%
Mean			0.9%

RE relative error, calculated as the ratio AE/EM, *AE* absolute error, *EM* endocranial module, computed after Holloway et al. (2004) as the mean of maximum hemispheric length (averaged hemispheric value), maximum endocranial width and basion-vertex height

for the left and right side in both physical and digital samples. Therefore, these results suggest that although some endocranial features can be perceived differently in the physical and digital reconstructions, these differences are associated with individual characteristics of the specimen and not to a systemic bias between the two different data sources.

The possible reasons behind the landmark discrepancy of the physical and digital data can be visualized by directly comparing the specimens (Fig. 9.8). The uncertainty in localizing landmarks can be influenced by specific anatomical features, like the presence of vessels or sutures, the degree of smoothness of the cortical surface and the morphology or degree of expression of the sulcal pattern. Hence, this study suggests that such individual features may be the main source of possible discrepancy, more than generalized factors associated with the endocranial geometry, cerebral asymmetries or with the physical/digital rendering. An idiosyncratic pattern of error was also mentioned by Schoenemann et al. (2007) concerning their comparison of digital and physical endocast reconstruction. Despite these differences, on both data types, the landmarks displaying greater variation are the same: the lateral sulcus and the supramarginal and angular gyri (Figs. 9.7 and 9.8), where references for an exact localization of the landmarks are less abundant.

Four main conclusions can be drawn from this survey. First, the error associated with locating cortical landmarks on virtual endocasts is acceptable at least when dealing with species-specific differences in hominoids, and these landmarks can be used for phylogenetic studies. Second, the effect of the different kind of data is minimal, with no patent differences between digital and physical landmarking. Third, the pattern of landmark discrepancy, rather than being associated with specific and shared anatomical factors, is largely related to the topographic characteristics of each specimen. Fourth, the parietal landmarks are the most problematic as they are located on areas that lack points of reference, and thus further caution should be taken when sampling landmarks to identify homologous areas in this region.

9.6 Conclusions and Future Challenges

In this chapter, we reviewed the challenges of landmark analysis when dealing with paleoneurological samples. The difficulties in landmarking endocasts are intrinsically associated with the quality of the endocast, the clearness of the imprints of brain features and the types of landmark we can sample. Indeed, the intra-observer error in landmarking the brain and endocasts is larger than that of landmarking skulls and bones. Nonetheless, intra-observer error on endocasts is still small and definitely acceptable at least when dealing with interspecific differences in humans and apes. Although digital samples may provide additional challenges when localizing landmarks, the effect seems to be marginal, and results are similar to those obtained from physical casts. Taking into account the relevant advantages of the digital replicas (in terms of reconstructions, geometrical modelling and calculation), working with virtual endocasts is definitely a convenient approach. Needless to say, an optimal option would be combining the two kinds of data or at least having the possibility to consider and evaluate the cortical features also on a physical mould. It is worth noting that the recognition of cortical features on an endocast is not only based on the feature itself but is better performed by evaluating the overall distribution of the cortical areas. That is, landmarking a specific cortical trait requires the evaluation of the morphology associated with that trait, plus an evaluation of the neighbouring areas, which delimitate and restrict the available positions of their contiguous elements. As such, 'reading endocasts' is something which should be performed taking into account the whole object and not its features separately (Bruner 2017).

Because of the limits associated with endocranial morphology, the evaluation of intra-observer errors should always be acknowledged in research papers using endocasts, in order to consider the actual range of resolution available.

Bosses representing areas show more uncertainty than sulcal references. In this case, the use of sliding landmarks can seriously add to the analysis. It must be however taken into consideration that sliding landmarks deal strictly with the geometry of the object, more than with its biological boundaries and proportions. Further issues on landmark uncertainties concern asymmetries, taking into account that

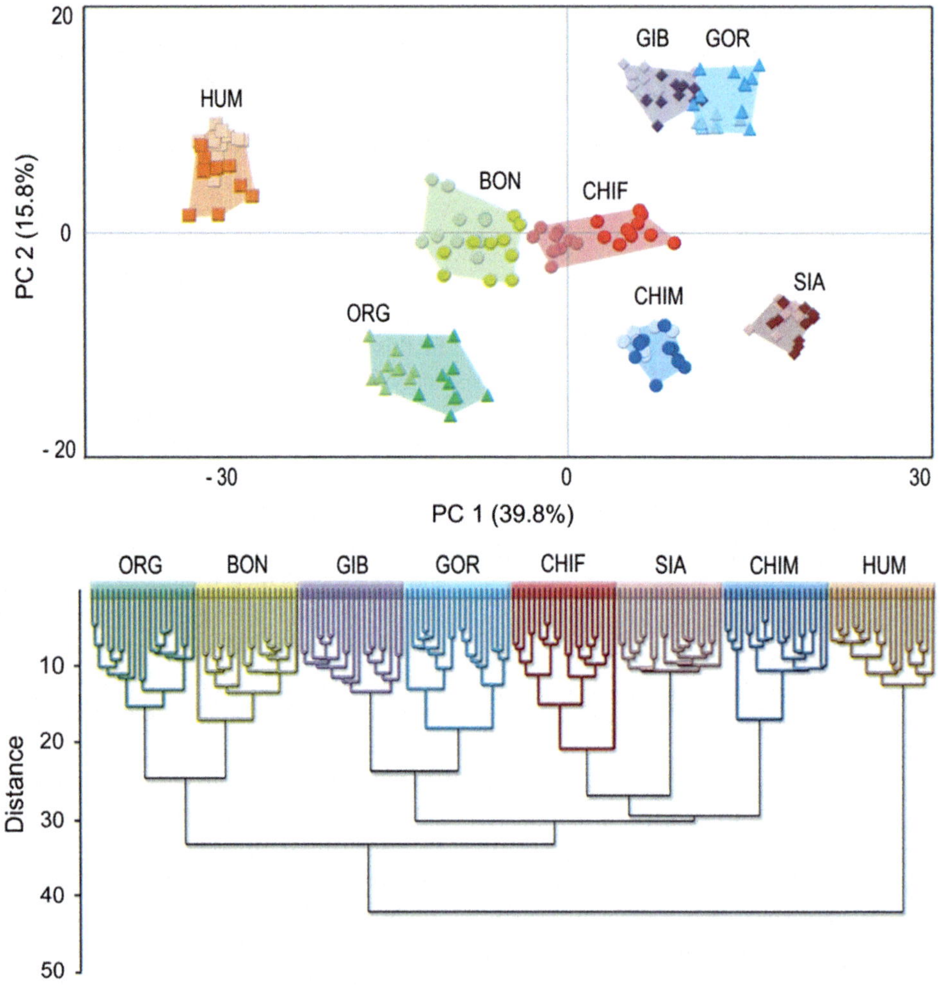

Fig. 9.6 After the Procrustes superimposition, the replicas from physical and digital samples, though with some separation, cluster together in shape space. The first two principal components represent together 55.6% of the variance. PC1 is associated with endocranial heights, separating taller vs flatter endocasts largely because of the position of the central sulcus, lateral sulcus and supramarginal gyrus. PC2 is associated with endocranial lengthening, mostly influenced by frontal proportions. The UPGMA phenogram (cophenetic correlation coefficient = 0.88) further illustrates the clustering, indicating that the discrepancy does not affect the shape analysis. Specimens are shown in Fig. 9.5

often one side (generally the left one) shows sharper and more shaped sulcal patterns.

In this discipline more than in others, the personal experience of the anatomist may play a determinant role when dealing with interobserver discrepancies (Ross and Williams 2008). In many fields, the computation of an interobserver error is a necessary step to add certainty and, most of all, to allow that data can be reliably shared among different researchers. Taking into account the complex anatomy of the brain, the lack of information on many anatomical aspects, the large individual variability and the partial and incomplete relationships between the brain and endocast, in this case (endocast landmarking), we simply suggest avoiding mixing data from different observers. However, digital sampling in morphometrics allows a new possibility: a multiple-observers option. Digital anatomical replicas are generally associated with a fixed spatial reference, and therefore the position of the landmarks can be established after multiple-observers agreement based on discussion, evaluation and different sessions aimed at reaching a common consensus between different anatomists (e.g. Bruner et al. 2014b). Finally, all these aspects indicate that there should be no mixing of data from different observers in different studies.

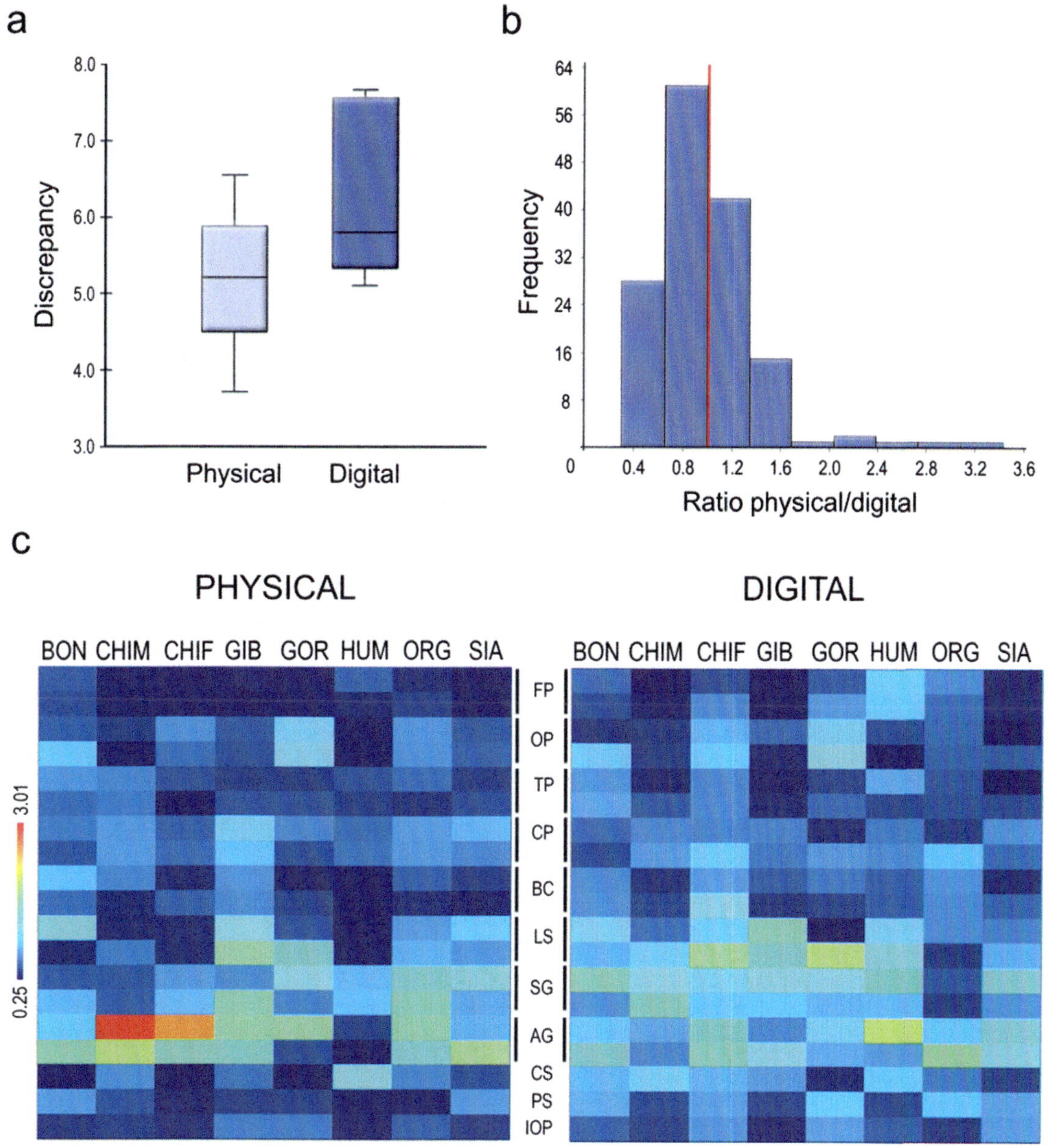

Fig. 9.7 Comparison between physical and digital discrepancy. Landmark variation was calculated on the Procrustes residuals with the same computation as in the absolute error. The overall variation for each specimen (physical and digital) was computed as the sum of its standard deviations of the first four principal components (whose percentage of explained variance is above 5%). Such discrepancy is larger for the digital sample than for the physical sample (**a**), although difference is not statistically significant. A cell-to-cell ratio (**b**) of physical/digital matrices (**c**) shows that the majority of landmarks have larger values on the digital sample (the same values give a result of 1, represented by the *thick line*). The matrices and the specimens are shown in Fig. 9.5

As a final remark, it is necessary to bear in mind that shape analysis of endocasts is just a preliminary step in paleoneurology. Endocranial morphology must be inevitably considered within a wider structural frame constituted at least by the overall cranial system. Primary neural changes must be separated from secondary changes due to cranial constraints (Bruner 2015). Furthermore, ultimately, endocranial morphological changes must be interpreted

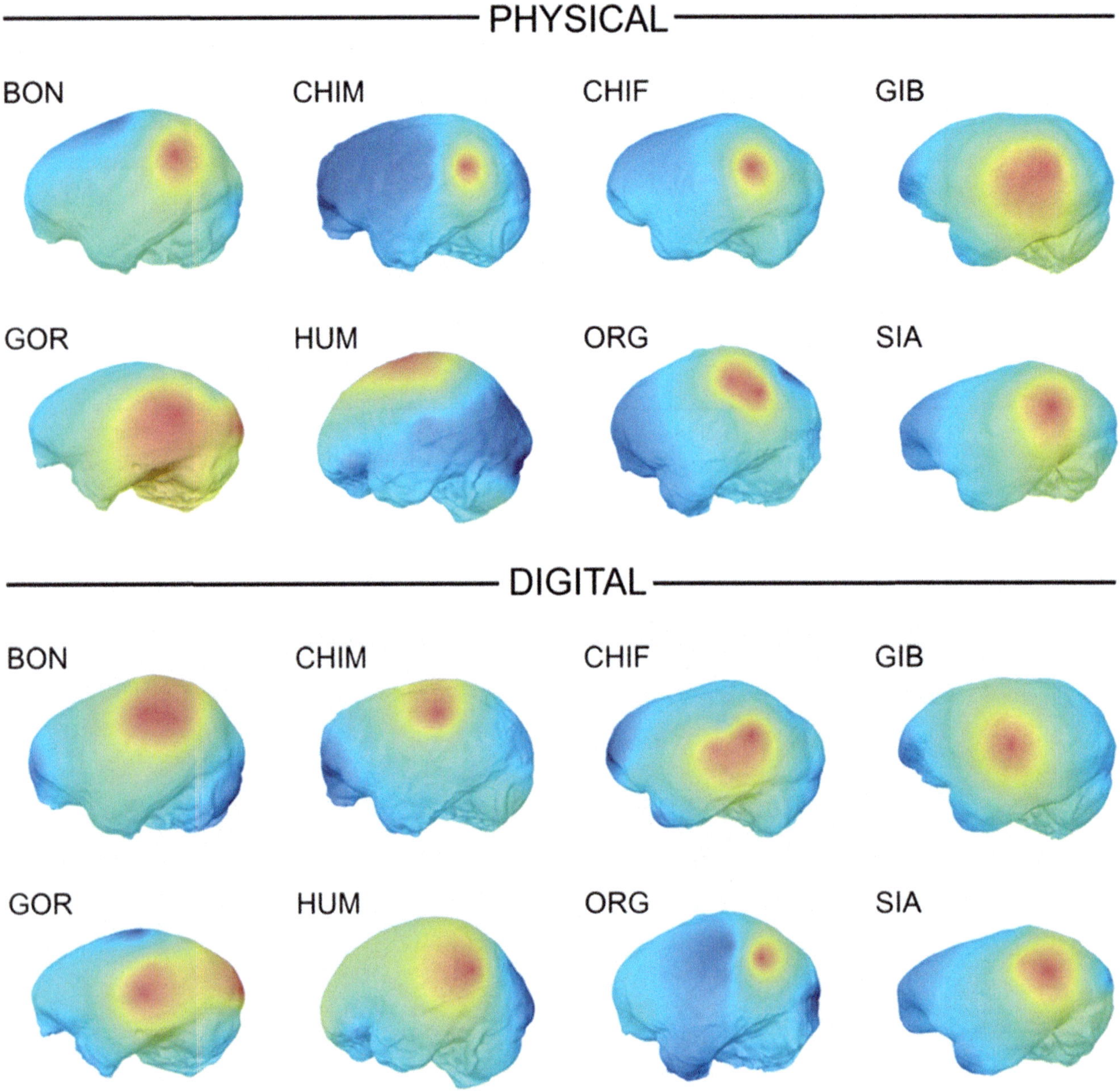

Fig. 9.8 The pattern of landmark discrepancy differs between the physical and digital samples, even when comparing specimen-by-specimen. For instance, on the human endocast, there is larger discrepancy on the central sulcus on the physical sample and on the angular gyrus on the digital sample. Nonetheless, in general, the larger discrepancies are located on the parietal region. Maps and specimens are shown in Fig. 9.5

according to the available information on the histological and cytological factors associated with the anatomical variability, as well as with their functional meaning. Concerning sulcal patterns, it must be taken into account that they can be the passive result of mechanical folding schemes, with no direct functional phylogenetic meaning (Toro 2012; Bayly et al. 2014; Tallinen et al. 2016). In this case, sulcal morphology can nonetheless provide information on the morphogenetic processes associated with the cortical development, depending on factors associated with time and rate of cortical growth, differential volumetric increases and tissue biomechanical properties. Without a proper biological context, the study of endocasts remains a descriptive study of phenotypic similarity that, although relevant, cannot be used to test specific hypotheses on brain evolution.

Acknowledgements This paper is funded by the Spanish Government (CGL2015-65387-C3-3-P). ASPP is also funded by the Atapuerca Foundation, Spain. We are grateful to Eugénia Cunha, Aida Gómez-Robles, José Manuel de la Cuétara, Markus Bastir, Hana Píšová and Gizéh Rangel de Lázaro for their help and suggestions.

References

Adams DC, Rohlf FJ, Slice DE (2004) Geometric morphometrics: ten years of progress following the "revolution". Ital J Zool 71: 5–16

Aldridge K (2011) Patterns of differences in brain morphology in humans as compared to extant apes. J Hum Evol 60: 94–105

Amano H, Kikuchi T, Morita Y, Kondo O, Suzuki H, Ponce de León MS, Zollikofer CPE, Bastir M, Stringer C, Ogihara N (2015) Virtual reconstruction of the Neanderthal Amud 1 cranium. Am J Phys Anthropol 158:185–197

Bayly PV, Taber LA, Kroenke CD (2014) Mechanical forces in cerebral cortical folding: a review of measurements and models. J Mech Behav Biomed Mater 29:568–581

Bienvenu T, Guy F, Coudyzer W, Gilissen E, Roualdès G, Vignaud P, Brunet M (2011) Assessing endocranial variations in great apes and humans using 3D data from virtual endocasts. Am J Phys Anthropol 145:231–246

Bookstein FL (1997) Landmark methods for forms without landmarks: morphometrics of group differences in outline shape. Med Image Anal 1(3):225–243

Bookstein FL (1991) Morphometric tools for landmark data: geometry and biology. Cambridge University Press, New York

Bookstein F, Schäfer K, Prossinger H, Seidler H, Fieder M, Stringer C, Weber GW, Arsuaga J-L, Slice DE, Rohlf FJ, Recheis W, Mariam AJ, Marcus LF (1999) Comparing frontal cranial profiles in archaic and modern *Homo* by morphometric analysis. Anat Rec 257:217–224

Bruner E (2004) Geometric morphometrics and paleoneurology: brain shape evolution in the genus *Homo*. J Hum Evol 47: 279–303

Bruner E (2015) Functional craniology and brain evolution. In: Bruner E (ed) Human paleoneurology, vol 3. Springer, Cham, pp 57–93

Bruner E (2017) The fossil evidence of human brain evolution. In: Kaas J (ed) Evolution of nervous systems 2e volume 4. Springer, Oxford, pp 63–92

Bruner E, Bastir M (2009) Landmarks could slide, brains cannot: interpreting models of shape variation. In: Abstracts in Memòria Especial 3 of the I Iberian Symposium on Geometric Morphometrics, Paleontologia I Evoluciò, Barcelona, 25–26 July 2009

Bruner E, Ripani M (2008) A quantitative and descriptive approach to morphological variation of the endocranial base in modern humans. Am J Phys Anthropol 137:30–40

Bruner E, Manzi G, Arsuaga JL (2003) Encephalization and allometric trajectories in the genus *Homo*: evidence from the Neandertal and modern lineages. Proc Natl Acad Sci 100: 15335–15340

Bruner E, De la Cuétara JM, Masters M, Amano H, Ogihara N (2014a) Functional craniology and brain evolution: from paleontology to biomedicine. Front Neuroanat 8:19

Bruner E, Rangel de Lázaro G, de la Cuétara JM, Martín-Loeches M, Colom R, Jacobs HIL (2014b) Midsagittal brain variation and MRI shape analysis of the precuneus in adult individuals. J Anat 224:367–376

Cheverud JM (1996) Developmental integration and the evolution of pleiotropy. Am Zool 36:44–50

Chollet MB, Aldridge K, Pangborn N, Weinberg SM, DeLeon VB (2014) Landmarking the brain for geometric morphometric analysis: an error study. PLoS ONE 9:e86005

Corner BD, Lele S, Richtsmeier JT (1992) Measuring precision of three-dimensional landmark data. J Quant Anthropol 3:347–359

Free SL, O'Higgins P, Maudgil DD, Dryden IL, Lemieux L, Fish DR, Shorvon SD (2001) Landmark-based morphometrics of the normal adult brain using MRI. NeuroImage 13:801–813

Gómez-Robles A, Hopkins WD, Sherwood CC (2014) Modular structure facilitates mosaic evolution of the brain in chimpanzees and humans. Nat Commun 5:4469

Goriely A, Geers MGD, Holzapfel GA, Jayamohan J, Jérusalem A, Sivaloganathan S, Squier W, Van Dommelen JAW, Waters S, Kuhl E (2015) Mechanics of the brains: perspectives, challenges, and opportunities. Biomech Model Mechanobiol 14:931–965

Gunz P, Mitteroecker P (2013) Semilandmarks: a method for quantifying curves and surfaces. Hystrix 24:103–109

Gunz P, Mitteroecker P, Bookstein FL (2005) Semilandmarks in three dimensions. In: Slice DE (ed) Modern morphometrics in physical anthropology. Kluwer Academic/Plenum Publishers, New York, pp 73–98

Gunz P, Mitteroecker P, Neubauer S, Weber GW, Bookstein FL (2009) Principles for the virtual reconstruction of hominin crania. J Hum Evol 57:48–62

Gunz P, Neubauer S, Maureille B, Hublin J-J (2010) Brain development after birth differs between Neanderthals and modern humans. Curr Biol 20:R921–R922

Hale AR, Honeycutt KK, Ross AH (2014) A geometric morphometric validation study of computed tomography-extracted craniofacial landmarks. J Craniofac Surg 25:231–237

Hammer Ø, Harper DAT (2006) Paleontological data analysis. Blackwell Publishing, Oxford

Hammer Ø, Ryan P, Harper DAT (2001) PAST: paleontological statistics software package for education and data analysis. Palaeontol Electron 4:9

Holloway RL (1975) The role of human social behavior in the evolution of the brain. James Arthur lecture on the evolution of the human brain. American Museum of Natural History, New York

Holloway RL (1981) Exploring the dorsal surface of hominoid brain endocasts by stereoplotter and discriminant analysis. Philos Trans R Soc Lond B 292:155–166

Holloway RL, Broadfield DC, Yuan MS (2004) The human fossil record, volume 3: brain endocasts, the paleoneurological evidence. Wiley-Liss, Hoboken

Huxley TH (1863) Evidence as to man's place in nature. Williams & Norgate, London

Klingenberg CP (2008) Novelty and "homology-free" morphometrics: what's in a name? Evol Biol 35:186–190

Klingenberg CP (2010) Evolution and development of shape: integrating quantitative approaches. Nat Rev Genet 11:623–635

Klingenberg CP (2013) Cranial integration and modularity: insights into evolution and development from morphometric data. Hystrix 24: 43–58

Kobayashi Y, Matsui T, Haizuka Y, Ogihara N, Hirai N, Matsumura G (2014) Cerebral sulci and gyri observed on macaque endocasts. In: Akazawa T et al (eds) Dynamics of learning in Neanderthals and modern humans volume 2: cognitive and physical perspectives, replacement of Neanderthals by modern humans series. Springer, Japan, pp 131–137

Lieberman DE, Pearson OM, Mowbray KM (2000) Basicranial influence on overall cranial shape. J Hum Evol 38:291–315

Lieberman DE, McBratney BM, Krovitz G (2002) The evolution and development of cranial form in *Homo sapiens*. Proc Natl Acad Sci U S A 99:1134–1139

Martin RD, Barbour AD (1989) Aspects of line-fitting in bivariate allometric analyses. Folia Primatol 53:65–81

Maudgil DD, Free SL, Sisodiya SM, Lemieux L, Woermann FG, Fish DR, Shorovon SD (1998) Identifying homologous anatomical landmarks on reconstructed magnetic resonance images of the human cerebral cortical surface. J Anat 193:559–571

Mitteroecker P, Gunz P (2009) Advances in geometric morphometrics. Evol Biol 36:235–247

Moss ML, Young RW (1960) A functional approach to craniology. Am J Phys Anthropol 18:281–292

Neubauer S, Gunz P, Hublin J-J (2009) The pattern of endocranial ontogenetic shape changes in humans. J Anat 215: 240–255

Neubauer S, Gunz P, Hublin J-J (2010) Endocranial shape changes during growth in chimpanzees and humans: a morphometric analysis of unique and shared aspects. J Hum Evol 59:555–566

O'Higgins P (2000) The study of morphological variation in the hominid fossil record: biology, landmarks and geometry. J Anat 197: 103–120

Oxnard C, O'Higgins P (2015) Biology clearly needs morphometrics. Does morphometrics need biology? Biol Theory 4:84–97

Oyen OJ, Walker A (1977) Stereometric craniometry. Am J Phys Anthropol 46:177–182

Ponce de León MS, Zollikofer CPE (2001) Neanderthal cranial ontogeny and its implications for late hominid diversity. Nature 412: 534–538

Richtsmeier JT, Paik CH, Elfert PC, Cole TM III, Dahlman HR (1995) Precision, repeatability, and validation of the localization of cranial landmarks using computed tomography scans. Cleft Palate Craniofac J 32:217–227

Richtsmeier JT, DeLeon VB, Lele SR (2002) The promise of geometric morphometrics. Am J Phys Anthropol S35:63–91

Richtsmeier JT, Aldridge K, DeLeon VB, Panchal J, Kane AA, Marsh JL, Yan P, Cole TM III (2006) Phenotypic integration of neurocranium and brain. J Exp Zool 306B:360–378

Rohlf FJ, Marcus LF (1993) A revolution in morphometrics. Trends Ecol Evol 8:129–132

Ross AH, Williams S (2008) Testing repeatability and error coordinate landmark data acquired from crania. J Forensic Sci 53: 782–785

Schoenemann PT, Gee J, Avants B, Holloway RL, Monge J, Lewis J (2007) Validation of plaster endocasts morphology through 3D CT image analysis. Am J Phys Anthropol 132(2):183–92

Scott N, Neubauer S, Hublin J-J, Gunz P (2014) A shared pattern of postnatal endocranial development in extant hominoids. Evol Biol 41:572–594

Spoor F, Jeffrey N, Zonneveld F (2000) Using diagnostic radiology in human evolutionary studies. J Anat 197:61–76

Tallinen T, Chung JY, Rousseau F, Girard N, Lefèvre J, Mahadevan L (2016) On the growth and form of cortical convolutions. Nat Phys 12:588–593

Thompson D'A (1942) On growth and form. Cambridge University Press, Cambridge

Toro R (2012) On the possible shapes of the brain. Evol Biol 39: 600–612

Valeri CJ, Cole TM, Lele S, Richtsmeier JT (1998) Capturing data from three-dimensional surfaces using fuzzy landmarks. Am J Phys Anthropol 107:113–124

Verheyen WN (1957). Bijdrage tot de craniometrie van Colobus badius (Kerr 1792). Annales du Musée Royal du Congo Belge 8/62, Tervuren

von Cramon-Taubadel N, Frazier BC, Lahr MM (2007) The problem of assessing landmark error in geometric morphometrics: theory, methods, and modifications. Am J Phys Anthropol 134:24–35

Weidenreich F (1941) The brain and its role in the phylogenetic transformation of the human skull. Trans Am Phil Soc NS XXXI: 321–442

Wiley DF, Amenta N, Alcantara DA, Ghosh D, Kil YJ, Delson E, Harcourt-Smith W, Rohlf FJ, St john K, Hamann B (2005) Evolutionary morphing. In: Proceedings of IEEE visualization, pp 431–438

Zelditch ML, Swidersky DL, Sheets HD, Fink WL (2004) Geometric morphometrics for biologists. Elsevier, San Diego

Zollikofer CPE, Ponce de León MS (2013) Pandora's growing box: inferring the evolution and development of hominin brains from endocasts: Pandora's growing box. Evol Anthropol 22:20–33

Zollikofer CPE, Ponce De León MS, Martin RD (1998) Computer-assisted paleoanthropology. Evol Anthropol 6:41–54

Comparing Endocranial Surfaces: Mesh Superimposition and Coherent Point Drift Registration

10

Ján Dupej, Gizéh Rangel de Lázaro, Ana Sofia Pereira-Pedro, Hana Píšová, Josef Pelikán, and Emiliano Bruner

Abstract

The endocranial cavity is a major source of information for the assessment of brain morphology in extinct species. Digital molds of the endocranium can be reconstructed from fossil remains. In paleoneurology, these so-called endocasts are examined using shape analysis and multivariate statistical methods to quantify differences among species and individuals. These surfaces are relatively smooth and offer few landmarks; as such, morphometric comparisons are not straightforward, and correspondence search algorithms are necessary to identify loci of equivalent anatomical variation. Many solutions to this so-called correspondence problem have been proposed, but these often require considerable manual input. Here, we present the application in paleoneurology of a correspondence search and symmetrization algorithm originally designed for facial and palatal scans. Homologous representations of surfaces were used to render a computed visualization of differences in shape between modern humans, Neanderthals, archaic humans, and chimpanzees.

Keywords

Endocasts • Surface analysis • Correspondence problem • Paleoneurology

10.1 Introduction

Paleoneurologists rely on endocranial casts to study cerebral morphology in extinct species, as the neurocranium offers some direct evidence about the shape and size of the brain. Volumes and impressions in the endocranial cavity can be used to create casts using computer graphics tools, which are commonly referred to as endocasts. Although these casts provide only partial information on brain anatomy, they are nonetheless the only direct anatomical evidence of variations

J. Dupej (✉)
Katedra software a výuky informatiky, Matematicko-fyzikální fakulta, Univerzita Karlova, Malostranské náměstí 25, 118 00 Praha 1, Czech Republic

Katedra antropologie a genetiky člověka Přírodovědecká fakulta, Univerzita Karlova, Viničná 7, 128 43 Praha 2, Czech Republic
e-mail: jdupej@cgg.mff.cuni.cz

G. Rangel de Lázaro
Área de Prehistoria, Universitat Rovira i Virgili, Avinguda Catalunya 35, 43002 Tarragona, Spain

Institut Català de Paleoecologia Humana i Evolució Social (IPHES), Zona Educacional 4, Campus Sescelades URV (Edifici W3), 43007 Tarragona, Spain
e-mail: gizeh.rangel@urv.cat

A.S. Pereira-Pedro • E. Bruner
Programa de Paleobiología, Centro Nacional de Investigación sobre la Evolución Humana, Paseo Sierra de Atapuerca 3, 09002 Burgos, Spain
e-mail: sofia.aspp@gmail.com; emiliano.bruner@cenieh.es

H. Píšová
Katedra antropologie a genetiky člověka Přírodovědecká fakulta, Univerzita Karlova, Viničná 7, 128 43 Praha 2, Czech Republic
e-mail: hana.pisova@natur.cuni.cz

J. Pelikán
Katedra software a výuky informatiky, Matematicko-fyzikální fakulta, Univerzita Karlova, Malostranské náměstí 25, 118 00 Praha 1, Czech Republic
e-mail: pepca@cgg.mff.cuni.cz

E. Bruner et al. (eds.), *Digital Endocasts*, Replacement of Neanderthals by Modern Humans Series,
DOI 10.1007/978-4-431-56582-6_10

in brain evolution. Because an endocast only captures the shape of the neurocranial space, surface methods are useful to analyze and quantify its geometry. Surfaces can be captured using surface scanners, although computed tomography is the most common approach (Gunz 2015; see Chap. 12). Traditional morphometrics including length, distance, and angle can provide some insight into endocranial variability (Holloway et al. 2004; Bruner and Holloway 2010; Bruner et al. 2011). However, given the complex geometry of endocranial casts, landmark-based geometric morphometrics can provide more useful quantitative data (Bruner 2004; Neubauer et al. 2009; Gunz et al. 2010).

Surfaces can be represented as triangle meshes—i.e., piecewise linear simplifications of the original surface that exhibit no homology and have variable vertex counts, to which geometric morphometrics cannot be directly applied. A simple solution to this problem is to select landmarks on the studied surface and analyze explicit correspondences between these points using landmark-based approaches. These have the advantage of being computationally simple and—for biologically relevant landmarks—of relying on specific anatomical elements. A surface consisting of tens or hundreds of thousands of points is typically reduced to a small number of landmarks. However, many surfaces and structures do not offer sufficient possibilities for landmarks owing to the lack of salient and reliable anatomical features. In such cases, computational geometry and computer graphics approaches can be applied that take into account the entire surface after standardized preprocessing transformations. The challenge in identifying corresponding vertices or other primitives in different datasets is known as the *correspondence problem.*

10.2 Correspondence Problem

Since the earliest applications of geometric morphometrics to neurocranial outlines, landmarks were selected using homologous or geometrical references (e.g., Lieberman et al. 2002; Bookstein et al. 2003; Bruner et al. 2004). *Generalized Procrustes Analysis* (GPA; Gower 1975) is used to normalize coordinate variation and to account for size, position, and orientation. Thin plate spline (TPS) interpolation function is used to decompose and visualize differences in shape (Bookstein 1991). The identification or generation of homologous vertices in sets of surface models is a most challenging problem. Many algorithms have been proposed that vary in their computational complexity, need for explicit correspondences, and parameters and other computational aspects. A common characteristic is that these methods generate samplings of the analyzed surfaces that can be considered "homologous" at least in geometrical terms and thus can be processed with morphometric or statistical methods. This is usually accomplished by transferring the topology of a template surface to all analyzed surfaces. The template is either an idealized surface supplied prior to processing or one of the surfaces from the sample. Alternatively, approaches such as sliding semilandmarks create a regular sampling of each surface (Bookstein 1997). A specific approach that represents surfaces and curves with semilandmarks has been developed in Gunz and Mitteroecker (2013) and Neubauer et al. (2009). Initial semilandmarks are distributed along a curve or surface, and then their position along those objects is iteratively refined until convergence. The refinement minimizes either Procrustes distance or TPS bending energy. This algorithm has also been successfully used to process endocast surfaces (Neubauer et al. 2009; Gunz 2015).

There are other methods that operate directly on meshes and use their vertices instead of landmarks. *Dense Correspondence Analysis* (DCA) is an algorithm originally designed for facial scans (Hutton et al. 2003). This relatively simple approach relies on a small number of explicit correspondences that have to be known prior to processing. These are given as sets of homologous landmarks for each surface. A template surface known as base mesh is also selected, whose topology is forced onto all remaining surfaces (known as floating meshes). The choice of base mesh has little or no influence on the statistical results, as long as it contains no large holes and is sampled with a regular coverage of vertices. In DCA, a TPS (Bookstein 1989) is fitted to deform each landmark configuration onto mean landmarks. Next, on each floating surface, the closest point to every vertex of the base mesh is identified. These closest points are considered homologous to the base mesh vertices, after the TPS deformation is reversed. Basically, this algorithm interpolates the given explicit correspondences to the entire surfaces.

Several approaches that perform correspondence search using conformal maps have been proposed. *Conformal maps* are bijective holomorphic functions that assign to each point on the surface a unique point on a plane, thereby unwrapping the surface into a two-dimensional domain. Furthermore, conformal maps locally preserve angles. Unlike algorithms that fit spatial deformations (Hutton et al. 2003), these methods exploit the fact that correspondences lie strictly on the studied surfaces. Thus, the typical workflow unwraps both studied surfaces onto a plane, registers them as planar representations, and transfers these correspondences into the original three-dimensional space. Algorithms belonging to this class usually differ in the way the planar representations are calculated (conformal maps, harmonic maps, geometric flows, etc.) and in the employed registration procedure.

Most of such algorithms are commonly applied to facial recognition (Szeptycki et al. 2010). Harmonic maps are created from the surfaces and registered using landmarks. Conformal maps have been created using Yamabe flow with landmark constraints (Zeng et al. 2014), an iterative procedure that progressively flattens a surface into the desired planar representation. A review and performance testing of several conformal mapping algorithms coupled with automatic feature detection are presented in Wang et al. (2006). It should also be noted that, while these methods might generate relatively accurate correspondences due to the lack of closest-point search, they also impose limitations on the geometry of the surfaces in question. Most methods require that the surfaces are single component, genus-0 with disc topology, i.e., surfaces with an outer boundary and no holes. This poses no problem for facial models, but the endocast surface is closed and thus the support for genus-0 spherical topology is needed. Some authors (Zeng et al. 2014) have suggested that generalization to other topologies is possible as well.

Surface parameterizations, of which conformal maps are a subclass, have also been applied to analyze endocasts, using quasiconformal spherical maps to parametrize the endocast surfaces with spherical coordinates (Specht et al. 2007). In this case, nonrigid registration is performed using landmarks that are manually placed on the surfaces. Because of spherical mapping, a common regular tessellation scheme was introduced to each surface, removing the need for selecting a template surface. Finally, a correspondence search procedure was proposed (Durrleman et al. 2012; Beaudet and Bruner 2017; Beaudet et al. 2016) that assumes growth will occur along surface tangents and interpolated growth patterns of bonobo and chimpanzee endocranial surfaces.

10.3 Coherent Point Drift

In this chapter we compare digital endocasts by using *Coherent Point Drift-Dense Correspondence Analysis* (CPD-DCA; Dupej et al. 2014), a modification of DCA that uses an automatic nonrigid registration algorithm to mitigate dependence on landmarks. This is an appropriate method when homologous landmarks cannot be placed in sufficient quantity and coverage, as in many paleontological cases. This algorithm also has a correspondence rejection stage that prevents incorrectly matched vertices from contaminating subsequent statistical analyses.

We have used four digital endocasts: one modern human, one Neanderthal (Saccopastore 1), one *Homo ergaster* (KNM-ER 3733), and a chimpanzee. All of these models are triangle meshes, each with approximately 25,000 vertices. Before rendering visualizations, vertex homology was enforced, and interindividual differences in position, orientation, and scaling were removed using CPD-DCA (Dupej et al. 2014). This algorithm is not overly sensitive to the precision of landmarks and imposes few restrictions on the topology of the studied surfaces. Furthermore, it is fully implemented in Morphome3cs (www.morphome3cs.com), a software for statistical processing and visualizations in geometric morphometrics. All processing and visualization have been performed using this software.

10.3.1 CPD-DCA

In the following text, we will be using the following notation. A triangle mesh $\boldsymbol{X} = (\boldsymbol{x}, \boldsymbol{t})$ comprises its vertices $\boldsymbol{x}_i \in \mathbb{R}^3$, $i = 1, \ldots m$ and vertex indices of the triangles $\boldsymbol{t}_j = \{1, ..m\}^3$, $j = 1, \ldots n$. We also refer to the set of vertices as vert$(\boldsymbol{X}) := \boldsymbol{x}$ and set of triangles' vertex indices as tri$(\boldsymbol{X}) := \boldsymbol{t}$.

This topology transfer algorithm is based on DCA, originally introduced by Hutton et al. (2003). Instead of landmark-fitted TPS deformations, an automatic nonrigid algorithm is used to deform the base mesh onto each of the studied floating meshes. Specifically, nonrigid coherent point drift (CPD; Myronenko and Song 2010) is considered a state-of-the-art algorithm and was chosen due to its favorable properties, including its robustness to outliers and noise and parameters that allow control over-regularization of deformation. Furthermore, CPD is formulated in a way that allows for easy implementation on modern graphics cards (GPUs) and can thus achieve high processing speeds, despite being computationally intensive. Even though CPD-DCA has been originally intended to work on larger samples of surfaces, it can equally be applied to map the differences between two individual surfaces.

CPD-DCA operates on the surfaces represented as triangle meshes $\boldsymbol{X}_i$, $i = 1, \ldots N$ and the corresponding landmark sets, as shown in Fig. 10.1. The first step is rigid preregistration. As in DCA, this is performed by fitting generalized Procrustes transformation (GPA) onto the landmark configurations manually placed on each mesh, obtaining the transformations f_i. These transformations are applied on entire meshes $\boldsymbol{Y}_i = (f_i(\boldsymbol{x}_i), \boldsymbol{t}_i)$. A complete GPA with size normalization was used, and the original size of each surface s_i was stored for later use. It should also be noted that this procedure is not intended for precise rigid alignment of the studied surfaces but to provide a good initial alignment of the surfaces and ultimately to assist in the convergence of the subsequent nonrigid step.

Nonrigid CPD was used to deform the base mesh onto each floating mesh. As previously noted (Hutton et al. 2003), the choice of the base mesh has little influence on the statistical results, which we will not generate here, and also on the visualizations, as long as the base surface contains no larger holes and is covered with vertices approximately evenly. Without loss of generality, we let $\boldsymbol{Y}_1$ be the base mesh. CPD works on point clouds, not surfaces; therefore it is used to align the vertices of base mesh onto the vertices of a floating mesh. This registration algorithm employs a probabilistic approach as it represents the data as Gaussian mixture models and fits one set to the other using an expectation-maximization procedure. The points are soft matched, i.e., one-to-many instead of one-to-one correspondences are generated, improving convergence properties. Regularization based on motion coherence theory and terms suppressing the influence of outliers are present in the algorithm. A

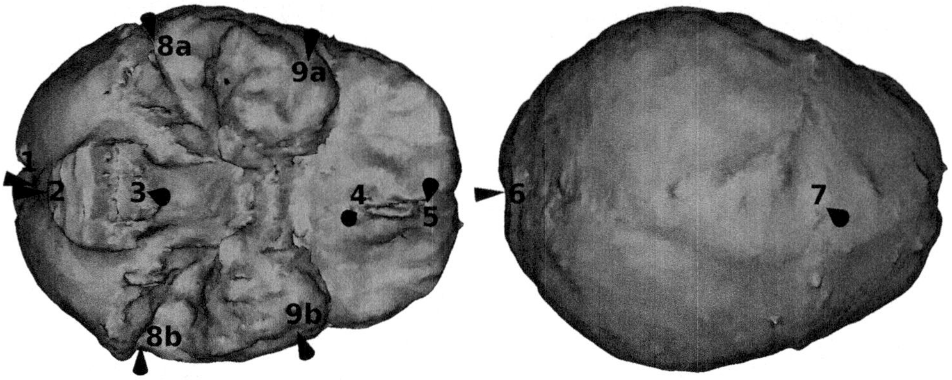

Fig. 10.1 The landmarks used in the correspondence search, shown on the modern human endocast: *1* internal occipital protuberance, *2* endopisthion, *3* endobasion, *4* planum sphenoideum, *5* foramen caecum, *6* lambda, *7* bregma, *8ab* pyramidal base-symmetric, *9ab* posterior border of the anterior fossa

comprehensive explanation of the procedure, including performance tests, is given in Myronenko and Song (2010). The alignment may not be perfect—that is, deformed base vertices do not necessarily lie on the floating surface; therefore, each transformed base vertex was snapped onto the floating surface, based on the closest-point principle. For each $i \neq 1$, we found $\tilde{z}_i = \mathrm{CPD}(\boldsymbol{y}_1 \rightarrow \boldsymbol{y}_i)$. Finally, z_i were found for each $i \neq 1$ as the closest points on $\boldsymbol{Y}_i$ to $\tilde{z}_i$. For $i = 1$, we set trivially $z_i = \boldsymbol{y}_1$, yielding the final remeshes $\boldsymbol{Z}_i = (z_i, \boldsymbol{t}_1)$.

A fast CPD initialization procedure was used (Dupej et al. 2015). The original implementation (Dupej et al. 2014) also featured a correspondence rejection step that discards vertices in the remeshes if the distances to mean mesh exceed a threshold or if the adjacent triangles are excessively resized (as a result of the remeshing). This creates holes in the surfaces (which have to be filled later) and is generally useful for statistical processing of a larger set of surfaces as vertices that were probably incorrectly matched are excluded from principal component analysis. In this way, much of the variability unrelated to the actual shape differences has no effect on the results. Because we only use CPD-DCA to prepare our data for visualizations, we do not use this part of the processing pipeline.

Until this point, the remeshes were in landmark-defined alignment. We therefore perform a final rigid registration step that suppresses the influence of user-supplied landmarks on the results. Furthermore, until this point, the size of the objects was determined solely on the basis of landmarks, which may not be accurate. We therefore perform GPA on all homologous vertices $z_{i,j}$, $1 \leq i \leq N$, $1 \leq j \leq m$ that were generated in the previous topology transfer stage. Again, full GPA is performed, and the size of each object is isolated as $\hat{s}_i$, leaving each surface scaled to a centroid size (from all vertices) of 1. Note that both size variables used here, $\hat{s}_i$ and s_i, are calculated as centroid size (CS), the former on all homologous vertices and the latter on landmarks.

$$\hat{s}_i = \sqrt{\sum_j \left\| z_{i,j} - \boldsymbol{t}_i \right\|^2}$$

where $\boldsymbol{t}_i$ is the vertex centroid of the i-th surface. On the other hand, some studies (Gunz 2015; Specht et al. 2007) have used the volume bounded by the closed surface as their metric of size.

10.3.2 Asymmetry

Endocasts exhibit a rough bilateral symmetry. This symmetry is, however, not perfect, and thus, for the purposes of visualization, the surfaces were artificially symmetrized. For the construction of symmetric specimens, we used a modification of CPD-DCA, similar to Dupej et al. (2013). In correspondence search, this algorithm was used to identify correspondences between base and floating meshes. Here, we applied it to find correspondences between a surface and its mirror reflection. First, paired landmarks were identified. A plane was fitted in the least-square sense to the non-paired landmarks (which were assumed to lie on the medial plane) and to midpoints of the paired landmarks. A mirror mesh was then created by reflecting the original mesh about this symmetry plane. Note that the mirror mesh and original mesh were in rigid alignment (landmark based), sufficient for a successful nonrigid registration using CPD.

The correspondence search algorithm is an enhancement of Dupej et al. (2013) and Krajíček et al. (2012), where correspondences were identified using the closest-point principle. We took advantage of the fact that a surface was being registered onto its mirror self, and thus, for each vertex, correspondences were found effectively twice. Indeed, we used this observation to improve the precision of symmetry correspondences at the cost of computation time. The procedure is schematically depicted in Fig. 10.2. We started with

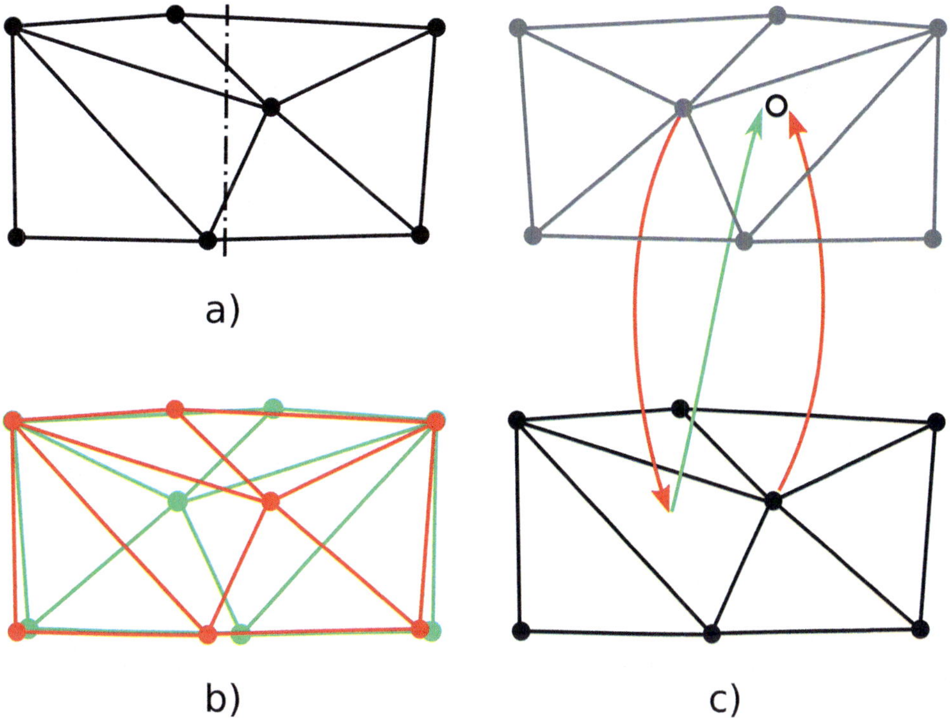

Fig. 10.2 A schematic illustration of correspondence search procedure in symmetric mesh generation phase. (**a**) A surface represented as a triangulation of vertices and its symmetry plane. (**b**) Original (*red*) and mirrored (*green*) surface after nonrigid registration. For clarity, elastic deformation is disregarded. (**c**) Closest point (empty *black circle*) in the mirror mesh (*gray*) is combined from two closest-point searches

two meshes $\boldsymbol{X}$ and $\boldsymbol{Y}$, representing the original and mirror object, respectively. We performed low-rank CPD, deforming $\boldsymbol{Y}$ onto $\boldsymbol{X}$, obtaining the deformed mirror mesh $\tilde{\boldsymbol{Y}}$. Then, for each vertex $\boldsymbol{x}_i \in \text{vert}(\boldsymbol{X})$, we found the closest point in $\tilde{\boldsymbol{Y}}$ and marked it $\tilde{\boldsymbol{p}}_0$. However, that point was located on the deformed mesh although its location on the nondeformed mirror surface was required. Therefore, we calculated the barycentric coordinates a , b , c of $\tilde{\boldsymbol{p}}_0$ within the triangle it was found in and calculated $\boldsymbol{p}_0 = a\boldsymbol{s} + b\boldsymbol{t} + c\boldsymbol{u}$, where $\boldsymbol{s}$, $\boldsymbol{t}$, and $\boldsymbol{u}$ are the vertices of the same triangle on the mirror mesh.

Note that this is the point previously used as the homologous mirrored point (Dupej et al. 2013; Krajíček et al. 2012). However, we took the point $\boldsymbol{y}_i \in \text{vert}(\tilde{\boldsymbol{Y}})$, which is the registered mirror counterpart of $\boldsymbol{x}_i$ and found the point $\tilde{\boldsymbol{p}}_1$ in $\boldsymbol{X}$ and, again, transferred that correspondence to the nondeformed mirror mesh using barycentric coordinates, obtaining $\boldsymbol{p}_1$. That correspondence is effectively the product of a reverse registration (original onto mirror) in contrast to $\boldsymbol{p}_0$, which resulted from mirror onto original fitting. The homologous correspondence point $\boldsymbol{p} = \frac{1}{2}(\boldsymbol{p}_0 + \boldsymbol{p}_1)$ was then the midpoint of the two correspondence points. Finally, the symmetrized vertices $z_i = \frac{1}{2}(\boldsymbol{p} + \boldsymbol{x}_i)$ were used to construct the symmetrized surface $\boldsymbol{Z} := (z, \text{tri}(\boldsymbol{X}))$ along with the topology of the original mesh.

It is important to note that the vertices of the symmetric meshes remain homologous. At this point, we can either visualize the differences between size-normalized symmetrized surfaces (i.e., shape) without any modification or scale them back by the factor $s_i \hat{s}_i$, thereby restoring their original size and allowing us to assess form in addition to shape.

10.3.3 Visualizations

Visualization of the distances between one surface and another can be constructed by color-coding the distances of corresponding vertices or the distances from a vertex to the closest point on the other surface. We refer to the former as *feature distance* and to the latter as *shell distance*. It is also important to visualize which surface is locally more prominent, i.e., show the *signed* distance. Given the surface normal $\boldsymbol{n}_i$ and the homologous pair of vertices $\boldsymbol{a}_i$ and $\boldsymbol{b}_i$, this can be calculated as follows:

$$d = \text{sgn}(\boldsymbol{n}_i \cdot (\boldsymbol{b}_i - \boldsymbol{a}_i))\|\boldsymbol{b}_i - \boldsymbol{a}_i\|$$

where $\cdot$ denotes dot product of vectors and sgn($\cdot$) is the sign function. If the angle of the vectors $\boldsymbol{b}_i - \boldsymbol{a}_i$ and $\boldsymbol{n}_i$ is lower than 90°, the value d will attain positive values. Conversely, if the angle is greater than 90°, the value d will be negative. The same equation can be used for shell distance, as well as for feature distance. In the presented visualizations, we preferred signed shell distance.

To facilitate the visual assessment of the produced distance maps, we applied these to the mean model, which was simply constructed from the centroids of aligned homologous vertices $\boldsymbol{M} = (\boldsymbol{m}, \text{tri}(\boldsymbol{X}))$, where $\boldsymbol{m}_i = \frac{1}{N}\sum_{j=1}^{N} z_{j,i}$.

Due to the symmetrization procedure, we could also visualize signed individual asymmetry of these surfaces, simply by taking $\boldsymbol{a}_i$ and $\boldsymbol{b}_i$ as the corresponding vertices of the symmetrized and correspondence meshes, respectively. This signed distance is then positive in the areas that are more prominent than they would be, if the surface were ideally symmetric.

Despite the nonrigid registration algorithm that does not use landmarks, our approach still partly relies on landmarks for rigid preregistration. Another limitation was the relative computational intensity of CPD. Morphome3cs performs the most intensive parts of CPD on modern graphics accelerators, which are uniquely suited for such tasks; however, such hardware must be installed in the used computer.

10.4 Comparing Endocranial Surfaces

We used the above procedure to compare a modern human endocast (adult male) with those of an early Neanderthal (Saccopastore 1, dated to 130–250 thousand years), a *Homo ergaster* (KNM-ER 3733, dated to about 1.5 million years), and a chimpanzee (*Pan troglodytes*, adult male). Color maps showed the signed shell distances on mean models (Fig. 10.3). Differences are based on size-normalized

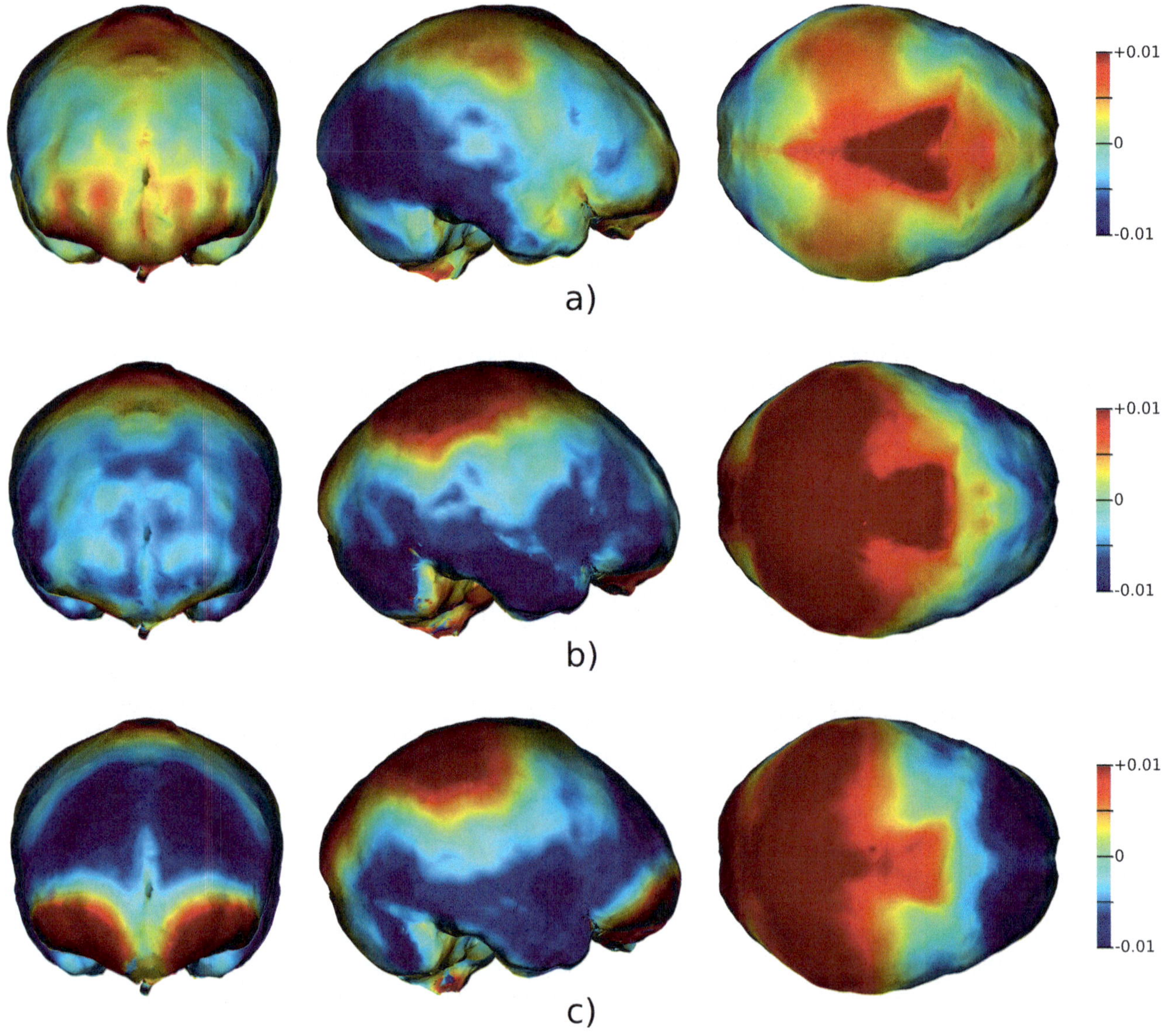

Fig. 10.3 Visualization of signed shell distances between the shapes of two endocasts. Positive values (*red*) indicate relative prominence of the surface capturing modern human; negative values (*blue*) indicate relative prominence of the other surface. All color maps are in the same scale and are shown on the mean surface. (**a**) Modern human vs. Saccopastore 1, (**b**) modern human vs. KNM-ER3733, (**c**) modern human vs. chimpanzee

models, and the endocranial surface of the modern human was arbitrarily selected as the base mesh.

According to these results, modern humans displayed a general enlargement of the parietal surface. Compared to the Neanderthal endocast, our modern specimen showed mid-sagittal expansion of the frontoparietal area and parasagittal expansion of the upper parietal volume. Differences relative to *H. ergaster* and *P. troglodytes* were even more pronounced, involving the whole parietal region. These patterns are in agreement with paleoneurological evidence suggesting lateral widening of the upper parietal surface in Neanderthals and a general expansion and longitudinal bulging of the entire upper parietal areas in modern humans (Bruner et al. 2003, 2014; see Chap. 15). Modern humans also displayed a relative enlargement of the fronto-orbital areas, even more pronounced compared to the chimpanzee. It is currently debated whether humans have larger frontal areas than living apes (e.g., Semendeferi et al. 1997; Rilling 2006; Barton and Venditti 2013; Smaers 2013; see Chap. 14); nonetheless there is evidence of a general change in proportion and lateral widening of these areas in modern humans and Neanderthals, associated with physical constraints in these two species due to increased physical contact between frontal lobes and orbits (Bruner and Holloway 2010; Beaudet and Bruner 2017). Differences between modern humans and Neanderthals are less apparent but include changes in the morphology of the anterior cranial fossa (Bastir et al. 2011).

It should be noted that any observed differences are relative due to normalization. Accordingly, relative increases in the size of some areas must be necessarily associated with a relative decrease in others. In modern humans, relatively larger parietal surfaces are associated with relatively smaller temporal, occipital, and frontal surfaces. Therefore, it is not possible to separate the enlargement of the parietal area from the reduction of other regions based solely on these comparisons. However, in this case we know that, apart from a general increase in brain size, modern humans have larger parietal lobes (Bruner 2004) and bones (Bruner et al. 2011) in absolute terms. We can therefore interpret the maps as a result of parietal enlargement and not of temporo-occipital reduction. Without this additional information, shape comparisons only reveal relative differences, and they cannot be used to discriminate between absolute enlargement of one area vs. reduction of another.

A second observation concerns the landmarks used to initialize surface matching. In this case cranial references have been used, instead of brain (cortical) references. Accordingly, the process may have been influenced by cranial boundaries to a greater degree than by the topology of the cerebral regions.

Additionally, we computed comparative maps for individual asymmetry of the same endocasts (Fig. 10.4). Asymmetries were more pronounced in humans than in chimpanzee.

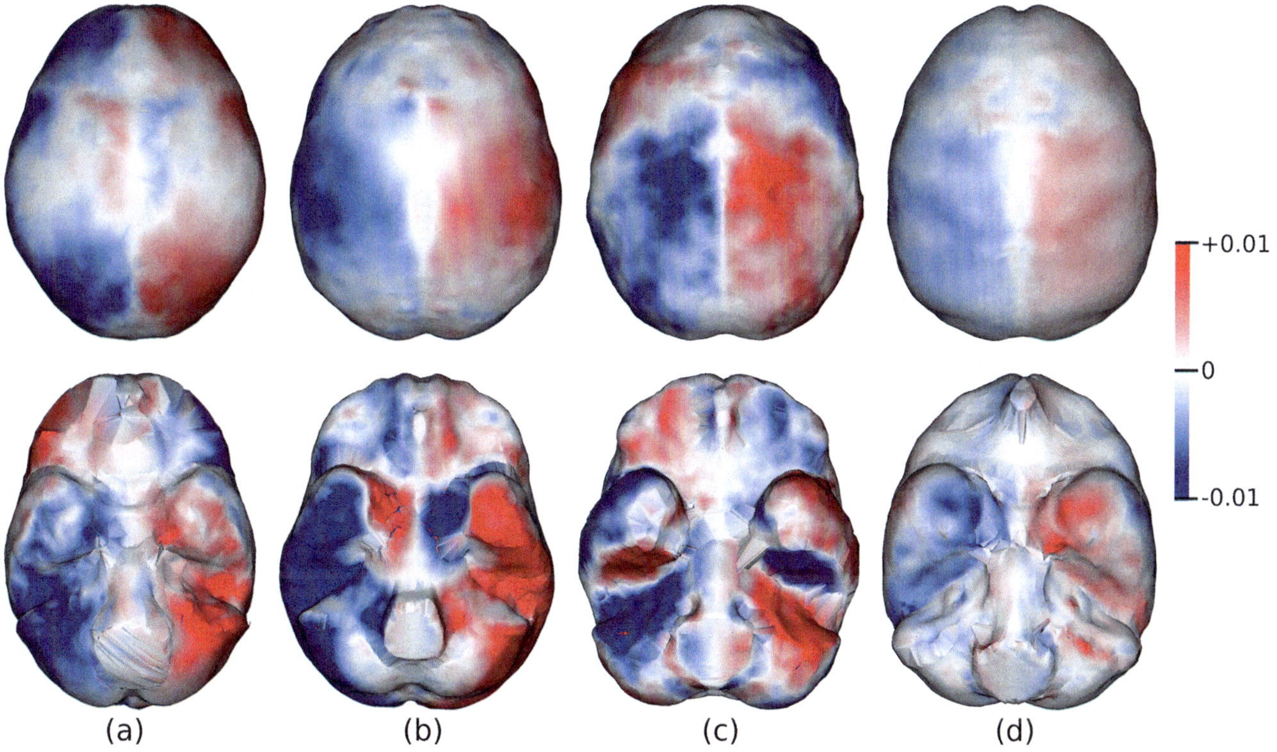

Fig. 10.4 Symmetrized surfaces of the endocasts with a color-mapped visualizations of individual asymmetry. Positive values shown in *red* indicate that the particular area is locally more prominent than it would be if the surface were ideally symmetric. (**a**) Modern human, (**b**) Saccopastore 1, (**c**) KNM-ER3733, (**d**) chimpanzee

Humans display larger and more frequent asymmetries than apes (Holloway and De la Coste-Lareymondie 1982), but species-specific features have not yet been described, and allometric effects due to brain size differences cannot be ruled out (Gómez-Robles et al. 2013). The frontoparietal surface is generally larger on the right side, while the temporal, occipital, and cerebellar areas are larger on the left hemisphere. Asymmetries have long been investigated in modern humans, fossil hominids, and living apes, through computed tomography and magnetic resonance, in terms of both volumes and surfaces (e.g., Barrick et al. 2005; Fournier et al. 2011; Balzeau et al. 2012), and the results presented here in this survey are in general agreement with the available data from literature. Minor local asymmetries require statistical samples to be evaluated in terms of intraspecific variation and are beyond the scope of the current work.

10.5 Conclusion

In this paper, we have reviewed selected methods for quantitative assessment and comparison of digitized surfaces. We have demonstrated that geometric morphometric methods, designed for facial and palatal surface scans, can be also applied for shape analysis of endocasts. In particular, we have used CPD-DCA, a correspondence search algorithm that is based on an automatic nonrigid registration algorithm, for homologous resampling of the studied surfaces. We also added an additional step in the CPD-DCA workflow that creates artificially symmetrized surfaces, enhancing visualization. These symmetrized samplings were then used to render visualizations of shape variation between pairs of species and specimens. Limitations of this approach include partial dependence on manually placed landmarks and computational demands. Nonetheless, we have developed a complete pipeline for quantitative analysis of endocast surfaces that is fully implemented and integrated in Morphome3cs. These visualizations can assist morphometric evaluations of form and shape differences in ontogeny and phylogeny, localizing areas undergoing specific volumetric changes and quantifying similarities and differences among living and extinct taxa.

Acknowledgments JD and JP are supported by the grant SVV-2016-260332. ASPP is funded by the Atapuerca Foundation, Spain. GRL is funded by the International Erasmus Mundus Doctorate in Quaternary and Prehistory consortium (IDQP). EB is supported by the Spanish Government (CGL2012-38434-C03-02).

References

Balzeau A, Grimaud-Hervé D, Détroit F, Holloway RL, Combès B, Prima S (2012) First description of the Cro-Magnon 1 endocast and study of brain variation and evolution in anatomically modern Homo sapiens. Bull Mém Soc Anthropol 25:1–18

Barrick TR, Mackay CE, Prima S, Maes F, Vandermeulen D, Crow TJ, Roberts N (2005) Automatic analysis of cerebral asymmetry: an exploratory study of the relationship between brain torque and planum temporale asymmetry. NeuroImage 24:678–691

Barton RA, Venditti C (2013) Human frontal lobes are not relatively large. PNAS 110:9001–9006

Bastir M, Rosas A, Gunz P, Peña-Melian A, Manzi G, Harvati K, Kruszynski R, Stringer C, Hublin JJ (2011) Evolution of the base of the brain in highly encephalized human species. Nat Commun 2:588

Beaudet A, Bruner E (2017) A frontal lobe surface analysis in three archaic African human fossils: OH 9, Buia, and Bodo. C R Palevol 16:499–507

Beaudet A, Dumoncel J, de Beer F, Duployer B, Durrleman S, Gilissen E, Hoffman J, Tenailleau C, Thackeray JF, Braga J (2016) Morphoarchitectural variation in South African fossil cercopithecoid endocasts. J Hum Evol 101:65–78

Bookstein FL (1989) Principal warps: thin-plate splines and the decomposition of deformations. Pattern Anal Mach Intell IEEE Trans 11:567–585

Bookstein FL (1991) Morphometric tools for landmark data: geometry and biology. Cambridge University Press, New York

Bookstein FL (1997) Landmark methods for forms without landmarks: morphometrics of group differences in outline shape. Med Image Anal 1:225–243

Bookstein FL, Gunz P, Mitterœcker P, Prossinger H, Schæfer K, Seider H (2003) Cranial integration in homo: singular warps analysis of the midsagittal plane in ontogeny and evolution. J Hum Evol 44:167–187

Bruner E (2004) Geometric morphometrics and paleoneurology: brain shape evolution in the genus homo. J Hum Evol 47:279–303

Bruner E, Holloway RL (2010) A bivariate approach to the widening of the frontal lobes in the genus homo. J Hum Evol 58:138–146

Bruner E, Manzi G, Arsuaga JL (2003) Encephalization and allometric trajectories in the genus homo: evidence from the Neandertal and modern lineages. PNAS 100:15335–15340

Bruner E, Saracino B, Ricci F, Tafuri M, Passarello P, Manzi G (2004) Midsagittal cranial shape variation in the genus homo by geometric morphometrics. Coll Antropol 28:99–112

Bruner E, De La Cuétara JM, Holloway R (2011) A bivariate approach to the variation of the parietal curvature in the genus homo. Anat Rec 294:1548–1556

Bruner E, Rangel de Lázaro G, de la Cuétara JM, Martín-Loeches M, Colom R, Jacobs HIL (2014) Midsagittal brain variation and MRI shape analysis of the precuneus in adult individuals. J Anat 224:367–376

Dupej J, Krajíček V, Velemínská J, Pelikán J (2013) Analysis of asymmetry in triangular meshes. Proceedings of the 33rd conference on geometry and graphics

Dupej J, Krajíček V, Velemínská J, Pelikán J (2014) Statistical mesh shape analysis with nonlandmark nonrigid registration. 12th symposium on geometry processing

Dupej J, Krajíček V, Pelikán J (2015) Low-rank matrix approximations for coherent point drift. Pattern Recogn Lett 52:53–58

Durrleman S, Pennec X, Trouvé A, Ayache N, Braga J (2012) Comparison of the endocranial ontogenies between chimpanzees and bonobos via temporal regression and spatiotemporal registration. J Hum Evol 62:74–88

Fournier M, Combès B, Roberts N, Keller S, Crow TJ, Hopkins W, Prima S (2011) Surface-based method to evaluate global brain shape asymmetries in human and chimpanzee brains. In: 2011 I.E. International Symposium on biomedical imaging: from nano to macro. IEEE, pp 310–316

Gómez-Robles A, Hopkins WD, Sherwood CC (2013) Increased morphological asymmetry, evolvability and plasticity in human brain evolution. Proc R Soc B 280:20130575

Gower JC (1975) Generalized procrustes analysis. Psychometrika 1:33–51

Gunz P (2015) Computed tools for paleoneurology. In: Bruner E (ed) Human paleoneurology. Springer, Cham, pp 177–208

Gunz P, Mitteroecker P (2013) Semilandmarks: a method for quantifying curves and surfaces. Hystrix 24:103–109

Gunz P, Neubauer S, Maureille B, Hublin J-J (2010) Brain development after birth differs between Neanderthals and modern humans. Curr Biol 20:R921–R922

Holloway RL, De La Coste-Lareymondie MC (1982) Brain endocast asymmetry in pongids and hominids: some preliminary findings on the paleontology of cerebral dominance. Am J Phys Anthropol 58:101–110

Holloway RL, Broadfield DC, Yuan MS (2004) Brain endocasts – the paleoneurological evidence. Wiley, Hoboken

Hutton TJ, Buxton BF, Hammond P, Potts HWW (2003) Estimating average growth trajectories in shape-space using kernel smoothing. IEEE Trans Med Imaging 22:747–753

Krajíček V, Dupej J, Velemínská J, Pelikán J (2012) Morphometric analysis of mesh asymmetry. J WSCG 20:65–72

Lieberman DE, McBratney BM, Krovitz G (2002) The evolution and development of cranial form in *Homo sapiens*. PNAS 99: 1134–1139

Myronenko A, Song X (2010) Point set registration: coherent point drift. IEEE Trans Pattern Anal Mach Intell 32:2262–2275

Neubauer S, Gunz P, Hublin JJ (2009) The pattern of endocranial ontogenetic shape changes in humans. J Anat 215:240–255

Rilling JK (2006) Human and nonhuman primate brains: are they allometrically scaled versions of the same design? Evol Anthropol 15:65–77

Semendeferi K, Damasio H, Frank R, Van Hoesen GW (1997) The evolution of the frontal lobes: a volumetric analysis based on three-dimensional reconstructions of magnetic resonance scans of human and ape brains. J Hum Evol 32:375–388

Smaers JB (2013) How humans stand out in frontal lobe scaling. PNAS 110:E3682–E3682

Specht M, Lebrun R, Zollikofer CPE (2007) Visualizing shape transformation between chimpanzee and human braincases. Vis Comput 23:743–751

Szeptycki P, Ardabilian M, Chen L, Zeng W, Gu D, Samaras D (2010) Conformal mapping-based 3D face recognition. Data Process Vis Transm Symp 3D:17–20

Wang S, Wang Y, Jin M, Xiangfeng G, Samaras D (2006) 3D surface matching and recognition using conformal geometry. 2006 I.E. Computer Society Conference on Computer Vision and Pattern Recognition (CVPR'06), 2006, pp 2453–2460

Zeng W, Lui LM, Kong H, Gu X (2014) Surface registration by optimization in constrained diffeomorphism space. 2014 I.E. Conference on Computer Vision and Pattern Recognition, Columbus, OH, 2014, pp 4169–4176

Reconstruction and Statistical Evaluation of Fossil Brains Using Computational Neuroanatomy 11

Takanori Kochiyama, Hiroki C. Tanabe, and Naomichi Ogihara

Abstract

To investigate differences in the cognitive abilities of fossil humans, it is important to be able to objectively infer possible differences in the anatomy and morphology of their brains from the insides of fossil crania. In this chapter, we present a new mathematical framework to virtually reconstruct fossil brains and to statistically evaluate possible morphological differences between fossil and extant human brains by means of computational neuroanatomy. Specifically, a fossil endocast was spatially deformed to a modern human endocast segmented from an MR image, and a mapping between these two endocast shapes was calculated. The modern human gray and white matter segmented from the MR images was then inversely transformed to reconstruct a virtual brain for the fossil cranium. Computational morphometry can then be used to statistically compare the reconstructed fossil brain with the brains of modern humans. The volume of each brain region can also be quantified by using neuroanatomical labels for the brain locations. To evaluate the accuracy of the reconstructed brain, the brains of modern subjects were reconstructed according to CT-derived endocasts, and were then compared with the subjects' true brains, as derived from the MRI. The overall shapes of the reconstructed brains were in good agreement with those of the corresponding true brains. Although some limitations certainly apply, the present brain reconstruction techniques are expected to contribute to an improved understanding of the evolution of the human brain.

Keywords

Skull • Brain anatomy • Virtual anthropology • Endocast • Neanderthal • Morphometry

11.1 Introduction

Acquisition of a large and complex brain is one of the fundamental hallmarks of human evolution. To understand the evolution of brain development in the human lineage, researchers have evaluated endocranial casts or endocast morphology, as the brain itself is not fossilized. Hence, endocast morphology is currently the most useful source of information for inferring morphological and functional differences between modern and fossil human brains. To explore possible differences in the cognitive abilities of fossil humans, previous studies have evaluated endocranial volume as a measure of brain size (e.g., Falk 2012) or have

T. Kochiyama (✉)
Department of Cognitive Neuroscience, Advanced Telecommunications Research Institute International, 2-2-2 Hikaridai Seika-cho, Sorakugun, Kyoto 619-0288, Japan

Brain Activity Imaging Center, ATR-Promotions, Kyoto, Japan
e-mail: kochiyam@atr.jp

H.C. Tanabe
Department of Psychology, Graduate School of Environmental Studies, Nagoya University, Furo-cho, Chikusa-ku, Nagoya, Aichi 464-8601, Japan
e-mail: htanabe@lit.nagoya-u.ac.jp

N. Ogihara
Department of Mechanical Engineering, Faculty of Science and Technology, Keio University, 3-14-1 Hiyoshi, Kohoku-ku, Yokohama 223-8522, Japan
e-mail: ogihara@mech.keio.ac.jp

E. Bruner et al. (eds.), *Digital Endocasts*, Replacement of Neanderthals by Modern Humans Series,
DOI 10.1007/978-4-431-56582-6_11

extrapolated local differences in brain shape on the basis of imprints of sulci and gyri on the endocasts (e.g., Holloway et al. 2004), as well as anatomical landmarks and neurocranial sutures (e.g., Bruner et al. 2003, 2014, 2015). However, endocranial volume is too coarse a measure to determine the neuroanatomical correlates of evolutionary changes. Functional neuroimaging research has revealed that cortical areas are specialized for certain aspects of perceptual, cognitive, and/or motor processing and that this specialization is anatomically segregated within the cortex (Marshall and Fink 2003). Furthermore, identification of cortical features according to the endocranial surface is known to be difficult for adult crania in both fossil and extant humans (Ogihara et al. 2015). Anatomical landmarks and neurocranial sutures do not necessarily correspond to the anatomical boundaries of brain lobes (Bruner et al. 2015). Pearce et al. (2013) estimated occipital lobe size using orbit size as a proxy, but such indirect estimation of brain organization also has limitations. In order to reveal the functional characteristics of the brains that were enclosed in fossil crania, it is essential to develop a new methodology to reconstruct the original brain from fossil endocranial morphology.

In the present chapter, we aim to present a new mathematical framework to virtually reconstruct fossil brains and then to use computational neuroanatomy to statistically infer possible morphological differences between fossil and extant human brains (Kochiyama et al. 2014; Tanabe et al. 2014; Ogihara et al. 2015). Computational neuroanatomy involves a series of computational techniques developed for quantitative descriptions and statistical comparisons of neural structures such as brains (Ashburner and Friston 2007). In the field of human neuroimaging, computational neuroanatomy techniques are frequently used for statistical comparisons of differences in the volume of brain regions and for patterns of brain activity obtained with functional magnetic resonance imaging (fMRI; Ogawa et al. 1990; Kwong et al. 1992; reviewed by Huettel et al. (2009)). One of the strengths of this methodology lies in the fact that it allows automatic segmentation of brains and extraction of brain subregions from MR images. Furthermore, the computational neuroanatomical framework provides a 3D image registration technique for quantitative comparisons of brains that are highly variable in size and shape (e.g., spatial patterns of sulci and gyri) between individuals; this is performed using a spatial transformation of one brain to another (a so-called standardized or template brain), with full probabilistic considerations. By making use of these computational neuroanatomical techniques, we hope to achieve objective, reproducible, and mathematically sound estimation of the antemortem appearance of a brain that would have been enclosed in a fossil cranium, thus allowing detailed comparisons of the brain morphology with that of modern humans and also other fossil humans.

This chapter comprises six sections. In the next section, we introduce our approach for the reconstruction of fossil brains using computational neuroanatomy techniques, with a special focus on spatial normalization and computational morphometry as principle methodologies for fossil brain reconstruction and statistical comparisons. The third section describes actual procedures for restoration of a fossil brain using computational neuroanatomy. In this example, we used CT-scanned crania of five modern Japanese participants to mathematically estimate the brain morphology inside each cranium and compared the restored brains with MRI-scanned (true) brains of the same subjects. In the fourth section, we evaluate the accuracy of the present reconstruction method. The fifth section deals with statistical techniques for comparing restored brain morphology between populations. In the sixth and last section, we briefly summarize the present brain restoration methodology, and discuss the strengths and limitations of the present method, as well as possible implications for future studies.

11.2 Strategies for Brain Reconstruction and Statistical Evaluation

To reconstruct brains from their fossil crania, we first made two assumptions. The first assumption is that there is a morphological correspondence between endocranial and brain morphology. This is a reasonable assumption under non-pathological conditions. If no correspondence existed between cranial and brain forms, the estimation of brain morphology from a fossil cranium would not be possible. The second assumption is that fossil human brains, such as those of Neanderthals and early modern humans, can be computationally reconstructed by deforming modern human brains based on endocranial shape, as they are phylogenetically close to each other. This may be a somewhat crude assumption, but the 3D brain structures of chimpanzees (*Pan troglodytes*) and bonobos (*Pan paniscus*) are reportedly similar, despite the fact that divergence between the two species is considered to have occurred approximately 1.5–2.1 million years ago (Stone et al. 2010). Since the divergence between Neanderthals and anatomically modern humans took place much more recently (approximately 0.6–0.8 million years ago; Meyer et al. 2016), we believe we can reasonably predict the morphology of fossil brains by deforming the brains of modern humans.

Under these assumptions, we attempt to establish a method to computationally reconstruct the 3D structure of the fossil brain using a mathematical framework from the field of computational neuroanatomy. In functional brain mapping methods, such as those used in fMRI, differences in activation are analyzed statistically for separate regions of the brain. However, the size and shape (e.g., spatial patterns of sulci and gyri) of each participant's brain vary considerably. To allow for statistical comparisons of the activity of each voxel within brain images, each brain must firstly be transformed onto a standardized template brain. This process is frequently termed spatial normalization. This spatial normalization process also allows examination of morphological differences between brains on the basis of computational morphometric techniques, such as voxel-based and deformation-based morphometry. Regional patterns in brain activity or morphology are statistically compared by assuming parametric statistical models at each voxel, with model parameters being estimated to make inferences and test hypotheses about regionally specific effects (Friston 2007).

The Functional Imaging Laboratory Group at the University College London created the statistical analysis package statistical parametric mapping or SPM (Friston 1997, http://www.fil.ion.ucl.ac.uk/spm/) to perform such techniques, and this has been a very popular software package within the neuroimaging research field. In the present study, we used the SPM analytical scheme, particularly the spatial normalization and computational morphometry, to reconstruct and statistically compare fossil brains.

Spatial normalization is an image registration technique involving the spatial transformation of images. Specifically, a source image is warped by spatial transformations to fit it to a target or template image, with the iterative procedure being driven by a cost function such as minimization of the mean-squared difference or maximization of the mutual information between the two images. Of the various spatial transformation techniques available, the recently developed large deformation diffeomorphic metric mapping (LDDMM) offers a promising approach (Avants and Gee 2004; Miller et al. 2005, 2006; Ashburner 2007; Wang et al. 2007; Ashburner and Friston 2011), with this high-dimensional nonlinear transformation offering several million degrees of freedom, as opposed to the one thousand or so in the original SPM nonlinear transformation (Friston 2007). During the optimization step, the LDDMM algorithm minimizes two metrics: the image intensity difference between the target and source image and the geodesic distance (squared distance measure of the deformations). Minimization of such metric distances between anatomical shapes is a distinct feature of this method, which thereby allows a high-quality registration. Diffeomorphic mappings like LDDMM are smooth and topology-preserving, which means that by using them, we are able to perform precise, smooth, and well-behaved transformations.

The SPM software has two implementations of LDDMM: the DARTEL toolbox (Ashburner 2007) and the geodesic shooting toolbox (Ashburner and Friston 2011). The DARTEL method we adopted for our scheme has the following characteristics: (1) it allows global one-to-one mapping (i.e., we can transform the data back and forth); (2) it is available in the pre-processing options of the SPM package and is regularly used in the neuroimaging research field; and (3) template images in standardized stereotaxic space are available (i.e., images can be transformed into this template image space). It thus becomes feasible to extract differences in brain morphology in fossil humans and to relate the relevant brain regions to specific cognitive abilities identified by functional neuroimaging studies on extant humans.

In the present study, we employ the spatial normalization technique provided in the SPM package as a core part of the methodology of the brain reconstruction. The reconstruction method in the present study can be outlined as follows (Fig. 11.1): a complete endocast with no missing surface is reconstructed from the fossil cranial CT image [see Amano

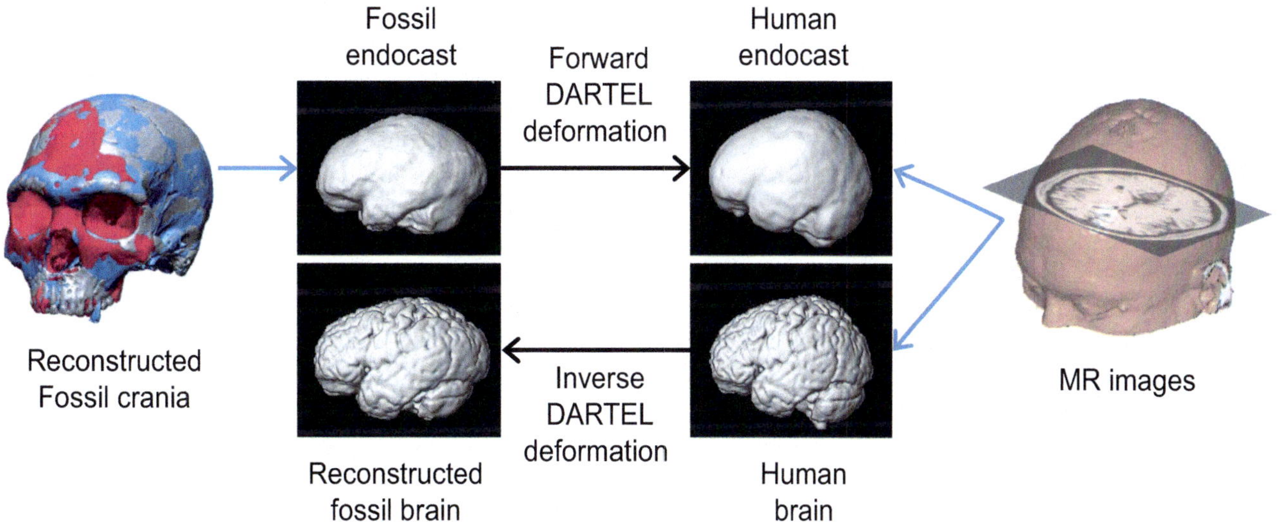

Fig. 11.1 Outline of fossil brain reconstruction methodology

et al. (2015) and Ogihara et al. (Chap. 2, this volume) for virtual reconstruction of fossil crania]. This is then spatially deformed to the modern human endocast segmented from an MR image. The DARTEL algorithm is used to estimate a global one-to-one smooth and continuous mapping between these two endocast shapes. Using the deformation or flow field in DARTEL, the modern human gray and white matter segmented from the MR image is then inversely transformed to reconstruct the virtual brain for the fossil cranium, which is consequently best fitted to its endocast shape. The volume of each brain region can also then be quantified using the neuroanatomical labels for the brain locations (Tzourio-Mazoyer et al. 2002).

The reconstructed brain shapes were then statistically evaluated using computational morphometry, a neuroimaging analysis technique that allows examination of focal differences in brain anatomy. Unlike manual region tracing approaches, it is fully automatic and performs voxel-by-voxel tests with high resolution and no requirement for a priori assumptions about regions of interest. There are two types of commonly used morphometric analysis in the field of neuroimaging: voxel-based morphometry (VBM) (Ashburner and Friston 2000; Good et al. 2001) and deformation-based morphometry (DBM) (Ashburner et al. 1998; Chung et al. 2001; Chung et al. 2003). VBM analysis is able to detect differences in the regional volumes of gray and white matter while discounting global differences in brain shape. The method uses a whole-brain mass-univariate statistical approach to examine differences at each voxel position. In contrast, DBM analysis is performed on the deformation fields used for the spatial transformation. As the deformation field is a vector field, the DBM approach uses multivariate statistics (Worsley et al. 2004; Chung et al. 2010). The advantage of this approach is that it is not only able to detect volumetric differences but also differences in local brain morphology (i.e., shrinkage and/or enlargement of brain regions and surfaces). By employing multiple morphometric techniques, differences in brain anatomy can be carefully examined.

11.3 Brain Reconstruction Procedure

In this section, we present an actual procedure for brain reconstruction based on endocranial morphology. While it would be more interesting to present brain reconstructions of Neanderthals or anatomically modern humans based on the CT scans of actual fossils, we herein used the CT scan data of five modern Japanese crania for reconstruction of the brains. This allowed evaluation of the accuracy of the reconstructed brains, as MRI scans of the subjects' true brains were available for comparison.

The following image processing routines were mainly performed using SPM12 revision 6225 (http://www.fil.ion.ucl.ac.uk/spm) implemented in MATLAB R2012b (MathWorks, MA, USA) and in-house MATLAB programs.

11.3.1 Target Endocasts

Cranial CT scans of five modern Japanese male participants were obtained (age range, 28–47 years). The bone regions were segmented from each stack of CT images, and the endocranial cavities were then extracted using ITK-SNAP software (Yushkevich et al. 2006; http://www.itksnap.org). A semiautomatic segmentation using a region competition algorithm was used. Each of the endocasts was written out as a NIfTI file, which is a standard image format used in neuroimaging, and is fully compliant with the SPM software package. The images were resampled to a resolution of 1.5 × 1.5 × 1.5 mm to conform to the MRI data. MRI scans of the same participants were also obtained for evaluation of the reconstructed brains, as described in the next section. This study was approved by the ethics committee of the Faculty of Science and Technology, Keio University, and informed consent was obtained from all of the participants.

11.3.2 Brains and Endocasts of the Modern Japanese Population

Cranial MRI scans of 512 healthy Japanese volunteers were obtained (256 females and 256 males, age range, 18–46 years). The whole-brain structural MRI data were obtained using a magnetization-prepared rapid-acquisition gradient-echo (MP-RAGE) sequence with 1 mm resolution. A total of 512 T1-weighted MR images of modern-day humans were used for further analysis. This MRI study was approved by the ethical committee of the National Institute for Physiological Sciences, Okazaki, Japan, and all of the participants provided written informed consent.

The T1 images of the modern-day humans ($N = 512$) were segmented into gray matter (GM), white matter (WM), cerebrospinal fluid (CSF), skull, and scalp using the unified segmentation–normalization procedure (Fig. 11.2). This is a standard automated segmentation method that is frequently used in neuroimaging research, and detailed descriptions of the method and their validation are presented elsewhere (Ashburner and Friston 2005; Malone et al. 2015). Briefly, unified segmentation–normalization is a Bayesian probabilistic framework that includes tissue probability maps (TPM) as priors and a mixture of Gaussian models for the MR signal intensity of each tissue class, in combination with bias correction (correction for intensity inhomogeneity in the MRI image) and spatial normalization (image alignment to the TPM). It therefore enables an accurate classification of the previously mentioned tissue types. The resulting segmented

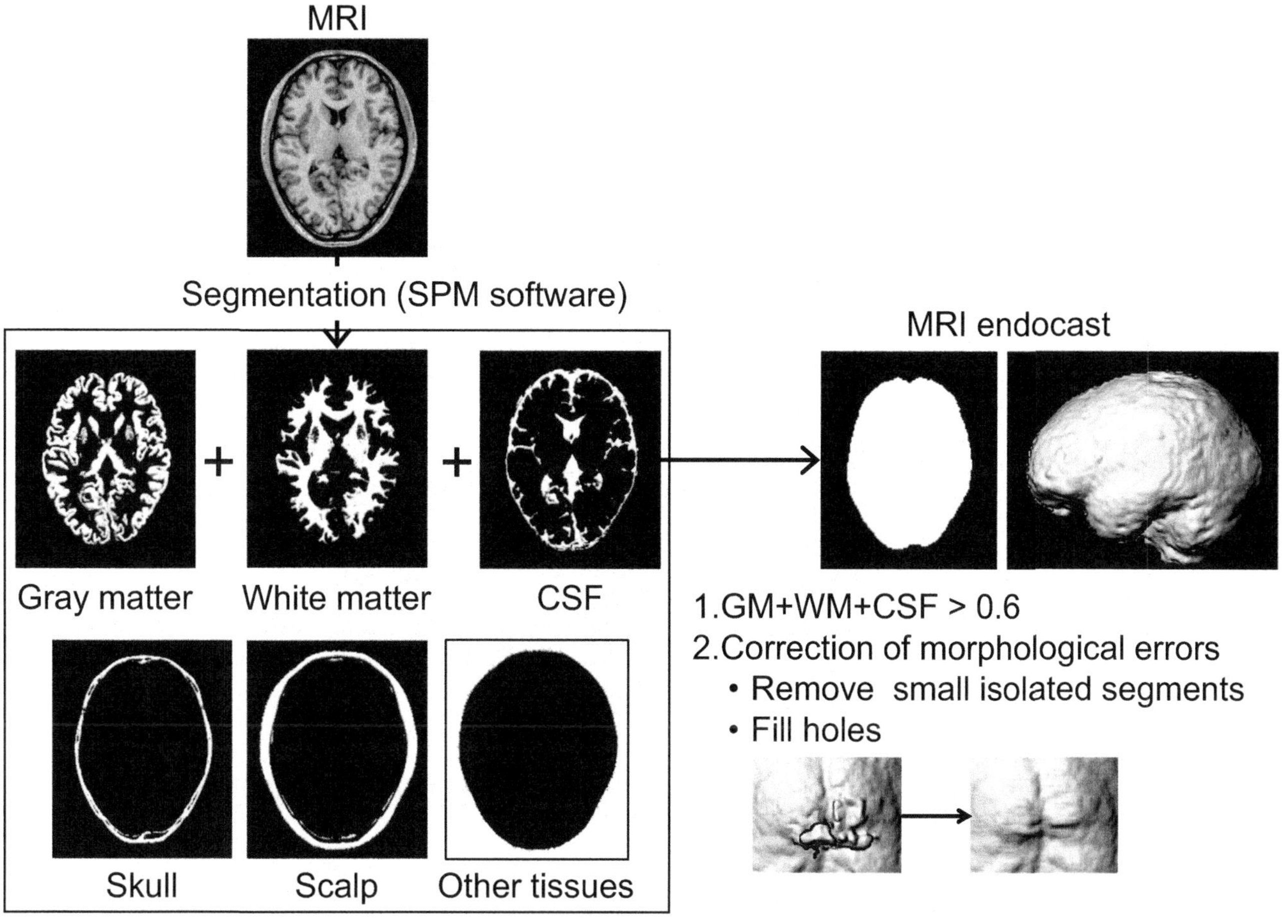

Fig. 11.2 Extraction of endocranial surfaces (endocasts) from MRI. The endocast was generated by thresholding the gray matter (GM), white matter (WM), and cerebrospinal fluid (CSF) regions segmented using the unified segmentation–normalization procedure. Surface cleanup and morphological corrections were also performed

images (GM, WM, and CSF) were resampled to a resolution of 1.5 × 1.5 × 1.5 mm.

The 3D structure of the human brain was generated as the sum of the GM and WM probability images. The voxel values of the segmented image represented the probability of each voxel belonging to each of the tissue classes. The sum image of the GM and WM images was binarized with a 50% threshold to display the brain surface using the surface extraction function in SPM software.

The corresponding endocasts were also generated by calculating the sum of the GM, WM, and CSF probability images and then thresholding this sum image to make a rough binary endocast image. The threshold probability was set to 0.6, which was determined by the supplementary analysis on independent data sets summarized in Fig. 11.3. To determine the optimal threshold for the sum of the intracranial tissue probability images, we evaluated the fraction of correctly and falsely overlapping voxels (i.e., the true- and false-positive rates) between the MRI-derived and CT-derived endocasts at various threshold settings (Fig. 11.3a, b) and plotted a receiver-operating characteristic (ROC) curve (Fig. 11.3c). The optimal threshold values were chosen to balance the sensitivity and specificity of the matching between the MRI- and CT-derived endocasts (Fig. 11.3c). The mean optimal threshold value across five participants was 0.58 ± 0.12. Figure 11.3d shows axial slices of the CT skull image overlaid with the MRI-derived endocast thresholded at a value of 0.6. The endocasts are virtually identical to each other, indicating that the endocast can be successfully extracted from MRI data using this method with the above threshold setting.

After obtaining the binarized endocast image, we performed surface cleanup and morphological corrections using both automatic and manual morphological operations in the MATLAB Image Processing Toolbox and MRICRON medical image viewer and analysis software (www.nitrc.org/projects/mricron). Firstly, we applied three-dimensional morphological

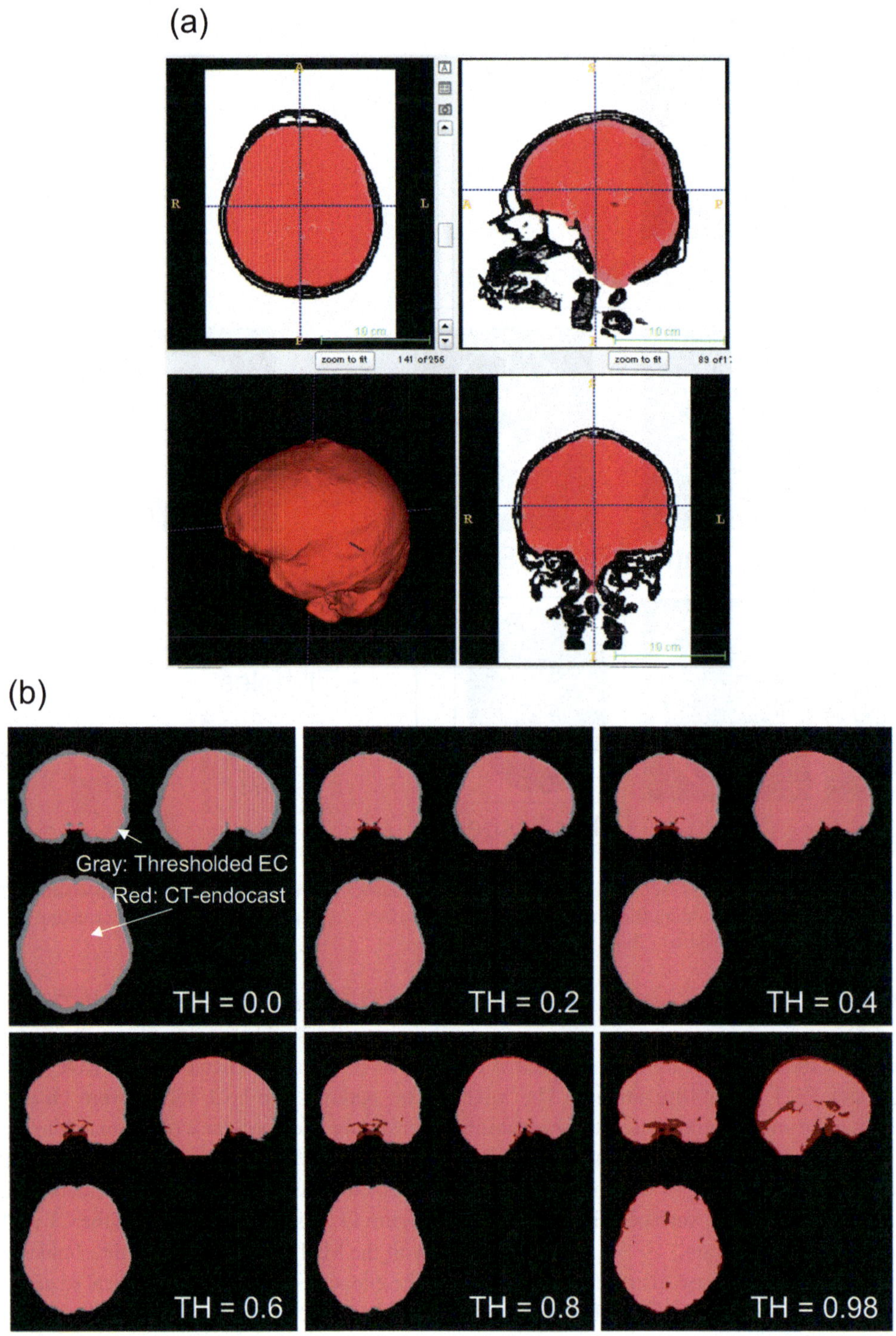

Fig. 11.3 Evaluation of the extracted endocast. (**a**) CT-derived endocast (*red*) of a representative participant used as a gold standard. (**b**) Effects of tissue probability threshold (TH) on the accuracy of the MRI-derived endocast (*gray*) compared with CT-derived endocast (*red*). (**c**) Receiver-operating characteristic graph. (**d**) Axial slices of the CT-scanned cranium with the MRI-derived endocast

erosion followed by dilation (i.e., morphological opening) to the endocast image to remove small connected and disconnected regions that were not a part of the endocast surface. In this processing stage, topological errors such as small bridges on the endocast surface were also corrected. Next, we applied a three-dimensional morphological dilation followed by erosion (i.e., morphological closing) to fill in surface pits and holes and structural gaps such as interhemispheric fissures. Finally, a

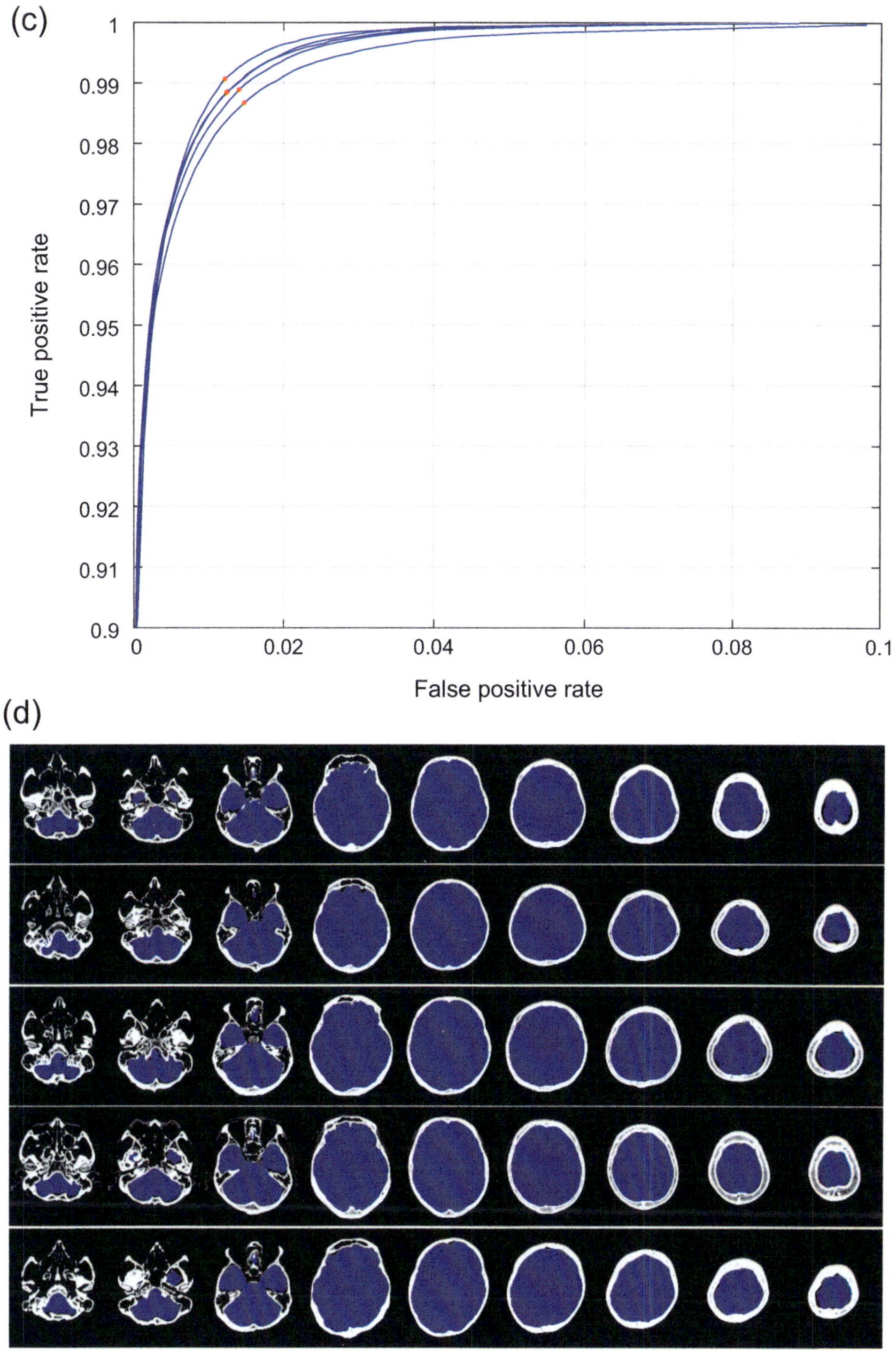

Fig. 11.3 (continued)

morphological filling procedure was used to fill in the internal voids of the endocast. After automatic processing, the quality of the resulting endocast was visually inspected, and when necessary, additional manual corrections were conducted to remove unwanted surface irregularities (bumps and pits).

11.3.3 Reconstruction of Brains Using Spatial Deformation

To reconstruct the brains of the target endocasts from the modern human brain, the deformation fields were defined

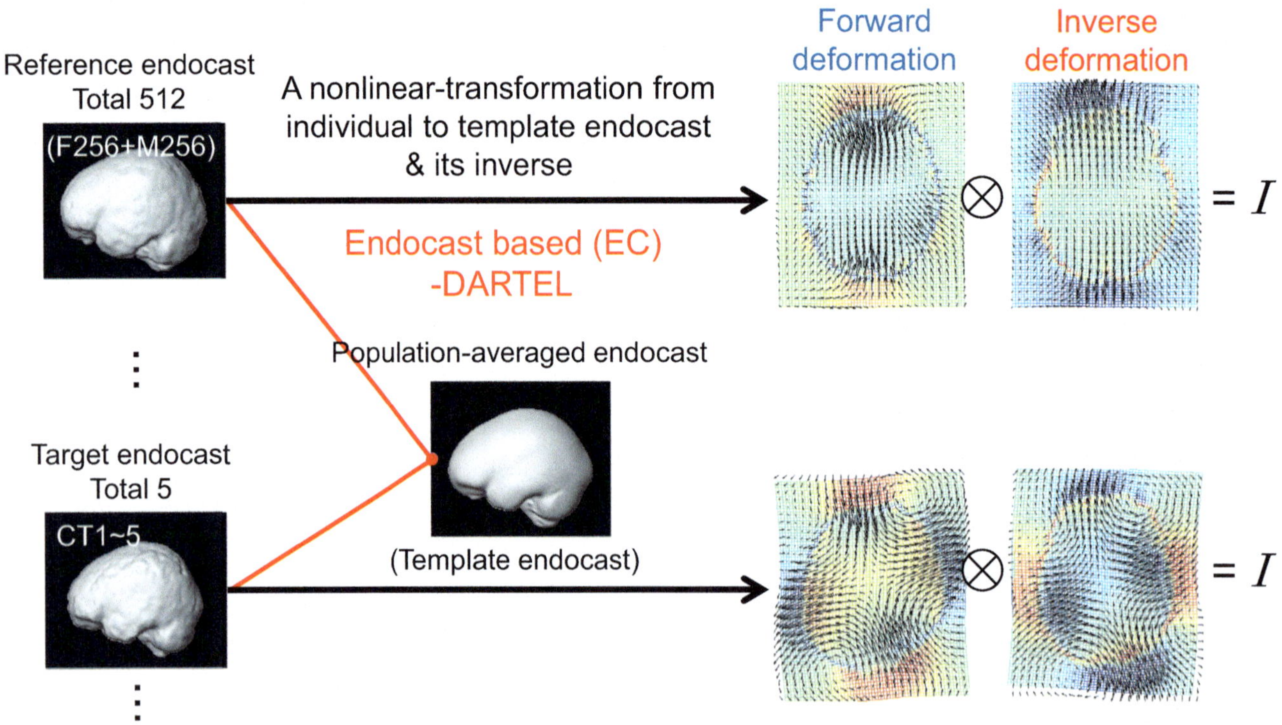

Fig. 11.4 Spatial deformation from an individual endocast to a population-averaged template endocast (forward deformation) and vice versa (inverse deformation) using DARTEL

using the endocasts. In brief, the endocast image reconstructed from the CT image of a target skull was spatially transformed to match the modern human endocast image segmented from the MRI. Using the deformation field from this spatial transformation, the modern human brain was inversely deformed to create the brain shape to fit inside the target cranium.

The analysis pipeline starts by creating a population-averaged endocast image (template) using DARTEL (EC-DARTEL; Fig. 11.4). The endocast images from different subjects were firstly registered to each other by a rigid-body transformation, so that all of the endocasts were in roughly the same spatial position and orientation. An initial population-averaged image was calculated by averaging all the aligned images. Each subject's endocast image was then spatially normalized by warping it to fit the initial average image created with DARTEL. The normalized images were then used to update the average image. By repeating this process, the population-averaged endocast image was generated. This process involved iterative minimization of the sum of squares of the difference between individual and template images using a Bayesian probabilistic formulation, with control (regularization) of the smoothness of the deformation field represented by an elastic energy (Ashburner 2007). This regularized iterative optimization allowed calculation of a fine-detailed and accurate average image, while avoiding the calculation becoming trapped in a local minimum. Finally, two types of deformation field were obtained from DARTEL (Fig. 11.4): a forward deformation field that allows an individual endocast image to be transformed to the average endocast image and an inverse deformation field that can be used to transform the average endocast image back to a subject-specific image.

The forward deformation field (forward EC-DART in Fig. 11.5) was applied to each reference brain of a modern human individual to create a spatially normalized brain. The normalized brains of all the modern humans were then averaged to create a population-averaged brain. As this average brain is defined by the DARTEL transformation, it could be transformed to each individual brain using the inverse transformation. Therefore, to reconstruct the brain of the target cranium, the average brain was transformed into the target endocast using the inverse deformation field (inverse EC-DART in Fig. 11.5).

In the present framework, no explicit information on anatomical information, such as digitized anatomical landmarks or imprints of sulci and gyri on the endocasts, was incorporated into the calculation of the deformation

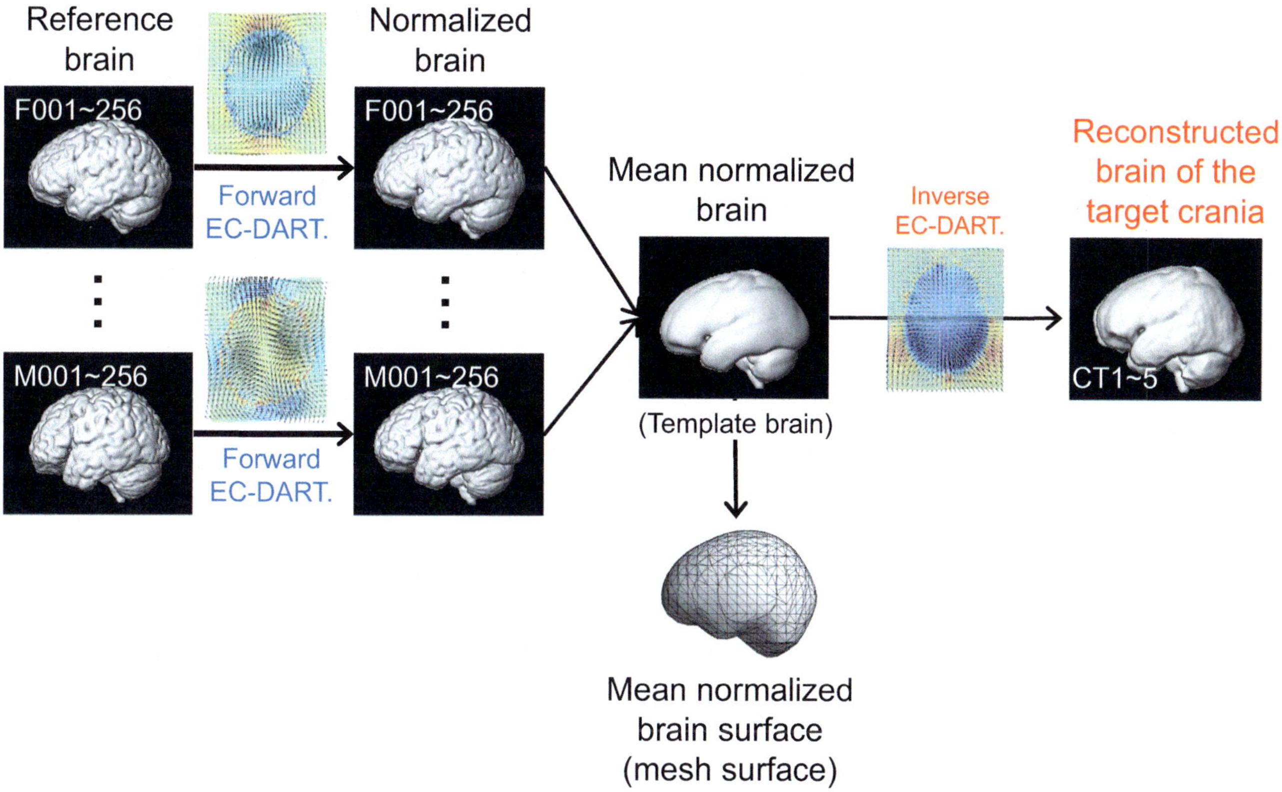

Fig. 11.5 Reconstruction of brains by sequentially applying the forward and inverse deformation fields

field. However, it should be technically possible to incorporate such additional information into the calculation of the deformation fields by introducing it as a form of constraint equation.

11.4 Evaluation of the Reconstructed Brains

Figure 11.6 shows the reconstructed and true brains of the five target crania. The overall shapes of the reconstructed brains were in good agreement with those of the corresponding true brains. We evaluated the accuracy of the brain reconstruction method by comparing the reconstructed and corresponding true brains. However, as the reconstructed brain is the DARTEL transformation of the population-averaged brain, individual sulci and gyri are not present on the reconstructed brain surface (Fig. 11.6). Therefore, for the comparisons, the deformation field for the transformation from each target brain (GM and WM) to the population-averaged (mean normalized) brain (GM and WM) was calculated using the DARTEL procedure (BR-DARTEL). The population-averaged brain with GM and WM information was inversely transformed using the inverse deformation field to obtain the true brain for each target cranium (Fig. 11.7).

Firstly, we calculated the Euclidean distance between the corresponding homologous vertices of the reconstructed (EC-DARTEL) and true brain (BR-DARTEL) surfaces (Fig. 11.8). This operation is justified, as the two surface meshes share the same topology. A surface deviation map was then created by averaging the Euclidean distances across all of the modern humans on a vertex-by-vertex basis and visualizing them on a normalized brain surface.

Secondly, we evaluated the overlapping of the GM between the reconstructed and true brains (Fig. 11.9). We employed a region of interest (ROI) approach to investigate regional dependencies in the overlap accuracy. The GM images were binarized with a 10% threshold in line with standard practice (e.g., http://dbm.neuro.uni-jena.de/vbm8/vbm8-manual.pdf). To estimate the volume of each brain region, we parcellated each individual brain into 25 structural regions (12 per hemisphere and the cerebellar vermis; Table 11.1) using the automated anatomical labeling (AAL) technique (Tzourio-Mazoyer et al. 2002). We

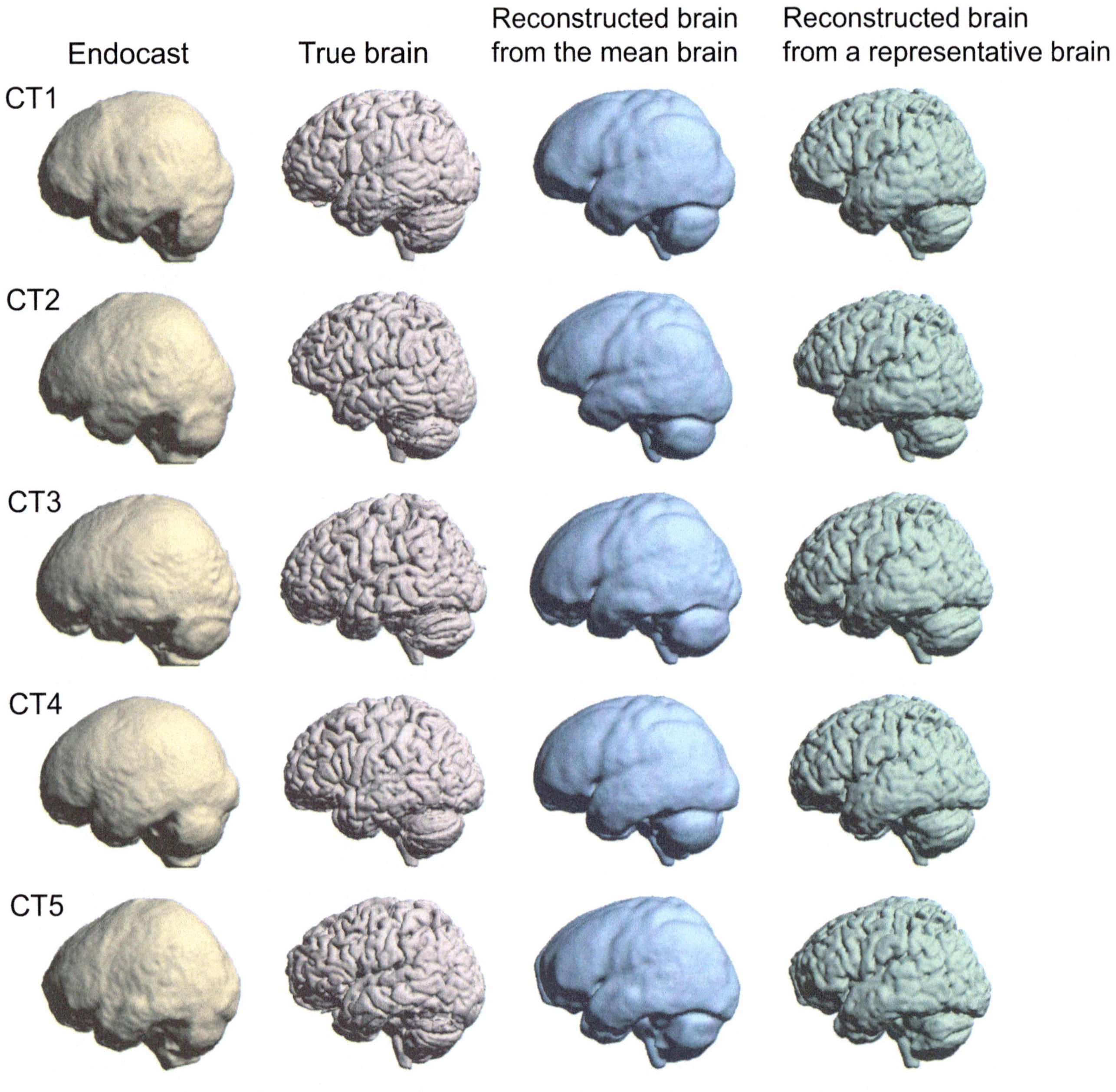

Fig. 11.6 Reconstructed brains. *Left two columns*: endocasts of the target crania and the corresponding true brains. *Right two columns*: reconstructed brains based on the population-averaged modern human brain and one representative modern human brain. The surface rendering was created using the surface extraction function of SPM software

included only AAL regions that appear on the surface, as the volume of deep brain structures such as the basal ganglia and limbic regions cannot be precisely estimated by our reconstruction method based solely on endocranial shape. For each of the 25 structural regions, we counted the number of correctly overlapping (true positive, TP), correctly non-overlapping (true negative, TN), falsely overlapping (false positive, FP), and falsely non-overlapping (false negative, FN) voxels within each ROI and calculated the accuracy as the ratio between the number of correctly identified voxels (TP + TN) and the total number of voxels (TP + FN + TN + FP) in the corresponding brain region.

Figure 11.10 shows a color map demonstrating the accuracy of the reconstructed brain surface. We observed only relatively small deviations over the whole brain (<3 mm), with the exception of the superior parietal lobule, where the largest morphological variability is observed among adult human brains (Bruner et al. 2014, 2017) and brain regions

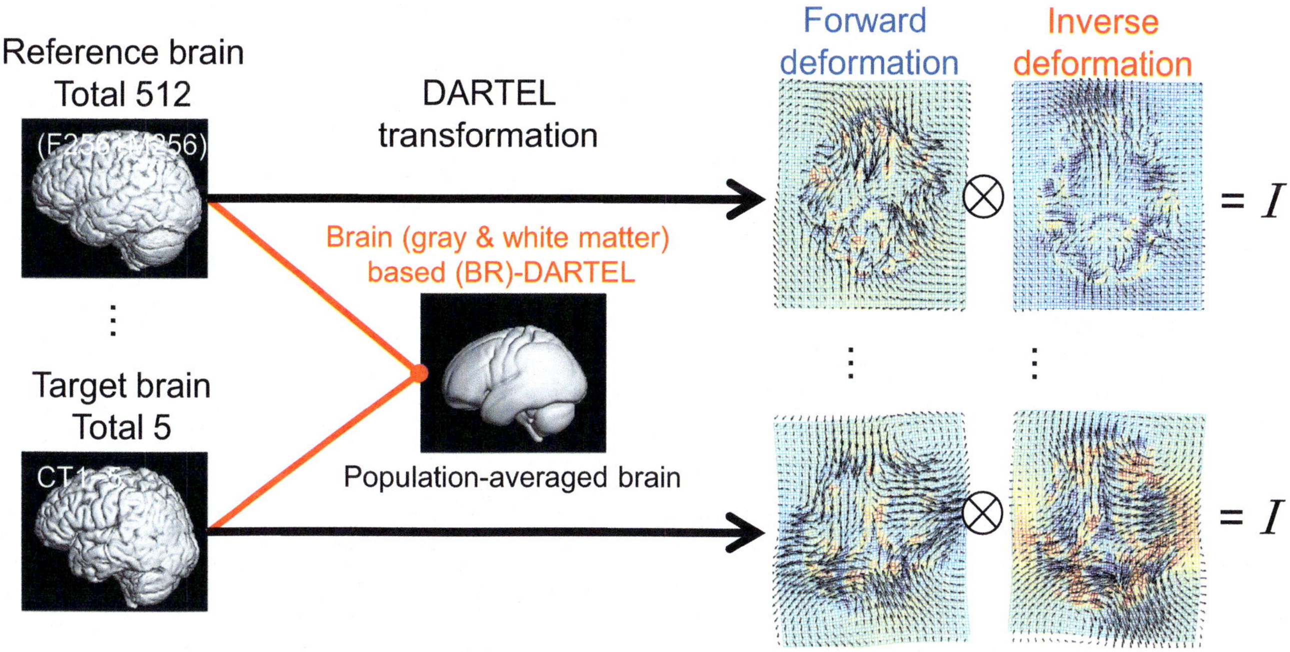

Fig. 11.7 Brain (GM and WM)-based DARTEL procedure for evaluation of the reconstructed brain. The deformation field from each brain to the population-averaged (mean normalized) brain was calculated using the DARTEL procedure (BR-DARTEL), and the population-averaged brain with GM and WM information was inversely transformed using the inverse of the deformation field to obtain the true brain for each target cranium for comparison purposes

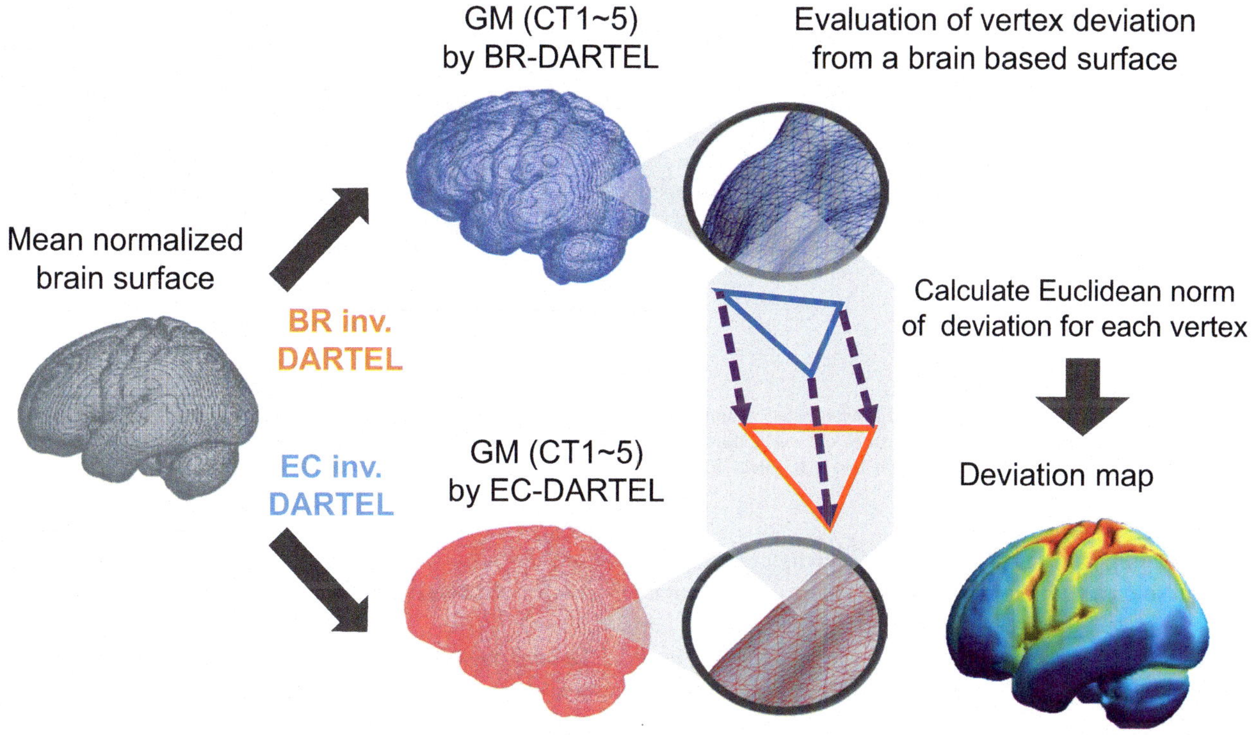

Fig. 11.8 Evaluation of the accuracy of a reconstructed brain surface. The surface deviation map shows the Euclidean deviation from the true to reconstructed brain

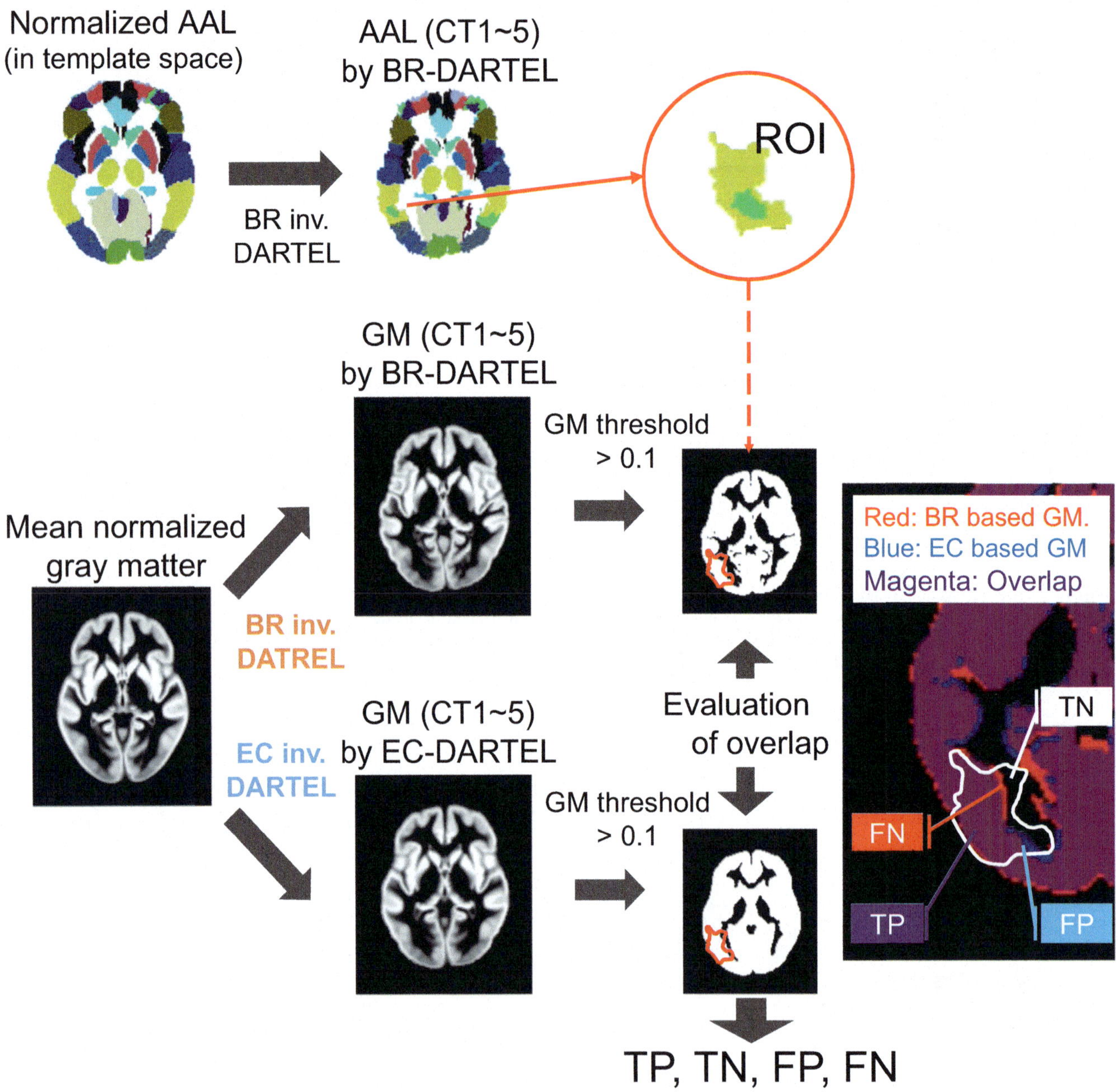

Fig. 11.9 Evaluation of the accuracy of the estimated volume of each parcellated brain region. Statistical measures of overlapping between the true and reconstructed brains within the anatomical ROIs were calculated. *ROI* region of interest, *TP* true positive, *TN* true negative, *FP* false positive, *FN* false negative

with a complicated pattern of folds and fissures (e.g., the central sulcus). Table 11.2 summarizes the volume accuracy, and again, we found decreased accuracy in the superior parts of the brain, although the accuracy was above 80% in the inferior parts, including the cerebellum. These results suggest that the proposed procedures are able to reconstruct the brain morphology with sufficient accuracy, despite the fact that it relies only on information from the endocast.

11.5 Statistical Comparisons of Brain Shape

To infer possible differences in reconstructed brain morphology between two groups such as Neanderthals and anatomically modern humans, statistical tests on the morphological brain differences between the two populations are necessary. To do this, we demonstrate three different statistical analyses using our

Table 11.1 Correspondence between the automated anatomical labeling (AAL) atlas and the 25 parcellated brain regions

Region	Subregions	Region name of AAL atlas				
Frontal	Superior/middle regions	Frontal_Sup	Frontal_Mid	Supp_Motor_Area		
	Inferior regions	Frontal_Inf_Oper	Frontal_Inf_Tri			
	Orbitofrontal regions	Frontal_Sup_Orb	Frontal_Mid_Orb	Frontal_Inf_Orb	Frontal_Med_Orb	Rectus
Sensory motor		Precentral	Postcentral	Paracentralobule		
Parietal	Superior/inferior regions	Parietal_Sup	Parietal_Inf	Precuneus		
	Temporoparietal junction	SupraMarginal	Angular			
Temporal	Superior/middle regions	Temporal_Sup	Temporal_Mid			
	Inferior/medial regions	Temporal_Inf	ParaHippocampal	Temporal_Pole_Sup	Temporal_Pole_Mid	
Occipital	Superior/middle regions	Calcarine	Cuneus	Occipital_Sup	Occipital_Mid	
	Inferior regions	Occipital_Inf	Lingual			
Cerebellum	Anterior parts	Cerebelum_3	Cerebelum_4_5			
	Posterior parts	Cerebelum_Crus1	Cerebelum_Crus2	Cerebelum_6	Cerebelum_7b	
		Cerebelum_8	Cerebelum_9	Cerebelum_10		
	Vermis	Vermis_1_2	Vermis_3	Vermis_4_5	Vermis_6	
		Vermis_7	Vermis_8	Vermis_9	Vermis_10	

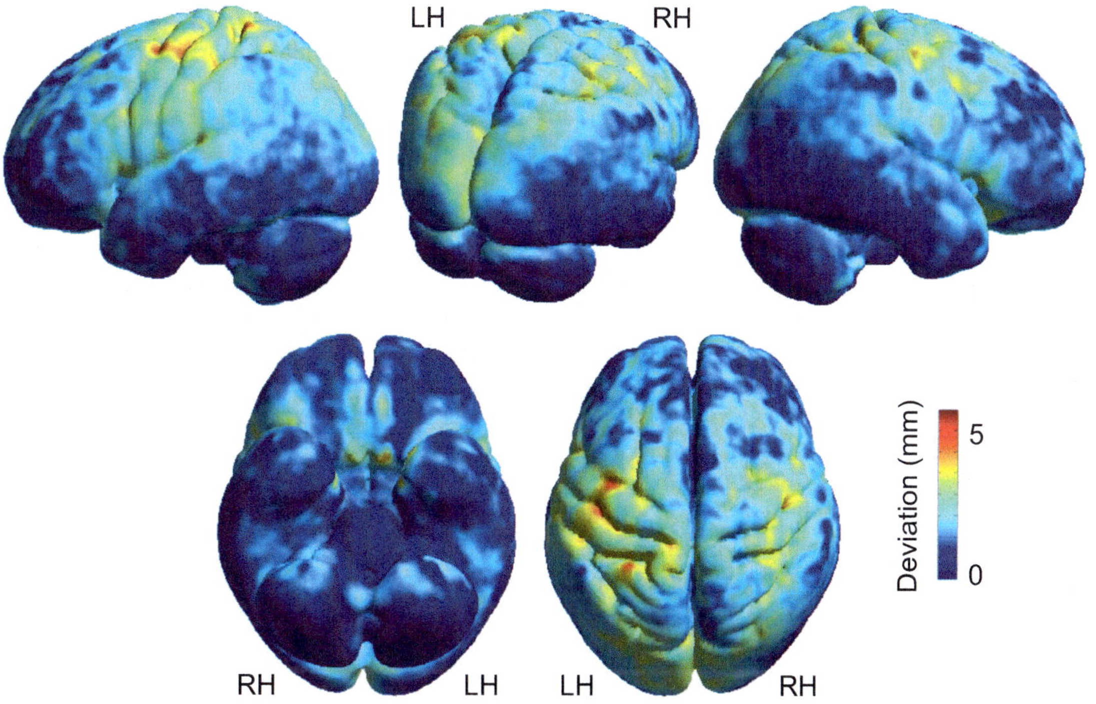

Fig. 11.10 The accuracy of a reconstructed brain surface. The color map represents the Euclidean deviation (mm) indicating the degree of accuracy

unified framework: surface displacement-based morphometry (SDBM), surface normal displacement-based morphometry (SNDBM), and voxel-based morphometry (VBM). While it would be more interesting to present comparisons of reconstructed fossil brain morphology using the above three statistical analyses, in this study, we introduce each of these approaches using an example that examines sex-related differences in brain shape in the modern Japanese population.

11.5.1 Surface Displacement-Based Morphometry Analysis

Surface displacement-based morphometry (SDBM) analysis is a variant of deformation-based morphometry (Ashburner et al. 1998; Chung et al. 2001; Chung et al. 2003) and enables statistical evaluation of the magnitude and direction of changes in the shape of the brain surface.

Table 11.2 Mean accuracy of the volume estimations for each parcellated brain region

Region	Subregions	LH	RH
Frontal	Superior/middle regions	85.8 ± 2.4	87.0 ± 1.8
	Inferior regions	86.5 ± 4.1	89.2 ± 2.4
	Orbitofrontal regions	91.3 ± 1.9	91.4 ± 1.6
Sensory motor		83.0 ± 2.9	84.2 ± 2.1
Parietal	Superior/inferior regions	89.0 ± 1.1	90.0 ± 1.6
	Temporoparietal junction	87.8 ± 1.3	89.3 ± 3.0
Temporal	Superior/middle regions	90.9 ± 1.7	90.6 ± 1.4
	Inferior/medial regions	92.6 ± 1.8	91.8 ± 1.1
Occipital	Superior/middle regions	86.7 ± 2.2	87.9 ± 1.5
	Inferior regions	93.5 ± 1.6	93.8 ± 0.9
Cerebellum	Anterior parts	94.4 ± 1.6	96.0 ± 1.2
	Posterior parts	94.6 ± 0.9	94.4 ± 0.7
	Vermis	91.0 ± 2.3	

Data are means ± s.d.%. *LH* left hemisphere, *RH* right hemisphere

The analytical workflow is illustrated in Fig. 11.11. Firstly, we created a template brain surface from the mean normalized GM image of modern humans using the surface extraction function in SPM. The resulting surface mesh contained approximately 52,000 vertices. The template brain surface was then inversely transformed into the individual brain surface using the deformation file derived previously and described above. Finally, we calculated the 3D displacement vector field required to transform the positions of the mesh vertices of the template brain to the corresponding homologous vertices of the individual brain. This operation was possible because the same topology was maintained throughout the process. The resulting displacement fields were entered into a one-way multivariate analysis of variance (MANOVA) to test for group differences in brain shape. The total endocast volume was added as a covariate of no interest to account for the different cranial sizes. The Surfstat toolbox (http://www.math.mcgill.ca/keith/surfstat/) was used to fit the statistical model and make inferences based on univariate and multivariate random field theory (Cao and Worsley 1999; Worsley et al. 2004; Chung et al. 2010). The surface statistical map was generated after computing Hotelling's T statistics on a vertex-by-vertex basis for each contrast of the group comparison. In this example case, the resulting map shows sex-related differences in the modern human brain shape. Regions demonstrating a shape difference between the groups (males and females in this case) were deemed significant if they exceeded a statistical threshold of $p < 0.05$, family-wise error (FWE) corrected for multiple comparisons at the peak level over the whole-brain surface using random field theory (Worsley et al. 1999). The results of the SDBM analysis are shown in the right side (statistical inference) of Fig. 11.11.

11.5.2 Surface Normal Displacement-Based Morphometry Analysis

While the displacement field is used to evaluate the magnitude of surface deformation in SDBM, it also allows a detailed characterization of the displacement by extraction of directional information. We developed a new method which we named *surface normal displacement-based morphometry* analysis (SNDBM). SNDBM can assess the regional expansion and shrinkage of the brain surface using the surface normal displacement map. This map consists of scalar-valued displacements from one group to another, limited to the surface normal direction.

The analytical workflow is illustrated in Fig. 11.12. The displacement vectors are projected onto the surface normal of the template brain surface on a vertex-by-vertex basis. The projections have scalar values, allowing them to be processed by univariate statistical models (e.g., the general linear model [GLM)] in SPM software). These projection values represent the signed normal displacement from the template to the individual brain, where a positive or negative value means surface expansion or shrinkage, respectively. Similar methods have been developed for subcortical shape analysis (c.f. Patenaude et al. 2011; Li et al. 2012).

The resulting surface normal displacement map is entered into the GLM (Friston et al. 1995), where the normal displacement between two or more different groups is statistically compared on a voxel-by-voxel basis. We employed a two-sample *t*-test in the present example case. The design matrix of this model had two regressors of interest, which were the estimates of the mean normal displacement from the template surface to the surface of each group. The total endocast volume was also added as a covariate of no interest, to correct for the different cranial sizes. For intergroup comparisons of the regional specific expansion or shrinkage of the brain surface, we examined the linear contrasts of the two groups. The resulting voxel-wise *t*-test for each contrast constituted a statistical parametric map of the *t*-statistics (SPM$\{t\}$). The cluster-forming threshold (height threshold) for the SPM$\{t\}$ was set at $p < 0.001$, and a value of $p < 0.05$ with FWE correction for multiple comparisons at the cluster level for the entire brain surface was then used (Friston et al. 1994; Worsley et al. 1996). The results of the SNDBM analysis are shown on the right side (statistical inference) of Fig. 11.12.

11.5.3 Morphometric Analysis of Regional Brain Volume

Finally, we introduce voxel-based morphometry (VBM; Wright et al. 1995; Ashburner and Friston 2000) to localize regions of intergroup shape differences estimated from the endocasts. VBM is a fully automated and well-established

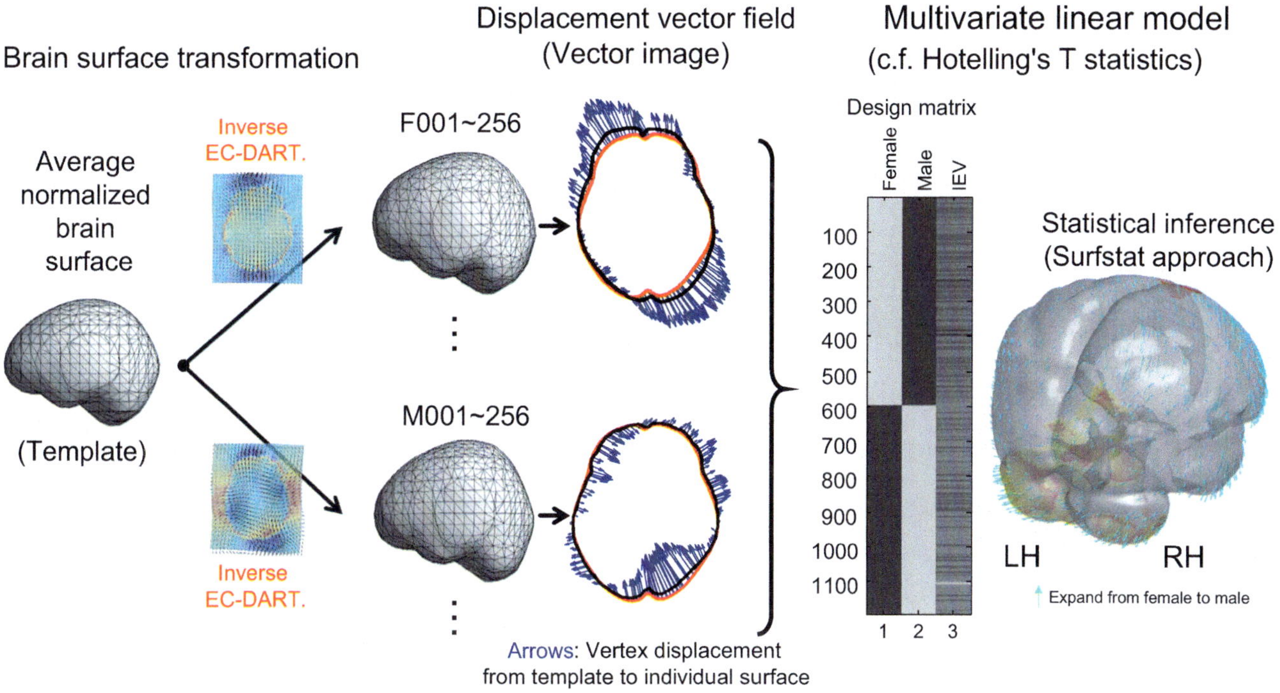

Fig. 11.11 Surface displacement-based morphometry analysis. Statistical parametric map of Hotelling's T statistics for female vs. male contrast overlaid on a surface rendering of the brain. *Red–yellow* indicates the significant morphological changes between females and males ($p < 0.05$, FWE-corrected for multiple comparisons at the cluster level). *Small cyan arrows* indicate the surface expansion/shrinkage from female to male

technique that allows quantitative examination of the whole-brain anatomical shape on a voxel-by-voxel basis.

The analytical workflow is illustrated in Fig. 11.13. This is the so-called optimized VBM approach (Good et al. 2001). In this example, we used the endocast images instead of gray and white matter images. Firstly, we calculated the Jacobian determinant of the deformation field derived from the DARTEL normalization. The Jacobian matrix of the deformation (i.e., the spatial derivatives of the deformation field) contains information about the local stretching, shearing, and rotations involved in the deformation. The determinant of the Jacobian matrix (Jacobian determinant) represents the relative volume of the individual endocast relative to the population-averaged endocast. A Jacobian determinant voxel value of greater than 1 means relative volume expansion of the individual image voxel in comparison with the average endocast image voxel, and vice versa. We then created the Jacobian-modulated normalized endocast image, which is the spatially normalized endocast image multiplied by the Jacobian determinant of the deformation field on a voxel-by-voxel basis. Our approach can therefore be referred to as (a simple form of) tensor-based morphometry (TBM), as it involves statistical analysis of a Jacobian determinant image (Ashburner and Friston 2004).

The pre-processed images were statistically compared on a voxel-by-voxel basis using the GLM (Friston et al. 1995). In the present example case, we used a two-sample t-test to compare sex-related differences in the anatomical brain structure of the modern humans. In order to account for different cranial sizes, the total endocast volume was added as a covariate of no interest. Significant volume changes for each contrast were assessed using one-tailed t-statistics on a voxel-by-voxel basis. The resulting set of voxel values for each contrast constituted the statistical parametric map of the t-statistics (SPM$\{t\}$). The cluster-forming threshold (height threshold) for the SPM$\{t\}$ was set at $p < 0.001$, with $p < 0.05$ with FWE correction for multiple comparisons at the cluster level used for the entire endocast (Friston et al. 1994; Worsley et al. 1996). The results of the VBM analysis are shown on the right side (statistical inference) of Fig. 11.13.

11.6 Discussion

The present study describes the development of a new mathematical framework to virtually reconstruct fossil brains and to statistically evaluate morphological differences between fossil and extant human brains by means of computational neuroanatomy and functional neuroimaging. Specifically, we presented a spatial normalization technique to computationally reconstruct brain morphology from modern human brains on the basis of endocranial shape. This allowed the

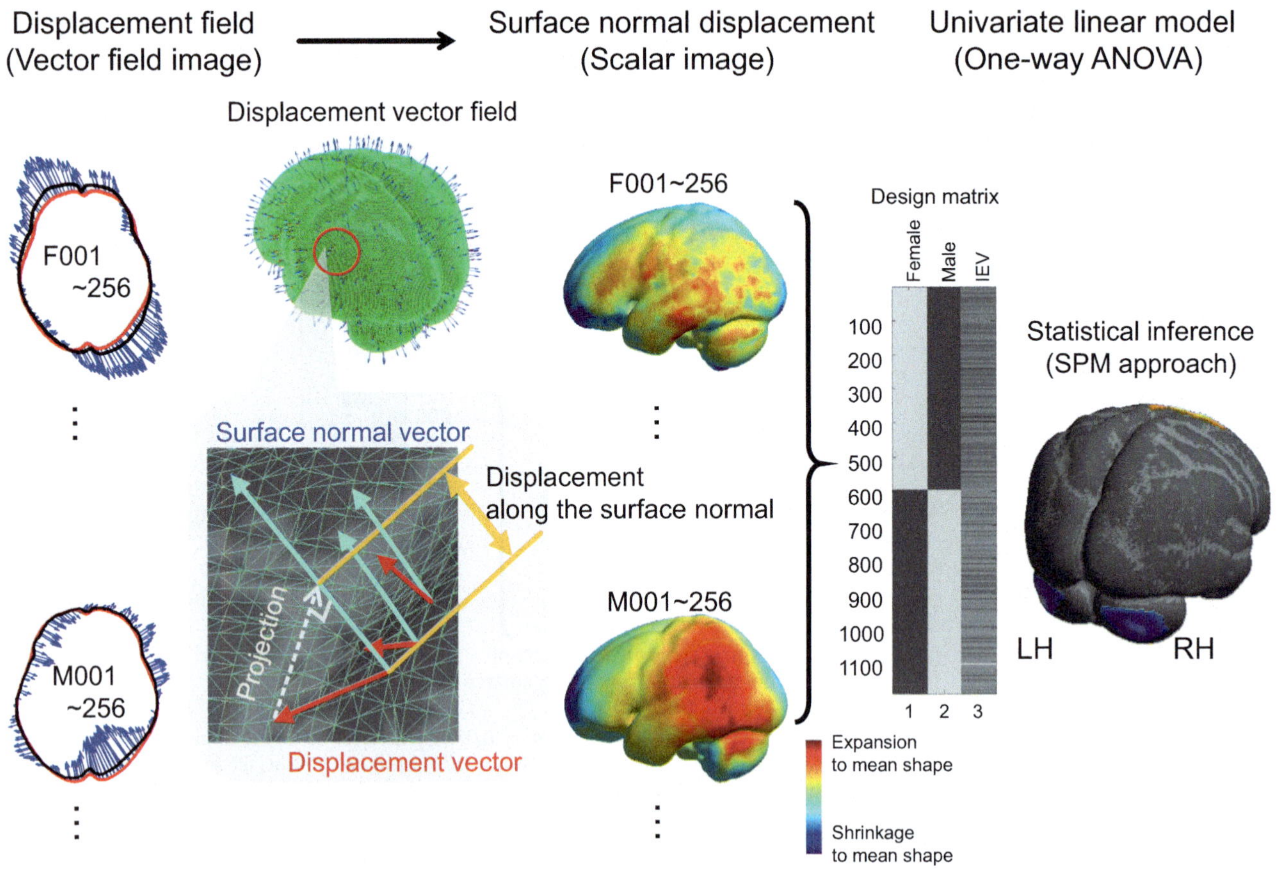

Fig. 11.12 Surface normal displacement-based morphometry analysis. A statistical parametric map of the *t*-statistics indicating where female brains were expanded compared with male brains (*red–yellow*) and vice versa (*blue–cyan*), overlaid on a surface rendering of the brain. The statistics are thresholded at $p < 0.05$ FWE-corrected for multiple comparisons at the cluster level, with a cluster-forming threshold of $p < 0.001$. *RH* right hemisphere, *LH* left hemisphere

estimation of cortical features and the volumes of various brain regions. We also presented methods for statistical comparisons of reconstructed brains using whole-brain morphometric analysis, SDBM and SNDBM analyses for evaluation of the location and direction of changes in the brain surface morphology among different populations, and endocast VBM analysis for estimation of the change in local volume from one population to another. These methods should be useful to identify morphological differences of the brain and infer possible functional differences between fossil and extant human brains in the fields of physical anthropology, archaeology, and paleoneurology.

Our unified reconstruction and evaluation approach offers several advantages in comparison with conventional paleoneurology methods. First, our analytical workflow starts from the reconstruction of the brain itself. The subsequent morphological measurements and evaluations are conducted on the reconstructed brain, not on the endocranial volume or shape. To the best of our knowledge, this is the first attempt to actually reconstruct 3D brain shape on the basis of endocranial shape. Such reconstruction of the brain shape allows more detailed investigations on the relationship between fossil cranial morphology. Second, our method allows automated whole-brain analysis. Conventional fossil ecto- and endocranial analyses typically use anatomical landmarks and sliding semi-landmarks, which are digitized onto cranial surfaces (e.g., Gunz et al. 2010, 2012; Bastir et al. 2011). However, although this approach is effective where landmarks are readily available, there are many morphometric features in cranial anatomy that may be difficult to quantify by visual inspection and, as a result, may be overlooked. To the contrary, the present whole-brain analysis could be described as using infinite-dimensional landmarks to represent the overall shape and is therefore free from manual landmark placement. Therefore, our analysis can provide unbiased and objective measures of brain morphology, even for a large number of datasets, although the automated analysis requires careful checking to ensure that homologous structures are compared.

The present reconstruction framework also allows quantitative statistical analysis of the reconstructed and living brains in anatomical standard space. Fossil studies are

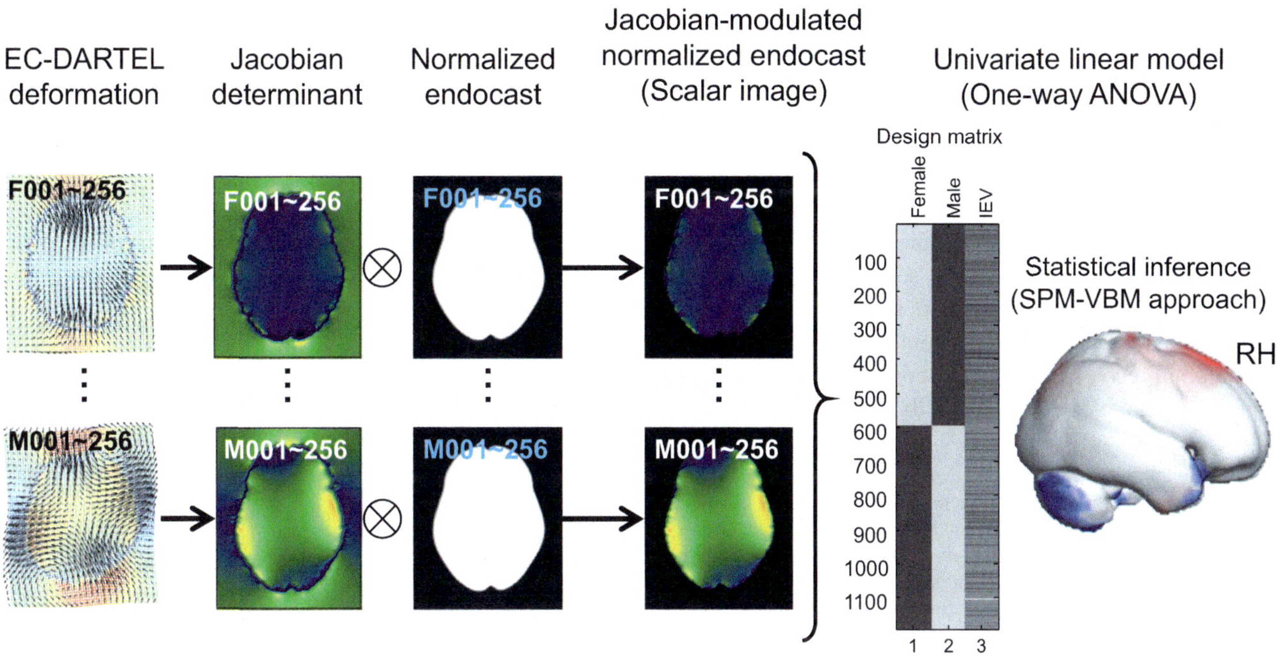

Fig. 11.13 Voxel-based morphometry analysis. Statistical parametric map of *t*-statistics for female brains indicating where they were larger than those of male subjects (*red*) and vice versa (*blue*), overlaid on a surface rendering of the brain. The statistics are thresholded at $p < 0.05$ FWE-corrected for multiple comparisons at the cluster level with a cluster-forming threshold of $p < 0.001$. *RH* right hemisphere

generally limited by their small sample size, which hampers correct estimation of confidence intervals. However, with our analysis method, when the test statistics (e.g., *t*-statistics) are constructed, the within-group variance can be drawn from a large modern human population, under the assumption of equivalent variance among species. The same approach is often used in an extreme case where a single specimen is compared with a sample (Sokal and Rohlf 1995). These techniques enable us to make comparisons not only between Neanderthals and modern humans but also between different fossil humans (e.g., Neanderthals and early modern humans). Using the SPM approach, all statistics were reported after correction for multiple comparisons, which has often been ignored in previous studies. Another merit of our statistical analysis is that the statistical model based on a GLM has enough flexibility to include additional factors (e.g., other types of fossil humans) and covariates of interest, such as gender and age. The same statistical model can quite easily be extended to incorporate various morphological quantities. Furthermore, the anatomical standardization is valuable, as it allows researchers discussing the functional significance of detected areas to refer to functional maps of not only individual functional MRI experiments but also the results of meta-analyses of functional neuroimaging studies. These advantages can only currently be realized by our unified platform.

One limitation of the present brain reconstruction methodology is the assumption that fossil human brains, such as those of Neanderthals and early modern humans, can be reconstructed by deforming modern human brains according to endocranial shape because of their phylogenetic closeness to extant humans. This means that what this framework actually presents as a reconstructed fossil brain is a deformed modern human brain corresponding to an endocast. Additionally, there may be a large variation in morphological correspondence between endocranial shape and brain shape. Furthermore, Neanderthals may have had features specific to their lineage that are absent in modern humans. In the present study, we tried to validate our methodology using modern human data. However, the applicability of the method to a fossil cranium can never actually be completely validated, as we do not have the real brain that belonged to the fossil cranium. Therefore, these limitations should be considered when interpreting the results and making inferences about possible morphological differences in the brain. It may be possible to validate our numerical inferences of the brain surface by correlations between brain surface areas and measurable morphological proxies obtained from comparative primate data, such as orbital height, as in Pearce et al.

(2013). However, such biological correlations have not been established for most brain regions, and therefore our technique may serve as a useful method to estimate the size of the internal brain areas of fossil humans. Although it is currently not possible to validate our methodology owing to the inherent nature of the problem, with the advent of fossil genome projects (Vernot et al. 2016), it might potentially be possible to infer differences in brain structure and organization on the basis of comparisons of genome sequences between *Homo neanderthalensis* and early modern *Homo sapiens*. Our computational reconstruction should hopefully be verified when brain reconstruction based on evolutionary and developmental genomics becomes possible in the near future.

Acknowledgments The authors express their sincere gratitude to T. Akazawa of Kochi Institute of Technology for giving the opportunity to participate in the research project "Replacement of Neanderthals by Modern Humans: Testing Evolutionary Models of Learning" and for lending his continuous guidance and support throughout the course of the study. The authors also thank N. Sadato of the National Institute for Physiological Sciences; O. Kondo and T. Suzuki of the University of Tokyo; H. Amano, T. Kikuchi, and Y. Morita of Keio University; K. Hasegawa of Nagoya University; H. Ishida, A. Yogi, and S. Murayama of the University of the Ryukyus; and M. Nakatsukasa of Kyoto University, for collaborations in this project. This project is supported by Scientific Research on Innovative Areas "Replacement of Neanderthals by Modern Humans: Testing Evolutionary Models of Learning" (#22101001, #22101006, #22101007) and "The Evolutionary Origin and Neural Basis of the Empathetic Systems" (#16H01486) from the Ministry of Education, Culture, Sports, Science and Technology of Japan (MEXT).

References

Amano H, Kikuchi T, Morita Y, Kondo O, Suzuki H, Ponce de Leon MS, Zollikofer CP, Bastir M, Stringer C, Ogihara N (2015) Virtual reconstruction of the Neanderthal Amud 1 cranium. Am J Phys Anthropol 158(2):185–197

Ashburner J (2007) A fast diffeomorphic image registration algorithm. NeuroImage 38:95–113

Ashburner J, Friston KJ (2000) Voxel-based morphometry-the methods. NeuroImage 11:805–821

Ashburner J, Friston KJ (2004) Morphometry. In: Frackowiak RSJ (ed) Human brain function. Academic Press, Boston, pp 707–722

Ashburner J, Friston KJ (2005) Unified segmentation. NeuroImage 26:839–851

Ashburner J, Friston KJ (2007) Computational anatomy. In: Friston KJ, Ashburner J, Kiebel SJ, Nichols TE, Penny WD (eds) Statistical parametric mapping: The analysis of functional brain images. Academic Press, New York, pp 49–98

Ashburner J, Friston KJ (2011) Diffeomorphic registration using geodesic shooting and Gauss-Newton optimisation. NeuroImage 55:954–967

Ashburner J, Hutton C, Frackowiak R, Johnsrude I, Price C, Friston K (1998) Identifying global anatomical differences: deformation-based morphometry. Hum Brain Mapp 6:348–357

Avants B, Gee JC (2004) Geodesic estimation for large deformation anatomical shape averaging and interpolation. NeuroImage 23 (Suppl 1):S139–S150

Bastir M, Rosas A, Gunz P, Peña-Melian A, Manzi G, Harvati K, Kruszynski R, Stringer C, Hublin JJ (2011) Evolution of the base of the brain in highly encephalized human species. Nat Commun 2:588

Bruner E, Manzi G, Arsuaga JL (2003) Encephalization and allometric trajectories in the genus Homo: evidence from the Neandertal and modern lineages. Proc Natl Acad Sci U S A 100(26):15335–15340

Bruner E, Rangel de Lázaro G, de la Cuétara JM, Martín-Loeches M, Roberto Colom R, Jacobs HIL (2014) Midsagittal brain variation and shape analysis of the precuneus in adult humans. J Anat 224:367–376

Bruner E, Amano H, de la Cuétara JM, Ogihara N (2015) The brain and the braincase: a spatial analysis on the midsagittal profile in adult humans. J Anat 227:268–276

Bruner E, Pereira-Pedro AS, Chen X, Rilling JK (2017) Precuneus proportions and cortical folding: a morphometric evaluation on a racially diverse human sample. Ann Anat 211:120–128

Cao J, Worsley KJ (1999) The detection of local shape changes via the geometry of Hotelling's T2 field. Ann Statist 27:925–942

Chung MK, Worsley KJ, Paus T, Cherif C, Collins DL, Giedd JN, Rapoport JL, Evans AC (2001) A unified statistical approach to deformation-based morphometry. NeuroImage 14:595–606

Chung MK, Worsley KJ, Robbins S, Paus T, Taylor J, Giedd JN, Rapoport JL, Evans AC (2003) Deformation-based surface morphometry applied to gray matter deformation. NeuroImage 18:198–213

Chung MK, Worsley KJ, Nacewicz BM, Dalton KM, Davidson RJ (2010) General multivariate linear modeling of surface shapes using SurfStat. NeuroImage 53:491–505

Falk D (2012) Hominin paleoneurology: where are we now? In: Hofman MA, Falk D (eds) Evolution of the PrimateBrain: from neuron to behavior. Elsevier, London, pp 255–272

Friston KJ (1997) Analyzing brain images: principles and overview. In: Frackowiak RSJ, Friston KJ, Frith CD, Dolan RJ, Price CJ, Zeki S, Ashburner J, Penny WD (eds) Human brain function. Academic, San Diego, pp 25–41

Friston KJ (2007) Statistical parametric mapping: the analysis of functional brain images, 1st edn. Academic, London

Friston KJ, Worsley KJ, Frackowiak RS, Mazziotta JC, Evans AC (1994) Assessing the significance of focal activations using their spatial extent. Hum Brain Mapp 1:210–220

Friston KJ, Holmes A, Poline JB, Frith CD, Frackowiak RSJ (1995) Statistical parametric maps in functional imaging: a general linear approach. Hum Brain Mapp 2:189–210

Good CD, Johnsrude I, Ashburner J, Henson RN, Friston KJ, Frackowiak RS (2001) Cerebral asymmetry and the effects of sex and handedness on brain structure: a voxel-based morphometric analysis of 465 normal adult human brains. NeuroImage 14:685–700

Gunz P, Neubauer S, Maureille B, Hublin JJ (2010) Brain development after birth differs between Neandertals and modern humans. Curr Biol 20:R921–R922

Gunz P, Neubauer S, Golovanova L, Doronichev V, Maureille B, Hublin JJ (2012) A uniquely modern human pattern of endocranial development. Insights from a new cranial reconstruction of the Neandertal newborn from Mezmaiskaya. J Hum Evol 62:300–313

Holloway RL, Broadfield DC, Yuan MS (2004) The human fossil record: brain endocasts: the paleoneurological evidence. Wiley, Hoboken

Huettel SA, Song AW, McCarthy G (2009) Functional magnetic resonance imaging, 2nd edn. Sinauer Associates, Sunderland

Kochiyama T, Tanabe HC, Ogihara N (2014) Reconstruction of the brain from skull fossils using computational anatomy. In: Akazawa T, Ogihara N, Tanabe HC, Terashima H (eds) Dynamics of learning in Neanderthals and modern humans. Volume 2: Cognitive and physical perspectives. Springer Japan, Tokyo, pp 191–200

Kwong KK, Belliveau JW, Chesler DA, Goldberg IE, Weisskoff RM, Poncelet BP, Kennedy DN, Hoppel BE, Cohen MS, Turner R,

Cheng HM, Brady TJ, Rosen B (1992) Dynamic magnetic resonance imaging of human brain activity during primary sensory stimulation. Proc Natl Acad Sci U S A 89:5675–5679

Li S, Wang Y, Xu P, Pu F, Li D, Fan Y, Gong G, Luo Y (2012) Surface morphology of amygdala is associated with trait anxiety. PLoS One 7:e47817

Malone IB, Leung KK, Clegg S, Barnes J, Whitwell JL, Ashburner J, Fox NC, Ridgway GR (2015) Accurate automatic estimation of total intracranial volume: a nuisance variable with less nuisance. NeuroImage 104:366–372

Marshall JC, Fink GR (2003) Cerebral localization, then and now. Neuroimage Suppl 1:S2–S7

Meyer M, Arsuaga JL, de Filippo C, Nagel S, Aximu-Petri A, Nickel B, Martínez I, Gracia A, de Castro JMB, Carbonell E, Viola B, Kelso J, Prüfer K, Pääbo S (2016) Nuclear DNA sequences from the Middle Pleistocene Sima de los Huesos hominins. Nature 531:504–507

Miller MI, Beg MF, Ceritoglu C, Stark C (2005) Increasing the power of functional maps of the medial temporal lobe by using large deformation diffeomorphic metric mapping. Proc Natl Acad Sci U S A 102:9685–9690

Miller MI, Trouve A, Younes L (2006) Geodesic shooting for computational anatomy. J Math Imaging Vis 24:209–228

Ogawa S, Lee TM, Kay AR, Tank DW (1990) Brain magnetic resonance imaging with contrast dependent on blood oxygenation. Proc Natl Acad Sci U S A 87:9868–9872

Ogihara N, Amano H, Kikuchi T, Morita Y, Hasegawa K, Kochiyama T, Tanabe HC (2015) Towards digital reconstruction of fossil crania and brain morphology. Anthropol Sci 123:57–68

Patenaude B, Smith SM, Kennedy DN, Jenkinson M (2011) A Bayesian model of shape and appearance for subcortical brain segmentation. NeuroImage 56:907–922

Pearce E, Stringer C, Dunbar RIM (2013) New insights into differences in brain organization between Neanderthals and anatomically modern humans. Proc R Soc B 280:2013168

Sokal RR, Rohlf FJ (1995) Biometry: the principles and practice of statistics in biological research, 3rd edn. Freeman, New York

Stone AC, Battistuzzi FU, Kubatko LS, Perry GH Jr, Trudeau E, Lin H, Kumar S (2010) More reliable estimates of divergence times in pan using complete mtDNA sequences and accounting for population structure. Philos Trans R Soc B 365:3277–3299

Tanabe HC, Kochiyama T, Ogihara N, Sadato N (2014) Integrated analytical scheme for comparing the Neanderthal brain to modern human brain using neuroimaging techniques. In: Akazawa T, Ogihara N, Tanabe HC, Terashima H (eds) Dynamics of learning in Neanderthals and modern humans. Volume 2: Cognitive and physical perspectives. Springer Japan, Tokyo, pp 203–207

Tzourio-Mazoyer N, Landeau B, Papathanassiou D, Crivello F, Etard O, Delcroix N, Mazoyer B, Joliot M (2002) Automated anatomical labeling of activations in SPM using a macroscopic anatomical parcellation of the MNI MRI single-subject brain. NeuroImage 15:273–289

Vernot B, Tucci S, Kelso J, Schraiber JG, Wolf AB, Gittelman RM, Dannemann M, Grote S, McCoy RC, Norton H, Scheinfeldt LB, Merriwether DA, Koki G, Friedlaender JS, Wakefield J, Pääbo S, Akey JM (2016) Excavating Neandertal and Denisovan DNA from the genomes of Melanesian individuals. Science. doi:10.1126/science.aad9416

Wang L, Beg F, Ratnanather T, Ceritoglu C, Younes L, Morris JC, Csernansky JG, Miller MI (2007) Large deformation diffeomorphism and momentum based hippocampal shape discrimination in dementia of the Alzheimer type. IEEE Trans Med Imaging 26:462–470

Worsley KJ, Marrett S, Neelin P, Vandal AC, Friston KJ, Evans AC (1996) A unified statistical approach for determining significant signals in images of cerebral activation. Hum Brain Mapp 4:58–73

Worsley KJ, Andermann M, Koulis T, MacDonald D, Evans AC (1999) Detecting changes in nonisotropic images. Hum Brain Mapp 8:98–101

Worsley KJ, Taylor JE, Tomaiuolo F, Lerch J (2004) Unified univariate and multivariate random field theory. NeuroImage 23(Suppl 1): S189–S195

Wright IC, McGuire PK, Poline JB, Travere JM, Murray RM, Frith CD, Frackowiak RS, Friston KJ (1995) A voxel-based method for the statistical analysis of gray and white matter density applied to schizophrenia. NeuroImage 2:244–252

Yushkevich PA, Piven J, Hazlett HC, Smith RG, Ho S, Gee JC, Gerig G (2006) User-guided 3D active contour segmentation of anatomical structures: significantly improved efficiency and reliability. NeuroImage 31:1116–1128

12 Endocasts and the Evo-Devo Approach to Study Human Brain Evolution

Simon Neubauer and Philipp Gunz

Abstract

The brain is a highly plastic organ and is shaped not only during prenatal but also during postnatal development. The analysis and comparison of ontogenetic patterns of endocranial size increase and endocranial shape changes can therefore add further evidence for the interpretation of hominin brain evolution. Here we focus on digital endocast data and the methodology used to document and compare developmental patterns of endocranial shape changes. We outline how geometric morphometrics of endocranial landmark data can be used in an evo-devo approach to human brain evolution, discuss how developmental simulations help to compare ontogenetic patterns among species, present different visualization techniques that help to interpret ontogenetic shape changes, provide an overview of our current knowledge, present new data on early postnatal shape changes in apes, and discuss open questions.

Keywords

Geometric morphometrics • Semilandmarks • Developmental simulation • Thin-plate spline network • Shape analysis • Brain development

12.1 Introduction

Fossil endocasts offer a glimpse into the evolution of hominin brain function, cognition, and behavior. They provide direct evidence about brain size, brain shape, and convolutional organization on the outer brain surface (Holloway 1978; Falk 1980, 1987; Bruner et al. 2003; Bruner 2004; Holloway et al. 2004; Neubauer et al. 2009; Neubauer 2014). Unfortunately, endocasts are mute about internal brain organization and internal maturation. The brain's neural network is critical for its function (Schenker et al. 2005; Sporns 2011; Bianchi et al. 2013; Buckner and Krienen 2013; van den Heuvel and Sporns 2013), and how this network is built during ontogeny plays a major role (Greenough et al. 1987; Casey et al. 2000, 2005; Sowell et al. 2003, 2004 Gogtay et al. 2004; Kramer et al. 2004; Nagy et al. 2004; Shaw et al. 2006; Sakai et al. 2011; Miller et al. 2012; Gómez-Robles et al. 2013, 2015). Modern humans are characterized by a pattern of brain development that is prolonged temporally. Humans therefore experience an enhanced impact of environmental stimuli and influences of social interactions on neural connectivity within a plastic brain, which are critical for the development of our cognitive and behavioral capabilities (Greenough et al. 1987; Kramer et al. 2004).

Based on the relationship of high brain plasticity, external stimuli, and cognitive development, the analysis of endocranial growth and development, i.e., the analysis of patterns of size increase and shape changes during ontogeny, can add further evidence for hominin brain evolution (Smith and Tompkins 1995; Zollikofer and Ponce de León 2010; Neubauer and Hublin 2012; Neubauer 2015; Hublin et al. 2015; Gunz 2016). Studying how development influences

S. Neubauer (✉) • P. Gunz
Department of Human Evolution, Max Planck Institute for Evolutionary Anthropology, Deutscher Platz 6, 04103 Leipzig, Germany
e-mail: simon.neubauer@eva.mpg.de; gunz@eva.mpg.de

E. Bruner et al. (eds.), *Digital Endocasts*, Replacement of Neanderthals by Modern Humans Series,
DOI 10.1007/978-4-431-56582-6_12

evolution as well as how developmental patterns evolved is necessary to integrate our understanding about the tightly interrelated evolution of our brain, behavior and culture, metabolic costs of large brains, energy allocation, life history patterns, and childbirth mechanisms (Martin 1983; Harvey et al. 1987; Aiello and Wheeler 1995; Smith and Tompkins 1995; Trevathan 1996; Rosenberg and Trevathan 2002; Leigh 2004, 2012; Leigh and Blomquist 2007; Ponce de León et al. 2008; Weaver and Hublin 2009; Fischer and Mitteroecker 2015). While other recent studies and reviews have discussed variation and evolutionary changes of ontogenetic endocranial size increases or brain mass increases (Leigh 2012; O'Connell and DeSilva 2013; Neubauer 2015; Hublin et al. 2015; Cofran and DeSilva 2015), we focus here on how digital endocasts and analytical methods can help to better understand the evolutionary developmental processes that have shaped the human brain. We provide an overview on how to use three-dimensional coordinate data and geometric morphometrics in an evo-devo approach and review our current knowledge about endocranial ontogenetic shape changes. Moreover, we discuss how developmental simulations can help to compare ontogenetic patterns among species and present different visualization techniques that help to interpret the results. Finally, we present new data on early postnatal shape changes in apes to discuss open questions.

12.2 How to Capture Endocranial Shape?

Geometric morphometrics comprise a set of techniques for the multivariate statistical analysis of shape based on Cartesian coordinates of homologous landmarks in which the geometric relationships between these measurement points are preserved throughout the analyses (Bookstein 1991; Slice 2007; Mitteroecker and Gunz 2009). Analyses of endocranial shape have been limited for a long time by the scarcity of well-defined, homologous landmarks on the smooth endocranial surface. Landmarks defined on brain convolutions (e.g., Bruner et al. 2003; Bruner 2004) make it possible to delimit various brain regions. The overall shape of the endocranium can be captured using sliding semilandmarks (Bookstein 1997; Gunz et al. 2005; Gunz and Mitteroecker 2013): a series of points is digitized along homologous curves and on homologous surfaces and then allowed to slide so as to minimize the bending energy of the thin-plate spline interpolation between each individual of a sample and the sample mean. The goal of this procedure is to achieve geometric correspondence among semilandmarks within this sample.

Based on semilandmarks, we have developed an endocranial landmark set that captures the overall endocranial shape, compartmentalized by semilandmarks on curves (Neubauer et al. 2009) so that, for example, the areas of the cerebrum and cerebellum can be distinguished (Neubauer et al. 2009, 2010, 2012; Gunz et al. 2010, 2011, 2012; Scott et al. 2014; Gunz 2015). Recently, Ponce de León et al. (2016) and Zollikofer et al. (2017) adopted a similar landmark set. Here, we introduce an updated version (Fig. 12.1) of our original landmark set (Neubauer et al. 2009) with a higher density of semilandmarks, comprising 29 anatomical landmarks, 110 curve semilandmarks, and 796 surface semilandmarks, in total a set of 935 points. Curve semilandmarks are distributed over two midplane curves (midsagittal cerebral curve and clivus curve) and four bilateral curves (sphenoid curve, hemi-foramen-magnum ridge, curve on the upper border of the transverse sinus continuing onto the petrous ridge, and curve on the lower

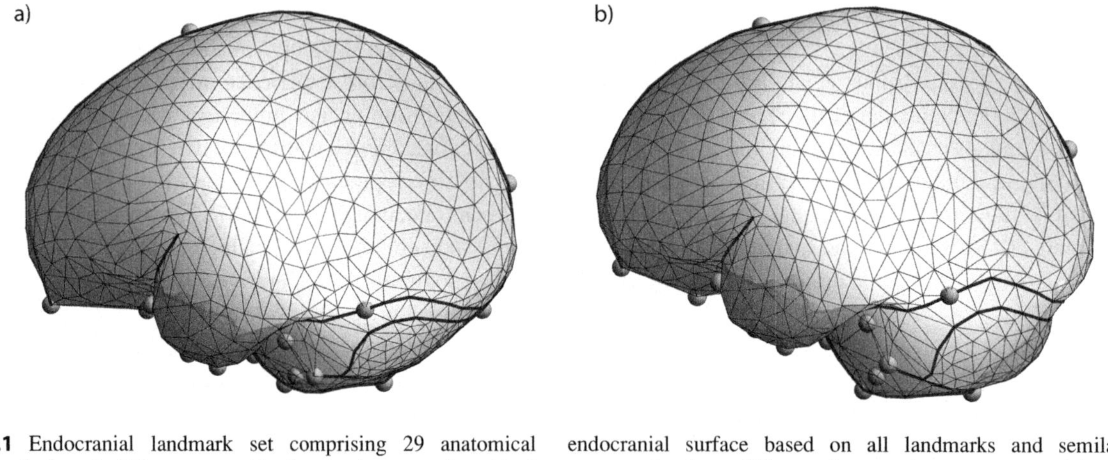

Fig. 12.1 Endocranial landmark set comprising 29 anatomical landmarks (*spheres*), 110 curve semilandmarks (connected as *black solid lines*), and 796 surface semilandmarks. The triangulated endocranial surface based on all landmarks and semilandmarks captures the shape difference between human neonates (**a**) and human adults (**b**) and can be used to visualize results

border of the transverse sinus continuing onto the border of the sigmoid sinus).

Using Procrustes superimposition (Gower 1975; Rohlf and Slice 1990), information about location and orientation is removed from the raw coordinates, and each specimen is scaled to unit centroid size. The latter is the size measure used in geometric morphometrics, computed as the square root of the summed up squared distances of all landmarks to the centroid of the landmark configuration (Bookstein 1991). In this way, the shape of differently sized age groups and species can be analyzed and compared using Procrustes shape variables, while size variation can be investigated separately based on centroid size. Figure 12.1 shows average shapes of human neonates and human adults and illustrates the advantages of such a landmark set: the geometric relationships between measurement points are kept throughout the analyses, and the triangulated versions of these dense point clouds are well suited to visualize results.

12.3 How to Analyze Endocranial Shape During Ontogeny?

There is a growing body of literature using geometric morphometrics to study ontogenetic patterns of craniofacial shape changes in extant and extinct hominins that ask if and how postnatal ontogeny contributes to adult morphological variation within and between groups (Ponce de León and Zollikofer 2001; Lieberman et al. 2002; Ackermann and Krovitz 2002; Cobb and O'Higgins 2004, 2007; Mitteroecker et al. 2004, 2005; Bastir and Rosas 2004a, b; Viðarsdóttir and Cobb 2004; Zollikofer and Ponce De León 2004; McNulty et al. 2006a; Bastir et al. 2006; Bulygina et al. 2006; Lieberman et al. 2007; Singleton et al. 2010; Neubauer et al. 2010; McNulty 2012; Gunz 2012; Freidline et al. 2012, 2013; Ponce de León et al. 2016; Zollikofer et al. 2017). Conceptually, these studies attempt quantifying shape changes during development, i.e., shape variation with increasing age. However, most times chronological age is available neither for fossils nor for comparative extant primate specimens. As a proxy for calendar age, many studies therefore use either size or age classes based on dental eruption.

However, there are important differences in the growth patterns among humans and apes as well as hominins (Schultz 1940, 1941; Count 1947; Holt et al. 1975; Jordaan 1976; Martin 1983; Jolicoeur et al. 1988; Smith and Tompkins 1995; Fragaszy and Bard 1997; Vrba 1998; Rice 2002; Leigh 2004, 2006; Coqueugniot et al. 2004; Fragaszy et al. 2004; Kennedy 2005; Vinicius 2005; DeSilva and Lesnik 2006; Hublin and Coqueugniot 2006; Ponce de León et al. 2008; Zollikofer and Ponce de León 2010; Coqueugniot and Hublin 2012; Leigh 2012; O'Connell and DeSilva 2013; Hublin et al. 2015; Cofran and DeSilva 2015) so that size increase may not be related with age in the same way in all the species. Furthermore, endocranial shape changes can occur after adult brain size has been attained. It is also difficult to choose an appropriate model for multivariate regression computations if shape changes are not linearly related to size. Using centroid size is appropriate when the focus is on allometry, i.e., the relationship between size increase and associated shape changes. In other cases, using size as a proxy for age can, however, complicate the comparison of ontogenetic shape trajectories.

Recent studies (Smith et al. 2010, 2015; Le Cabec et al. 2015) have shown that age at death estimations of fossils based on dental eruption and calcification patterns of living primates is not as straightforward as previously thought. These studies found that the pace for dental development might differ from somatic development. This for example leads to an overestimation of the actual age at death in Neandertals (Smith et al. 2010) and in early *Homo* as well as australopiths (Smith et al. 2015). Instead of converting dental eruption and calcification stages to calendar ages, dental ontogenetic stages can be used to define age groups. Such age groups are reproducible and homologous among hominin groups at least in some way.

Defining ontogenetic trajectories as a vector of regression coefficients resulting from multivariate regression of shape data onto dental age groups (Frost et al. 2003; McNulty et al. 2006b) or, alternatively, the series of vectors between consecutive age group means, or in other words several segments of a trajectory, (Neubauer et al. 2009, 2010; Scott et al. 2014) allows describing the overall large-scale shape changes during postnatal ontogeny and proved to be appropriate to compare ontogenetic patterns of different groups.

To analyze ontogenetic patterns of shape changes, so-called developmental simulations can be used (e.g., McNulty et al. 2006b; Neubauer et al. 2010; Gunz et al. 2010; Gunz 2012; Freidline et al. 2012, 2013; Scott et al. 2014). They help to assess the validity of ontogenetic patterns in fossils when fossil samples are small and provide information about differences of ontogenetic patterns between species. For example, applying the ontogenetic trajectory of one species to the neonates of another species, i.e., simulating individuals that grew up according to another species' ontogenetic pattern, results in simulated adults that can be compared to the actual adults of this species. Furthermore, the pattern of developmental changes, i.e., the shape of the ontogenetic trajectories, can be compared in this way, which helps to understand how similar or different the ontogenetic patterns are (McNulty et al. 2006b; Neubauer et al. 2010). Ponce de León et al. (2016) argued that "evolutionary developmental shifts between species cannot be modeled straightforwardly by shifts of entire developmental trajectories in shape space." Instead, these authors suggested comparing the direction of trajectory segments. However, if only a few individuals (e.g., in a cross-sectional sample of a fossil species) define such a trajectory segment, the direction may be unstable just because of individual shape variation not related to age (see discussion below). Moreover, developmental simulations among humans and extant apes (Neubauer et al. 2010; Scott et al. 2014) show that the developmental patterns of hominids are highly conserved. In those cases where differences are found, developmental simulations are informative about evolutionary developmental shifts.

12.4 Shared and Unique Aspects of Endocranial Ontogenetic Shape Changes

In our previous work, we have used a semilandmark-based landmark set and developmental simulations in a geometric morphometrics framework to analyze and compare endocranial ontogenetic patterns of humans, chimpanzees, gorillas, orangutans, gibbons, and Neandertals (Neubauer et al. 2009, 2010; Gunz et al. 2010, 2011, 2012; Scott et al. 2014). We have found shared but also unique aspects of postnatal shape changes elucidating how ontogeny affects and generates morphological differences among adults. Figure 12.2 revisits our previous findings using the updated landmark set and shows new data for early postnatal shape changes that are discussed below. The principal component analysis illustrates the major variation driven by the shape differences among species and the ontogenetic shape changes. The ontogenetic trajectories are represented by the series of vectors between consecutive age group means, variation within age groups by convex hulls. Note that there is an overlap of shape variation among consecutive age groups due to the cross-sectional nature of the samples.

The endocranial ontogenetic trajectory of humans was found to be nonlinear with distinct phases of different shape changes: (1) early shape changes occur from age group 1 (individuals without any erupted teeth) to age group 2 (individuals with incomplete deciduous dentition) and (2) shape changes occurring after this first phase (Neubauer et al. 2009). The ontogenetic trajectory of humans does not overlap with those of apes, showing that at no time during postnatal development, humans and apes have similar endocranial shapes (Neubauer et al. 2010). Prenatal ontogenetic patterns therefore cause differences of neonatal shape and largely affect adult morphological differences between humans and apes. Among apes, trajectories are also separated but partly overlap (Scott et al. 2014). Zollikofer et al.'s (2017) analyses based on samples of different individuals support this finding.

Based on developmental simulations, we found that differences in postnatal ontogeny contribute to the adult differences between humans and chimpanzees in addition to the prenatally established ones (Neubauer et al. 2010). However, these postnatal species-specific differences are found mainly in the shape changes from age group 1 to 2. In contrast, ontogenetic patterns from age group 2 until adulthood are similar. In a subsequent study, we found

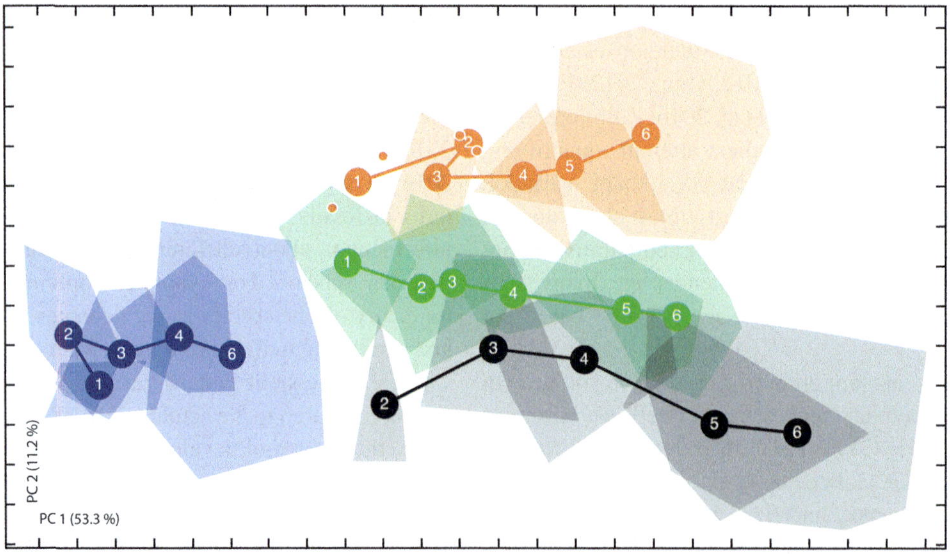

Fig. 12.2 Endocranial shape space: principal component analysis of endocranial shape data (PC 1 versus PC 2, together about 65% of total shape variation) of humans (*blue*), chimpanzees (*green*), gorillas (*black*), and orangutans (*orange*). Variation within age groups is represented by convex hulls. Age group averages labeled with numbers are connected with lines to build an ontogenetic trajectory. For age groups 1 and 2 of orangutans, single individuals (two for each group) are shown instead of convex hulls

similar patterns of shape changes in gorillas, orangutans, and gibbons from age group 2 onward (Scott et al. 2014) and suggested that these shared developmental shape changes are conserved in hominoids. Zollikofer et al. (2017) confirmed these results independently. The shared shape changes include "an expansion of the anterior and posterior cranial fossae, an overall reduction of superior regions, a superolateral rotation of the temporal poles and a posterior angling of the foramen magnum that is accompanied by an inferior expansion of the clivus region and a superior rotation of the cribriform plate" (Scott et al. 2014).

However, differences in the amounts of shared shape changes and potentially some minor deviations from this overall hominoid pattern contribute to adult morphologies in addition to shape differences that have been accumulated before the eruption of the first teeth (Neubauer et al. 2010; Scott et al. 2014; Neubauer 2014, 2015; Gunz 2016). The timing of the process of development itself seems to be variable among hominids: gorillas, e.g., show more pronounced shape changes than chimpanzees along the shared pattern during early ontogeny with variable size-shape change association (Scott et al. 2014). In other words, these data support the notion that postnatal species-specific ontogenetic trajectories after the eruption of the deciduous dentition were not affected by *major* evolutionary shifts of developmental patterns but "only" by fine-tuning of directions, amounts, and timing of shape changes.

These shape changes occurring later in ontogeny probably are less informative about the maturing brain. This is due to the fact that brain size and shape are integrated with the size and shape of the adjacent cranial base and face (Martínez-Abadías et al. 2012; Bruner 2015; Zollikofer et al. 2017). Brain size increase slows down with increasing age and ceases well before adult endocranial shape is attained, while facial size and shape continue to change (Bastir and Rosas 2006; Bastir et al. 2006, 2010; Neubauer et al. 2009).

In humans, the first postnatal phase comprises shape changes that make the endocast more globular (Neubauer et al. 2009, 2010). We therefore called this early period the "globularization phase" (Neubauer et al. 2010; Gunz et al. 2010). During this period, the cranial bones are not yet fused, and the brain grows rapidly, generating tension along the endocranial surface via the dura mater as the direct connection between the brain and the bones including the *falx cerebri* and the tentorium cerebelli. This activates osteoblast deposition, drift, and endochondral growth (functional matrix hypothesis; Moss and Young 1960; Duterloo and Enlow 1970; Lieberman et al. 2000a). It is therefore reasonable to assume that the growing brain drives most endocranial shape changes during this phase.

In this context, endocranial shape reflects interrelationships of the tempo and mode of brain development and the development of surrounding connective tissues and cranial bones with decreasing influence of the brain when growth rates decelerate later in ontogeny. The "spatial packing hypothesis" posits that brain enlargement causes a spatial packing problem solved by cranial base flexion and coronal orientation of the petrous bones (Biegert 1963; Gould 1977; Ross and Ravosa 1993; Lieberman et al. 2000a, b, 2008). Our data are partially supportive of this idea as they demonstrate that the listed changes of the cranial base go along with a globularization of the bony braincase. This is also consistent with recent findings by Zollikofer et al. (2017). However, endocranial shape change is not exclusively driven by size increase as (1) the transition from the globularization phase to the shared phase is not linked to abrupt changes in growth rates and (2) endocranial shape continues to change after endocranial size has been attained (Neubauer et al. 2009).

Our sample included neonates only of humans and chimpanzees but not of gorillas and orangutans. Based on a comparison between chimpanzees and humans, we have argued that humans are derived in having a globularization phase. However, Zollikofer et al. (2017) found early endocranial shape changes that are reminiscent of the human globularization phase in gorillas and orangutans but not (or to a far lesser degree) in chimpanzees and bonobos. These authors therefore suggested that the genus *Pan* (chimpanzees and bonobos) is the one being derived in having no globularization phase. Zollikofer et al. (2017) used partial least squares (PLS) analyses to study the covariation patterns between the shape of the endocranium and the shape of the cranial base and face. It is possible that early endocranial shape changes described in their study are not synonymous with the globularization phase described in our previous work (Neubauer et al. 2009, 2010; Gunz et al. 2010, 2011, 2012). Below we present new data to further discuss this issue.

Intriguingly, the endocranial shape differences between modern humans and Neandertals (Bruner et al. 2003; Bruner 2004) resemble the ontogenetic shape changes during the globularization phase in humans. Using developmental simulations based on virtual reconstructions of subadult and adult Neandertals (including two Neandertal neonates from Le Moustier and Mezmaiskaya), we were able to confirm the hypothesis that Neandertals also followed the shared hominid ontogenetic pattern, but unlike modern humans they did not undergo a postnatal globularization phase (Gunz et al. 2010, 2011, 2012). Using multiple reconstructions based on different plausible assumptions and reference individuals, we made sure that this conclusion is not influenced by the uncertainty of the required reconstructions. We argued that the modern human globularization phase is a derived developmental phase in early ontogeny that underlies one of the two autapomorphic features of the modern human cranium (Lieberman et al. 2002). Ponce de León et al. (2016) challenged this conclusion and suggested that postnatal brain development was similar in modern humans and Neandertals (and apes with the exception of the genus *Pan*; Zollikofer et al. 2017). According to these authors, adult shape differences result predominantly from differences in prenatal development.

In the next two sections, we describe different techniques to visualize developmental shape changes, and we present new data and analyses of early postnatal ontogeny in apes before further discussing the human globularization phase and the ontogenetic pattern in Neandertals.

12.5 Visualization Techniques for Endocranial Ontogenetic Shape Changes

Intuitive visualizations of three-dimensional shape differences or shape changes for smooth surfaces like the endocranial globularization phase are not an easy endeavor. Using vector fields (Fig. 12.3b) between two landmark configurations or superposition of warped surfaces corresponding to these landmark sets can be helpful but is sometimes misleading because they are affected by the Procrustes superimposition procedure. Thin-plate spline (TPS) grids between two landmark configurations, on the other hand, represent actual shape changes without the influence of the Procrustes fit. The information of shape variation is represented in the deformation of such grids that are initially constructed as perfectly parallel lines with right angles and the same distance to each other in the x and y direction. They can be computed even between landmark configurations that are not superimposed. For three-dimensional data, a TPS deformation grid can be used to illustrate shape changes in a particular plane of interest (Fig. 12.3c). Using the same thin-plate spline algebra, a three-dimensional surface can be warped: shape changes (i.e., development) are then depicted as an animation by a series of interpolating steps from one age group to the next age group. Multiple still frames of such animations can be used for print publication (Fig. 12.3a). However, these still images are not as intuitive as an animation. Zollikofer and Ponce de León (2002) proposed to decompose TPS displacement vectors relative to the orientation of the surface into a normal component visualized as a color-coded surface and a tangential component visualized as a vector field. Unfortunately, these visualizations are not easy to read.

Here, we discuss alternatives to visualizing shape changes or differences for print that attempt to translate the idea of intuitive two-dimensional thin-plate spline grids into three dimensions. One option is to create a network from a series of 3D grids that are also interconnected along the third dimension (Fig. 12.3d). Such TPS networks are not influenced by Procrustes superimposition but can result in confusing, illegible plots, when the network is not oriented well in relationship to the viewpoint. Another option is to construct cubes centered at each of the landmarks and semilandmarks (Figs. 12.3e). These cubes are warped according to the thin-plate spline function between two landmark configurations. The amount of shearing and position of these cubes illustrate the local shape changes just like the TPS network (Fig. 12.3d). However, the initial cubes are already less regularly arranged compared to the TPS network so that the shape difference conveyed by the shearing and position of the warped cubes in relationship to the initial cubes might be harder to interpret.

Another challenge in describing shape changes and differences (especially for smooth, round objects without clear and delimited structures like endocasts) is the very concept of shape itself. If, for example, a region is larger in one shape than the other, it is *relatively* larger and need not be *absolutely* larger. It is often difficult, however, to interpret a relative enlargement of a region when the shape comprises other regions that might also change in size, all affecting overall shape. This problem also affects color-coding of the endocranial surface due to positive versus negative allometric expansion (Zollikofer and Ponce de León 2002; Ponce de León et al. 2016; Zollikofer et al. 2017).

To visualize localized allometric surface expansion (Fig. 12.3f), we utilize the fact that our landmark set consists of a large number of relatively evenly distributed landmarks and semilandmarks that can be triangulated for proper visualization of the endocranial surface, thereby not requiring interpolation of endocranial surface points but only using actual data points. The endocranial surface is color-coded according to local expansion (positive allometry) or contraction (negative allometry). This figure might be misleading, however, because during growth none of the regions actually decrease in size. We can also compute the difference of the *absolute* surface areas of all the triangles between average forms of two age groups and (because not all triangles have the same surface area) color-code the percentage of surface area increases from the younger to the older age group (Fig. 12.3g). Since humans increase more in size than apes, using the same scale for the color codes for the entire range of variation in all species would result in unused areas of the color scale in apes and thereby less differentiation in visualizing shape changes in all species. We therefore adapt the ranges of color codes according to the overall range of surface increases in each species and according to frequencies and adjacency of similar percentage values to highlight specific regions with similar size increase within one species (Figs. 12.4 and 12.8). For example, all triangles corresponding to the cerebellar hemispheres in human age group 2 have surface areas approximately 2–2.84 times

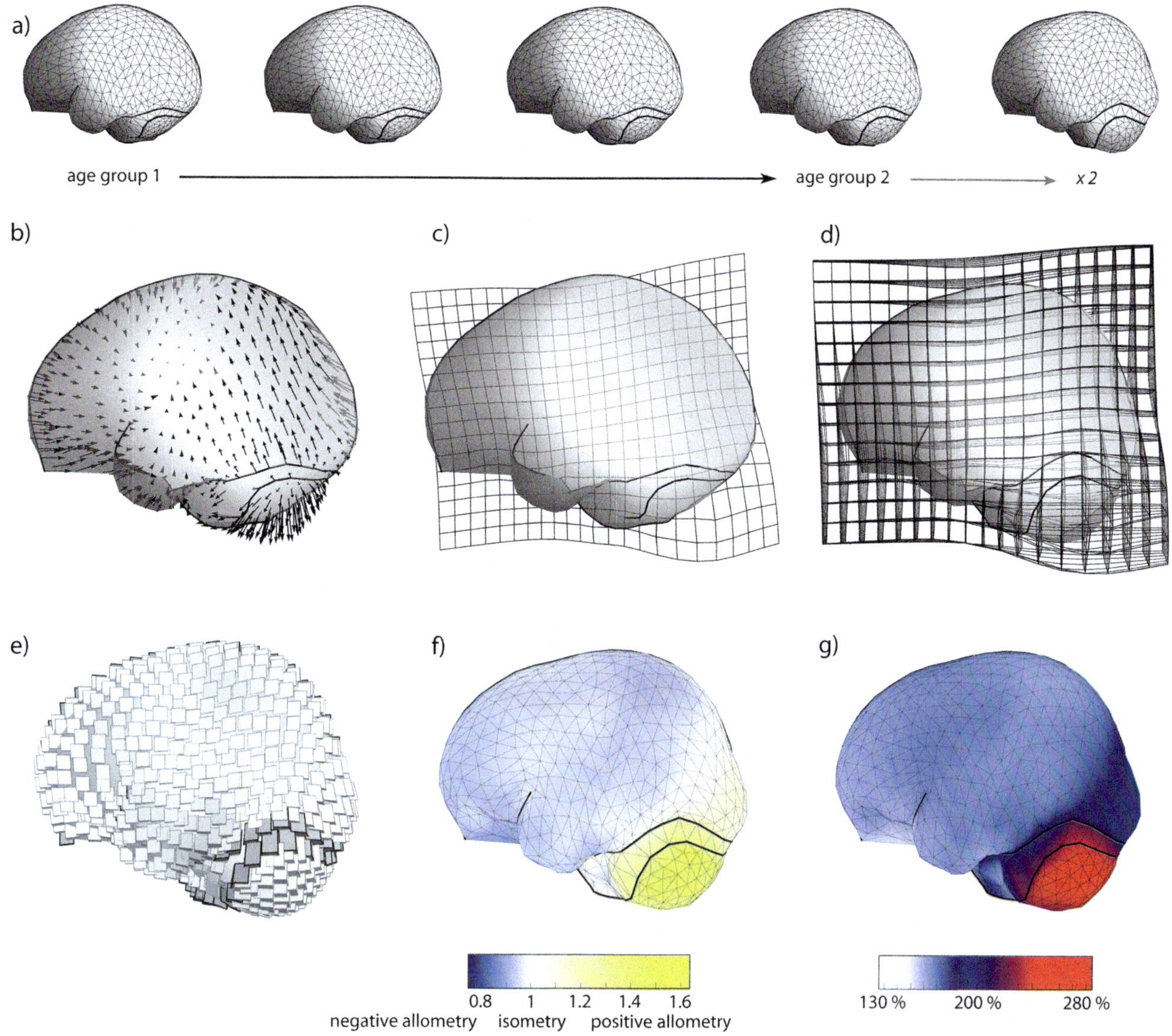

Fig. 12.3 Different visualization techniques shown for the human globularization phase with an exaggeration factor of 2: (**a**) series of warped surfaces from age group 1 to 2, (**b**) vector field, (**c**) TPS grid in the midsagittal plane, (**d**) TPS network, (**e**) TPS cubes, (**f**) color-coded surface according to local allometric expansion of the endocranial surface, (**g**) color-coded surface according to absolute local surface increases. (**b, c**) include a semitransparent endocranial surface for age group 1, (**d–g**) include an endocranial surface according to the exaggerated shape changes

larger than (200–284% of) the ones in age group 1. Size changes in all other triangles of the endocranial surface are substantially smaller. So as to visualize the more subtle signals on the cerebral surface (appearing almost uniformly blue in Fig. 12.3g), we set a threshold and color-code the range of surface area increase in cerebellar triangles with one color (red) and use a gradient from white to blue to represent the smaller values (Fig. 12.4). Identifying regions by adjacent triangles with similar absolute surface area increase and delimiting these regions from others with other ranges of absolute surface area increase help visualizing some key aspects of shape changes. Due to differences of absolute size increases among species and the described color range adaptations, the scales of color-coding are not directly comparable between species. These color ranges therefore need to be adapted to highlight the similarities as well as the differences between species.

All of the mentioned options for visualization have advantages and disadvantages. In the following discussion about early ontogenetic shape changes in humans, chimpanzees, and orangutans, we use (1) vector fields that are informative about shape change directions, (2) TPS networks oriented to standard views that are informative about local shape changes, and (3) color-coded surface representations as described above that are informative about absolute size increases in specific regions. We suggest that each of these three techniques alone

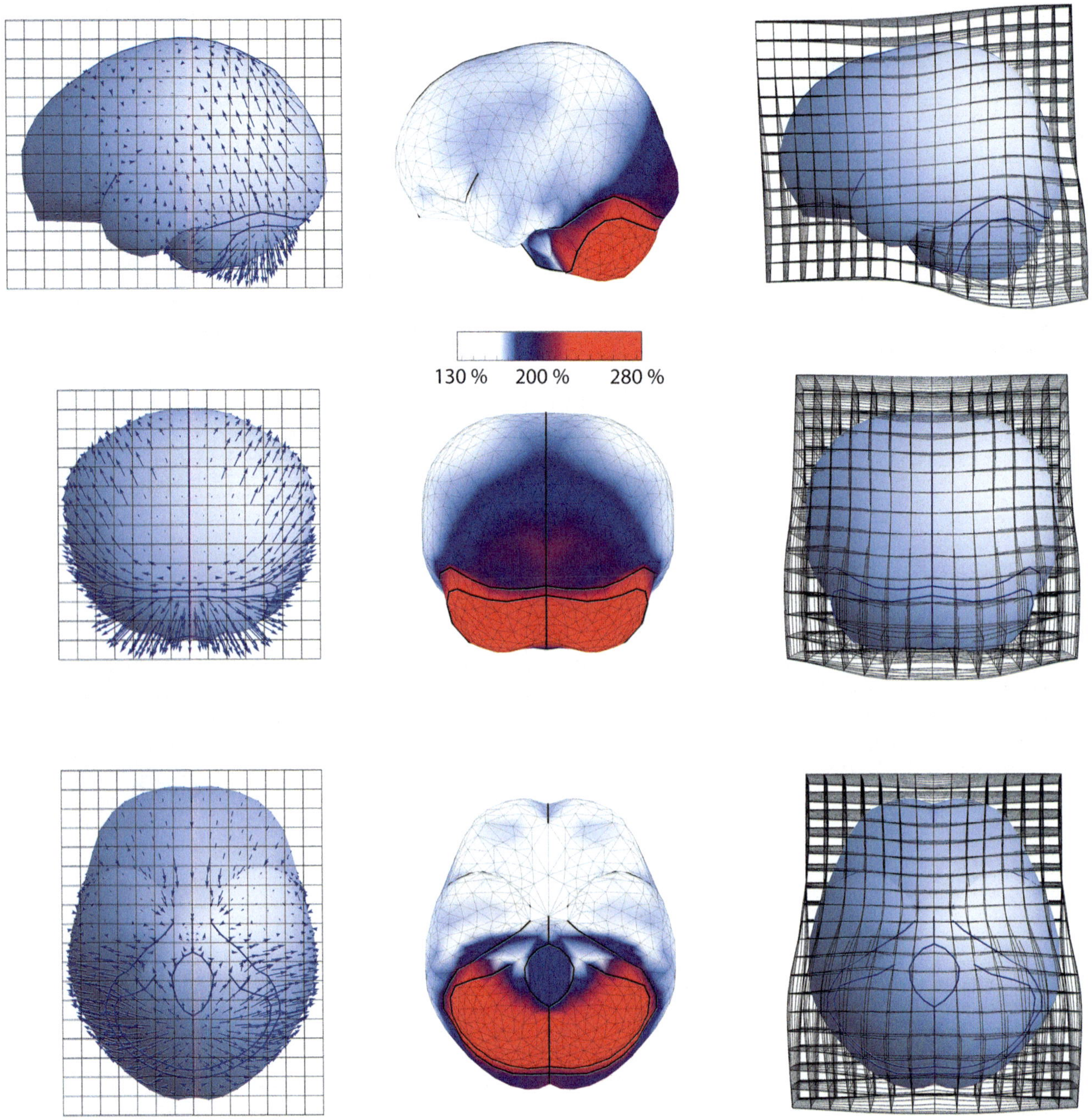

Fig. 12.4 Visualization of human early endocranial shape changes (globularization phase) in different views. The *left column* shows the surface of the average shape of age group 1 with vector fields and the TPS network before TPS warping to age group 2. The *middle column* shows the endocranial surface color-coded according to absolute surface area increase, exaggerated by a factor of 2. The *right column* shows the TPS network to illustrate shape changes. Legend illustrates color-coding for minimal and maximal increases of local surface areas from 126 to 284%

might miss some relevant information but that their combination allows comprehensive descriptions of shape changes or shape differences between two configurations. To emphasize developmental patterns and to make them better comparable between different species, we exaggerate visualizations to show the same amount of shape changes along the respective species-specific trajectory segments. To this end, we compute the lengths of the vectors between age groups 1 and 2 in each species, representing the amounts of shape changes. The trajectory segment from age group 1 to 2 is largest in orangutans and smallest in chimpanzees. The length of the human vector is intermediate. According to the lengths of species-specific trajectory segments, we therefore exaggerate shape changes from age group 1 to 2 in humans by a factor of 2, in chimpanzees by a factor of 2.3, and in orangutans by a factor of 1.6.

Figure 12.4 shows the shape changes during the human globularization phase. The cerebellar hemispheres bulge, become more convex, and expand as seen from the direction of vectors, deformation of the TPS network, and the extensive surface increase (surface area in red grows by a factor of about 2–2.84). The foramen magnum reorients, its posterior border moving anteroinferiorly. The vectors and the sheared TPS network also capture that the transverse sinuses elevate laterally and become more arched. The cerebral region as a whole becomes more globular. The occipital region flattens because regions directly anterior (above) of the transverse sinus extend more posteriorly as seen from the sheared TPS network. The surface area of the occipital regions nearly doubles (dark blue regions). Following the size increase of the cerebellum, posterior temporal regions bulge outward, and the temporal poles rotate supero-posteromedially so that the interpetrosal angle increases. Interestingly, the transition from the occipital region to more anteriorly located regions of less surface area increase (lighter blue) appears to correspond roughly to the lambdoid suture. Parietal regions, especially the parietal bosses, bulge as seen from vectors and the sheared TPS network. Interestingly, the parietal bosses are among the regions with the least absolute surface area increase (under 160% and therefore shown in white). The frontal region just anteriorly of the bulged parietal regions flattens. Frontal poles develop, and the sheared TPS network reveals some bulging of the frontal eminences that like the parietal eminences as well as the frontal bec, orbital regions, the hypophyseal fossa region, and the temporal poles belong to the regions with the least surface area increase (white areas in inferior view). Together these shape changes make the endocast more globular (Neubauer et al. 2009, 2010). The techniques of visualization used here reveal a correspondence of areas of low and high endocranial surface area increase and the location of sutures and ossification centers.

Extensive work has demonstrated that adult modern humans compared to other members of the genus *Homo* have absolutely and relatively larger parietal bones (Bruner et al. 2011) as well as absolutely and relatively larger parietal lobes (Bruner et al. 2003; Bruner 2004, 2010) and that parietal bone size is correlated with parietal lobe size among human adults (Bruner et al. 2015a, b). Parietal bulging in modern humans is related to large shape variation in the precuneus (Bruner et al. 2015a, b) which is larger in humans than in chimpanzees (Bruner et al. 2016). Besides a pronounced cerebellar growth, our data highlight a parietal bulging but not so much relative increases in parietal areas during the globularization phase in early postnatal development. However, some of the cited literature is based on 2D data of the midsagittal plane, while developmental parietal bulging is especially expressed parasagittally. Expansion of the midsagittal parietal area is intermediate during the globularization phase, being higher than in other parietal, frontotemporal pole, and orbital areas but lower than in the occipital and cerebellar areas. We therefore suggest that the developmental globularization phase is not only responsible for the variation of parietal bulging but also for the variation in parietal size within modern humans and among individuals of the genus *Homo*. Additional work is required to substantiate the precise relationships.

12.6 New Data on Early Ontogenetic Shape Changes

As mentioned above, based on data from gorilla and orangutan neonates, Zollikofer et al. (2017) challenged our previous interpretation that the globularization phase is unique to modern humans (Neubauer et al. 2010; Gunz et al. 2010, 2011, 2012). These authors argued that key aspects of the postnatal globularization phase are absent only in chimpanzees and bonobos and therefore suggested that postnatal endocranial development in the genus *Pan* is derived, whereas humans, gorillas, and orangutans retained the ancestral pattern of endocranial development.

Here, we present new analyses on early postnatal shape changes including an expanded sample of chimpanzee neonates ($n = 12$) and chimpanzees of age group 2 (defined as individuals with incomplete deciduous dentition; $n = 7$) and four orangutan babies, two of which are neonates and the other two being classified to dental age group 2. The sample size for young *Pongo* individuals is small, but we can approximate early postnatal shape changes that correspond (according to dental age groups) to the human globularization phase. We use PCA, developmental simulations, and the visualization techniques detailed above to compare early ontogenetic patterns of endocranial shape changes in humans and apes.

One of the orangutan neonatal crania is shown in Fig. 12.5. This individual has an endocranial volume of 153 ml. It is worth noting here that at the time of birth, it already has a very globular neurocranium (like the second neonate in our sample, 132 ml). Given that orangutan adults do not have globular neurocrania, this fairly globular shape of the orangutan neonates makes it unlikely that further globularization and cranial base flexion occur (early) postnatally (as implied by the analyses of Zollikofer et al. 2017).

The principal component analysis of endocranial shape shown in Fig. 12.2 is driven by species differences among apes and humans and the shared ontogenetic shape changes from age group 2 onward. Human shape changes from age group 1 to 2 nonetheless appear different to later shape changes. A bend in the trajectory at the transition from the early postnatal period to the following shared phase like in humans cannot be observed in orangutans and chimpanzees.

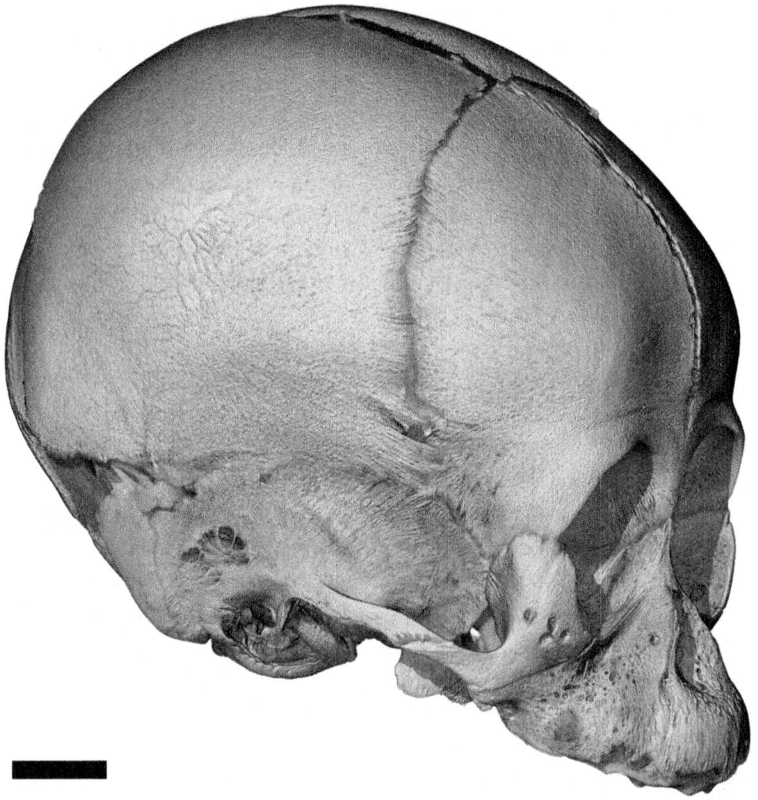

Fig. 12.5 An orangutan neonate has a very globular neurocranium so that further globularization until the first deciduous teeth erupt is rather unlikely (scale bar is 1 cm)

For gorillas, we unfortunately do not have neonatal data (age group 1).

To highlight shape changes in early human ontogeny, we computed a between-group principal component analysis (Mitteroecker and Bookstein 2011) of human age groups 1 and 2 and the adult mean shape of apes, seeking a morphospace in which shape variation associated to the human globularization phase and the morphological differences between humans and apes is maximized. As expected, the first component of such a morphospace (Fig. 12.6) captures the shape difference between humans and apes and the shape changes associated to the shared phase, while the second component describes human early postnatal shape changes. Neither chimpanzees nor orangutans have a trajectory segment from age group 1 to 2 that is parallel or approximately parallel to the one in humans, or in other words, they do not show shape changes from age group 1 to age group 2 that are similar to the human globularization phase. It is worth mentioning here that the direction of the trajectory segment in orangutans is based only on four individuals. However, it is evident that the early shape changes in both chimpanzees and orangutans differ from their later ontogenetic shape changes during the shared phase. Chimpanzee and orangutan neonates might undergo shape changes involving *some* aspects of the human globularization phase, given that the PC scores of age groups 1 and 2 along bgPC 2 increase like in humans albeit to a lesser degree (Fig. 12.6). It is important to note, however, that shape changes that are specific to early ontogeny in chimpanzees or orangutans might be underrepresented in such a morphospace.

To shed further light on early ontogenetic shape changes, we therefore computed standard principal component analyses for each species separately so that PCs are not influenced by the shape differences among species (Fig. 12.7). We furthermore performed developmental simulations to check if ontogenetic trajectories of apes include a human globularization phase. To this end we applied the human trajectory segment, that is, the vector from age group 1 to age group 2, to the neonates of the other species. We then checked whether those simulated endocranial shapes look like the actual individuals of the species' age group 2. As our sample does not contain gorilla neonates, we performed backward simulations for gorillas, applying the reversed human globularization phase but also the early chimpanzee shape changes, to age group 2 of gorillas, thereby simulating neonatal shape of gorillas along human- and chimpanzee-like development.

Applying the globularization phase to human neonates (Fig. 12.7a) illustrates how the developmental simulations

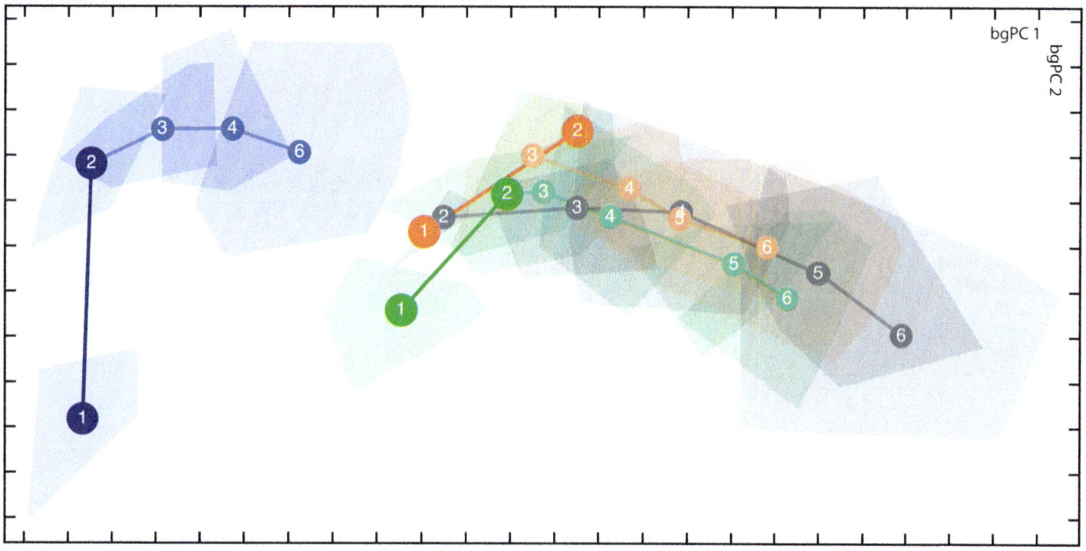

Fig. 12.6 Endocranial morphospace: between-group PCA (bgPCA) to maximize shape variation associated with the human globularization phase and the main difference between humans and apes. Labels like in Fig. 12.2

work: as expected, simulated individuals (Hh2) plot with the measured individuals of age group 2. In contrast, using the segment from chimpanzee age group 1 to 2, simulated individuals (Hc2) do not fall on the human ontogenetic trajectory. Likewise when chimpanzee neonates "develop" along the human vector (Fig. 12.7b), these simulated infants (Ch2) do *not* cluster with chimpanzees of age group 2 or with chimpanzees of other age groups. It follows that the endocranial shape changes during this developmental phase differ between modern humans and chimpanzees or, in other words, that chimpanzees clearly do not exhibit a humanlike globularization phase, in line with our previous results as well as Zollikofer et al.'s (2017).

A bend in the chimpanzee trajectory (Fig. 12.7b) suggests that early postnatal shape changes are different to the later shared phase. Indeed, visualization of early shape changes in chimpanzees (Fig. 12.8b) reveals that similar endocranial regions as during the human globularization phase are affected (Fig. 12.8a). Cerebellar regions expand the most (surface areas of triangles reaching between about 175% and 213%, shown in red) but in different directions than in humans as is evident from the vector field and the sheared TPS network. While the cerebellar hemispheres become more convex, the cerebellar region as a whole flattens because the cerebellar hemispheres rotate outward in relation to the midplane and the foramen magnum shifts. The upper cerebral region as a whole flattens, especially the occipital and posterior frontal regions. Posterior temporal regions bulge and rotate superolaterally, and the temporal poles rotate superiorly. The interpetrosal angle increases. The surface areas of occipital regions, the temporal poles, and midsagittal brainstem regions increase by a factor of about 1.5–1.7 (blue) with other regions of the endocranial surface growing less. Anterior parietal regions flatten less than the surrounding areas. In chimpanzees this region is located more anteriorly than the human parietal bulging area. Parietal and frontal regions grow the least (white surface denotes increases of under 140%). A constriction approximately along the coronal suture causes the anterior frontal region to appear relocated in relation to more posterior regions. The frontal region becomes more pointed and the frontal bec well developed. While some of these shape changes appear to be similar to shape changes during the human globularization phase, the combination of all shape changes makes the endocast of chimpanzees more *elongated* during early development. Developmental simulations and comparison of visualizations therefore support the hypothesis that early endocranial shape changes differently between humans and chimpanzees from age group 1 to 2. It is important to note here that using only color-coding of absolute local surface increases (Fig. 12.8) or only color-coding of positive versus negative allometric expansion (Ponce de León et al. 2016; Zollikofer et al. 2017), early shape changes in humans and chimpanzees might appear more similar than they actually are. It is therefore critical to also visualize and consider the direction of shape changes (Fig. 12.8).

Since our sample unfortunately does not contain gorilla neonates, we can only perform backward simulations, applying the reversed early shape changes of humans and chimpanzees to gorilla individuals of age group 2 (Fig. 12.7c) and evaluate the simulated neonatal shape (Fig. 12.7e, f). Both simulated neonates (Gh1 and Gc1; Fig. 12.7c, e, f) yield plausible (biologically and morphologically conceivable) endocranial shapes of a hypothetical gorilla neonate. These

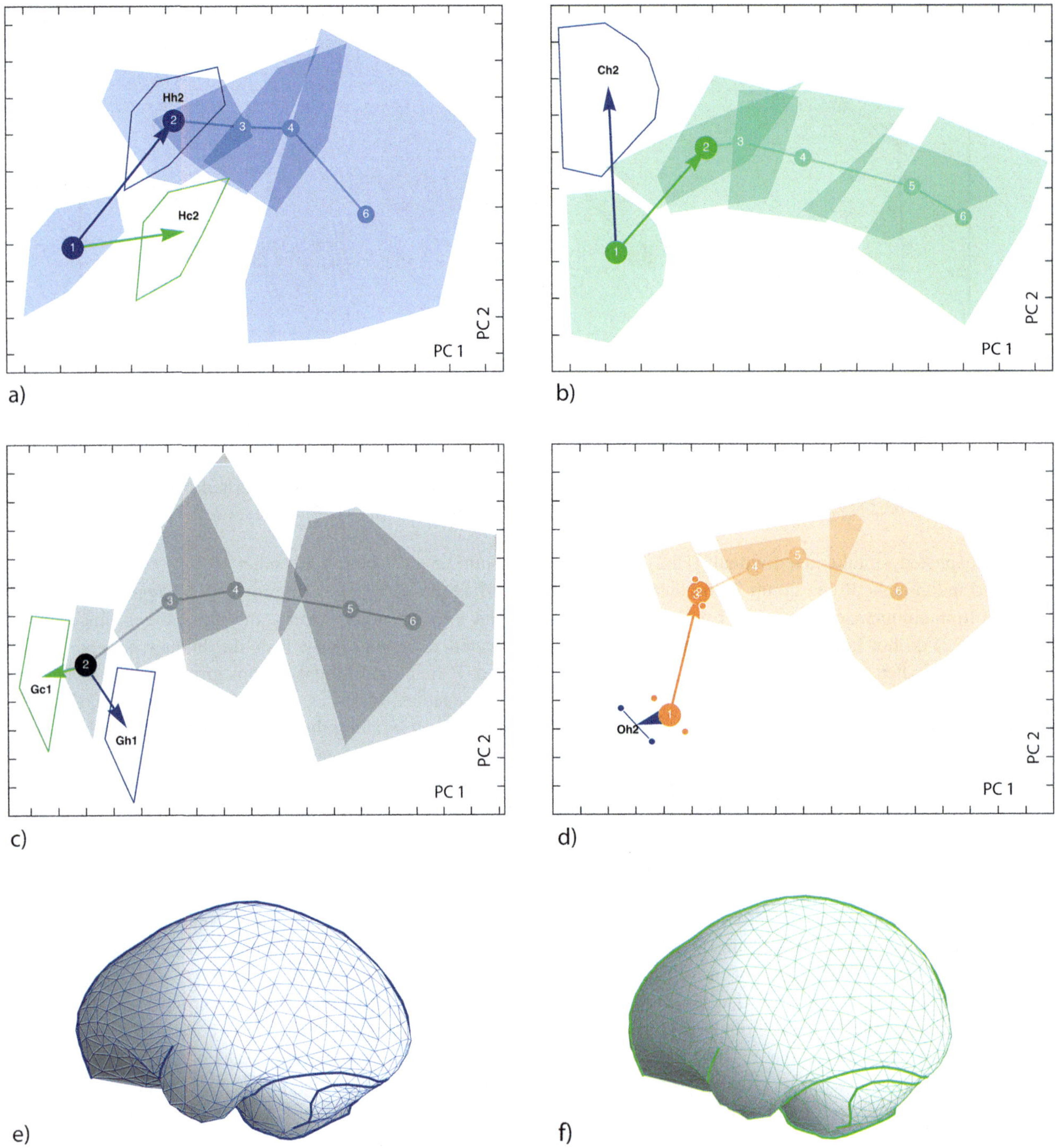

Fig. 12.7 Endocranial shape spaces computed for each species separately showing developmental simulations of the first segment of the ontogenetic trajectories (age group 1 to 2). Labels like in Figs. 12.2 and 12.6. The first (capital) letter denotes species that another species' trajectory segment is applied to, and the second (lowercase) letter denotes the other species' pattern of shape changes. (**a**) Humans, (**b**) chimpanzees, (**c**) gorillas, (**d**) orangutans, (**e**) average shape of gorilla neonates simulated according to a humanlike development, and (**f**) a chimpanzee-like development (backward simulations as depicted in (**c**))

backward simulations therefore cannot determine whether gorillas undergo a humanlike globularization phase, or if they develop similar to chimpanzees. In contrast, our previous work demonstrated that Neandertals did not undergo a globularization phase to the extent of modern humans, because a simulated Neandertal neonate assuming a humanlike globularization would have an endocast that morphologically is not conceivable in having no cerebellum (Gunz et al. 2010, 2012).

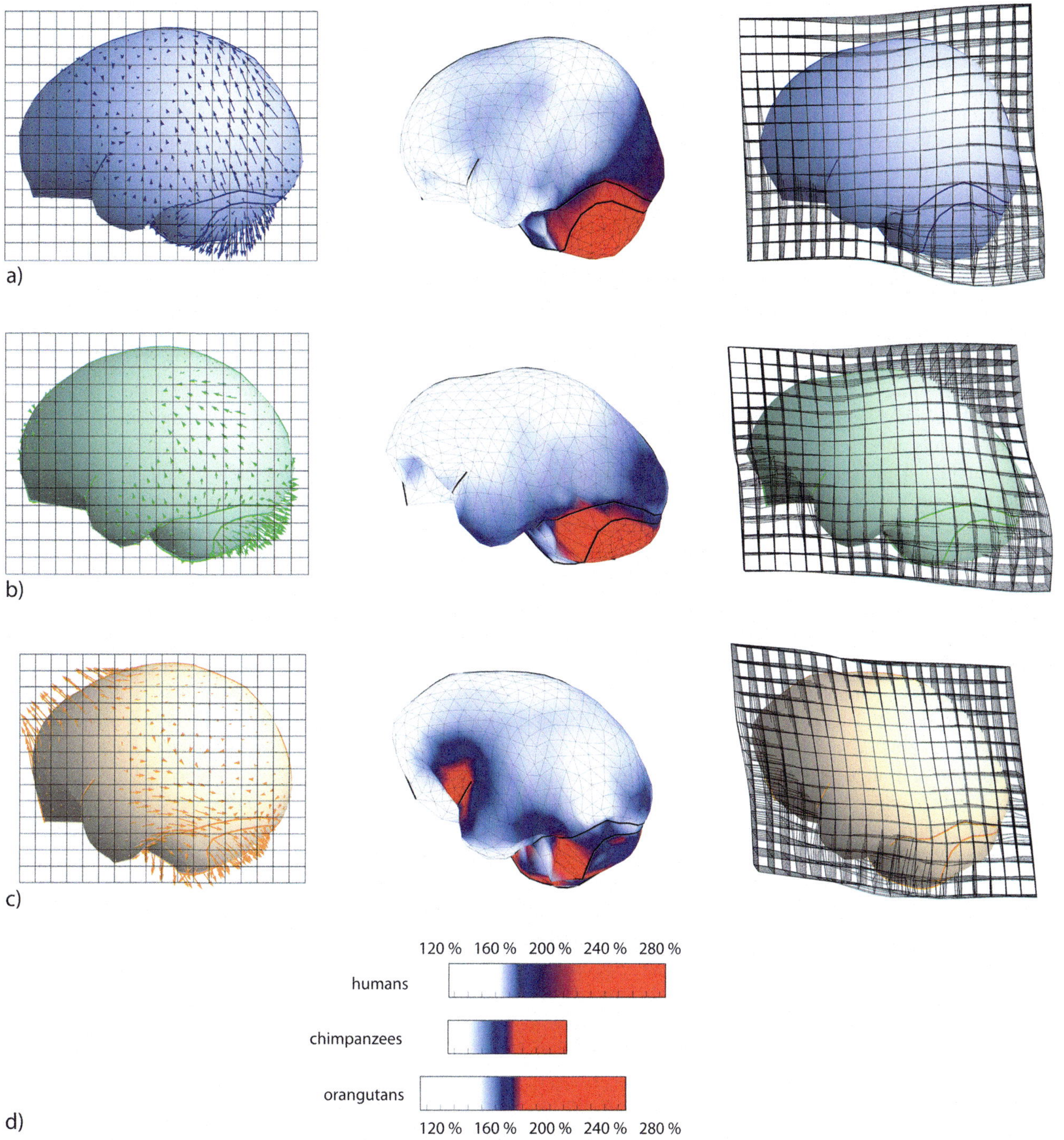

Fig. 12.8 Visualization of early endocranial shape changes in humans (**a**), chimpanzees (**b**), and orangutans (**c**) shown in lateral view. (**d**) Legends for color-coding in each species. For further explanations, see Fig. 12.4

Orangutans, as seen in a separate principal component analysis, show an unexpected trajectory (Fig. 12.7d). The two individuals of age group 2 plot within the variation of age group 3. This pattern is also apparent in the PCA plots of the pooled samples (Figs. 12.2 and 12.6) in which the average shape of age group 2 appears to be even more advanced along the ontogenetic trajectory than the average of age group 3. Most likely this is a random effect of the small sample size, and the two individuals capture one extreme of the shape variation of age group 2 that overlaps with the variation of age group 3. Interestingly, the orangutan trajectory also includes a strong bend between the first and the following segments of the ontogenetic trajectory. Developmental simulations (Fig. 12.7d) and visualization of shape changes (Fig. 12.8c) reveal, however, that the pattern in early orangutan ontogeny does not correspond to a humanlike globularization phase nor the pattern found in chimpanzees. *Pongo* neonates simulated to develop along the human pattern (Oh2) do not look like individuals of age groups 2 (or 3). The first two principal components of the orangutan shape space do not capture the shape variation related to the human globularization phase resulting in a very short, globularization vector (Fig. 12.7d). Visualization of these shape changes reveals yet another pattern of development during this developmental period (Fig. 12.8c). However, sample sizes of young orangutan individuals have to be improved to substantiate this claim.

As captured with the current sample, orangutan early shape changes are completely different from the pattern found in humans (Fig. 12.8c). The cerebellar hemispheres become slightly more convex, but the surface area of the cerebellar region increases only by a factor of about 1.7, while most surface area increase (over 190% and up to 260%, shown in red) occurs in other less wide-stretched regions: the midsagittal clivus/brain stem region, the region between the sigmoid sinus and petrous ridge, the region around the confluences of sinuses, and the lateral extension of the sphenoid curve around the opening of the Sylvian fissure (lateral sulcus). The interpetrosal angle barely changes. Posterior temporal regions rotate superolaterally. The cerebral region as a whole flattens. The frontal region appears more pointed, and a well-expressed frontal bec develops and rearranges the prefrontal and orbital regions. Regions with comparable increase of surface area to the cerebellar region are frontotemporal regions and the occipital poles. Regions with the least increase in surface area (under 145%) are posterior temporal regions, frontal poles, and the frontal bec. Interestingly, regions of lower increase in surface area on the vault could correspond to the coronal and lambdoid sutures. While these data further support the hypothesis that the globularization phase is derived in humans and thereby contributing to the human autapomorphic globular neurocranium, they raise the question what causes variation in early postnatal endocranial shape changes. Better samples of very young ape individuals are required to further investigate this question.

12.7 Neandertal Endocranial Ontogenetic Trajectory

Ponce de León et al. (2016) argued for a similar (postnatal) brain development in humans and Neandertals and suggested that Neandertals also undergo a globularization phase. In an attempt to explain the differences between their and our study, they questioned our visualization techniques to describe the globularization phase and the validity of developmental simulations as described above. Furthermore, Ponce de León et al. (1) used a dense mesh of surface semilandmarks, (2) added two important Neandertal individuals to age group 2 (Dederiyeh 1 and 2) that were not available to us but dropped another one that we used in our analyses (Pech de l'Azé) and had only one Neandertal neonate available while we have used two, and (3) generated more alternative reconstructions for the Mezmaiskaya neonate.

Above, we have shown the validity of our visualization techniques and demonstrated that developmental simulations are useful in comparing ontogenetic trajectories. All results shown in this chapter are based on a denser endocranial landmark set with approximately the same number of (semi)landmarks as the landmark set used by Ponce de León et al. (2016) and Zollikofer et al. (2017). This does not change our previous results and conclusions.

Concerning Neandertal sample composition and reconstruction uncertainties, we want to add some important comments here. While we agree with Ponce de León et al.'s critique that Pech de l'Azé seems to have an endocranial shape at the border of the range of variation of age group 2 in the overlap zone with age group 1 (see Fig. 8 in Gunz et al. 2012), Dederiyeh 1 and 2 seem to have an endocranial shape at the border of the range of variation of age group 2 in the overlap zone with age group 3 (see Fig. 1 in Ponce de León et al. 2016). It seems that using Pech de l'Azé therefore supports the absence of a globularization phase and using Dederiyeh 1 and 2 the presence of a globularization phase. An average shape based on all three individuals should therefore better represent this age group's average shape to define the ontogenetic trajectory. On the other hand, Ponce de León et al. (2016) treated multiple reconstructions of the Mezmaiskaya neonate as if they were different individuals and computed an average shape for Neandertal age group 1 to define the first trajectory segment. However, some of their multiple reconstructions support an absence and some a presence of a globularization phase in different magnitudes.

This discussion illustrates that the computation of ontogenetic trajectories is sensitive to random sampling effects when based on only a few fossil specimens. We therefore encourage the usage of developmental simulations that are not prone to small sample sizes and reconstruction uncertainties when based on appropriate reference species as described in our previous work (Neubauer et al. 2010; Gunz et al. 2010, 2011, 2012). We therefore maintain that early postnatal endocranial development differs between modern humans and Neandertals.

12.8 Conclusions

We reviewed here state-of-the-art digital data and methodology to document and compare developmental patterns of endocranial shape changes. To summarize, we argue that the postnatal globularization phase is indeed uniquely human, not occurring in chimpanzees, orangutans, and Neandertals. Based on PCA analysis and various visualization techniques, we documented variation in ontogenetic shape changes during early development (trajectory segment from age group 1 to 2) among apes. However, in apes shape changes during this developmental phase seem to contribute less to species-specific adult variation than the human globularization phase (see Figs. 12.2 and 12.6). For the future, samples sizes of young ape individuals need to be increased so that several open questions can be discussed. These include how facial shape changes interact reciprocally with endocranial shape changes especially in early development (using appropriate methods like partial least squares analysis, e.g., Bookstein et al. 2003; Mitteroecker and Bookstein 2007, 2008, 2009; Mitteroecker et al. 2012; Zollikofer et al. 2017), how the speed of brain growth is related to specific patterns of endocranial shape changes, how this might be different among species, and what causes differences in early endocranial shape changes. It seems clear, however, that the absolute size difference and the difference in growth rates between apes and humans are not exclusively related to the globular neurocranium of modern humans, as Neandertals have very similar brain sizes and growth rates in early ontogeny comparable to modern humans.

Acknowledgments We thank Emiliano Bruner, Naomichi Ogihara, and Hiroki Tanabe for the invitation to contribute to this book and Jean-Jacques Hublin for stimulating discussions. We are grateful to the constructive comments by Emiliano Bruner and two anonymous referees that substantially improved the manuscript. Thanks to H. Coqueugniot, C. Feja, C. Funk, B. Herzig, J.-J. Hublin, J.L. Kahn, F. Mayer, D. Plotzki, F. Renoult, P. Schoenefeld, U. Schwarz, K. -Spanel-Borowski, F. Spoor, H. Temming, F. Veillon, G.W. Weber, A. Winter, and A. Winzer, the Kyoto University Primate Research Institute (via the Digital Morphology Museum), the University of Zurich, the Ivorian authorities (especially the Ministry of the Environment and Forests and the Ministry of Research), the Swiss Centre for Scientific Research Abidjan, and the Phyletic Museum Jena for curating of and access to specimens and acquisition of CT data and N. Scott, C. Roeding, S. Stelzer, M. Baeuchle, C. Moore, and A. Strauss for data organization and preprocessing. Our research was supported by EU FP6 Marie Curie Actions grant MRTN-CT-2005-019564 "EVAN" and the Max Planck Society.

References

Ackermann RR, Krovitz GE (2002) Common patterns of facial ontogeny in the hominid lineage. Anat Rec 269:142–147

Aiello LC, Wheeler P (1995) The expensive-tissue hypothesis: the brain and the digestive system in human and primate evolution. Curr Anthropol 36:199–221

Bastir M, Rosas A (2004a) Comparative ontogeny in humans and chimpanzees: similarities, differences and paradoxes in postnatal growth and development of the skull. Ann Anat 186:503–509

Bastir M, Rosas A (2004b) Facial heights: evolutionary relevance of postnatal ontogeny for facial orientation and skull morphology in humans and chimpanzees. J Hum Evol 47:359–381

Bastir M, Rosas A (2006) Correlated variation between the lateral basicranium and the face: a geometric morphometric study in different human groups. Arch Oral Biol 51:814–824

Bastir M, Rosas A, O'Higgins P (2006) Craniofacial levels and the morphological maturation of the human skull. J Anat 209: 637–654

Bastir M, Rosas A, Stringer C, Manuel Cuétara J, Kruszynski R, Weber GW, Ross CF, Ravosa MJ (2010) Effects of brain and facial size on basicranial form in human and primate evolution. J Hum Evol 58:424–431

Bianchi S, Stimpson CD, Bauernfeind AL, Schapiro SJ, Baze WB, McArthur MJ, Bronson E, Hopkins WD, Semendeferi K, Jacobs B, Hof PR, Sherwood CC (2013) Dendritic morphology of pyramidal neurons in the chimpanzee neocortex: regional specializations and comparison to humans. Cereb Cortex 23:2429–2436

Biegert J (1963) The evaluation of characteristics of the skull, hands and feet for primate taxonomy. In: Washburn SL (ed) Classification and human evolution. Aldine De Gruyter, Chicago, pp 116–145

Bookstein FL (1991) Morphometric tools for landmark data: geometry and biology. Cambridge University Press, Cambridge

Bookstein FL (1997) Landmark methods for forms without landmarks: morphometrics of group differences in outline shape. Med Image Anal 1:225–243

Bookstein FL, Gunz P, Mitteroecker P, Prossinger H, Schaefer K, Seidler H (2003) Cranial integration in *Homo*: singular warps analysis of the midsagittal plane in ontogeny and evolution. J Hum Evol 44:167–187

Bruner E (2004) Geometric morphometrics and paleoneurology: brain shape evolution in the genus *Homo*. J Hum Evol 47:279–303

Bruner E (2010) Morphological differences in the parietal lobes within the human genus. Curr Anthropol 51:77–88

Bruner, E. 2015. Functional craniology and brain evolution. Human Paleoneurology. Springer, p. 57–94

Bruner E, Manzi G, Arsuaga JL (2003) Encephalization and allometric trajectories in the genus *Homo*: evidence from the Neandertal and modern lineages. Proc Natl Acad Sci U S A 100:15335–15340

Bruner E, De La Cuétara JM, Holloway R (2011) A bivariate approach to the variation of the parietal curvature in the genus homo. Anat Rec 294:1548–1556

Bruner E, Amano H, de la Cuétara JM, Ogihara N (2015a) The brain and the braincase: a spatial analysis on the midsagittal profile in adult humans. J Anat 227:268–276

Bruner E, Román FJ, de la Cuétara JM, Martin-Loeches M, Colom R (2015b) Cortical surface area and cortical thickness in the precuneus of adult humans. Neuroscience 286:345–352

Bruner E, Preuss TM, Chen X, Rilling JK (2016) Evidence for expansion of the precuneus in human evolution. Brain Struct Funct 1-8

Buckner RL, Krienen FM (2013) The evolution of distributed association networks in the human brain. Trends Cogn Sci 17:648–665

Bulygina E, Mitteroecker P, Aiello L (2006) Ontogeny of facial dimorphism and patterns of individual development within one human population. Am J Phys Anthropol 131:432–443

Casey BJ, Giedd JN, Thomas KM (2000) Structural and functional brain development and its relation to cognitive development. Biol Psychol 54:241–257

Casey BJ, Tottenham N, Liston C, Durston S (2005) Imaging the developing brain: what have we learned about cognitive development? Trends Cogn Sci 9:104–110

Cobb SN, O'Higgins P (2004) Hominins do not share a common postnatal facial ontogenetic shape trajectory. J Exp Zoolog B Mol Dev Evol 302:302–321

Cobb SN, O'Higgins P (2007) The ontogeny of sexual dimorphism in the facial skeleton of the African apes. J Hum Evol 53:176–190
Cofran Z, DeSilva JM (2015) A neonatal perspective on Homo erectus brain growth. J Hum Evol 81:41–47
Coqueugniot H, Hublin JJ (2012) Age-related changes of digital endocranial volume during human ontogeny: results from an osteological reference collection. Am J Phys Anthropol 147:312–318
Coqueugniot H, Hublin JJ, Veillon F, Houët F, Jacob T (2004) Early brain growth in *Homo erectus* and implications for cognitive ability. Nature 431:299–302
Count EW (1947) Brain and body weight in man – their antecedents in growth and evolution – a study in dynamic somatometry. Ann N Y Acad Sci 46:993–1122
DeSilva J, Lesnik J (2006) Chimpanzee neonatal brain size: implications for brain growth in *Homo erectus*. J Hum Evol 51:207–212
Duterloo HS, Enlow DH (1970) A comparative study of cranial growth in *Homo* and *Macaca*. Am J Anat 127:357–368
Falk D (1980) Hominid brain evolution: the approach from paleoneurology. Yearb Phys Anthropol 23:93–107
Falk D (1987) Hominid paleoneurology. Annu Rev Anthropol 16: 13–28
Fischer B, Mitteroecker P (2015) Covariation between human pelvis shape, stature, and head size alleviates the obstetric dilemma. Proc Natl Acad Sci U S A 112:5655–5660
Fragaszy DM, Bard K (1997) Comparison of development and life history in *Pan* and *Cebus*. Int J Primatol 18:683–701
Fragaszy DM, Visalberghi E, Fedigan LM (2004) The complete capuchin: the biology of the genus *Cebus*. Cambridge University Press, Cambridge
Freidline SE, Gunz P, Harvati K, Hublin J-J (2012) Middle pleistocene human facial morphology in an evolutionary and developmental context. J Hum Evol 63:723
Freidline SE, Gunz P, Harvati K, Hublin JJ (2013) Evaluating developmental shape changes in homo antecessor subadult facial morphology. J Hum Evol 65:404–423
Frost SR, Marcus LF, Bookstein FL, Reddy DP, Delson E (2003) Cranial allometry, phylogeography, and systematics of large-bodied papionins (primates: cercopithecinae) inferred from geometric morphometric analysis of landmark data. Anat Rec A Discov Mol Cell Evol Biol 275:1048–1072
Gogtay N, Giedd JN, Lusk L, Hayashi KM, Greenstein D, Vaituzis AC, Nugent TF, Herman DH, Clasen LS, Toga AW, Rapoport JL, Thompson PM (2004) Dynamic mapping of human cortical development during childhood through early adulthood. Proc Natl Acad Sci U S A 101:8174–8179
Gómez-Robles A, Hopkins D, Sherwood C (2013) Increased morphological asymmetry, evolvability and plasticity in human brain evolution. Proc R Soc B Biol Sci 280:20130575
Gómez-Robles A, Hopkins WD, Schapiro SJ, Sherwood CC (2015) Relaxed genetic control of cortical organization in human brains compared with chimpanzees. Proc Natl Acad Sci U S A 112: 14799–14804
Gould SJ (1977) Ontogeny and phylogeny. Belknap Press of Harvard University Press, Cambridge
Gower JC (1975) Generalized procrustes analysis. Pyschometrika 40: 33–51
Greenough WT, Black JE, Wallace CS (1987) Experience and brain development. Child Dev 58:539–559
Gunz P (2012) Evolutionary relationships among robust and Gracile Australopiths: an "Evo-devo" perspective. Evol Biol 39:472
Gunz P 2015 Computed tools for paleoneurology. Human paleoneurol. Springer, p. 39-55
Gunz P, Mitteroecker P (2013) Semilandmarks: a method for quantifying curves and surfaces. Hystrix,. The. Italian. J Mammal 24:–7
Gunz P, Mitteroecker P, Bookstein FL (2005) Semilandmarks in three dimensions. In: Slice DE (ed) Modern Morphometrics in physical anthropology. Kluwer Academic/Plenum Publishers, New York, pp 73–98
Gunz P, Neubauer S, Golovanova L, Doronichev V, Maureille B, Hublin JJ (2012) A uniquely modern human pattern of endocranial development. Insights from a new cranial reconstruction of the Neandertal newborn from Mezmaiskaya. J Hum Evol 62:300–313
Gunz P, Neubauer S, Maureille B, Hublin JJ (2010) Brain development after birth differs between Neanderthals and modern humans. Curr Biol 20:R921–R922
Gunz P, Neubauer S, Maureille B, Hublin JJ (2011) Virtual reconstruction of the Le Moustier 2 newborn skull. Implications for Neandertal ontogeny. PALEO 22:155–172
Gunz P (2016) Growing up fast, maturing slowly: the evolution of a uniquely modern human pattern of brain development. In: Rolian C Boughner JA (eds) Developmental Approaches to Human Evolution. Wiley, p. 261–283
Harvey PH, Martin RD, Clutton-Brock TH (1987) Life histories in comparative perspective. In: Smuts BB, Cheney DL, Seyfarth RM, Wrangham RW, Struhsaker TT (eds) Primate Societies. University of Chicago Press, Chicago, pp 181–196
Holloway RL, Broadfield DC, Yuan MS (2004) The human fossil record: brain Endocasts, The Paleoneurological Evidence. Wiley-Liss, Hoboken
Holloway RL (1978) The relevance of endocasts for studying primate brain evolution. In: Noback CR (ed) Sensory Systems of Primates. Plenum Press, New York, pp 181–200
Holt AB, Cheek DB, Mellits ED, Hill DE (1975) Brain size and the relation of the primate to the nonprimate. In: Cheek DB (ed) Foetal and postnatal cellular growth: hormones and nutrition. Wiley, New York, pp 23–44
Hublin JJ, Coqueugniot H (2006) Absolute or proportional brain size: that is the question. A reply to Leigh's (2006) comments. J Hum Evol 50:109–113
Hublin JJ, Neubauer S, Gunz P (2015) Brain ontogeny and life history in pleistocene hominins. Philos Trans R Soc Lond Ser B Biol Sci 370
Jolicoeur P, Baron G, Cabana T (1988) Cross-sectional growth and decline of human stature and brain weight in 19th-century Germany. Growth Dev Aging 52:201–206
Jordaan HV (1976) Newborn: adult brain ratios in hominid evolution. Am J Phys Anthropol 44:271–278
Kennedy GE (2005) From the ape's dilemma to the weanling's dilemma: early weaning and its evolutionary context. J Hum Evol 48:123–145
Kramer AF, Bherer L, Colcombe SJ, Dong W, Greenough WT (2004) Environmental influences on cognitive and brain plasticity during aging. J Gerontol A Biol Sci Med Sci 59:M940–M957
Le Cabec A, Tang N, Tafforeau P (2015) Accessing developmental information of fossil hominin teeth using new synchrotron microtomography-based visualization techniques of dental surfaces and interfaces. PLoS One 10:e0123019
Leigh SR (2004) Brain growth, life history, and cognition in primate and human evolution. Am J Primatol 62:139–164
Leigh SR (2006) Brain ontogeny and life history in *Homo erectus*. J Hum Evol 50:104–108
Leigh SR (2012) Brain size growth and life history in human evolution. Evol Biol 39:587–599
Leigh SR, Blomquist GE (2007) Life history. In: Campbell CJ, Fuentes A, MacKinnon KC, Panger M, Bearder SK (eds) Primates in perspective. Oxford University Press, Oxford, pp 396–407
Lieberman DE, Carlo J, Ponce de León M, Zollikofer CP (2007) A geometric morphometric analysis of heterochrony in the cranium of chimpanzees and bonobos. J Hum Evol 52:647–662
Lieberman DE, Hallgrímsson B, Liu W, Parsons TE, Jamniczky HA (2008) Spatial packing, cranial base angulation, and craniofacial shape variation in the mammalian skull: testing a new model using mice. J Anat 212:720–735

Lieberman DE, McBratney BM, Krovitz G (2002) The evolution and development of cranial form in *Homo sapiens*. Proc Natl Acad Sci U S A 99:1134–1139
Lieberman DE, Pearson OM, Mowbray KM (2000a) Basicranial influence on overall cranial shape. J Hum Evol 38:291–315
Lieberman DE, Ross CF, Ravosa MJ (2000b) The primate cranial base: ontogeny, function, and integration. Am J Phys Anthropol Suppl 31:117–169
Martin RD (1983) Human brain evolution in an ecological context. 52nd James Arthur lecture on the evolution of the human brain. American Museum of Natural History, New York
Martínez-Abadías N, Esparza M, Sjøvold T, González-José R, Santos M, Hernández M, Klingenberg CP (2012) Pervasive genetic integration directs the evolution of human skull shape. Evolution 66:1010–1023
McNulty KP (2012) Evolutionary development in Australopithecus africanus. Evol Biol 39:488
McNulty KP, Frost SR, Strait DS (2006a) Examining affinities of the Taung child by developmental simulation. J Hum Evol 51: 274–296
McNulty KP, Frost SR, Strait DS (2006b) Examining affinities of the Taung child by developmental simulation. J Hum Evol 51:274–296
Miller DJ, Duka T, Stimpson CD, Schapiro SJ, Baze WB, McArthur MJ, Fobbs AJ, Sousa AM, Sestan N, Wildman DE, Lipovich L, Kuzawa CW, Hof PR, Sherwood CC (2012) Prolonged myelination in human neocortical evolution. Proc Natl Acad Sci U S A 109: 16480–16485
Mitteroecker P, Bookstein F (2007) The conceptual and statistical relationship between modularity and morphological integration. Syst Biol 56:818–836
Mitteroecker P, Bookstein F (2008) The evolutionary role of modularity and integration in the hominoid cranium. Evolution 62:943–958
Mitteroecker P, Bookstein F (2009) The ontogenetic trajectory of the phenotypic covariance matrix, with examples from craniofacial shape in rats and humans. Evolution 63:727–737
Mitteroecker P, Bookstein F (2011) Linear discrimination, ordination, and the visualization of selection gradients in modern Morphometrics. Evol Biol 38:1–15
Mitteroecker P, Gunz P (2009) Advances in geometric Morphometrics. Evol Biol 36:235–247
Mitteroecker P, Gunz P, Bernhard M, Schaefer K, Bookstein FL (2004) Comparison of cranial ontogenetic trajectories among great apes and humans. J Hum Evol 46:679–697
Mitteroecker P, Gunz P, Bookstein FL (2005) Heterochrony and geometric morphometrics: a comparison of cranial growth in *Pan paniscus* versus *Pan troglodytes*. Evol Dev 7:244–258
Mitteroecker P, Gunz P, Neubauer S, Müller G (2012) How to explore morphological integration in human evolution and development? Evol Biol 39:536–553
Moss ML, Young RW (1960) A functional approach to craniology. Am J Phys Anthropol 18:281–292
Nagy Z, Westerberg H, Klingberg T (2004) Maturation of white matter is associated with the development of cognitive functions during childhood. J Cogn Neurosci 16:1227–1233
Neubauer S (2014) Endocasts: possibilities and limitations for the interpretation of human brain evolution. Brain Behav Evol 84:117–134
Neubauer S (2015) Human brain evolution: ontogeny and phylogeny. Human Paleoneurol. Springer International Publishing, p. 95–120
Neubauer S, Gunz P, Hublin JJ (2009) The pattern of endocranial ontogenetic shape changes in humans. J Anat 215:240–255
Neubauer S, Gunz P, Hublin JJ (2010) Endocranial shape changes during growth in chimpanzees and humans: a morphometric analysis of unique and shared aspects. J Hum Evol 59:555–566
Neubauer S, Hublin JJ (2012) The evolution of human brain development. Evol Biol 39:568–586
Neubauer S, Gunz P, Weber GW, Hublin JJ (2012) Endocranial volume of Australopithecus africanus: new CT-based estimates and the effects of missing data and small sample size. J Hum Evol 62: 498–510
O'Connell CA, DeSilva JM (2013) Mojokerto revisited: evidence for an intermediate pattern of brain growth in Homo erectus. J Hum Evol 65:156–161
Ponce de León MS, Golovanova L, Doronichev V, Romanova G, Akazawa T, Kondo O, Ishida H, Zollikofer CP (2008) Neanderthal brain size at birth provides insights into the evolution of human life history. Proc Natl Acad Sci U S A 105:13764–13768
Ponce de León MS, Zollikofer CP (2001) Neanderthal cranial ontogeny and its implications for late hominid diversity. Nature 412: 534–538
Ponce de León MS, Bienvenu T, Akazawa T, Zollikofer CP (2016) Brain development is similar in Neanderthals and modern humans. Curr Biol 26:R665–R666
Rice SH (2002) The role of heterochrony in primate brain evolution. In: Minugh-Purvis N, McNamara KJ (eds) Human evolution through developmental change. The John Hopkins University Press, Baltimore, pp 154–170
Rohlf FJ, Slice D (1990) Extensions of the Procrustes method for the optimal superimposition of landmarks. Syst Zool 39:40–59
Rosenberg K, Trevathan W (2002) Birth, obstetrics and human evolution. BJOG 109:1199–1206
Ross CF, Ravosa MJ (1993) Basicranial flexion, relative brain size, and facial kyphosis in nonhuman primates. Am J Phys Anthropol 91: 305–324
Sakai T, Mikami A, Tomonaga M, Matsui M, Suzuki J, Hamada Y, Tanaka M, Miyabe-Nishiwaki T, Makishima H, Nakatsukasa M, Matsuzawa T (2011) Differential prefrontal white matter development in chimpanzees and humans. Curr Biol 21:1397–1402
Schenker NM, Desgouttes AM, Semendeferi K (2005) Neural connectivity and cortical substrates of cognition in hominoids. J Hum Evol 49:547–569
Schultz AH (1940) Growth and development of the chimpanzee. Contrib Embryol 28:1–63
Schultz AH (1941) The relative size of the cranial capacity in primates. Am J Phys Anthropol 28:273–287
Scott, N., Neubauer, S., Hublin, J.-J., Gunz, P., (2014) A shared pattern of postnatal Endocranial development in extant hominoids. Evol Biol 1–23
Shaw P, Greenstein D, Lerch J, Clasen L, Lenroot R, Gogtay N, Evans A, Rapoport J, Giedd J (2006) Intellectual ability and cortical development in children and adolescents. Nature 440:676–679
Singleton M, Mcnulty KP, Frost SR, Soderberg J, Guthrie EH (2010) Bringing up baby: developmental simulation of the adult cranial morphology of rungwecebus kipunji. Anat Rec 293:388–401
Slice DE (2007) Geometric morphometrics. Annu Rev Anthropol 36: 261–281
Smith TM, Tafforeau P, Cabec AL, Bonnin A, Houssaye A, Pouech J, Moggi-Cecchi J, Manthi F, Ward C, Makaremi M, Menter CG (2015) Dental ontogeny in Pliocene and early pleistocene hominins. PLoS One 10:e0118118
Smith TM, Tafforeau P, Reid DJ, Pouech J, Lazzari V, Zermeno JP, Guatelli-Steinberg D, Olejniczak AJ, Hoffman A, Radovcic J, Makaremi M, Toussaint M, Stringer C, Hublin JJ (2010) Dental evidence for ontogenetic differences between modern humans and Neanderthals. Proc Natl Acad Sci U S A 107: 20923–20928
Smith BH, Tompkins RL (1995) Toward a life history of the Hominidae. Annu Rev Anthropol 24:257–279
Sowell ER, Peterson BS, Thompson PM, Welcome SE, Henkenius AL, Toga AW (2003) Mapping cortical change across the human life span. Nat Neurosci 6:309–315
Sowell ER, Thompson PM, Leonard CM, Welcome SE, Kan E, Toga AW (2004) Longitudinal mapping of cortical thickness and brain growth in normal children. J Neurosci 24:8223–8231

Sporns O (2011) The human connectome: a complex network. Ann N Y Acad Sci 1224:109–125
Trevathan WR (1996) The evolution of bipedalism and assisted birth. Med Anthropol Q 10:287–290
van den Heuvel MP, Sporns O (2013) Network hubs in the human brain. Trends Cogn Sci 17:683–696
Vinicius L (2005) Human encephalization and developmental timing. J Hum Evol 49:762–776
Viðarsdóttir US, Cobb S (2004) Inter- and intra-specific variation in the ontogeny of the hominoid facial skeleton: testing assumptions of ontogenetic variability. Ann Anat 186:423–428
Vrba ES (1998) Multiphasic growth models and the evolution of prolonged growth exemplified by human brain evolution. J Theor Biol 190:227–239
Weaver TD, Hublin JJ (2009) Neandertal birth canal shape and the evolution of human childbirth. Proc Natl Acad Sci U S A 106: 8151–8156
Zollikofer CPE, Ponce de León MS (2002) Visualizing patterns of craniofacial shape variation in Homo sapiens. Proc R Soc B Biol Sci 269:801–807
Zollikofer CPE, Ponce De León MS (2004) Kinematics of cranial ontogeny: heterotopy, heterochrony, and geometric morphometric analysis of growth models. J Exp Zool B Mol Dev Evol 302:322–340
Zollikofer CP, Ponce de León MS (2010) The evolution of hominin ontogenies. Semin Cell Dev Biol 21:441–452
Zollikofer CP, Bienvenu T, Ponce de León MS (2017) Effects of cranial integration on hominid endocranial shape. J Anat 230: 85–105

Networking Brains: Modeling Spatial Relationships of the Cerebral Cortex

13

Emiliano Bruner, Borja Esteve-Altava, and Diego Rasskin-Gutman

Abstract

Brain mapping has always been a priority in neurobiology and evolutionary neuroanatomy. In the last century, methodological issues and technical advances have generated a vivid debate on the parcellation and functions of the cortical territories. Brain structure is generally analyzed by considering the network of connections associated with neural pathways. Nonetheless, there is still a major debate on the recognition of the spatial and geometrical components of the cerebral cortex. The maps produced by Korbinian Brodmann in the early twentieth century on the basis of histological patterns represented a pioneering and decisive step in this sense, being a reference until the present day. Network models allow a numerical analysis of the spatial relationships among anatomical elements, supplying a quantitative tool to evaluate their reciprocal geometrical organization. This approach is able to analyze the spatial parameters associated with an anatomical system, characterized by the relationships of its elements. The network analysis of the spatial contiguity of Brodmann's areas approximately describes the major cerebral lobes. A frontal cluster includes only the prefrontal areas. There is a large parieto-occipital block including also the precentral and paracentral cortex. The cortical areas identified by the model match different areas of craniocerebral relationships, namely, the anterior fossa influenced by the upper face (prefrontal cortex), the middle fossa influenced by cranial base and mandibular integration (temporal cortex), and the vault which is characterized by more linear brain-bone dynamics (parieto-occipital cortex). The maps of Brodmann, after one century of contributions, are now replaced by finer parcellations obtained with new technical approaches based on histology, biochemistry, and metabolism, enhanced by advances in brain imaging and digital biology. Besides issues associated with cognitive processing, structural factors can influence geometrical and mechanical properties of the cerebral morphology. Network theory, applied to alternative parcellation schemes or to specific brain districts, can provide essential information on evolutionary factors channeling or constraining the evolution of the brain spatial organization.

Keywords

Brain maps • Network modeling • Brodmann's areas • Brain lobes • Evolutionary neuroanatomy

E. Bruner (✉)
Programa de Paleobiología, Centro Nacional de Investigación sobre la Evolución Humana, Paseo Sierra de Atapuerca 3, 09002 Burgos, Spain
e-mail: emiliano.bruner@cenieh.es

B. Esteve-Altava
Structure & Motion Laboratory, Department of Comparative Biomedical Sciences, The Royal Veterinary College, Hawkshead Lane, Hatfield AL9 7TA, UK

Department of Anatomy, Howard University College of Medicine, 520 W Street NW, Washington, DC 20059, USA
e-mail: boresal@gmail.com

D. Rasskin-Gutman
Theoretical Biology Research Group, Cavanilles Institute of Biodiversity and Evolutionary Biology, University of Valencia, c/ Catedratico Jose Beltran Martinez 2, CP, 46980 Paterna, Valencia, Spain
e-mail: diego.rasskin@uv.es

E. Bruner et al. (eds.), *Digital Endocasts*, Replacement of Neanderthals by Modern Humans Series,
DOI 10.1007/978-4-431-56582-6_13

13.1 Mapping the Brain Cortex

Since the earliest modern studies in neuroanatomy, researchers have tried to understand a possible association between cortical areas and functions, probably reflecting an expectation which is rooted in far more ancient popular beliefs. According to this view, specific brain areas may be responsible of specific behavior or cognitive processes. The main evidence came from observations of altered behaviors in impaired and injured individuals, with historical case studies ranging from the language areas (Broca's and Wernicke's areas) to the famous case of Phineas Gage (Goldenberg 2004). Such view of the brain as a compartmentalized computer formed by interacting but specialized areas found an extreme expression in the positivistic approach of phrenologists, trying to associate every mental attitude or capacity with a specific and determined cerebral area. This is somehow similar to some current approaches to genes and molecules, aimed at associating specific biological traits with specific physiological or behavioral conditions. A large debate, still in vogue, was then developed contrasting the localized view against holistic perspectives aimed at highlighting the importance of the entire brain system over its specific parts.

In craniology, bones refer to units which have a clear structural, embryological, and generally homologous roles, and that can be defined according to clear borders and landmarks. In contrast, in neuroanatomy, lobes and sulci are terms without such firm biological characterization and whose boundaries are not strictly defined. Thus, traditional neuroanatomical terminology refers to elements – lobes, sulci, and gyri – which do not represent real biological units but conventional areas which have been named to supply a shared and convenient language. Generally, lobes have been defined according to generalized functional associations or to raw and imprecise anatomical borders. This is why, very early in the story of the field, neuroanatomists have tried to find units based on objective biological features, beyond the general and irregular appearance of the sulcal schemes.

Many different methods and techniques were used in the last century to supply alternative maps of the brain cortex, but without any doubt, the most popular and distinguished are the ones proposed by Korbinian Brodmann (see Zilles and Amunts 2010). Brodmann (1868–1918), a German histologist, was influenced by Oskar and Cécile Vogt, dedicating his life to analyze the distribution of different cell types on the cortical surface of the brain in many different mammal species (Pearce 2005; Annese 2009). By using the stain procedure developed by Franz Nissl, he identified 52 brain areas (Fig. 13.1), publishing a seminal book in 1909 (Šimić and Hof 2015). All along his career, he further revised his maps and areas, introducing questions and issues which are still open (Judaš et al. 2012).

Brodmann did not conclude his work; he had many personal and professional difficulties during his life, dying prematurely, apparently because of an infection contracted during an autopsy. An important part left out of his work was the recompilation of a visual atlas, necessary to display the cytoarchitectural features and criteria he used to put forward his parcellation of the brain cortex.

There are at least three main issues with brain maps: *variability*, *correspondence*, and *homology*. Anatomical and morphological differences associated with individual variations can be notable, and mapping requires a statistical approach to distributions and sample variability (Eickhoff et al. 2005; Van Essen and Dierker 2007). Correspondence between anatomical and architectural elements is also very variable, and gross anatomy (sulcal pattern) is hardly associated with strict histological or functional areas (Amunts et al. 1999; Amunts and Zilles 2012). Homology among mammals or among primates is often scarcely known at functional, histological, and anatomical levels. For example, Brodmann recognized in humans only 43 areas of the 52 described in other species (Zilles and Amunts 2010).

Brain mapping can deal with functional and structural aspects of the cerebral anatomical networks (Raichle 2010; Alexander-Bloch et al. 2013; Craddock et al. 2013). In terms of *function*, it can represent the system of elements involved in the underlying cerebral processes and associated with physiological and cognitive mechanisms. In this case, the network formed by these elements influences and is influenced by the functional result associated with the system (e.g., cognition, metabolism, etc.). Generally, functional networks are based on co-activation during specific processes. In terms of *structure*, mapping is aimed at representing the system of elements involved in the spatial organization of cerebral anatomy. From a structural perspective, we can identify two different targets. In neuroanatomy, structural relationships generally refer to axonal connections between neural areas. In this case structural networks are defined in terms of shared neural pathways. In morphology, structural relationships generally refer to shared morphogenetic and biomechanical influences, due to spatial requirements and geometric properties associated with growth, development, allometric rules, and physical constraints. These contexts, the functional and the structural, are associated with different factors, involved in different activities, and often analyzed through different methods. Nonetheless, in terms of biology and evolution, they are the two sides of the same coin; they must be integrated, and such integration is the actual combination of traits and processes evaluated by natural selection.

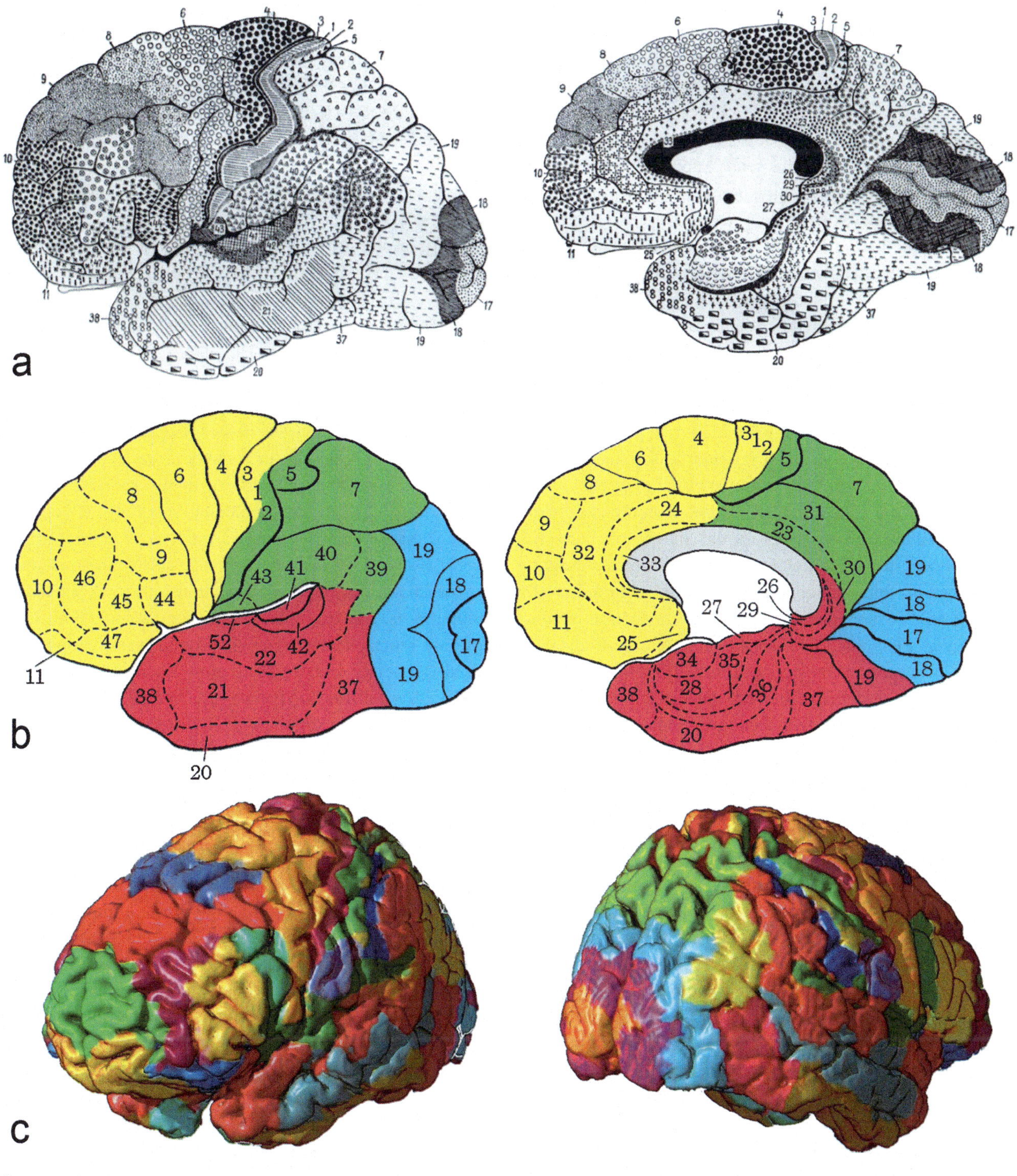

Fig. 13.1 Brodmann's areas: (**a**) original drawings by Brodmann; (**b**) a traditional schematic map of Brodmann's areas, as used in this study to compute network modeling based on spatial contiguity, with colors referring to frontal (*yellow*), parietal (*green*), occipital (*blue*), and temporal (*red*) lobes; (**c**) 3D view of Brodmann's areas (After Mark Dow, University of Oregon; Wikimedia Commons Public Domain)

13.2 Anatomical Network Analysis

Numerical modeling can be used to quantify and analyze different anatomical aspects, including metric covariance among morphological components, functional processes associated with morphological variation, or evolutionary and phylogenetic factors. In biology, network modeling concerns the relationships among the components of a biological system, and it is aimed at evaluating possible underlying rules which generate the actual organization of the system (in an organism, its phenotype). In neurobiology, network approaches have been widely used to investigate neural connectivity (e.g., Sporns et al. 2004; Meunier et al. 2010). The functioning of the brain strictly depends on its patterns of neural connections, so these schemes are a crucial issue to investigate the organization or the cerebral system. Furthermore, current advances in digital imaging have provided essential tools in this sense, supplying reliable morphological reconstruction of the neural fibers (Rilling 2008).

Nonetheless, network models can be also used to investigate the spatial arrangement of the anatomical elements. Such spatial organization is the final result of a morphogenetic process that is influenced by both intrinsic and extrinsic factors, in which a genetic program is executed within a specific physical environment, constrained by a system of forces (pressures and tensions) due to neighboring anatomical elements. Hence, in this case, we consider not the neural connectivity of the brain areas but the physical spatial connectivity, that is, the set of relationships associated with the direct spatial contiguity among cortical districts.

To generate a spatial model, we use anatomical network analysis (AnNA), a framework for the analysis of connectivity relations in a morphological system. Conceptually, it relies on an old anatomical adagio, the "Principe des connexions," which identifies physical connections among anatomical elements (i.e., bones, muscles, cartilages) as carriers of important biological information, often more so than their size and shape. Indeed, this assumption is at the foundation of comparative anatomy itself; it was championed by the French anatomist Etienne Geoffroy St. Hilaire in the XIX century, and it has been the focus of attention for comparative anatomy ever since. Geoffroy recognized that the shape and size of the same anatomical elements in different organisms vary greatly; so much that, in order to correctly identify them, it was more useful to analyze how they were connected to their anatomical surroundings. The study of connections, although intuitively sound, lacked until very recently a suitable mathematical framework to codify, manipulate, and analyze the patterns underlying these physical relationships in a meaningful way. AnNA has been developed in the past decade to fill this gap, applying and conceptualizing mathematical tools from graph theory and network analysis in a morphological context (for a review on AnNA, see Rasskin-Gutman and Esteve-Altava 2014 and references therein).

To begin a study with AnNA, one must first understand the biological meaning of the connectivity patterns it is about to analyze. This is a first, paramount step that determines the results and the usefulness of the conclusions drawn. But, why focusing on connections? Besides its classical appeal mentioned above, it is worth noting that any anatomical system can be teased apart on different levels of morphological organization. Such a division could encompass four related but semi-independent levels: (1) proportions, (2) connections, (3) orientations, and (4) articulations (Rasskin-Gutman and Buscalioni 2001). Of these four levels, the level of proportions (level 1) is the most studied one because it is related to size and shape and can be analyzed comparatively by using traditional morphometric tools with size and shape measurements or landmark-based geometric morphometrics using Cartesian coordinates. Other levels need different types of formalisms. AnNA is useful at the connection level (level 2), where the formalism is a codification of the physical connection among elements; this codification results on an adjacency matrix, filled by 1 s (representing connections) or 0 s representing the absence of connections. The other two levels can be formalized with a set of angles (orientation, level 3) and tables of motion range, to account for articulation (level 4).

The assumptions that we make about how morphological data represents function, development, or evolution determine the kind of conclusions we finally are able to draw. Connections describe the topological relations between anatomical parts, that is, their arrangement in a morphological system. Connections might also capture the presence of functional and developmental relationships (codependences) among parts. For example, connections among skull bones not only represent the topological boundaries among bones but also primary sites of bone growth and remodeling and sites of stress diffusion (Esteve-Altava et al. 2013).

Network theory supplies all the mathematical tools for the analysis of network models. A network is the combination of two sets: a set of nodes and a set of links; each link has two endpoints, that is, it represents a connection between two nodes. In this mathematical abstraction, the nodes stand for anatomical elements, and the links stand for interactions among elements. The most common representation of a network is a drawing of dots joined by lines: a line connecting two nodes indicates the presence of a mutual relation. Direct links indicate nonreciprocal relations, while weighted links indicate the strength of the interaction. Notice that all network representations are equivalent as long as the same links between nodes are kept. For simplicity we will describe only undirected (reciprocal) and unweighted networks.

The *adjacency matrix* ($A_{i,j}$) codifies the connections among the nodes of the network, that is, the number and the particular distribution of links between nodes. For undirected, unweighted networks, this is a symmetric binary matrix of size NxN, where 1 indicates the presence and 0 indicates the absence of connection. Thus, the adjacency matrix defines the neighborhood, the connectivity context of each node as all the nodes to which it connects. An adjacency matrix is the main source of data in many programs used to analyze networks, but it is not the only one. For example, a list of edges is also a very common source: a list in which each row indicates the origin and the destination of a link.

Some important descriptors and parameters for individual nodes and the whole matrix are listed below. While node descriptors are very useful to study the properties of individual elements in relation to others, system descriptions are useful to compare whole networks.

Node Degree sum of links a specific node has to other nodes in the network:

$$k_i = \sum\nolimits_j A_{i,j}$$

Clustering Coefficient ratio between the total number of links connecting its nearest neighbors and the total number of all possible links between all these nearest neighbors:

$$C_i = \frac{\sum t_i}{k_i(k_i - 1)}$$

where t_i is the number of links between the neighbors of node i.

Shortest Path Length between two nodes: their shortest distance measured as number of links to go from one node to the other:

$$l_{i,j} = d(n_i, n_j)$$

where $d(n_i,n_j)$ is the minimum distance in number of links to connect nodes i and j. Note that more than one path might have the shortest length.

Density total number of existing links (K) divided by the maximum number of possible links for a given number of nodes (N):

$$D = \frac{2K}{N(N - 1)}$$

Average Clustering Coefficient arithmetic mean of the clustering coefficient of all nodes in the network:

$$C = \frac{1}{N}\sum C_i$$

Average Shortest Path Length arithmetic mean of the shortest path length between all pairs of nodes:

$$L = \frac{1}{N - 1}\sum l_{i,j}$$

Degree Distribution frequency of occurrence of nodes with a given number of links:

$$P(k) = \frac{N_k}{N}$$

Clustering Coefficient Distribution clustering coefficient mean of all nodes with k links:

$$C(k) = \frac{\sum C_{i,k}}{N}$$

In addition, the organization of the network can be informative about its properties for a given function. For example, networks are often seen as scale-free, hierarchical, and/or small world, depending on the values of some of the parameters we just listed above. The presence of a community structure, or modules inside the network, is also very important in AnNA.

A network with a hierarchical organization shows a stratification of connections in various nested layers. The P(k) and the C(k) help assess the presence of a hierarchical organization in a network. The functional form of these distributions (e.g., uniform, Poisson, or power law) characterizes the organization of connections among the nodes. In general, a power-law distribution in both parameters indicates that the neighborhoods of low-degree nodes are highly clustered, forming blocks, while those of high-degree nodes are sparsely connected, which suggest that high-degree nodes are acting as connectors between blocks. The hierarchical organization of a network is defined as opposed to a random or a scale-free organization. In the former, the P(k) fits a Poisson function; in the latter, it fits a power-law function; in both the C(k) fits a discrete uniform function. A hierarchical organization is commonly observed in anatomical networks with a community structure.

A network with a small-world organization has a higher C and a lower or similar L to that of a random network, as a consequence of the presence of shortcut nodes. These nodes connect other nodes that would otherwise be far apart (i.e., high shortest path length). The presence of a small world is assessed by measuring the values of C and L and then comparing them to those of random equivalent networks (i.e., networks with the same number of nodes and links but randomly rewired). A common problem in anatomical networks is the small number of nodes ($N < 100$), which can hamper statistical comparisons to random models like this. A method to circumvent this problem has been proposed by Humphries and Gurney: a network is small world if $[(C/C_{\mathrm{rand}})/(L/L_{\mathrm{rand}})] \geq 0.012 \times N^{1.11}$. A small-world organization is common in anatomical networks and is related to the identification of a community structure.

Small-World Networks have a special kind of organization between regularity and randomness; their low shortest path length (L) gives them special dynamic relationships among nodes, and their high clustering coefficient (C) provides them with distinctive structural features. Having a low L means that the communication of any kind of properties among nodes (e.g., stress forces among bones) is more efficient, thanks to shortcut links; having a high C means that there are many clusters or associations between nodes, which can be putative modules.

Hierarchical Networks take their name from a very specific idea about hierarchy: nodes are organized as clusters within clusters; thus, C is also high in these types of networks, where the organization somehow depends on the existence of these clusters that become necessarily modular. Hierarchical networks are also scale-free, which means that its structure is preserved at any scale of observation; in addition, these networks always host highly connected nodes or hubs.

What does this mean in terms of the structural architecture of the brain? Hierarchical and small-world organizations are characteristic of biological systems that are integrated and, at the same time, able to maintain groupings of nodes tightly connected. Networks that are either hierarchical or small world can be said, thus, to hold structures that can be, at the same time, modular and integrated.

A network with a community structure is divided into groups of nodes that are more densely connected within the group than to nodes outside the group. A community, or connectivity module, is then a group of nodes with more links among them than to other nodes outside the module. Due to the enormous number of ways to divide networks into modules, there are many different methods and algorithms to find communities in networks, as well as to estimate the quality of different community structures in order to decide between the many possible. Fortunato (2010) has recently compiled the many methods available in the specialized literature in a systematic way.

The parameters associated with a specific network represent the way we can quantify the properties of the network, with three main scopes: to compare groups, to compare hypotheses, and to correlate functions. *Comparing groups* means testing differences between different networks. *Comparing hypotheses* requires a priori (hypothetical) models based on theoretical assumptions, which can be contrasted against the observed (real) networks. *Correlating functions* means to investigate the covariation between parameters and whatever biological or ecological variable. In all cases, the anatomical systems are described and compared by virtue of the structure and organization of the relationships among their elements.

13.3 Brodmann's Network

Cortical morphology is the results of a complex morphogenetic process, in which biomechanical factors are crucial in shaping the final cerebral form, at local and global level (e.g., Van Essen 1997; Hilgetag and Barbas 2005; Toro and Burnod 2005; Bayly et al. 2014; Tallinen et al. 2016). Therefore, the spatial organization of the cerebral areas supplies an interesting case study to apply AnNA and to investigate possible rules and constraints associated with brain parcellation in terms of spatial proximity and topology. In this example, we rely on the most basic and comprehensive criterion, modeling the spatial relationships between Brodmann's areas with a network based on the physical contiguity between areas, namely, considering whether or not two parts are in direct physical contact (Fig. 13.2). This criterion is simplistic, but it provides a preliminary survey on the issue and a direct example of application of network modeling to cortical spatial arrangement. The criterion is, thus, based on the assumption of structural interaction due to direct physical contact. Each node represents one individual area, and its connections represent their topological contiguity according to Brodmann's graphic scheme. We calculated its degree distribution, density, mean clustering coefficient, and mean shortest path length. We tested the fit of the degree distribution to four distribution functions: Poisson, log normal, exponential, and power law. Parameters were estimated by maximum likelihood and different functions compared using the negative log likelihood (nLLV), the Akaike information criterion (AIC), and the weighted Akaike information criterion (wAIC). The presence of a small-world effect in the network organization has been assessed by comparing the mean clustering coefficient and mean shortest path length of the network to that of 1000 random equivalent networks, which were generated by randomly rewiring the network connections among nodes keeping the original degree distribution. We calculated the small-world-ness (sw) of the network as the ratio between its clustering and path length and that of random equivalent networks $\left(\frac{C}{C_{\text{rand}}}/\frac{L}{L_{\text{rand}}}\right)$. A network is small world if sw $\geq 0.012n^{1.11}$ (Humphries and Gurney 2008).

We used a community detection algorithm to find hierarchical, overlapping modules in Brodmann's network. This algorithm was created by Shen et al. (2009) and implemented in *R* by Esteve-Altava (2015). It comprises the following steps:

1. Find all maximal cliques. A maximal clique is a subset of nodes in a network that is completely connected and is not a subset of another clique. Every maximal clique and subordinate node (i.e., a node that does not belong to any maximal clique) form the initial modules. We consider

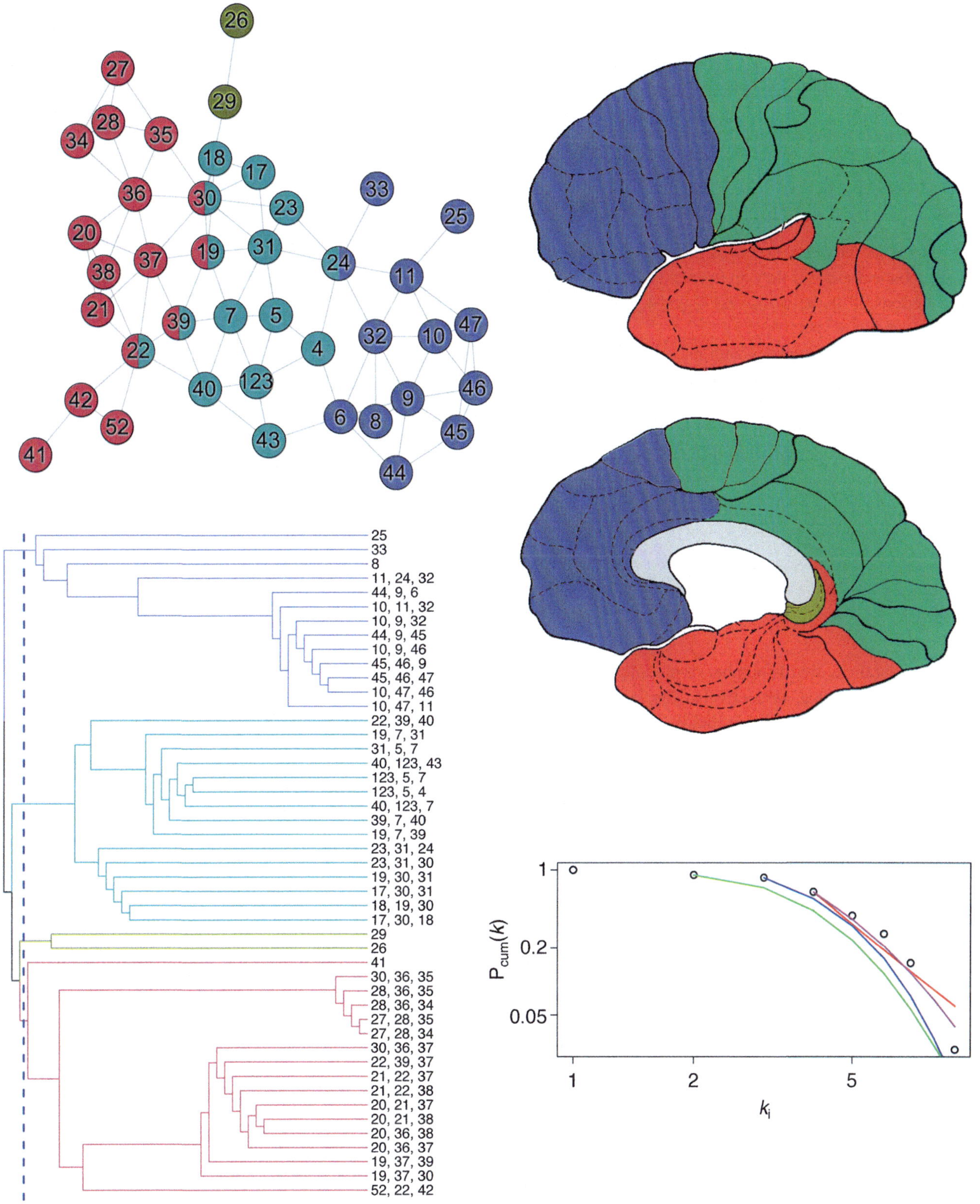

Fig. 13.2 Network used in this study, with nodes colored according to the module they belong following the results of the analysis. Numbers refer to Brodmann's areas (areas 1, 2, and 3 are considered as a single node, as in the original figure). Nodes belonging to more than one module (i.e., overlapping) are indicated with two colors. The colors of the clusters are used to evidence the groups onto Brodmann's maps. The dendrogram shows clusters composed by maximal cliques equal or larger than three nodes. Colors indicate modules identified according to the maximal extended modularity value calculated for the optimal partition (*dashed line*). The plot shows the cumulative degree distribution of Brodmann's network, showing the fit to log-normal (*green*), Poisson (*blue*), exponential (*purple*), and power-law (*red*) distributions

maximal cliques of length equal or greater than three nodes.

2. Calculate similarity between modules. The higher the number of connections between modules, the higher the similarity:

$$S = \frac{1}{2m} \sum_{v \in C_i, w \in C_j, v \neq w} \left(A_{vw} - \frac{k_v k_w}{2m} \right)$$

where A_{vw} is the element v,w of the adjacency matrix (1 if nodes connect and 0 if they do not), m is the total number of connections of the network, and k_v and k_w are the degree of nodes v and w, respectively.

3. Find and merge modules with maximal similarity. Select the pair of initial modules with the highest similarity and merge them into a new module. Repeat this step until there is only one module: the resulting grouping can be visualized as a dendrogram (Fig. 13.2).
4. Calculate the quality of each potential partition. For each branching event of the resulting dendrogram, we calculate its extended modularity:

$$QE = \frac{1}{2m} \sum_{i} \sum_{v \in C_i, w \in C_i} \frac{1}{O_v O_w} \left(A_{vw} - \frac{k_v k_w}{2m} \right)$$

where O_v and O_w are the number of modules to which node v and node w belong, respectively. The higher the QE, the better the partition of nodes in highly connected modules.

The resulting network of Brodmann's areas has 41 nodes and 87 connections ($D = 0.106$). This is a highly clustered, efficient network ($C = 0.472$; $L = 3.483$) that shows a small-world organization ($sw = 4.106$). The degree distribution of the network fits better to a log-normal decay (Table 13.1; Fig. 13.2).

According to the spatial contiguity criterion and to the threshold of extended modularity, we can identify four groups (Fig. 13.2). One group includes all the areas of the prefrontal cortex. A second group includes the parietal cortex, the occipital cortex, and the precentral gyrus/paracentral lobule. A third cluster includes the temporal cortex. A small fourth group includes the retrosplenial cortex, isolating area 26 and area 29 because of a scarce triangulation with the other areas.

This simple application of network modeling to the contiguity scheme of Brodmann's areas may suggest that the brain lobes, although representing conventional regions, are nonetheless spatially arranged in a way that generates a modular structure grouping the areas into larger units. If these four modules are real spatial units, we must consider the possibility that their spatial arrangements influenced the functional organization of the cortical areas. Of course we must also evaluate the opposite hypothesis that functional associations may have oriented their spatial arrangements due to intrinsic or extrinsic structural factors (like wiring or cranial constraints) or functional reasons (neural efficiency). In both cases, such arrangement was partially recognized by our anatomical terminology.

The frontal cluster only includes the prefrontal cortex. Despite the many studies on the topic, there is still a general debate on whether or not humans display, beyond a larger absolute size, specific frontal features when compared with other hominids (Bruner and Holloway 2010) or with living apes (Rilling 2006; Sherwood and Smaers 2013; see Chap. 14). There is no doubt that frontal areas changed their spatial relationships during human evolution: in Neanderthals and modern humans, they are positioned onto the orbital roof, introducing some constraints associated with the relationships between the brain and the upper facial block (Bruner et al. 2014). The precentral gyrus is rather grouped with the parieto-occipital cluster, instead than with the frontal areas. Actually, also in terms of functions, the motor areas are necessarily integrated with the sensorial and visuospatial cortex, namely, with the postcentral cortex and with the superior parietal lobules (Ackerley and Kavounoudias 2015). The parietal and occipital areas are clustered in a single large block. A morphological association between parietal and occipital cortex has been long recognized, also in terms of histological organization (Eidelberg and Galaburda 1984). Because of their noticeable contiguity, the parietal and occipital volumes are often analyzed together (e.g., Semendeferi and Damasio 2000). Also in terms of evolutionary variation of the braincase, these two districts are strongly integrated, suggesting shared morphogenetic patterns (Gunz and Harvati 2007). There is apparently an inverse relationship at an evolutionary level: modern humans are supposed to display larger parietal areas and smaller occipital areas (Bruner et al. 2003; De Sousa et al. 2010; see Chapt. 15). Nonetheless, when considering the volumetric variations in adult humans, there is no correlation between the parietal and occipital lobes, being the former inversely correlated to the

Table 13.1 Degree distribution fits to four functional functions

	Estimated parameters	nLLV[1]	AIC[2]	wAIC[3]
Poisson	$\lambda = 3.96$, for $k \geq 3$	−60.520	−58.52	4.40e-02
Log normal	$\mu = 1.46$, $\sigma = 0.354$, for $k \geq 2$	−68.768	−64.768	1.00e + 00
Exponential	$\lambda = 0.555$, for $k \geq 4$	−41.616	−39.616	3.45e-06
Power law	$\alpha = 3.682$, for $k \geq 4$	−41.616	−41.729	9.93e-06

frontal and temporal volumes (Allen et al. 2002). The temporal lobes are relatively larger in humans when compared with living apes' allometric patterns (Rilling and Seligman 2002). When compared with extinct human species, in modern humans, a more anterior tip of the temporal pole was hypothesized to be due to a specific increase of the temporal lobe volume (Bastir et al. 2008). According to the network clusters, the isolation of the retrosplenial cortex is also interesting, being these areas in contact with subcortical regions not included in this study. These areas are also associated with an allometric stretching influencing the morphology of the midsagittal brain and of the corpus callosum, tentatively interpreted as a mechanical effect of the tension exerted by the tentorium cerebelli (Bruner et al. 2010, 2012).

13.4 Networks and Evolution

13.4.1 Brains and Geometry

The pioneering work by Brodmann integrated histology and phylogeny, opening an essential methodological perspective in neurobiology and evolutionary neuroanatomy. The technical advances in the last decades have allowed an outstanding development of tools and approaches in digital and physical neuroscience (Preuss 2011; Rilling 2014). Cytoarchitectural studies are getting more and more specific with neural mapping, multiplying the number of areas, and adding different principles and criteria (e.g., Toga et al. 2006). Brain mapping is today also developed using information on biochemical elements (neurotransmitters and receptors) and on connectivity among areas. Modules and submodules of the brain are probably arranged with nodes and hubs as to integrate spatial and functional issues, with local nodes coordinating specific areas and global nodes coordinating together different areas (Meunier et al. 2010). Such schemes linking different elements are the results of genetic, physiological, and anatomical factors, and the resulting patterns of association are essential in normal ontogenetic processes as well as for pathological conditions (Alexander-Bloch et al. 2013). These same schemes are also the prime matter for any evolutionary change. Brain anatomy is probably organized on small-scale factors, modular organization, and local spatial interactions, which can facilitate evolutionary changes because of the degree of independence among areas (Gómez-Robles et al. 2014). Structural and functional networks share some important topological features (Hagmann et al. 2008), and in this sense, some crucial areas of integration between the two systems, like the precuneus, belong to districts which have undergone important morphological changes in our species (Bruner et al. 2014; Bruner et al. 2017). Interestingly the frontoparietal network, which is hypothesized to represent a relevant cognitive system, shows many similarities between humans and nonhuman primates (Caminiti et al. 2015) suggesting that evolutionary changes may be subtle, or associated with a matter of degree and reuse of plesiomorphic processes and structures, more than of brand-new elements.

Network modeling can be a useful tool to integrate multiple evidences from brain and cranial morphology, combining information from geometry and brain mapping (Fig. 13.3). These methods can be applied to cranial, endocranial, and brain elements, separating the brain and braincase or else describing their reciprocal relationships. Results can be used to describe and quantify the relationships within these anatomical systems or to match data from other kinds of brain mapping principles. This approach can reveal underlying patterns of structural organization, as well as phylogenetic differences. Most analyses on brain network modeling concern the organization of the neural connections (e.g., Sporns et al. 2004; Meunier et al. 2010). These studies are opening an exciting brand-new perspective in neuroscience, where functions are investigated by numerical models associated with spatial properties of the fiber arrangements. Brain functions are strongly based on wiring schemes, so connections are clearly a main issue. Nonetheless, from the pioneering works by D'Arcy Thompson on spatial functions and evolution (1942) to the seminal book by Stephen Jay Gould on ontogeny and phylogeny (1977), until the last advances in shape analysis and computed morphometrics (e.g., Mitteroecker and Bookstein 2008; Mitteroecker and Gunz 2009), we are further aware that spatial organization is also essential to channel the variation of the anatomical systems. Allometric rules, spatial constraints, and mechanical relationships are sensitive to geometrical factors underlying the phenotypic plasticity and the selective processes associated with its evolutionary success or failure. Therefore, beyond the scheme of neural connections, network analysis can be useful to investigate the spatial properties of the anatomical elements in terms of geometrical relationships among their parts. Such quantitative analyses can reveal underlying schemes and phenotypic patterns constraining evolutionary and functional processes.

The structure of the brain organization is supposed to be the consequence of selective forces optimizing costs and efficiency of the neural networks (Bullmore and Sporns 2012). An evolutionary pressure in this sense is likely, most of all when considering the ecological and metabolic costs of brain management. Nonetheless, we must always bear in mind that selection works only on characters influencing the reproductive rates and that many characters are integrated by polygenic and pleiotropic effects. For most features, optimization is therefore relative and secondary to

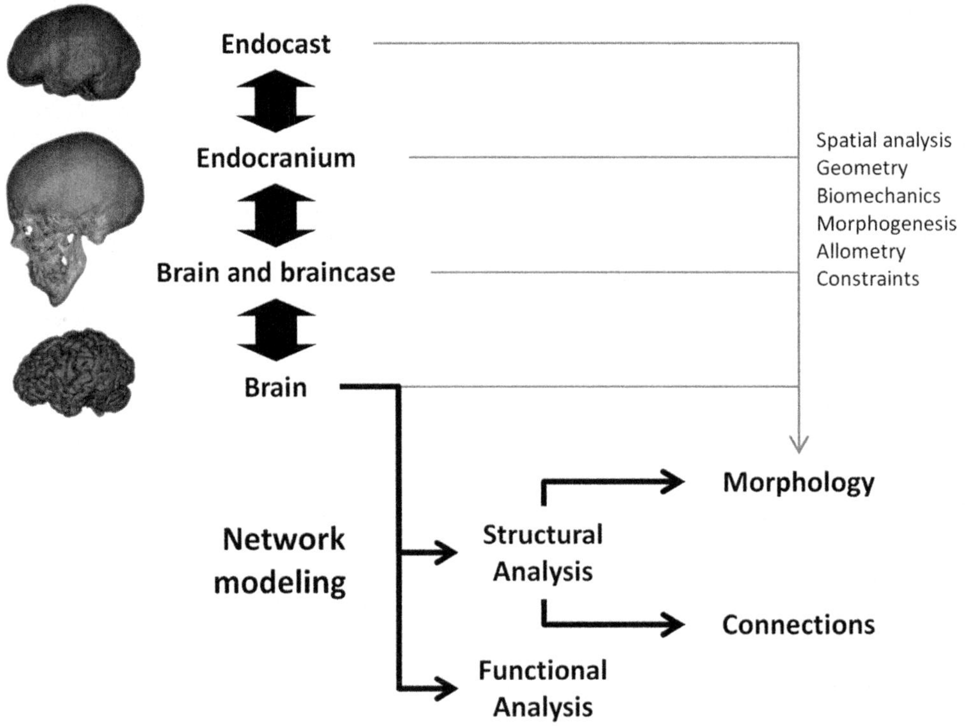

Fig. 13.3 Network modeling can be used to investigate functional and structural brain organization. Structural analyses concern the neural connections as well as the spatial and geometrical properties of the brain, which must take into account different levels of the brain-braincase system

a set of limits and constraints intrinsic to the biology of a species, channeled by rules and a given degree of phenotypic plasticity.

Spatial and functional parameters are necessarily related, but they are also influenced and constrained by distinct factors. In particular, many behavioral and cognitive aspects are currently interpreted in terms of functional imaging. The old popular view of a compartmentalized brain has permeated functional imaging, giving sometimes a modern appearance to past phrenological perspectives. It is necessary, hence, to take into considerations that brain functional imaging is based on biochemical and metabolic markers, whose relationships with the underlying cognitive processes are, to date, largely unknown (Raichle 2003). Therefore, apart from any possible shared mechanism associating spatial and functional data, mapping is generally the result of a biological distribution, while cognitive issues require an interpretation based on processes which are, at present, largely ignored.

13.4.2 Sulcal Patterns and Brain Structure

The limited correspondence between sulcal elements and cytoarchitecture and the complex relationships between cortical areas and cognitive functions further advise against using sulcal patterns for phylogenetic or cognitive inferences, at least as traditional evolutionary characters. Nonetheless, brain gross morphology still represents a major source of neuroanatomical information in those fields in which soft tissues are not available (like in paleoneurology) or in which histological studies are not feasible for physical, economic, or logistic reasons (as in living human samples and other medical contexts). Sulcal patterns can supply at least three kinds of information (Fig. 13.4): it can reveal changes in relative volumes and proportions, it can provide geometrical references to analyze spatial variations, and it can disclose underlying genetic programs and morphogenetic mechanisms.

Changes in relative volumes of different cortical areas can be evidenced analyzing the position and proportions of the cortical elements (gyri and sulci). Although with a lower resolution when compared with histological studies, the sulcal pattern can provide a direct quantitative evidence of relative volumetric differences between species or individuals. Dealing with fossil species, such information is the only direct evidence available in this sense (Bruner 2015).

Concerning brain form, the sulcal pattern can supply information on the spatial reciprocal organization of the brain elements, about the spatial relationships between brain and braincase, and about any functional factor statistically correlated with geometry. Beyond a strict shape comparison, the geometrical relationships among brain elements are relevant for all those functional issues associated with

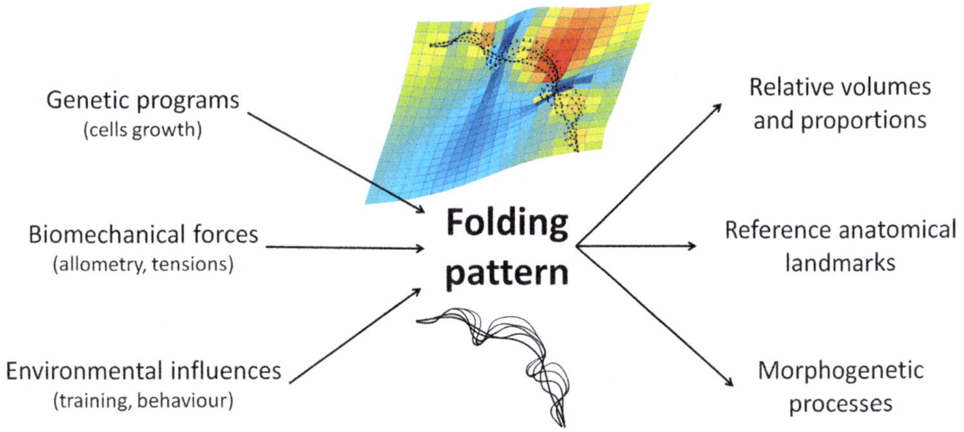

Fig. 13.4 Folding pattern is the result of genetic programs, biomechanical strains, and environmental effects. The analysis of its organization and geometry can therefore supply information on the cortical proportions, on the spatial variations, and on the underlying morphogenetic processes

spatial organization, such as connectivity. As a matter of fact, using information on spatial contiguity of the cortical element in fossil species can extend network analysis into a paleoneurological perspective, especially when dealing with higher taxonomic ranks (i.e., comparative analyses among genera, families, and orders). Among functional factors that can be investigated indirectly through their correlation with geometry, it is important to mention also thermoregulation and metabolism. Although brain thermoregulation mostly depends upon vascular physiology, heat dissipation patterns are also influenced by size and shape (Bruner et al. 2011). Human species displayed notable brain form differences, and these variations influenced the distribution of cortical heat. These patterns can be numerically simulated, linking spatial and functional information (Bruner et al. 2012). Furthermore, the spatial relationship between the brain and braincase is a relevant issue in neurosurgery and medicine (Ribas et al. 2006; Richtsmeier et al. 2006; Bruner et al. 2015).

Finally, sulcal morphology can also reveal underlying growth and developmental patterns associated with genetic pathways and morphogenetic constraints. Actually, there is a consistent relationship between genetic expression and cortical blocks (Chen et al. 2012). There is also a direct association between genes and the development of large cortical regions (Rakic 2004), and a correlation between sulcal morphology and genetic ancestry has been also recently described (Fan et al. 2015). Nonetheless, the genetic programs are probably aimed only at regulating the time and rate of cell growth, while the sulcal morphology is likely to be the result of intrinsic and automatic mechanical folding processes based on allometric responses, strain distributions, and surface adjustments (Tallinen et al. 2016). The spatial organization of the brain cortex is the result of differential growth of its areas, influenced by geometrical rules and energetic constraints (Hofman 2012). Beyond direct genetic programs influencing cell proliferation and diversification, macro- and microanatomical elements are linked through shared morphogenetic mechanisms (Van Essen 1997; Hilgetag and Barbas 2005 2006). Neurons can act themselves as biomechanical tensors redistributing growth forces and contributing to the final sulcal morphology. In this case, the final phenotype will depend also upon mechanical properties of the neurons as cortical mechanical units (Toro and Burnod 2005; Bayly et al. 2014) and upon their influence along growth trajectories (Toro 2012). Therefore, beyond the spatial and geometric information, it is possible to use sulcal patterns to evaluate indirectly changes and parameters of the underlying morphogenetic processes. Evolutionary changes in the sulcal pattern can reveal changes in relative proportions of the cortical areas or changes in the general morphogenetic sequence leading to that specific folding scheme. In this sense, the sulcal variation and arrangement are not relevant per se but as witness of an underlying structural difference (relative volumes, tissue mechanical properties, developmental forces, folding sequences).

According to the results of this introductory example on Brodmann's maps, it is interesting that the contiguity among areas is able to reveal three blocks that approximately correspond to the frontal, temporal, and parieto-occipital districts. If this is not by chance, it means that our conventional "lobes" may represent actual spatial and structural units. Modularity is often a matter of degree and hierarchical inclusive blocks, more than of absolute isolation between morphological regions. Nonetheless, it may reveal consistent groups of anatomical elements influenced by reciprocal or shared factors. In this case, it is worth noting that the three blocks described in this survey correspond also to different cranial districts and different kinds of relationships with the cranial morphogenetic system (Bruner 2015). The prefrontal cortex is housed in the anterior cranial fossa, which is structurally constrained by the upper facial block (Bruner et al. 2014). The temporal lobes are housed in the middle cranial fossa, constrained by the midface, the endocranial base, and the mandibular biomechanics (Lieberman et al. 2000; Bastir et al. 2004; Bastir and Rosas 2005, 2006). The parieto-occipital block is the largest component of the cranial vault, free from cranial constraints except for the spatial relationships between bones and suture. The three mentioned endocranial areas are quite independent in terms of morphological variation, probably because they are influenced by independent factors (Bruner and Ripani 2008). Actually, within the human genus, different morphological changes have been described for the prefrontal (Bruner and Holloway 2010), parieto-occipital (Bruner 2004; Gunz and Harvati 2007), and temporal (Bastir et al. 2008) areas. If these three cortical blocks based on contiguity of Brodmann's areas are real structural units, it remains therefore to be evaluated whether their internal cohesion is a cause or a consequence of the different relationships with the corresponding cranial districts. It is likely that a joint analysis on the cerebral and cranial elements can further add to these structural models, taking into consideration their reciprocal spatial relationships and consequent mechanical influences (Ribas et al. 2006; Bruner et al. 2015; Goriely et al. 2015).

13.4.3 Extending Networks

A final note concerns the relationships between brain morphology and environmental influences. Many current cognitive theories are giving more importance to nonneural components, like the whole body and the environment (Clark 2007, 2008). According to hypotheses on extended cognition and embodiment, the body and the environment are active parts of the cognitive experience (Malafouris 2010, 2013). The interaction between body and tools strongly influences the organization of the neural circuits, inducing micro- and macroanatomical changes in the brain (Iriki and Sakura 2008; Quallo et al. 2009; Iriki and Taoka 2012). Visuospatial integration is a clear example of cognitive functions in which biological and cultural factors can interact to generate feedbacks and integrative dynamics between the brain, body, and environment (Bruner and Iriki 2016). In this sense, selection can operate on specific traits or on the sensitivity of those traits to undergo biological responses after environmental influences. Phenotypic plasticity and evolvability can be targeted by selective forces, promoting or demoting the capacity of a biological component to respond to changes or training (Crispo 2007). It is therefore critical to investigate further to what extent cortical morphology is due to genetic, epigenetic, and environmental influences. A recent study suggests that modern humans show an apparent heritability for brain size and dimensions but, contrary to apes, less genetic constraints on the sulcal patterns (Gómez-Robles et al. 2015). Such phenotypic plasticity can be the result of a selective process increasing the environmental sensitivity of the brain structure. In all cases, a proper knowledge of the "brain geography," accounting for its neural groups and clusters, is a mandatory step, necessary to reveal the distribution of the spatial factors involved in the neural responses.

13.5 More Networks

In this study, we have shown how network modeling can be applied to brain spatial mapping taking into consideration contiguity among different areas, being the physical contact a factor relevant to structural analysis in ontogeny and phylogeny and a crucial aspect when dealing with morphological integration and local effects. Network approaches are often used to evaluate brain connectivity, but, in this case, we used network modeling to evaluate possible associations and constraints due to spatial proximity between adjacent areas. In this example, we used Brodmann's areas, a parcellation scheme which has been long applied in the last century. The results suggest that spatial contiguity generates a network which approximately separates the main lobe and matches different areas of brain-braincase relationships. Nonetheless, brain mapping is currently a proficient field of investigation, and different methods and criteria are at present providing different schemes and perspectives (Glasser et al. 2016). The same approach used here can be used with different kind of parcellations or considering different kind of spatial elements. Working with endocasts, spatial contiguity can be investigated as to evidence whether major evolutionary changes have influenced the underlying organization of brain morphology. Among human species, gross anatomical differences are more subtle, but among primates or mammals, variations in the position or composition of the brain elements have been more apparent. Also, in this introductory analysis, we consider only contiguity in terms of the presence or absence of physical contact. Future studies should weight such contact in terms of absolute and relative extension of the physical interfaces. Furthermore, such networks should be extended including the relationships between cortical and subcortical areas and between the brain (soft tissues) and braincase (hard tissues). Allometric and spatial constraints, as well as vascular and metabolic components, will be probably essential to provide consistent models able to evidence modular and/or integrated levels of organization associated with the brain spatial arrangement.

Acknowledgments EB is funded by the Spanish Government (CGL2015-65387-C3-3-P). BEA is funded by the European Union's Horizon 2020 research and innovation program under the Marie Skłodowska-Curie grant agreement No. 654155. DRG is funded by Spanish MINECO/FEDER (BFU2015-70927-R).

References

Ackerley R, Kavounoudias A (2015) The role of tactil afference in shaping motor behaviour and implications for prosthetic innovation. Neuropsychologia 79:192–205

Alexander-Bloch A, Giedd JN, Bullmore E (2013) Imaging structural co-variance between human brain regions. Nat Rev Neurosci 14: 322–336

Allen JS, Damasio H, Grabowski TJ (2002) Normal neuroanatomical variation in the human brain: an MRI-volumetric study. Am J Phys Anthropol 118:341–358

Amunts K, Zilles K (2012) Architecture and organizational principles of Broca's region. Trends Cogn Sci 16:418–426

Amunts K, Schleicher A, Burgel U, Mohlberg H, Uylings HBM, Zilles K (1999) Broca's region revisited: cytoarchitecture and intersubject variability. J Comp Neurol 412:319–341

Annese J (2009) In retrospect: Brodmann's brain map. Nature 461:884

Bastir M, Rosas A (2005) Hierarchical nature of morphological integration and modularity in the human posterior face. Am J Phys Anthropol 128:26–34

Bastir M, Rosas A (2006) Correlated variation between the lateral basicranium and the face: a geometric morphometric study in different human groups. Arc Oral Biol 51:814–824

Bastir M, Rosas A, Kuroe K (2004) Petrosal orientation and mandibular ramus breadth: evidence for an integrated petroso-mandibular developmental unit. Am J Phys Anthropol 123:340–350

Bastir M, Rosas A, Lieberman DE, O'Higgins P (2008) Middle craneal fossa and the origin of modern humans. Anat Rec 291:130–140

Bayly PV, Taber LA, Kroenke CD (2014) Mechanical forces in cerebral cortical folding: a review of measurements and models. J Mech Behav Biomed Mater 29:568–581

Bruner E (2004) Geometric morphometrics and paleoneurology: brain shape evolution in the genus homo. J Hum Evol 47:279–303

Bruner E (2015) Functional craniology and brain evolution. In: Bruner E (ed) Human paleoneurology. Springer, Cham, pp 57–94

Bruner E, Holloway R (2010) Bivariate approach to the widening of the frontal lobes in the genus homo. J Hum Evol 58:138–146

Bruner E, Iriki A (2016) Extending mind, visuospatial integration, and the evolution of the parietal lobes in the human genus. Quat Int 405: 98–110

Bruner E, Ripani M (2008) A quantitative and descriptive approach to morphological variation of the endocranial base in modern humans. Am J Phys Anthropol 137:30–40

Bruner E, Manzi G, Arsuaga JL (2003) Encephalization and allometric trajectories in the genus *Homo*: evidence from the Neanderthal and modern lineages. Proc Natl Acad Sci U S A 100: 15335–15340

Bruner E, Martin-Loeches M, Colom R (2010) Human midsagittal brain shape variation: patterns, allometry and integration. J Anat 216:589–599

Bruner E, Mantini S, Musso F, de la Cuétara JM, Ripani M, Sherkat S (2011) The evolution of the meningeal vascular system in the human genus: from brain shape to thermoregulation. Am J Hum Biol 23:35–43

Bruner E, De la Cuétara M, Musso F (2012) Quantifying patterns of endocranial heat distribution: brain geometry and thermoregulation. Am J Hum Biol 24:753–762

Bruner E, de la Cuétara JM, Masters M, Amano H, Ogihara N (2014) Functional craniology and brain evolution: from paleontology to biomedicine. Front Neuroanat 8:19

Bruner E, Amano H, de la Cuétara JM, Ogihara N (2015) The brain and the braincase: a spatial analysis on the midsagittal profile in adult humans. J Anat 227:268–276

Bruner E, Preuss T, Chen X, Rilling J (2017) Evidence for expansion of the precuneus in human evolution. Brain Struct Funct 222: 1053–1060

Bullmore E, Sporns O (2012) The economy of brain network organization. Nat Rev Neurosci 13:336–349

Caminiti R, Innocenti GM, Battaglia-Mayer A (2015) Organization and evolution of parieto-frontal processing streams in macaque monkeys and humans. Neuosci Biobehav Rev 56:73–96

Chen CH, Gutierrez ED, Thompson W, Panizzon MS, Jernigan TL, Eyler LT, Fennema-Notestine C, Jak AJ, Neale MC, Franz CE, Lyons MJ, Grant MD, Fischl B, Seidman LJ, Tsuang MT, Kremen WS, Dale AM (2012) Hierarchical genetic organization of human cortical surface area. Science 335:1634–1636

Clark A (2007) Re-inventing ourselves: the plasticity of embodiment, sensing, and mind. J Med Philos 32:263–282

Clark A (2008) Supersizing the mind. Embodiment, action, and cognitive extension. Oxford University Press, Oxford

Craddock RC, Jbabdi S, Yan CG, Vogelstein JT, Castellanos FX, Di Martino A, Kelly C, Heberlein K, Colcombe S, Milham MP (2013) Imaging human connectomes at the macroscale. Nat Methods 10: 524–539

Crispo E (2007) The Baldwin effect and genetic assimilation: revisiting two mechanisms of evolutionary change mediated by phenotypic plasticity. Evolution 61:2469–2479

De Sousa AA, Sherwood CC, Mohlberg H, Amunts K, Schleicher A, MacLeod CE, Hof PR, Frahm H, Zilles K (2010) Hominoid visual brain structure volumes and the position of the lunate sulcus. J Hum Evol 58:281–292

Eickhoff S, Stephan KE, Mohlberg H, Grefkes C, Fink GR, Amunts K, Zilles K (2005) A new SPM toolbox for combining probabilistic cytoarchitectonic maps and functional imaging data. NeuroImage 25:1325–1335

Eidelberg D, Galaburda AM (1984) Inferior parietal lobule. Arch Neurol 41:843–852

Esteve-Altava B (2015) An R implementation of an algorithm to detect hierarchical, overlapping modules in networks. Applications to the human skull. Figshare, https://dx.doi.org/10.6084/m9.figshare.1577668.v1

Esteve-Altava B, Marugán-Lobón J, Bastir M, Botella H, Rasskin-Gutman D (2013) Grist for Riedl's mill: a network model perspective on the integration and modularity of the human skull. J Exp Zool B 320:489–500

Fan CC, Bartsch H, Schork AJ, Chen CH, Wang Y, Lo MT, Brown TT, Kuperman JM, Hagler DJ Jr, Schork NJ, Jernigan TL, Dale AM, the Pediatric Imaging, Neurocognition, and Genetics Study (2015) Modeling the 3D geometry of the cortical surface with genetic ancestry. Curr Biol 25:1988–1992

Fortunato S (2010) Community detection in graphs. Phys Rep 486: 75–174

Glasser MF, Coalson TS, Robinson EC, Hacker CD, Harwell J, Yacoub E, Ugurbil K, Andersson J, Beckmann CF, Jenkinson M, Smith SM, Van Essen DC (2016) A multi-modal parcellation of human cerebral cortex. Nature 536:171–178

Goldenberg G (2004) The life of Phineas gage – stories and realities. Cortex 40:552–555

Gómez-Robles A, Hopkins WD, Sherwood CC (2014) Modular structure facilitates mosaic evolution of the brain in chimpanzees and humans. Nat Commun 5:4469

Gómez-Robles A, Hopkins WD, Schapiro SJ, Sherwood CC (2015) Relaxed genetic control of cortical organization in human brains compared with chimpanzees. Proc Natl Acad Sci U S A 112: 14799–804

Goriely A, Geers MGD, Holzapfel GA, Jayamohan J, Jérusalem A, Sivaloganathan S, Squier W, Van Dommelen JAW, Waters S, Kuhl E (2015) Mechanics of the brains: perspectives, challenges, and opportunities. Biomech Model Mechanobiol 14:931–965

Gould SJ (1977) Ontogeny and phylogeny. Belknap Press, Harvard University Press, Cambridge MA

Gunz P, Harvati K (2007) The Neanderthal "chignon": variation, integration, and homology. J Hum Evol 52:262–274

Hagmann P, Cammoun L, Gigandet X, Meuli R, Honey CJ, Wedeen VJ, Sporns O (2008) Mapping the structural core of human cerebral cortex. PLoS Biol 6:e159

Hilgetag CC, Barbas H (2005) Developmental mechanics of the primate cerebral cortex. Anat Embryol 210:411–417

Hilgetag CC, Barbas H (2006) Role of mechanical factors in the morphology of the primate cerebral cortex. PLoS Comput Biol 2: e22

Hofman MA (2012) Design principles of the human brain: an evolutionary perspective. Progr Brain Res 195:373–390

Humphries MD, Gurney K (2008) Network 'small-worldness': a quantitative method for determining canonical network equivalence. PLoS One 3:e0002051

Iriki A, Sakura O (2008) The neuroscience of primate intellectual evolution: natural selection and passive and intentional niche construction. Phil Trans R Soc London B 363:2229–2241

Iriki A, Taoka M (2012) Triadic (ecological, neural, cognitive) niche construction: a scenario of human brain evolution extrapolating tool use and language from the control of reaching actions. Phil Trans R Soc Lond B 367:10–23

Judaš M, Cepanec M, Sedmak G (2012) Brodmann's map of the human cerebral cortex – or Brodmann's maps? Translat Neurosci 3:67–74

Lieberman DE, Ross CF, Ravosa MJ (2000) The primate cranial base: ontogeny, function, and integration. Y Phys Anthropol 43:117–169

Malafouris L (2010) The brain-artefact Interface (BAI): a challenge for archaeology and cultural neuroscience. Soc Cogn Affect Neurosci 5:264–273

Malafouris L (2013) How things shape the mind: a theory of material engagement. MIT Press, Cambridge

Meunier D, Lambiotte R, Bullmore ET (2010) Modular and hierarchically modular organization of brain networks. Front Neurosci 4:200

Mitteroecker P, Bookstein F (2008) The evolutionary role of modularity and integration in the hominoid cranium. Evolution 62:943–958

Mitteroecker P, Gunz P (2009) Advances in geometric morphometrics. Evol Biol 36:235–247

Pearce JMS (2005) Brodmann's cortical maps. J Neurol Neurosurg Psychiatry 76:259

Preuss TM (2011) The human brain: rewired and running hot. Ann N Y Acad Sci 1225(S1):E182–E191

Quallo MM, Price CJ, Ueno K, Asamizuya T, Cheng K, Lemon RN, Iriki A (2009) Gray and white matter changes associated with tool-use learning in macaque monkeys. Proc Natl Acad Sci U S A 106: 18379–18384

Raichle ME (2003) Functional brain imaging and human brain function. J Neurosci 23:3959–3962

Raichle ME (2010) Two views of brain functions. Trends Cogn Sci 14: 180–190

Rakic P (2004) Genetic control of cortical convolutions. Science 303: 1983–1984

Rasskin-Gutman D, Buscalioni AD (2001) Theoretical morphology of the Archosaur (Reptilia: Diapsida) pelvic girdle. Paleobiology 27: 59–78

Rasskin-Gutman D, Esteve-Altava B (2014) Connecting the dots: anatomical network analysis in morphological EvoDevo. Biol Theor 9:178–193

Ribas GC, Yasuda A, Ribas EC, Nishikuni K, Rodrigues AJ (2006) Surgical anatomy of microneurosurgical sulcal key-points. Neurosurgery 59:S177–S208

Richtsmeier JT, Aldridge K, de Leon VB, Panchal J, Kane AA, Marsh JL, Yan P, Cole TM (2006) Phenotypic integration of neurocranium and brain. J Exp Zool 306B:360–378

Rilling JK (2006) Human and non-human primate brains: are they allometrically scaled versions of the same design? Evol Anthropol 15:65–67

Rilling JK (2008) Neuroscientific approaches and applications within anthropology. Yrb Phys Anthropol 51:2–32

Rilling JK (2014) Comparative primate neuroimaging: insights into human brain evolution. Trends Cogn Sci 18:45–55

Rilling JK, Seligman RA (2002) A quantitative morphometric comparative analysis of the primate temporal lobe. J Hum Evol 42:505–533

Semendeferi K, Damasio H (2000) The brain and its main anatomical subdivisions in living hominoids using magneticresonance imaging. J Hum Evol 38:317–332

Shen H, Cheng X, Cai K, Hu MB (2009) Detect overlapping and hierarchical community structure in networks. Phys A Stat Mech Appl 388:1706–1712

Sherwood CC, Smaers JB (2013) What's the fuss over human frontal lobe evolution? Trends Cogn Sci 17:432–433

Šimić G, Hof PR (2015) In search of the definitive Brodmann's map of cortical areas in humans. J Comp Neurol 523:5–14

Sporns O, Chialvo DR, Kaiser M, Hilgetag CC (2004) Organization, development and functions of complex brain networks. Trend Cogn Sci 8:418–425

Tallinen T, Chung JY, Rousseau F, Girard N, Lefèvre J, Mahadevan L (2016) On the growth and form of cortical convolutions. Nat Phys 12:588–593

Thompson D'AW (1942) On growth and form. Cambridge University Press, Cambridge

Toga AW, Thompson PM, Mori S, Amunts K, Zilles K (2006) Towards multimodal atlases of the human brain. Nat Rev Neurosci 7:952–966

Toro R (2012) On the possible shapes of the brain. Evol Biol 39: 600–612

Toro R, Burnod Y (2005) A morphogenetic model for the development of cortical convolutions. Cereb Cortex 15:1900–1913

Van Essen DC (1997) A tension-based theory of morphogenesis and compact wiring in the central nervous system. Nature 385:313–318

Van Essen DC, Dierker DL (2007) Surface-based and probabilistic atlases of primate cerebral cortex. Neuron 56:209–225

Zilles K, Amunts K (2010) Centenary of Brodmann's map – conception and fate. Nat Rev Neurosci 11:139–145

The Evolution of the Frontal Lobe in Humans

14

Ashley N. Parks and Jeroen B. Smaers

Abstract

In 1912, Korbinian Brodmann suggested that the "regio frontalis" (i.e., the prefrontal cortex) of the human brain was exceptionally large in comparison to other primates. His observations sparked over a century of neuroscientific inquiry into the frontal lobe and the prefrontal cortex in particular. Later work describing the role of the prefrontal cortex in human intelligence drove anthropologists and evolutionary neuroscientists to study its evolution as a means of revealing the evolutionary history of unique cognitive capacities of humans. Here we discuss the results of investigations into the frontal cortex from the perspectives of multiple disciplines: paleoneurology, comparative neuroanatomy, and phylogenetic comparative neuroanatomy. We will describe the different pieces of the puzzle that each of these disciplines contributes to forming a detailed picture of the evolution of the human frontal lobe. We then hone in on phylogenetic comparative approaches in order to investigate changes in frontal lobe scaling across anthropoids. We find that human frontal lobe enlargement is driven specifically by an expansion of the prefrontal cortex, not the frontal motor areas. These results are confirmed by comparisons of regions within the frontal lobe that indicate the human prefrontal cortex has expanded drastically in comparison to frontal motor areas. Furthermore, evolutionary rate analyses reveal that the rate of evolution of the prefrontal cortex size is higher than for the relative sizes of the frontal lobe or the frontal motor cortex. Overall, phylogenetic comparative analyses converge on the observation that different areas of the frontal lobe evolved at different rates of evolution, favoring exceptional enlargement of the prefrontal cortex, but not necessarily the frontal lobe as a whole. These perspectives thus confirm that the human brain is more than a scaled-up version of the monkey brain and that the putative unique expansion of the "regio frontalis" is indeed an important feature that may support human's unique cognitive abilities.

Keywords

Brain evolution • Prefrontal cortex • Frontal lobe • Primates • Comparative neuroanatomy

14.1 Introduction

The search for the neural substrate of human intelligence is a prevailing topic in the neurosciences. Ever since the landmark cytoarchitectonic mapping of the cerebral cortex by Brodmann (1912), a particular focus has been on the "regio frontalis." Brodmann noted that the prefrontal cortex is

A.N. Parks (✉)
Interdepartmental Doctoral Program in Anthropological Sciences, Stony Brook University, Circle Road, Stony Brook, NY 11794-4364, USA
e-mail: ashley.parks@stonybrook.edu

J.B. Smaers
Department of Anthropology, Stony Brook University, Circle Road, Stony Brook, NY 11794-4364, USA
e-mail: jeroen.smaers@stonybrook.edu

E. Bruner et al. (eds.), *Digital Endocasts*, Replacement of Neanderthals by Modern Humans Series,
DOI 10.1007/978-4-431-56582-6_14

disproportionately larger in humans compared to nonhuman primates, suggesting this region may have been subject to an exceptional evolutionary expansion. At the time, the functional underpinnings of the prefrontal cortex were, however, piecemeal. Subsequent neuroscientific work has demonstrated that this region is associated with a plethora of behavioral features that contribute to measures of general intelligence in humans (e.g., language, decision-making, theory of mind, reaching higher level goals (Asplund et al. 2010), planning (Rowe et al. 2001), introspection (Fleming et al. 2010), imagination, social information processing (Adolphs 2009)).

Considering the central role of the frontal lobe, and the prefrontal cortex in particular, for human intelligence, anthropologists and evolutionary neuroscientists have sought to study its evolution in the hope of unraveling the evolutionary history of humans' exceptional cognitive capacities (Passingham 1973). Here we discuss the findings of this endeavor as approached from different disciplines. Comparative neuroscientists utilize neuroimaging techniques to detail differences among humans, chimpanzees, and macaques (Avants et al. 2006; Van Essen and Dierker 2007; Glasser and Van Essen 2011). Paleoneurologists use fossil endocasts to track putative differences among hominin species (Neubauer 2014). Lastly, phylogenetic comparative neurobiologists study differences across a wide sample of extant species in order to map detailed patterns of change along individual lineages of the tree of life (Smaers and Soligo 2013).

In paleoneurological and phylogenetic comparative studies, the study of the prefrontal cortex is often proxied by the frontal lobe (the larger neuroanatomical region that subsumes the prefrontal cortex) (Bush and Allman 2004; Bruner et al. 2013; Falk 2014). To make an adequate distinction between the frontal lobe and prefrontal cortex, a brief anatomical and functional description of the frontal lobe and its constituent regions will first be provided. Throughout, a clear distinction will be maintained between studies that focus on the frontal lobe and those that focus on the prefrontal cortex.

14.2 Anatomy and Function of the Frontal Lobe

The human frontal lobe comprises the most anterior portion of the neocortex. It extends from the frontal pole anteriorly to the central sulcus posteriorly. It borders posteriorly with the postcentral gyrus of the parietal lobe, and it is separated from the temporal lobe by the lateral sulcus. The boundary between the primary motor cortex (area 4) of the frontal lobe and the somatosensory cortex (area 3) of the parietal lobe is also distinguished by clear differences in cytoarchitecture (cellular structure). The primary motor cortex is agranular, contains large pyramidal neurons, and is marked by generally diffuse lamination, whereas the primary somatosensory cortex (area 3) has a granular layer IV and is clearly differentiated from the frontal lobe by its sharply defined layers (Bucy 1937). The frontal lobe is functionally and structurally heterogeneous, as it contains multiple subdivisions that are the structural basis of different aspects of both motor and higher cognitive processing. The general functional subdivisions of the frontal lobe are the primary motor cortex, the premotor and supplementary motor areas, and the prefrontal cortex (Fig. 14.1).

The primary motor area occupies a strip of cortical tissue in the precentral gyrus, primarily in the anterior wall of the central sulcus. Its distinct cytoarchitecture is marked by the presence of large pyramidal neurons located in layer V called Betz motor cells. The primary motor cortex contains many cells of origin of descending motor pathways that are involved in the initiation of voluntary movements. Before the topographic organization of motor representation in area 4 had been confirmed using electrical studies, the mid-nineteenth century neurologist Hughlings Jackson predicted the pattern in which movements are mapped on to the primary motor cortex based on his observations of the predictable spread of tremors in epileptic patients (Jackson 1867). Jackson observed that partial seizures produced abnormal movements that progressed in a predictable manner from one part of the body to the next, i.e., from fingers to the hand, arm, shoulder, and, eventually, face. This sequence corresponds to the motor homunculus—a physical representation of the human body located in the precentral gyrus (Sira and Mateer 2014). Stimulation of the primary motor area elicits contralateral contraction in the muscles of the corresponding anatomical area (Fuster 2002; Sira and Mateer 2014).

Electrical stimulation of the premotor cortex, or Brodmann's area 6, also elicits muscle contraction, albeit at a higher threshold. The premotor area and supplementary motor areas are located anterior to the primary motor area. Originally, Brodmann (1909) determined that the premotor area and the primary motor cortex both lacked an internal granular layer IV and were thus determined to be architectonically agranular. However, it was later discovered that the premotor cortex is *dysgranular*, as it contains a faint granular layer IV. Unlike the sharp boundary between Brodmann's areas 3 and 4, the border between the premotor cortex (area 6) and the primary motor cortex (area 4) is somewhat more diffuse and is marked by the absence of Betz cells (Bucy 1937) as well as a faint granular layer IV in area 6. Both the number and size of Betz cells taper toward the anterior boundary of area 4, yet they remain larger than the pyramidal cells found in the premotor cortex.

The prefrontal cortex occupies the most anterior portion of the frontal lobe, although its precise delineation is a matter of contention. This region was originally coined the

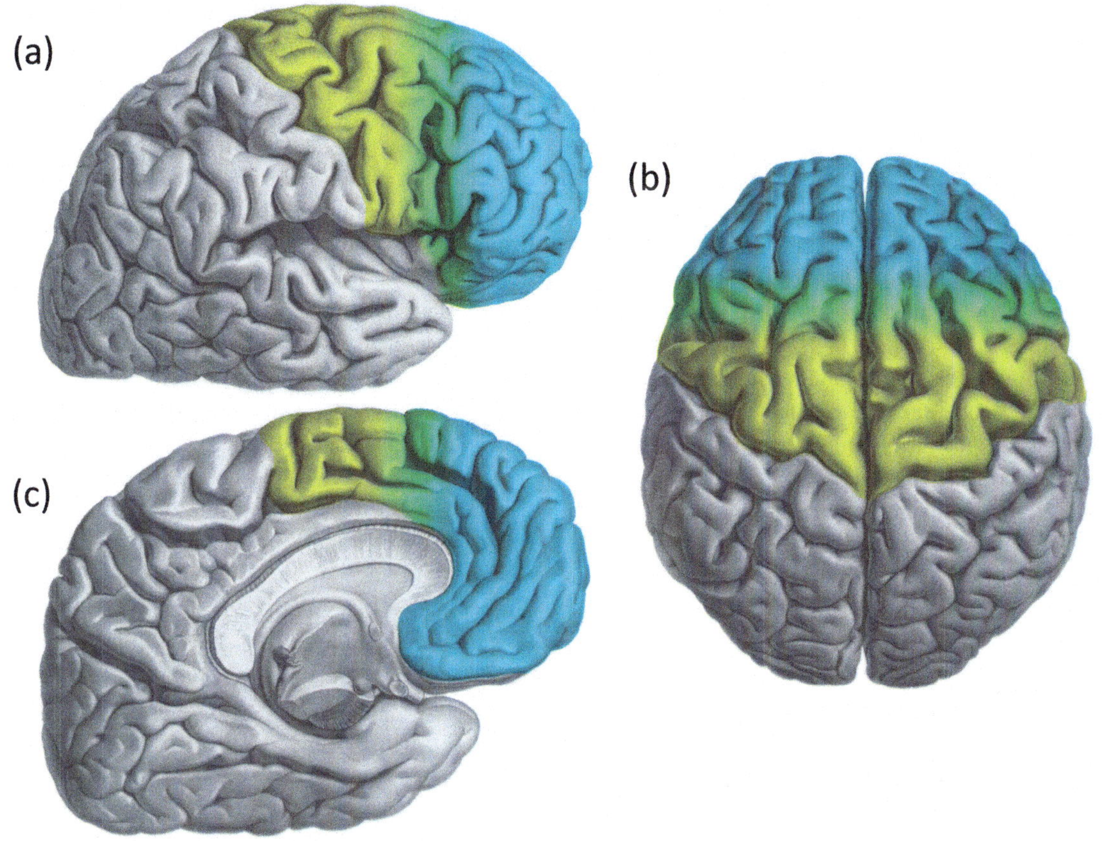

Fig. 14.1 Lateral (**a**), dorsal (**b**), and medial (**c**) view of the human brain illustrating the regions under consideration. Colors illustrate the approximate boundaries of the primary motor cortex (*yellow*), premotor cortex and supplementary motor cortex (*green*), and prefrontal cortex (*blue*) (Figure adjusted from Foville 1864)

"granular frontal cortex" or *regio frontalis*, by Brodmann, and was later referred to as the "prefrontal cortex" by subsequent researchers (Preuss 1995), although Brodmann originally reserved the term "prefrontal cortex" for a single area (area 11) within this region. In the mammalian brain, the prefrontal cortex is conventionally defined based on a combination of cytoarchitectonic and connectivity criteria, including a prominent granular layer IV and reciprocal connectivity with the mediodorsal nucleus (MD) of the thalamus (Rose and Woolsey 1948; Fuster 2002). While some researchers maintain that the use of MD projection and granularity for delineating the border of the prefrontal cortex ultimately converges on similar anatomical definitions of the region, these criteria are challenged in later works for their lack of diagnostic power (Preuss 1995). In primates, the prefrontal cortex has three major anatomical aspects: the lateral, medial, and ventral or orbital prefrontal cortex. Each prefrontal region is further divided into functionally and cytoarchitectonically distinct areas, such that there is considerable structural and functional variance within the prefrontal cortex itself. Each of these prefrontal areas plays a distinct role in the organization and control of behavioral, linguistic, and higher cognitive functions associated with intelligence (for more information, see Fuster 2008; Passingham and Wise 2012; Passingham et al. 2017). Unfortunately, no broad-scale comparative dataset of cytoarchitectonically distinct prefrontal subdivisions currently exists, prohibiting a detailed phylogenetic comparative analysis (though see Semendeferi et al. 2001 for an analysis of area 10 in apes).

14.3 A Paleoneurological Perspective

Paleoneurologists study the evolution of the frontal lobe in the hominin lineage by examining variation in such features as sulcal patterns, curvature of the frontal bone, and breadth of the anterior cranial fossae of the endocranial cavities of fossil specimens. Using endocasts, measurements of frontal gyri are derived from the imprints of sulcal patterns. The degree of sulcal preservation is, however, impacted by such variables as species, age of the individual, and geological conditions, such that sulcal imprints are typically poorly preserved in hominin fossils

(Falk 2014). In addition to sulcal patterns, several studies quantify the shape and proportions of the anterior cranial fossa, with particular regard to both the anterior curvature of the frontal bone and the diameter of the anterior cranial fossa at its widest aspect. Although frontal "bulging" and lateral widening have been used as proxies for frontal lobe expansion, the relationship between structural changes in the anterior cranial fossa and the underlying neural tissue is not straightforward, leading to conclusions that such features cannot provide unequivocal information on frontal lobe expansion (Bruner 2017). Nevertheless, three endocranial traits in particular have been discussed in the context of frontal lobe expansion: sulcal pattern variation, frontal bulging, and lateral widening.

Sulcal patterns in the frontal lobe have been a principal focus in the study of hominin endocasts. In the case of the genus *Homo*, every specimen displays sulcal patterns associated with the inferior frontal gyrus (including an outward protrusion at Broca's area) (Tobias 1987; Bruner and Holloway 2010). Frontal sulcal patterns in earlier specimens of *Homo*, such as *Homo habilis* and *Homo erectus*, have been reported to be markedly similar to those of modern *Homo sapiens* (Tobias 1987; Bruner and Holloway 2010). However, it is not clear how these patterns contrast with australopithecines due to sample size constraints. Indeed, evidence of frontal gyri in australopithecine endocasts are often observed in single, fragmentary specimens that are difficult to interpret in a broader phylogenetic context (Bruner 2017). For example, it has been suggested that a convexity in the anterior portion of the frontal lobe of specimen MH1 (*Australopithecus sediba*) represents the initial evolutionary stages of a more humanlike inferior frontal gyrus (Carlson et al. 2011; Falk 2012, 2014; Falk et al. 2012). More specimens are needed in order to characterize frontal sulcal patterns across australopithecine genera.

Additionally, modern human crania exhibit a characteristic anterior bulging of the frontal bone (Bruner et al. 2013). Although there is overlap in the degree of "frontal bulging" across a wide range of earlier and modern species of *Homo* (Bookstein et al. 1999), it has been suggested that the accentuated curvature of the frontal bone reflects underlying changes in cortical tissue (Lieberman et al. 2002). However, both the face and the vault of the skull contribute to formation of frontal bone morphology; thus, causality of frontal bulging is obscured by structural interactions between the neurocranial and splanchnocranial elements (Bookstein et al. 1999; Bruner et al. 2013; Bruner 2017). Indeed, the frontal bone comes into direct contact with the orbits, rendering frontal bone morphology susceptible to vertical constraints on facial growth (Enlow 1990). Thus, it is plausible that anterior bulging of the frontal bone is a structural by-product of spatial constraints that stem from changes to the hominin facial shape (Bruner 2017).

Lastly, the anterior cranial fossae of modern humans and Neanderthals have undergone a lateral widening (Bruner and Holloway 2010). In comparison to *Homo erectus* and australopithecines, modern humans and Neanderthals exhibit an evolutionary grade shift in the proportion of the anterior cranial fossa width relative to the width of the posterior portion of the cranium (Bruner and Holloway 2010). The frontal lobes are absolutely and relatively wider in Neanderthals and modern humans than in more archaic species. As Neanderthal and modern humans are the only two species whose frontal lobes lie directly on top of the orbits, it is possible that cranial constraints from direct contact with the orbits have caused a shift in proportions toward a greater maximum brain width (Bruner 2017).

Three hypotheses have been suggested to account for changes in the form and proportion of the anterior cranial fossa (Bruner and Holloway 2010; Bruner 2017): (1) Neanderthals and modern humans underwent a redistribution of cortical volume as a secondary consequence of structural constraints from having the frontal lobes lie directly on top of the orbits (Bookstein et al. 1999; Bruner and Manzi 2005; Bruner 2017); (2) underlying changes in cortical organization, specifically via the expansion of Broca's area and the evolution of language and complex cognition, have caused a lateral expansion of the anterior endocranial cavity; and (3) cranial constraints caused lateral widening, providing the spatial dimensions that would be exapted for new neural functions (i.e., new connections to Broca's area in association with language). Evidence thus far seems to favor that a geometric reconfiguration of frontal cortical mass was largely a secondary by-product of structural changes to the face and skull (Bruner and Holloway 2010; Bruner 2017). In sum, the fossil record does not provide evidence that directly addresses frontal expansion (Bruner 2017). Moreover, because internal brain reorganization cannot be deduced from shifts in the gross proportions of the cranium, the expansion of the human frontal lobe and consequent changes in cognition evade fossilization in the paleoanthropological record. Insights from other disciplines are necessary in order to address purported changes in human frontal lobe evolution.

14.4 A Comparative Neuroanatomical Perspective

Elucidating the nature of differences between the prefrontal cortex of humans and other animals has been an enduring question driving comparative neuroanatomical enquiry for over a century (Preuss and Goldman-Rakic 1991; Fuster 2002; Sherwood and Smaers 2013). Brodmann's (1909, 1912) seminal work highlighted the significance of the "regio frontalis" in primates, with a specific emphasis on the unique qualities of the human prefrontal cortex. Brodmann conducted a broad survey of mammalian cytoarchitecture and concluded that the granular frontal cortex (the area now referred to widely as the prefrontal cortex) is unique to primates and that the human prefrontal cortex is disproportionately large in comparison to that of nonhuman primates. Brodmann's work instigated an ongoing stream of research regarding the evolutionary significance of the prefrontal cortex, including whether or not new regions have evolved within the frontal lobe throughout the course of primate brain evolution.

Brodmann's first major hypothesis was that the granular frontal cortex (the prefrontal cortex) is unique to primates and is rudimentary or absent in all mammals. In support of Brodmann's hypothesis, several researchers have argued that the evolution of the frontal lobe in primates involved the addition of new functionally and cytoarchitectonically distinct areas that comprise the prefrontal cortex (Sanides 1964, 1970; Pandya et al. 1988). However, arguments against the evolutionary distinctiveness of the primate prefrontal cortex have been made by emphasizing similarities in connectivity patterns across mammals (Rose and Woolsey 1948). Specifically, the bidirectional connectivity of the mediodorsal (MD) nucleus projects similarly to the granular portion of the frontal lobe (the prefrontal cortex) in primates and to the nongranular cortex in other mammals. As an extension of this hypothesis, it has been argued that the MD-projection cortex in nonprimates is homologous to the prefrontal cortex of primates (Akert 1964). Lesion studies of the MD-projection cortex of rats provided support for the homology in the functional organization of the prefrontal cortex (Eichenbaum et al. 1983; Kolb 1984). It is now generally accepted that homologues of the orbital and cingulate portions of the prefrontal cortex exist in some nonprimate mammals (Ongür and Price 2000), while the dorsolateral prefrontal regions are not found outside of primates (Preuss 1995).

Brodmann's second major hypothesis was that the prefrontal cortex underwent considerable expansion throughout the course of human evolution, including the possible addition of new areas within the prefrontal cortex. Brodmann remarked that the prefrontal regions were not identical in human and nonhuman primates, noting that homologies between the monkey and human prefrontal regions were unclear. While many early works regarded the primate prefrontal cortex as a homogenous unit that lacked internal functional subdivisions (Lashley and Clark 1946; Von Bonin 1948), it is now well established that there is a high degree of functional and structural parcellation within the prefrontal cortex (Cavada and Goldman-Rakic 1989; Preuss and Goldman-Rakic 1989, 1991; Seltzer and Pandya 1989). However, whether or not the monkey prefrontal cortex contained a full complement of structurally and functionally distinct regions as the human prefrontal cortex was less clear. Walker (1940) created new cytoarchitectonic maps of the macaque brain in which he described new areas on the posterior orbital surface of the monkey prefrontal cortex. Modifications to Brodmann's original parcellation of the human prefrontal cortex were proposed to account for areas that Brodmann originally only identified in humans (Beck 1949; Petrides and Pandya 1994). It is now generally accepted on the basis of cytoarchitectonic evidence that the human orbital prefrontal cortex is homologous to the orbital prefrontal cortex of nonhuman primates (Semendeferi et al. 1998; Ongür and Price 2000). Similarly, the lateral prefrontal cortex of the macaque monkey has been argued to contain the same complement of cytoarchitectonic regions as that found in humans (Petrides 2005).

Whether or not the prefrontal cortex of humans is exceptionally enlarged, as suggested by Brodmann, or is to be expected for a primate of human body size, has been addressed in several comparative neuroimaging studies (Avants et al. 2006; Van Essen and Dierker 2007; Glasser and Van Essen 2011). For example, neuroimaging methods have been used to demonstrate that the relative size of the prefrontal cortex of humans is twice as large in comparison to that of chimpanzees (Avants et al. 2006). Furthermore, differences in the relative size of regions within the prefrontal cortex have been documented, such as the exceptional expansion of the human lateral prefrontal cortex in comparison to macaques (Van Essen and Dierker 2007). Distinctions in the connectivity patterns of the prefrontal cortex in humans and other primates have also been demonstrated using neuroimaging techniques. For example, unlike in chimpanzees and macaques, the human left ventral premotor cortex is strongly connected with the left middle and inferior temporal gyrus by means of the arcuate fasciculus (Rilling et al. 2008). Additionally, humans, but not macaques, exhibit strong functional connectivity between the anterior prefrontal cortex (aPFC) and the inferior parietal lobe (Mars et al. 2011).

In sum, comparative neuroanatomists have used direct comparisons of neuroanatomical variation between species in order to advance our knowledge of the evolution of the human frontal lobe and the prefrontal cortex in particular. Firstly, comparative neuroanatomical studies largely reject Brodmann's notion that the prefrontal cortex is unique to primates among mammals with the exception of the primate dorsolateral prefrontal cortex. The orbital and cingulate portions of the prefrontal cortex, on the other hand, are found in many other mammalian species. Secondly, the human prefrontal cortex has undergone considerable expansion in comparison to nonhuman primates, although it is generally accepted that the prefrontal cortex of humans and nonhuman primates contains the same complement of cytoarchitectonic regions. Thus, it is unlikely that the human prefrontal cortex expanded by the addition of new regions that are not found in other primates.

14.5 A Macroevolutionary Perspective

In order to make inferences about the evolutionary context in which the human brain evolved, comparative neuroanatomists often rely upon direct comparisons between humans and our closest living relatives, chimpanzees and bonobos. Implicit in this comparison is the notion that the neural architecture of the chimpanzee has not changed throughout the course of the evolution of *Pan*, such that the chimpanzee brain can function as a stand-in for the brain of the last common ancestor of chimpanzees and humans. This underlying assumption is problematic considering that both chimpanzees and humans have evolved for around 6 million years since their last common ancestor. This leaves the possibility that chimpanzees may also have evolved traits that are unique to their lineage. Several recent studies have indeed shown this to be the case (Sayers et al. 2012; Almécija et al. 2015). It is clear that any derived morphological traits of the chimpanzee brain will confound the results of an evolutionary analysis based off direct comparisons between the two species.

Additionally, the direct comparison of chimpanzee and human brains fails to take allometry into account, confounding interpretations of proportionality (Passingham 1973). Comparisons of proportions assume that variables scale at a ratio of one-to-one (i.e., isometry). It is, however, well established that many neural structures do not scale in a ratio of one-to-one but, rather, are found to scale allometrically with size (Finlay and Darlington 1995). Allometric scaling trends are critically important to the understanding of the evolution of human neural architecture, as shared allometries may reflect shared functional, genetic, and developmental constraints (Smaers et al. 2017). Departures from shared allometries, in particular, are evolutionarily informative because they reveal deviations from integration that highlight shifts in the functional, genetic, and/or developmental bauplan of animals. Such information is crucial to identify instances where brain organization shifts away from the generally integrated building plan of the vertebrate brain (Sylvester et al. 2010). A renewed effort to collect data for comparisons across a broad range of species (e.g., MacLeod et al. 2003; Sherwood et al. 2005a; Smaers et al. 2010, 2011b, 2013; Bauernfeind et al. 2013), in conjunction with ongoing advancements in comparative methods, permits a macroevolutionary account of brain evolution that is able to characterize allometric scaling trends (and species- or clade-specific deviations from allometry) in brain structures throughout a vast array of species in a primate phylogenetic tree.

In order to develop a macroevolutionary context of the human brain, statistical methods are used that incorporate the phylogenetic relatedness among the species under study in techniques that answer questions regarding the coevolution of traits, the scaling patterns of traits, and the tempo and mode of evolution of traits (Venditti et al. 2011; Khabbazian et al. 2016; Smaers and Rohlf 2016; Smaers et al. 2016). These approaches are of particular relevance to elucidating the evolution of the human frontal lobe because they address issues of allometry and putative human "uniqueness." It is clear, however, that these methods require information from a wide range of species. One advantage of comparing trait variation across a broad comparative sample is that it alleviates the issues presented by apomorphies in the direct comparison of two species.

A principal issue limiting the collection of prefrontal information across a wide range of species is that there is no standard method for delineating prefrontal boundaries across species (Sherwood and Smaers 2013). Different delineations of the prefrontal and frontal regions have been used across datasets, confounding the interpretability of comparisons of these prefrontal measurements. Some studies rely on gross anatomical landmarks, such as the genu of the corpus callosum, in order to delimit the boundaries of the prefrontal cortex (McBride et al. 1999; Schoenemann et al. 2005), while other researchers insist upon the use of cytoarchitectonic criteria for an accurate demarcation of prefrontal boundaries (Semendeferi et al. 2001; Sherwood et al. 2005b). Of the available datasets, only two provide information on the prefrontal cortex based on cytoarchitectonic criteria (Brodmann 1912; Smaers et al. 2011a). Brodmann (1912) provides information for 13 species (including humans) on total area of the granular frontal cortex, agranular cortex, neocortex, and striate cortex. Brodmann's delineation of the granular frontal cortex (i.e., the prefrontal cortex) is defined as all subregions of the frontal lobe that contain a prominent granular layer IV (including areas 8, 9, 10, 11, 44, 45, 46, and 47 in the human brain). Brodmann's cytoarchitectonic maps can be regarded as the most reliable dataset, with subsequent cytoarchitectonic maps providing highly similar impressions (von Economo and Koskinas 1925; Bailey et al. 1950; Bailey and Von Bonin 1951), including recent maps based on the difference in myelination between association cortex and other areas (Glasser and Van Essen 2011). Smaers et al. (2011a) provided a proxy of prefrontal volume across a wider range of species ($N = 19$) by employing a volumetric bootstrapping procedure along the frontal pole relative to the cytoarchitectonic borders between the frontal lobe and the parietal lobe. First, the borders between the frontal and parietal lobes were delineated based on cytoarchitectonic criteria (using 20 section intervals). Then, the cumulative volume of the first five anterior sections along the frontal pole were considered as a proxy for prefrontal volume, while the last five posterior sections were considered as a proxy of frontal motor area volume. For more information, see Smaers et al. (2011a, 2012, 2017). While the Brodmann dataset provides a more accurate delineation of the prefrontal cortex as defined by the presence of a prominent granular layer IV, the Smaers dataset provides a proxy for prefrontal size that underestimates possible exceptional expansion in great apes and humans (see more information in Smaers et al. 2017). Here we discuss the Brodmann and Smaers datasets only because they are the only two datasets that provide prefrontal information based on cytoarchitectonic criteria.

In order to test whether the human frontal lobe deviates significantly from allometric predictions, we performed a phylogenetic analysis of covariance (Smaers and Rohlf 2016). This procedure tests whether different groups in the sample indicate significant differences in the slope and intercept of the regression. Specifically, this procedure evaluates whether a model that includes separate slopes or intercepts for different groups provides a better fit for the data than a model with a single intercept. A phylogenetic ANCOVA test for differences in intercepts between groups for frontal lobe size versus brain size hereby constitutes a test on whether these different groups have significantly different values for relative frontal lobe size. Results (Table 14.1) indicate that human frontal cortex expansion (relative to the rest of the neocortex) is approaching significance in humans for both datasets (P is just below 0.05 for the Smaers data and non-significant for the Brodmann data). Similarly, human prefrontal cortex expansion relative to the rest of neocortex size nears significance ($P = 0.058$ for the Smaers data and $P < 0.014$ for the Brodmann data). Frontal motor area expansion is, however, not significant. Importantly, prefrontal expansion relative to frontal motor expansion is highly significant ($P < 0.01$ for the Smaers data and $P < 0.001$ for the Brodmann data).

These results support the conclusion that frontal motor areas have not significantly expanded in the human brain but, rather, that the prefrontal cortex has (Passingham 1975; Buckner and Krienen 2013; Glasser et al. 2014; Smaers et al. 2017). Results reveal that any enlargement in the frontal lobe is due to expansion in the prefrontal cortex, not the frontal motor areas. This is confirmed in a comparison of prefrontal to frontal motor areas indicating that the human prefrontal cortex has expanded dramatically relative to frontal motor areas. Together, our results suggest that the expansion of the frontal lobe is due to the exceptional expansion of the prefrontal cortex. Thus, a grade shift within the human frontal lobe is evident (frontal motor areas have not expanded significantly, even decreased in relative size, while prefrontal cortex size has expanded significantly). It should be noted that the current measure of frontal motor areas does not differentiate between primary motor areas and premotor areas. In alignment with previous suggestions (Blinkov and Glezer 1968; Passingham and Ettlinger 1974; Preuss 2004),

Table 14.1 Results from a phylogenetic analysis of covariance. Results relate to tests of differences in intercept among groups with the slope held constant. "Others" refers to all non-great ape primates in the sample. The analysis includes the comparison of multiple treatment groups (group a "v" group b) to a control group ("|" group c)

<table>
<tr><th></th><th></th><th colspan="7">PGLS ANCOVA</th></tr>
<tr><th></th><th></th><th></th><th colspan="3">Smaers data</th><th colspan="3">Brodmann data</th></tr>
<tr><th>Y</th><th>X</th><th>Grouping</th><th>F</th><th>P</th><th></th><th>F</th><th>P</th><th></th></tr>
<tr><td>Frontal</td><td>Rest of the neocortex</td><td>Among groups</td><td>2.651</td><td>0.103</td><td></td><td>3.384</td><td>0.118</td><td></td></tr>
<tr><td></td><td></td><td>Humans v others|great apes v others</td><td>5.085</td><td>0.040</td><td>*</td><td>4.160</td><td>0.097</td><td></td></tr>
<tr><td></td><td></td><td>Humans v great apes|others</td><td>4.789</td><td>0.045</td><td>*</td><td>0.202</td><td>0.672</td><td></td></tr>
<tr><td></td><td></td><td>Great apes v others|humans</td><td>1.380</td><td>0.258</td><td></td><td>4.015</td><td>0.101</td><td></td></tr>
<tr><td>Prefrontal</td><td>Frontal motor</td><td>Among groups</td><td>4.842</td><td>0.024</td><td>*</td><td>24.911</td><td>0.003</td><td>**</td></tr>
<tr><td></td><td></td><td>Humans v others|great apes v others</td><td>9.098</td><td>0.009</td><td>**</td><td>49.820</td><td>0.001</td><td>***</td></tr>
<tr><td></td><td></td><td>Humans v great apes|others</td><td>5.051</td><td>0.040</td><td>*</td><td>23.080</td><td>0.005</td><td>**</td></tr>
<tr><td></td><td></td><td>Great apes v others|humans</td><td>5.994</td><td>0.027</td><td>*</td><td>2.316</td><td>0.189</td><td></td></tr>
<tr><td>Prefrontal</td><td>Rest of the neocortex</td><td>Among groups</td><td>2.109</td><td>0.156</td><td></td><td>7.110</td><td>0.035</td><td>*</td></tr>
<tr><td></td><td></td><td>Humans v others|great apes v others</td><td>4.207</td><td>0.058</td><td></td><td>13.678</td><td>0.014</td><td>*</td></tr>
<tr><td></td><td></td><td>Humans v great apes|others</td><td>3.349</td><td>0.087</td><td></td><td>3.895</td><td>0.105</td><td></td></tr>
<tr><td></td><td></td><td>Great apes v others|humans</td><td>1.710</td><td>0.211</td><td></td><td>5.995</td><td>0.058</td><td></td></tr>
<tr><td>Frontal motor</td><td>Rest of the neocortex</td><td>Among groups</td><td>0.930</td><td>0.416</td><td></td><td>1.772</td><td>0.262</td><td></td></tr>
<tr><td></td><td></td><td>Humans v others|great apes v others</td><td>1.257</td><td>0.280</td><td></td><td>1.425</td><td>0.286</td><td></td></tr>
<tr><td></td><td></td><td>Humans v great apes|others</td><td>0.345</td><td>0.566</td><td></td><td>3.458</td><td>0.122</td><td></td></tr>
<tr><td></td><td></td><td>Great apes v others|humans</td><td>1.768</td><td>0.204</td><td></td><td>1.383</td><td>0.293</td><td></td></tr>
</table>

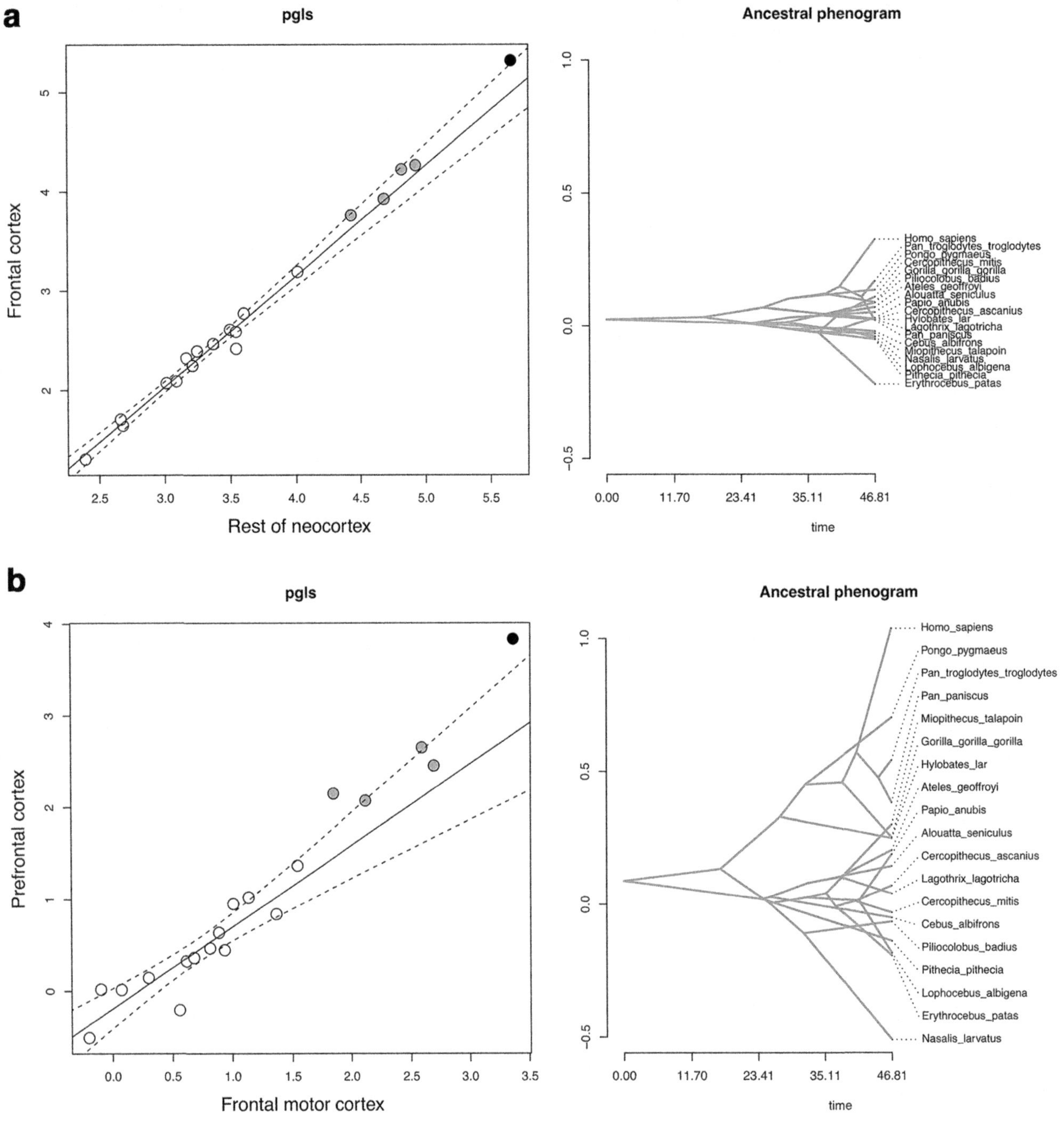

Fig. 14.2 Phylogenetic generalized least-squares analysis of the Smaers data. Confidence intervals (*dashed lines*) indicate the uncertainty in the estimation of the scaling parameters (i.e., the slope and intercept of the regression). Regression parameters are based on the scaling pattern of the non-hominoid sample. *Black circles* represent human values, whereas *gray circles* reflect great ape values. All other primates are indicated by *white circles*. *Scaling patterns* are displayed for four comparisons: (**a**) scales the frontal cortex to the rest of the neocortex (defined as the neocortex minus the frontal cortex); (**b**) the prefrontal cortex to frontal motor cortex; (**c**) the prefrontal cortex to the rest of the neocortex (defined as the neocortex minus the frontal cortex); (**d**) frontal motor cortex to the rest of the neocortex (defined as the neocortex minus the frontal cortex). *** indicates $P < 0.001$, ** $P < 0.01$, *$P < 0.05$. Ancestral phenograms depict estimations of trait evolution across independent lineages of a phylogeny (Smaers et al. 2016), such that both increases and decreases in the rate of evolution of relative brain structure volumes can be visualized

it is expected that the non-enlargement of human frontal motor areas applies particularly to the primary motor cortex, not necessarily to the premotor areas.

In addition to testing for scaling patterns, phylogenetic comparative methods can also be used to map the evolution of biological traits. Figures 14.2 and 14.3 display the scaling patterns of the frontal lobe and prefrontal cortex and the ancestral phenograms of relative frontal lobe and prefrontal cortex size. Ancestral phenograms display a best estimate of how relative frontal lobe and prefrontal cortex size have

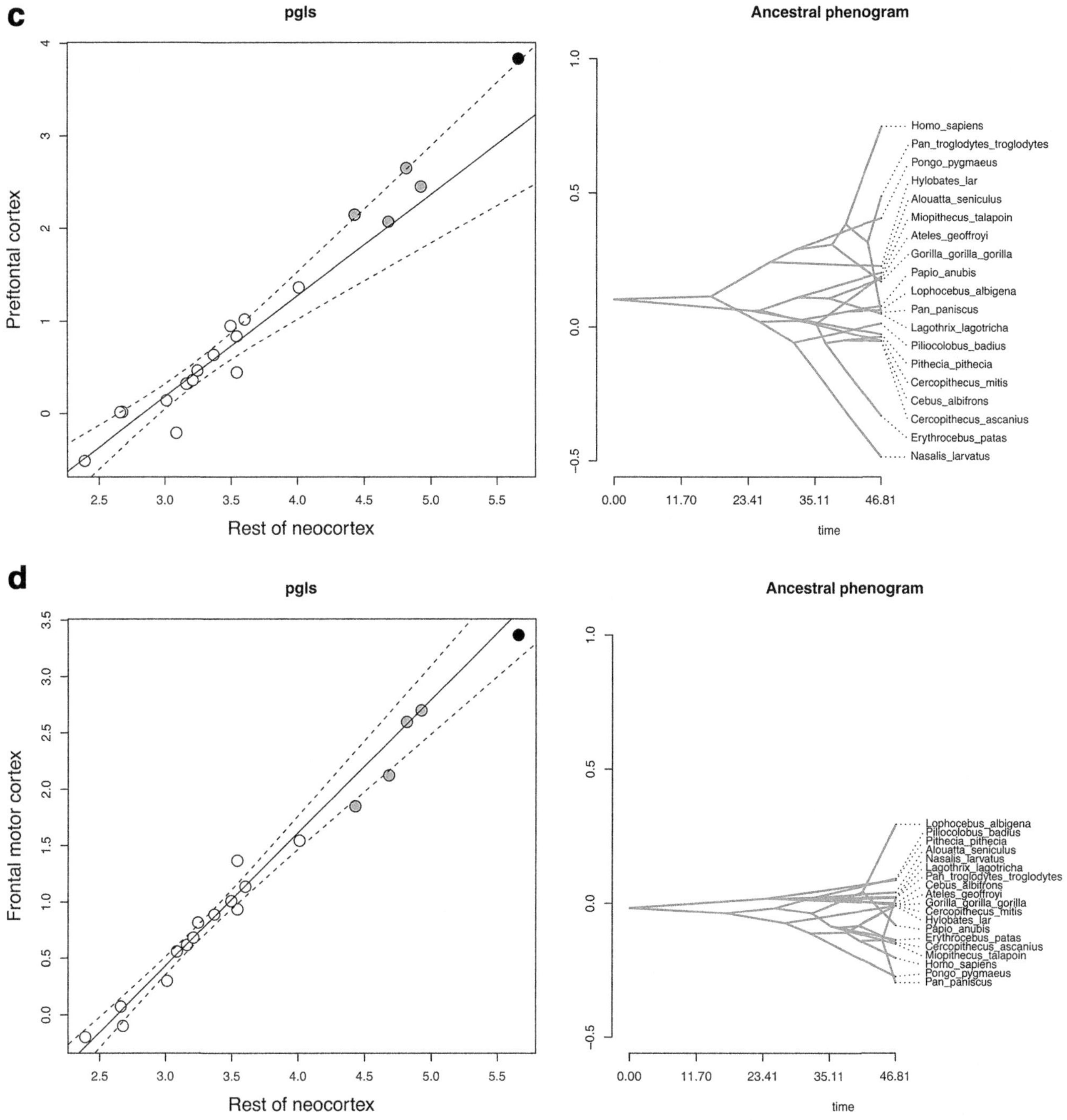

Fig. 14.2 (continued)

changed through time across the individual lineages of the phylogenetic tree. Results are indicated for both the Smaers (Fig. 14.2) and Brodmann (Fig. 14.3) datasets and lead to the same conclusion with regard to frontal and prefrontal cortex evolution. The specific measure of relative size matches the analyses performed in (a–d). Ancestral phenograms were computed using a multiple-variance Brownian motion procedure (Smaers et al. 2016) (equivalent results were obtained using a constant-variance Brownian motion procedure).

From the ancestral phenograms, it is clear that most of the cross-species variation that occurs in the frontal lobe is accounted for by changes in relative prefrontal cortex size. This is further confirmed by a rate analysis demonstrating that the rate of evolution of relative prefrontal cortex size is higher than that in the relative sizes of either frontal cortex or frontal motor areas (Fig. 14.4). The rate is hereby quantified as the Brownian motion rate parameter (σ^2) of a multiple-variance Brownian motion model (also here equivalent results were obtained using a constant-variance Brownian

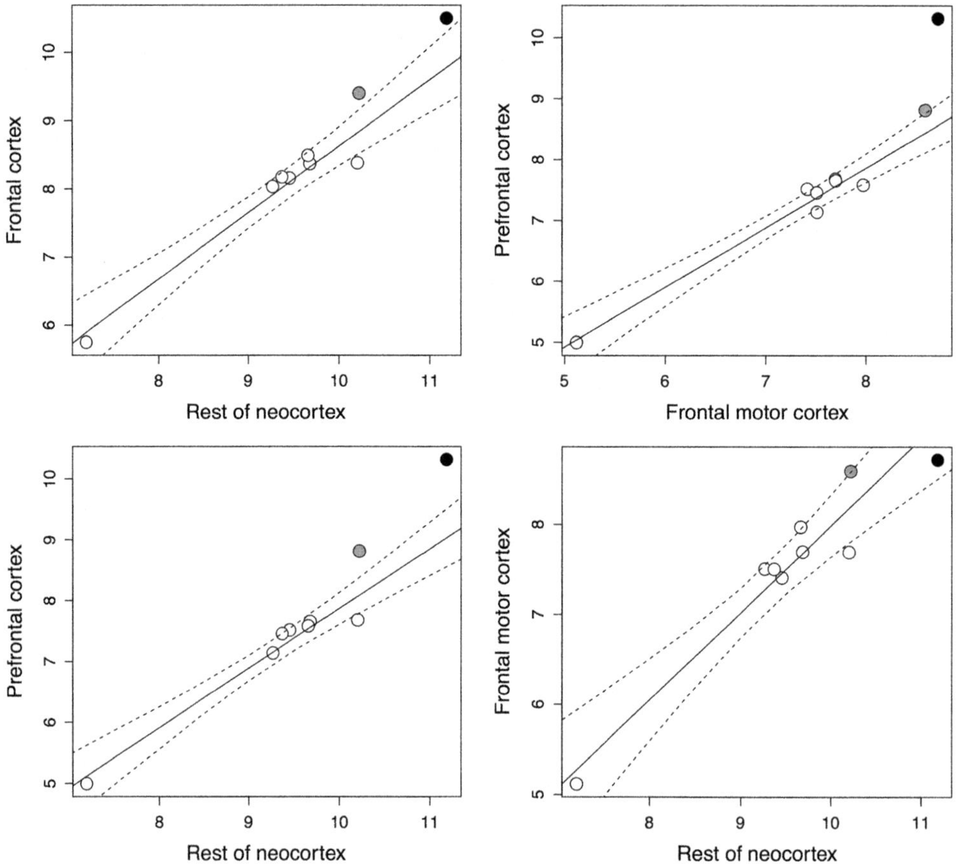

Fig. 14.3 Phylogenetic generalized least-squares analysis of the Brodmann data. Conventions as in Fig. 14.2

motion model) and calculated using a Bayesian MCMC procedure (with 10^6 iterations and a 20% burnin). The Brownian motion rate parameter is directly related to the amount of observed trait variation within a given time span. Traits with a higher rate of evolution are traditionally interpreted as being under a higher selective pressure. These results again confirm that comparative variation in frontal lobe volume is primarily a matter of variation in prefrontal volume.

Overall, phylogenetic comparative analysis of the frontal and prefrontal cortex thus demonstrates a grade shift within the frontal lobe toward more than predicted prefrontal cortex expansion. The traditional view that humans are an extension of the nonhuman primate allometric trend in terms of frontal (or prefrontal) evolution is not supported. Humans, great apes, and non-hominoid primates thus form three distinct grades in frontal lobe evolution (Passingham and Smaers 2014; Smaers et al. 2017). Evidence thus suggests that non-allometric expansion of the prefrontal cortex occurred at the dawn of great apes (~19–15 mya), such that selective pressures for higher cognitive functions underlie frontal lobe organization in both great apes and humans (Smaers et al. 2017). Exceptional expansion of the prefrontal cortex converges with functional data from cognitive neuroscience and primatology indicating that both great apes and humans are characterized by complex social cognition (Adolphs 2003, 2009) and the concomitant evolution of cultural traditions (van Schaik et al. 2003).

14.6 Discussion

Unpacking the biological basis of distinctly human cognitive and behavioral capacities is a major force of compelling scientific inquiry. Korbinian Brodmann's cytoarchitectonic maps of the cerebral cortex highlighted the significance of the "regio frontalis" in humans. His conclusion that the prefrontal cortex was especially enlarged in humans compared to nonhuman primates inspired over a century of neuroscientific research. Due to its fundamental role in human intelligence, it is no wonder that the investigation into the evolutionary history of the prefrontal cortex has involved a multidisciplinary approach. When insights from paleoneurology, comparative neuroanatomy, and phylogenetic comparative methods are taken in summation, a much more complete picture of the evolution of the frontal lobe, and the prefrontal cortex in particular, begins to emerge.

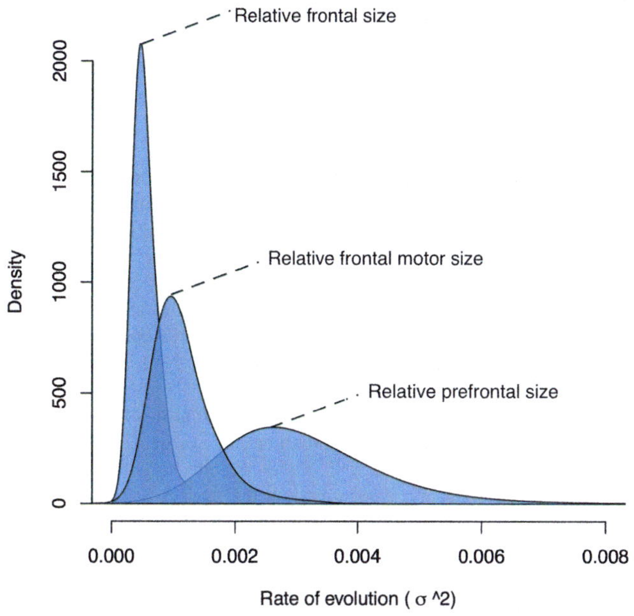

Fig. 14.4 Results from a rate analysis on the Smaers dataset. Rates are quantified as σ^2 (i.e., the Brownian motion rate parameter). Traits with a higher rate are commonly interpreted as being under a higher selective pressure. Rates are displayed for the expansion of the frontal cortex, prefrontal cortex, and frontal motor areas, each relative to the rest of the neocortex (defined as the neocortex minus the frontal cortex)

Paleoneurological studies provide the only direct evidence of changes in brain structure since our last common ancestors with chimpanzees. While evidence strongly suggests that there have been changes in the form and proportions of the anterior cranial fossa, it is unclear whether these structural differences reflect variations in underlying cortical organization. Modern human crania exhibit (a) enhanced curvature of the frontal bone and (b) a larger diameter of the anterior cranial fossa (i.e., "lateral widening"). it is not currently possible to determine whether or not these changes are driven specifically by the expansion of the frontal lobe. Moreover, internal changes in the relative sizes of structures within the frontal lobe are not revealed by gross changes in the form of the anterior cranial fossa. Thus, insights from other perspectives must be integrated in order to elucidate the nature of changes in underlying cortical organization. For example, genetic evidence suggests that the human prefrontal cortex underwent the majority of its exceptional expansion only since the emergence of anatomically modern humans (Shulha et al. 2012; Somel et al. 2014). More multidisciplinary evidence is needed to interpret structural changes in the anterior cranial fossa and whether this might reflect exceptional expansion of specific components within the frontal lobe since the last common ancestor of chimpanzees and humans.

Phylogenetic comparative studies provide a broader evolutionary picture, as they place the evolution of the hominin frontal lobe in a broader macroevolutionary context. Such studies analyze phylogenetic trees in association with data from many species in order to investigate the tempo, mode, and scaling patterns that underlie variation in the frontal lobe across a wide range of different species. Results of macroevolutionary analyses presented here demonstrate that the frontal cortex is expanded in comparison to the rest of the neocortex, but that this is due principally to the extraordinary expansion of the human prefrontal cortex. Comparisons of prefrontal and frontal motor areas confirm these results, demonstrating that the human prefrontal cortex has expanded dramatically relative to frontal motor areas. This is further confirmed by a rate analysis demonstrating that the rate of evolution of relative prefrontal cortex size is far higher than that in the relative sizes of either frontal cortex as a whole and frontal motor areas. These results also are in accordance with previous phylogenetic comparative analyses that demonstrate how humans, great apes, and other primates form three distinct grade shifts in prefrontal cortex expansion among primates that differ significantly from each other (Passingham and Smaers 2014; Smaers et al. 2017). Thus, phylogenetic comparative methods indicate that there has been significant reorganization within the frontal lobe throughout the course of human evolution that is driven primarily by prefrontal expansion.

Thus, multiple lines of evidence converge upon the observation that different areas of the human frontal lobe have evolved at different rates favoring prefrontal expansion. The different perspectives discussed here reveal unique pieces of this puzzle. Paleoneurology is unique in its ability to demonstrate structural changes of the anterior cranial fossa within the human lineage, comparative neuroanatomy provides a detailed picture of differences in cytoarchitecture and connectivity between humans and other primates, and phylogenetic comparative methods place these neuroanatomical differences in a larger macroevolutionary context. Together, these perspectives indicate that the human brain is more than a scaled-up version of the monkey brain and that the putative unique expansion of the "regio frontalis" is indeed an important feature that may support human's unique cognitive abilities. Future analyses would benefit from a continued effort to expand available datasets on cytoarchitectonically distinct areas of the cerebral cortex (Zilles et al. 2011). Furthermore, given the functional contribution of the frontoparietal (Genovesio et al. 2014; Caminiti et al. 2015) and cortico-cerebellar systems (Kelly and Strick 2003; Ramnani 2006; Koziol et al. 2014) and their contribution to explaining the evolution of brain organization in primates (Smaers et al. 2011b, 2013; Smaers and Soligo 2013; Smaers 2014), future research should also look to emphasize prefrontal connectivity and the targets of its various projections.

References

Adolphs R (2003) Cognitive neuroscience of human social behaviour. Nat Rev Neurosci 4:165–178

Adolphs R (2009) The social brain: neural basis of social knowledge. Annu Rev Psychol:693–716

Akert K (1964) In: Warren JM, Akert K (eds) The frontal granular cortex and behavior. McGraw-Hill, New York

Almécija S, Smaers JB, Jungers WL (2015) The evolution of human and ape hand proportions. Nat Commun 6:7717

Asplund CL, Todd JJ, Snyder AP, Marois R (2010) A central role for the lateral prefrontal cortex in goal-directed and stimulus-driven attention. Nat Neurosci 13:507–512

Avants BB, Schoenemann PT, Gee JC (2006) Lagrangian frame diffeomorphic image registration: morphometric comparison of human and chimpanzee cortex. Med Image Anal 10:397–412

Bailey P, Von Bonin G (1951) The isocortex of man. University of Illinois Press, Urbana

Bailey P, Von Bonin G, McCulloch W (1950) The isocortex of the chimpanzee. University of Illinois Press, Urbana

Bauernfeind AL, de Sousa AA, Avasthi T, Dobson SD, Raghanti MA, Lewandowski AH, Zilles K, Semendeferi K, Allman JM, Craig AD, Hof PR, Sherwood CC (2013) A volumetric comparison of the insular cortex and its subregions in primates. J Hum Evol 64: 263–279

Beck E (1949) A cytoarchitectural investigation into the boundaries of cortical areas 13 and 14 in the human brain. J Anat 83:147–157

Blinkov SM, Glezer II (1968) The human brain in figures and tables: a quantitative handbook. Basic Books, New York

Bookstein F, Schafer K, Prossinger H, Seidler H, Fieder M, Stringer C, Weber GW, Arsuaga JL, Slice DE, Rohlf FJ, Recheis W, Mariam AJ, Marcus LF (1999) Comparing frontal cranial profiles in archaic and modern homo by morphometric analysis. Anat Rec 257:217–224

Brodmann K (1909) Vergleichende Lokalisationslehre der Gro hirnrinde. Verlag von Ambrosius Barth, Leipzig

Brodmann K (1912) Neue ergebnisse uber die vergleichende histologische lokalisation der grosshirnrinde mit besondere berucksichtigung des stirnhirns. Anat Anzeiger 41:157–216

Bruner E, Holloway RL (2010) A bivariate approach to the widening of the frontal lobes in the genus homo. J Hum Evol 58:138–146

Bruner E, Manzi G (2005) CT-based description and phyletic evaluation of the archaic human calvarium from Ceprano, Italy. Anat Rec A: Discov Mol Cell Evol Biol 285:643–658

Bruner E, Athreya S, de la Cuétara JM, Marks T (2013) Geometric variation of the frontal squama in the genus homo: frontal bulging and the origin of modern human morphology. Am J Phys Anthropol 150:313–323

Bruner E (2017) The fossil evidence of human brain evolution. In: Kaas J (ed) Evolution of nervous systems, vol 4, 2nd edn. Elsevier, Oxford, pp 63–92

Buckner RL, Krienen FM (2013) The evolution of distributed association networks in the human brain. Trends Cogn Sci 17:648–665

Bucy PC (1937) A comparative cytoarchitectonic study of the motor and premotor areas in the primate cortex. J Nerv Ment Dis 85:343

Bush EC, Allman JM (2004) The scaling of frontal cortex in primates and carnivores. Proc Natl Acad Sci U S A 101:3962–3966

Caminiti R, Innocenti GM, Battaglia-Mayer A (2015) Organization and evolution of parieto-frontal processing streams in macaque monkeys and humans. Neurosci Biobehav Rev 56:73–96

Carlson KJ, Stout D, Jashashvili T, de Ruiter DJ, Tafforeau P, Carlson K, Berger LR (2011) The endocast of MH1, Australopithecus sediba. Science 333:1402–1407

Cavada C, Goldman-Rakic PS (1989) Posterior parietal cortex in rhesus monkey: I. Parcellation of areas based on distinctive limbic and sensory corticocortical connections. J Comp Neurol 287:393–421

Eichenbaum H, Clegg RA, Feeley A (1983) Reexamination of functional subdivisions of the rodent prefrontal cortex. Exp Neurol 79: 434–451

Enlow DH (1990) Facial growth, 3rd edn. W B Saunders, Philadelphia

Falk D (2012) Hominin paleoneurology: where are we now? In: Hofman MA, Falk D (eds) 1st edn. Elsevier B.V., Amsterdam

Falk D (2014) Interpreting sulci on hominin endocasts: old hypotheses and new findings. Front Hum Neurosci 8:134

Falk D, Zollikofer CPE, Morimoto N, Ponce de León MS (2012) Metopic suture of Taung (Australopithecus africanus) and its implications for hominin brain evolution. Proc Natl Acad Sci 109: 8467–8470

Finlay BL, Darlington RB (1995) Linked regularities in the development and evolution of mammalian brains. Science 268:1578–1584

Fleming SM, Weil RS, Nagy Z, Dolan RJ, Rees G (2010) Relating introspective accuracy to individual differences in brain structure. Science 329:1541–1543

Foville M (1864) L'anatomie de la physique et de la pathologie du syste' me nerveux ce re bro-spinal. Fortin, Masson et Companie, Paris

Fuster JM (2002) Frontal lobe and cognitive development. J Neurocytol 31:373–385

Fuster JM (2008) Anatomy of the Prefrontal Cortex. In: The Prefrontal Cortex, 4th edn. Academic, San Diego, pp 9–62

Genovesio A, Wise SP, Passingham RE (2014) Prefrontal-parietal function: from foraging to foresight. Trends Cogn Sci 18:72–81

Glasser MF, Van Essen DC (2011) Mapping human cortical areas in vivo based on myelin content as revealed by T1- and T2-weighted MRI. J Neurosci 31:11597–11616

Glasser MF, Goyal MS, Preuss TM, Raichle ME, Van Essen DC (2014) Trends and properties of human cerebral cortex: correlations with cortical myelin content. NeuroImage 93(Pt 2):165–175

Jackson HJ (1867) Remarks on the disorderly movements of chorea and convulsion, and on localisation. Med Times Gaz II:669–670

Kelly RM, Strick PL (2003) Cerebellar loops with motor cortex and prefrontal cortex of a nonhuman primate. J Neurosc: Off J Soc Neurosci 23:8432–8444

Khabbazian M, Kriebel R, Rohe K, Ane C (2016) Fast and accurate detection of evolutionary shifts in Ornstein-Uhlenbeck models. Methods Ecol Evol:811–824

Kolb B (1984) Functions of the frontal cortex of the rat: a comparative review. Brain Res 320:65–98

Koziol LF, Budding D, Andreasen N, D'Arrigo S, Bulgheroni S, Imamizu H, Ito M, Manto M, Marvel C, Parker K, Pezzulo G, Ramnani N, Riva D, Schmahmann J, Vandervert L, Yamazaki T (2014) Consensus paper: the Cerebellum's role in movement and cognition. Cerebellum 13:151–177

Lashley KS, Clark G (1946) The cytoarchitecture of the cerebral cortex of Ateles; a critical examination of architectonic studies. J Comp Neurol 85:223–305

Lieberman DE, Mcbratney BM, Krovitz G (2002) The evolution and development of cranial form in Homo sapiens. Proc Natl Acad Sci U S 99:1134–1139

MacLeod CE, Zilles K, Schleicher A, Rilling JK, Gibson KR (2003) Expansion of the neocerebellum in Hominoidea. J Hum Evol 44: 401–429

Mars RB, Jbabdi S, Sallet J, O'Reilly JX, Croxson PL, Olivier E, Noonan MP, Bergmann C, Mitchell AS, Baxter MG, Behrens TEJ, Johansen-Berg H, Tomassini V, Miller KL, Rushworth MFS (2011) Diffusion-weighted imaging Tractography-based Parcellation of the human parietal cortex and comparison with human and macaque resting-state functional connectivity. J Neurosci 31: 4087–4100

McBride T, Arnold SE, Gur RC (1999) A comparative volumetric analysis of the prefrontal cortex in human and baboon MRI. Brain Behav Evol 54:159–166
Neubauer S (2014) Endocasts: possibilities and limitations for the interpretation of human brain evolution. Brain Behav Evol 84:117–134
Ongür D, Price JL (2000) The organization of networks within the orbital and medial prefrontal cortex of rats, monkeys and humans. Cereb Cortex 10:206–219
Pandya DN, Seltzer B, Barbas H (1988) Input-output organization of the primate cerebral cortex. In: Steklis HD, Erwin J (eds) Comparative primate biology, vol 4. Alan R. Liss, New York, pp 39–80
Passingham RE (1973) Anatomical differences between the neocortex of man and other primates. Brain Behav Evol 7:337–359
Passingham RE (1975) Changes in the size and organisation of the brain in man and his ancestors. Brain Behav Evol 11:73–90
Passingham RE, Ettlinger G (1974) A comparison of cortical functions in man and the other primates. Int Rev Neurobiol 16:233–299
Passingham RE, Smaers JB (2014) Is the prefrontal cortex especially enlarged in the human brain? Allometric relations and remapping factors. Brain Behav Evol 84:156–166
Passingham RE, Wise SP (2012) The neurobiology of the prefrontal cortex. Oxford University Press, Oxford
Passingham RE, Smaers JB, Sherwood CC (2017) Evolutionary specializations of the human prefrontal cortex. In: Kaas JH (ed) Evolution of nervous systems, vol 4, 2nd edn. Elsevier, Oxford, pp 207–226
Petrides M (2005) Lateral prefrontal cortex: architectonic and functional organization. Philos Trans R Soc Lond 360:781–795
Petrides M, Pandya DN (1994) Comparative architectonic analysis of the human and the macaque frontal cortex. In: Boller F, Grafman J (eds) Handbook of neuropsychology. Elsevier, Amsterdam, pp 17–58
Preuss TM (1995) Do rats have prefrontal cortex? The rose-Woolsey-Akert program reconsidered. J Cogn Neurosci 7:1–24
Preuss TM (2004) What is it like to be human? In: Gazzaniga MS (ed) The cognitive neurosciences III, 3rd edn. MIT Press, Cambridge, pp 5–22
Preuss TM, Goldman-Rakic PS (1989) Connections of the ventral granular frontal cortex of macaques with perisylvian premotor and somatosensory areas: anatomical evidence for somatic representation in primate frontal association cortex. J Comp Neurol 282: 293–316
Preuss TM, Goldman-Rakic PS (1991) Myelo- and cytoarchitecture of the granular frontal cortex and surrounding regions in the strepsirhine primate Galago and the anthropoid primate Macaca. J Comp Neurol 310:429–474
Ramnani N (2006) The primate cortico-cerebellar system: anatomy and function. Nat Rev Neurosci 7:511–522
Rilling JK, Glasser MF, Preuss TM, Ma X, Zhao T, Hu X, Behrens TEJ (2008) The evolution of the arcuate fasciculus revealed with comparative DTI. Nat Neurosci 11:426–428
Rose JE, Woolsey CN (1948) The orbitofrontal cortex and its connections with the mediodorsal nucleus in rabbit, sheep and cat. Res Publ Assoc Res Nerv Ment Dis 27(1 vol):210–232
Rowe JB, Owen AM, Johnsrude IS, Passingham RE (2001) Imaging the mental components of a planning task. Neuropsychologia 39: 315–327
Sanides F (1964) The cyto-myeloarchitecture of the human frontal lobe and its relation to phylogenetic differentiation of the cerebral cortex. J Hirnforsch 7:269–282
Sanides F (1970) Functional architecture of motor and sensory cortices in primates in the light of a new concept of neocortex evolution. Primate Brain: Adv Primatol 1:137–201
Sayers K, Raghanti MA, Lovejoy CO (2012) Human evolution and the chimpanzee referential doctrine. Annu Rev Anthropol 41:119–138
Schoenemann PT, Sheehan MJ, Glotzer LD (2005) Prefrontal white matter volume is disproportionately larger in humans than in other primates. Nat Neurosci 8:242–252
Seltzer B, Pandya DN (1989) Frontal lobe connections of the superior temporal sulcus in the rhesus monkey. J Comp Neurol 281:97–113
Semendeferi K, Armstrong E, Schleicher A, Zilles K, Van Hoesen GW (1998) Limbic frontal cortex in hominoids: a comparative study of area 13. Am J Phys Anthropol 106:129–155
Semendeferi K, Armstrong E, Schleicher A, Zilles K, Van Hoesen GW (2001) Prefrontal cortex in humans and apes: a comparative study of area 10. Am J Phys Anthropol 114:224–241
Sherwood CC, Smaers JB (2013) What's the fuss over human frontal lobe evolution? Trends Cogn Sci 17:432–433
Sherwood CC, Holloway RL, Semendeferi K, Hof PR (2005a) Is prefrontal white matter enlargement a human evolutionary specialization? Nat Neurosci 8:537–538. author reply 538
Sherwood CC, Hof PR, Holloway RL, Semendeferi K, Gannon PJ, Frahm HD, Zilles K (2005b) Evolution of the brainstem orofacial motor system in primates: a comparative study of trigeminal, facial, and hypoglossal nuclei. J Hum Evol 48:45–84
Shulha HP, Crisci JL, Reshetov D, Tushir JS, Cheung I, Bharadwaj R, Chou HJ, Houston IB, Peter CJ, Mitchell AC, Yao WD, Myers RH, Fan CJ, Preuss TM, Rogaev EI, Jensen JD, Weng Z, Akbarian S (2012) Human-specific histone methylation signatures at transcription start sites in prefrontal neurons. PLoS Biol 10
Sira CS, Mateer CA (2014) Frontal lobes. In: Aminoff M, Daroff RB (eds) Encyclopedia of the neurological sciences, 2nd edn. Elsevier, San Diego, pp 358–365
Smaers JB (2014) Modeling the evolution of the cerebellum. From macroevolution to function. 1st edn. Elsevier B.V., Amsterdam
Smaers JB, Rohlf FJ (2016) Testing species' deviation from allometric predictions using the phylogenetic regression. Evolution 70: 1145–1149
Smaers JB, Soligo C (2013) Brain reorganization, not relative brain size, primarily characterizes anthropoid brain evolution. Proc R Soc B Biol Sci 280:20130269
Smaers JB, Schleicher A, Zilles K, Vinicius L (2010) Frontal white matter volume is associated with brain enlargement and higher structural connectivity in anthropoid primates. PLoS One:5
Smaers JB, Steele J, Case CR, Cowper A, Amunts K, Zilles K (2011a) Primate prefrontal cortex evolution: human brains are the extreme of a lateralized ape trend. Brain Behav Evol 77:67–78
Smaers JB, Steele J, Zilles K (2011b) Modeling the evolution of cortico-cerebellar systems in primates. Ann N Y Acad Sci 1225: 176–190
Smaers JB, Mulvaney PI, Soligo C, Zilles K, Amunts K (2012) Sexual dimorphism and laterality in the evolution of the primate prefrontal cortex. Brain Behav Evol 79:205–212
Smaers JB, Steele J, Case CR, Amunts K (2013) Laterality and the evolution of the prefronto-cerebellar system in anthropoids. Ann N Y Acad Sci 1288:59–69
Smaers JB, Mongle CS, Kandler A (2016) A multiple variance Brownian motion framework for estimating variable rates and inferring ancestral states. Biol J Linn Soc 118:78–94
Smaers JB, Gomez-Robles A, Parks AN, Sherwood CC (2017) Exceptional evolutionary expansion of prefrontal cortex in great apes and humans. Curr Biol 27:1–7
Somel M, Rohlfs R, Liu X (2014) Transcriptomic insights into human brain evolution: acceleration, neutrality, heterochrony. Curr Opin Genet Dev 29:110–119
Sylvester JB, Rich CA, YHE L, van Staaden MJ, Fraser GJ, Streelman JT (2010) Brain diversity evolves via differences in patterning. Proc Natl Acad Sci U S A 107:9718–9723
Tobias PV (1987) The brain of Homo habilis: a new level of organization in cerebral evolution. J Hum Evol:741–761

Van Essen DC, Dierker DL (2007) Surface-based and probabilistic atlases of primate cerebral cortex. Neuron 56:209–225

van Schaik CP, Ancrenaz M, Borgen G, Galdikas B, Knott CD, Singleton I, Suzuki A, Utami SS, Merrill M (2003) Orangutan cultures and the evolution of material culture. Science 299:102–105

Venditti C, Meade A, Pagel M (2011) Multiple routes to mammalian diversity. Nature 479:393–396

Von Bonin G (1948) The frontal lobe of primates; cytoarchitectural studies. Res Publ Assoc Res Nerv Ment Dis 27(1 vol.):67–83

von Economo C, Koskinas GN (1925) Die Cytoarchitektonik der Hirnrinde des Erwachsenen Menschen: Textband und Atlas mit 112 Mikrophotographischen Tafeln. Springer, Vienna

Walker AE (1940) A cytoarchitectural study of the prefrontal area of the macaque monkey. J Comp Neurol 73:59–86

Zilles K, Amunts K, Smaers JB (2011) Three brain collections for comparative neuroanatomy and neuroimaging. In: Johnson JI, Zeigler HP, Hof PR (eds) Resources and technological advances for studies of neurobehavioral evolution. Ann N Y Acad Sci, New York, pp E94–104

The Evolution of the Parietal Lobes in the Genus *Homo*

15

Emiliano Bruner, Hideki Amano, Ana Sofia Pereira-Pedro, and Naomichi Ogihara

Abstract

The parietal areas have always been of major interest in paleoneurology because of their remarkable variation among hominids. In neuroanatomy, difficulties in defining their blurred boundaries and their complex functions have delayed a proper quantitative study of their organization and evolution. Paleoneurological evidence indicates the upper parietal cortex, including its deep medial folds, as a probable area of evolutionary change. In modern humans, the intraparietal sulcus shows species-specific features when compared with other primates. The size and proportions of the precuneus represent a determining factor of variability among adults, and a major difference between human and chimpanzee midsagittal brain morphology. This medial element is a relevant connectivity hub of the whole brain, is central for the frontoparietal system, and has an important role in the default mode network. When compared with extinct human species, modern humans display a marked enlargement of the parietal bone and of the parietal lobes, inducing the longitudinal bulging of the whole parietal surface. This morphological change is very similar to the pattern associated, among living humans and between humans and apes, with the size variation of the precuneus. It remains to be understood to what extent such evolutionary variations are due to genetic selection or to environmental and physiological factors. These cortical areas are involved in many complex cognitive functions, but most of all they are central for visuospatial integration, coordinating body management, the eye-hand system, the interaction between body and environment, and the integration between body and inner cognitive levels including self-awareness, egocentric memory, social perception, and mental imagery.

Keywords

Paleoneurology • Parietal cortex • Precuneus • Visuospatial integration

15.1 Introduction

The parietal areas of the human brain were long neglected in neuroanatomy, at least when compared with other districts like the frontal or the temporal ones. There may be at least three main reasons for such a delay in the attention toward the parietal cortex. First, its boundaries are blurred and not strictly defined. In general, lobes are but a matter of conventional definition and not necessarily real functional or evolutionary units. Lobes are formed by elements which are

E. Bruner (✉) • A.S. Pereira-Pedro
Programa de Paleobiología, Centro Nacional de Investigación sobre la Evolución Humana, Paseo Sierra de Atapuerca 3, 09002 Burgos, Spain
e-mail: emiliano.bruner@cenieh.es; sofia.aspp@gmail.com

H. Amano • N. Ogihara
Department of Mechanical Engineering, Faculty of Science and Technology, Keio University, Yokohama 223-8522, Japan
e-mail: hideki_amano_0307@yahoo.co.jp; ogihara@mech.keio.ac.jp

E. Bruner et al. (eds.), *Digital Endocasts*, Replacement of Neanderthals by Modern Humans Series,
DOI 10.1007/978-4-431-56582-6_15

heterogeneous in terms of structure and functions, the association between anatomical surfaces and cytoarchitectural areas is not strictly consistent, the boundaries among the different components are blurred, and individual variability may be considerable. Most quantitative neuroanatomy has been developed on volume calculations and statistics, and for many areas different results due to different definitions or operational decisions have led to disagreements and open debates (see Allen et al. 2002; Sherwood and Smaers 2013). These limits are similar for all the brain districts, but for the parietal lobes things were even more difficult, and the few available evolutionary comparisons were computed on the parieto-occipital block because of the technical difficulties in isolating the parietal cortex with reliable homologous limits (Semendeferi and Damasio 2000). A second problem associated with the parietal lobes regards their importance as an integrative center. Apart from their crucial role in some specific cognitive tasks, the parietal areas integrate many different districts and processes. Therefore, they are involved in many cognitive procedures, which are often even difficult to categorize in terms of specific behavioral abilities. Because of such difficulties when dealing with cognitive labels, they have been tagged ever since with the general expression "associative cortex." Of course, the whole cortex is associative, and the term probably does not help that much. A third limitation concerns the spatial location of major parietal elements, hidden in the depth of the brain volume. The precuneus and the intraparietal sulcus represent a large percentage of the parietal cortex, and they are positioned in the core of the cerebral space. Many functional studies in neuroanatomy have been based on cortical damage (like strokes or accidents), associating damaged areas and impaired functions. The position of the deep parietal cortex is generally less sensitive to observable functional impairment, because damage in those areas is less compatible with the survival of the individual. It must be also added that, until recently, the main source of neuroanatomical information was studies on cadavers, generally limited by small samples and by the alterations associated with the study of an anatomical system out of its functional (living) conditions. Also in this case, deep elements were more complicated to analyze than outer areas.

Despite all these difficulties, many pioneering studies in neurobiology were available at the end of the past century, and there was plenty of evidence on the functional relevance of the parietal cortex (Mountcastle 1995). Biomedical imaging and digital analysis induced a step forward in functional and evolutionary neuroanatomy, improving the size of the samples and the anatomical resolution and most of all making it possible to analyze brain morphology in living individuals (Rilling 2008; Preuss 2011). After advances in brain mapping based on cytoarchitecture, connectivity, and neurotransmitters, the basic anatomical parcellation of the parietal cortex proposed by Brodmann in 1914 has now been replaced with a complex system of areas and subareas (Caspers et al. 2006, 2011; Scheperjans et al. 2008a, b; Mars et al. 2011). Such microanatomical diversity makes subtle differences between species very difficult to identify, because of the small-scale functional factors and because of uncertainties about the homology in different taxa.

15.2 A Glimpse Inside the Parietal Lobes

15.2.1 General Morphology

Despite the complex parcellation, in terms of gross anatomy what we conventionally call "parietal lobe" can be divided into four major parts: the anterior district (postcentral gyrus), the superior lobule, the inferior lobule, and the deep medial cortex (Figs. 15.1 and 15.2) (see Wild et al. 2017). The anterior district is limited to the postcentral gyrus, which is involved in somatosensory integration. The rest of the parietal lobe is dedicated to the integration of multiple processes, sometimes being labeled as a whole with the term "posterior parietal cortex." According to Allen et al. (2002), human parietal lobe volume is on average 137 cc for males and 120 cc for females. This difference (female value is 88% of the male figure) is statistically significant and in agreement with the mean difference for the rest of the brain (85–90%). Women generally display larger gray matter/white matter ratios for the parietal areas, suggesting a lesser degree of connectivity (Im et al. 2006; Luders et al. 2006; Salinas et al. 2012). Parietal lobes account for 25% of the hemispheric

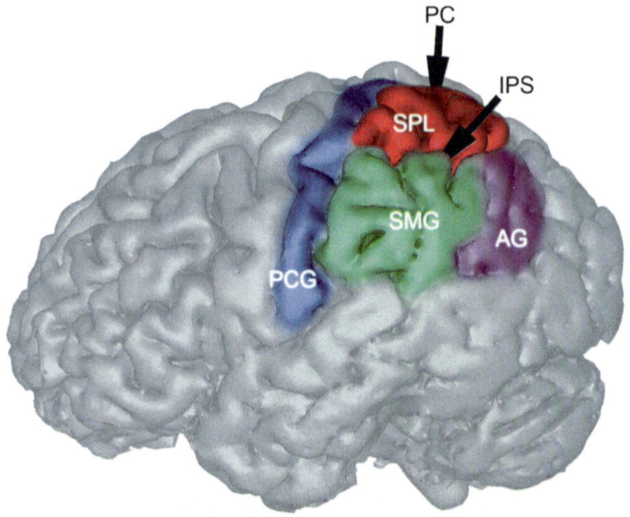

Fig. 15.1 Main regions of the parietal lobe: postcentral gyrus (PCG), supramarginal gyrus (SMG), angular gyrus (AG), superior parietal lobule (SPL). The arrows show the hidden positions of the precuneus (PC) and intraparietal sulcus (IPS)

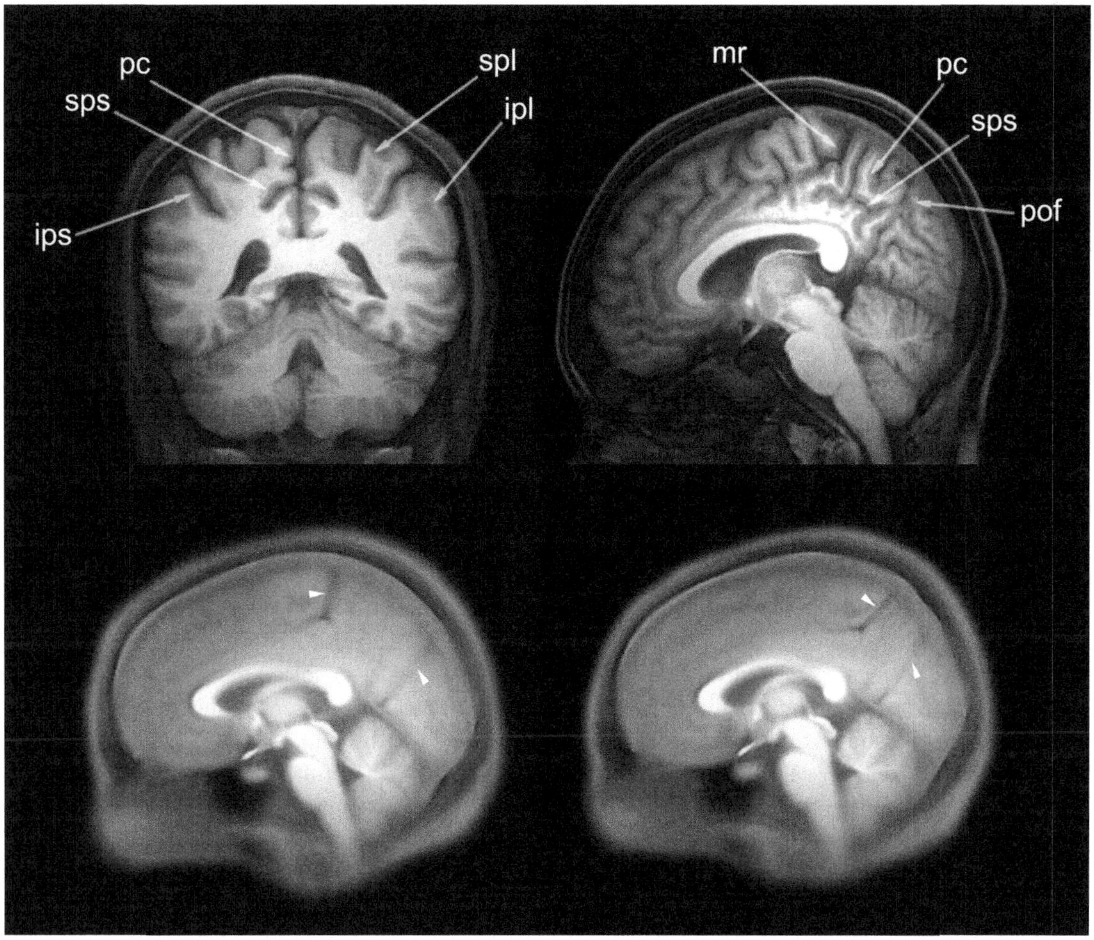

Fig. 15.2 *Above*: coronal and sagittal section of an adult human brain showing the main parietal elements, *ipl* inferior parietal lobule, *ips* intraparietal sulcus, *mr* marginal ramus of the cingulate sulcus, *pc* precuneus, *pof* parieto-occipital fissure, *spl* superior parietal lobule, *sps* subparietal sulcus. *Below*: superimposed MR images showing the main morphological pattern of variation in adult humans, associated with the longitudinal size of the precuneus (Modified from Bruner et al. 2014a)

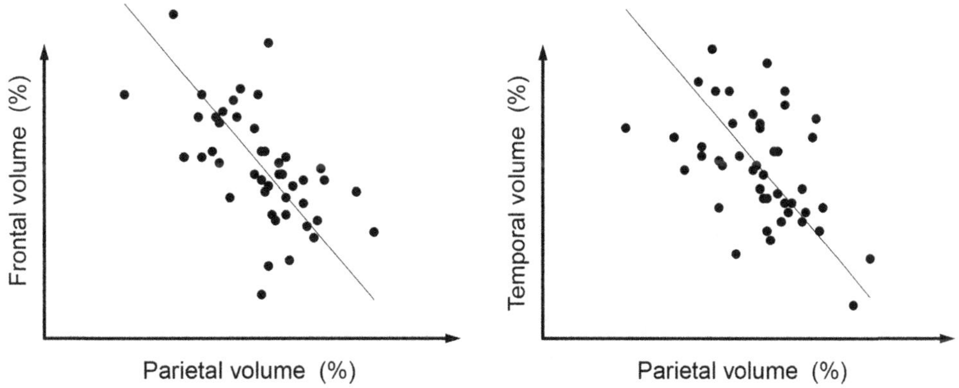

Fig. 15.3 In adult modern humans, the relative volume of the parietal lobe is inversely correlated with the relative volume of the frontal ($R = -0.69$) and temporal ($R = -0.49$) lobes (Original data from Allen et al. 2002)

volume, in both males and females. The lower areas are generally more pronounced on the left side, while the upper areas are larger on the right hemisphere. Nonetheless, quantitative evidence is not consistent: males do not show noticeable asymmetries, while in females the right side may be slightly larger. The relative volume of the parietal lobe is inversely correlated with the volume of the frontal lobes: the larger the former, the smaller the latter (Fig. 15.3). An inverse correlation can be also seen between the parietal and temporal lobes.

The upper parietal areas are more homogeneous than the lower ones and fade more gradually into the occipital cortex (Eidelberg and Galaburda 1984). These areas are associated with macroanatomical changes during human evolution and therefore will be described in more detail in this review. The precuneus is the medial surface of the upper parietal lobule, facing the midsagittal plane and in contact with the *falx cerebri*, a connective sheet which is formed by the invagination of the meninges between the two hemispheres. The precuneus displays a remarkable individual variability in terms of extension and cortical surface area, with no robust degree of asymmetry (Bruner e al. 2014a, 2015a, 2017a). It is separated from the paracentral lobule (somatosensory cortex) by the marginal ramus of the cingulate sulcus, and from the occipital lobule (visual cortex) by the parieto-occipital fissure. In the lower area, the precuneus fades into the posterior cingulate cortex. Although the subparietal sulcus is generally used as a macroanatomical reference to mark this separation, the sulcal pattern displays an extraordinary variability, hampering the recognition of constant morphological patterns (Kacar et al. 2015; Pereira-Pedro and Bruner 2016). Additional folding is slightly associated with larger precuneus dimensions (Bruner et al. 2017a). On the coronal plane, the height of the precuneus generally ranges between 33 and 39 mm in adult humans and has a visible influence on the vault outline (Pereira-Pedro and Bruner 2016; Fig. 15.4). The width and morphology of the subparietal sulcus can vary substantially, generally spanning 13–15 mm, but its dimensions do not seem to affect the outer parietal morphology.

The second deep parietal region is the intraparietal sulcus, a large cortical fold separating the upper and lower parietal lobules. It has a complex cytoarchitecture and it is formed by distinct areas (Ebeling and Steinmetz 1995; Choi et al. 2006). The intraparietal sulcus in humans is larger and far more complex than in nonhuman primates, with several branches (Grefkes and Fink 2005). The homology of these areas with nonhuman primates is still debated, but specific human features and functions have been described associated with this cortical region (Eidelberg and Galaburda 1984; Posner et al. 1984; Vanduffel et al. 2002; Orban et al. 2006). Also in this case, the sulcal scheme displays remarkable individual variation (Choi et al. 2006).

The inferior lobules are formed by the supramarginal gyrus and by the angular gyrus, largely outfolded on the brain surface. Their gross morphology is more homogeneous among individuals, although with some minor differences of the sulcal pattern. The supramarginal gyrus bridges the postcentral gyrus with the first temporal circumvolutions, and the angular gyrus separates the supramarginal gyrus from the occipital lobe.

Despite the macroanatomical differences of the parietal lobes between humans and nonhuman primates, there is a large correspondence between their respective neuroanatomical organization, suggesting conservative evolutionary schemes (Culham and Kanwisher 2001; Rushworth et al. 2001; Sereno et al. 2001; Caminiti et al. 2015). Apart from their direct contact with the somatosensory and visual areas, the parietal lobules are well connected with the cerebellum and with the temporal lobes, but the main connections are with the prefrontal cortex (Wise et al. 1997; Caminiti et al. 2015). The inferior lobules are connected with the dorsolateral prefrontal areas and the superior lobules with the dorsomedial prefrontal areas.

15.2.2 Functions

Each part of the parietal lobe is actually formed by several distinct areas (Caspers et al. 2006, 2011; Scheperjans et al. 2008a, b; Mars et al. 2011). The homology of these areas among primates is not fully understood. They are all crucial nodes for many different cognitive functions. The complex scheme of parcellation, and the fact that those functions are involved in different and multiple processes, often makes any strict association between areas and specific cognitive processes too general and misleading.

The postcentral gyrus is mainly involved in somatosensory processes, namely, integrating stimuli (in particular tactile ones) from the whole body (Ackerley and Kavounoudias 2015). The rest of the parietal lobe (the "posterior parietal cortex") has generally been divided into superior and inferior lobules. In the inferior areas, the supramarginal gyrus is largely involved in language comprehension (Deschamps et al. 2014), as well as in attention and sensory integration (Nejad et al. 2015), and abstract categorization (Leshinskaya and Caramazza 2016). The angular gyrus is often associated with calculation capacity, although it is also involved in attention and decision-making processes (Seghier 2013; Studer et al. 2014). Apart from the angular gyrus, numerical cognition is also associated with the intraparietal sulcus (Cantlon et al. 2006; Ansari 2008).

The superior areas are a bridge between the body as represented by the somatosensory areas and the world outside as represented by the visual models coded at the occipital lobe (see Bruner 2010a for a review). In this sense, the parietal areas integrate the self and the environment (Sakata et al. 1997). Visuospatial integration deals with the relationship between body and space, proprioception, handling, and eye-hand coordination. Reaching and grasping are particularly associated with the intraparietal areas (Battaglia-Mayer et al. 2000, 2003) and directly involved in toolmaking and tool use (Stout and Chaminade 2007; Stout et al. 2000). Such spatial management is not strictly a mechanical issue

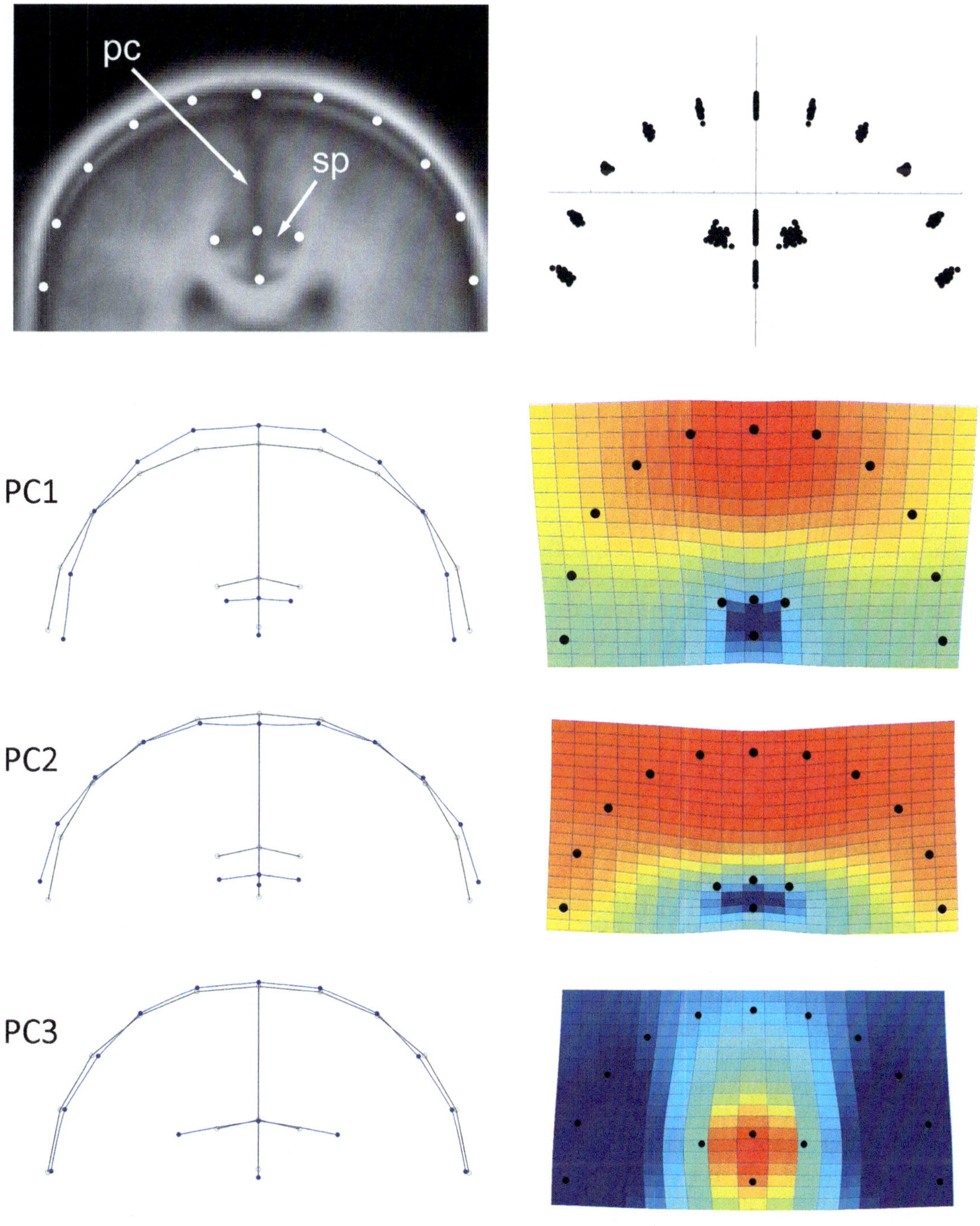

Fig. 15.4 In coronal section, the precuneus (pc) appears like a vertical structure with a main lateral fold, the subparietal sulcus (sp). The area below this sulcus is generally included in the posterior cingulate cortex. In this image, the base of the medial parietal area and the margin of the subparietal sulcus are landmarked along with the midsagittal point and 10 equally distant semilandmarks from the midsagittal plane to the lateral outer parietal profile in 50 adult humans from the Oasis sample (Marcus et al. 2007). Landmarks were registered and symmetrized by Procrustes superimposition (Bookstein 1991; *top right*) and a principal component analysis of the residual shape variables computed with MorphoJ 1.06a (Klingenberg 2011) and PAST 2.17c (Hammer et al. 2001). Shape variation is structured on three components (*left*, wireframes; *right*, thin-plate spline deformation grids and expansion maps; *red*, dilation; *blue*, contraction). PC1 (39%) is associated with vertical stretching of the parietal outline through an increase of the precuneal height. PC2 (27%) is associated with vertical changes of precuneal and posterior cingulate reciprocal proportions. PC3 (15%) is associated with the width of the subparietal sulcus (Data from Pereira-Pedro and Bruner 2016)

between body and space, but it is intermingled with awareness, abstract representations, attention, intention, and simulation (Andersen et al. 1997; Rushworth et al. 2001; Yantis et al. 2002; Andersen and Buneo 2002; Wardak et al. 2005; Tunik et al. 2007). Even language is perceived in terms of activation of the corresponding neural areas involved in the body actions associated with the meaning of the contents (Hauk et al. 2004; Buccino et al. 2005; Klepp et al. 2015).

The precuneus is a critical bridge between body and vision and between visuospatial integration and egocentric memory, being also crucial in attention, self-awareness, and self-centered mental imagery (Fletcher et al. 1995; Lundstrom et al. 2005; Cavanna and Trimble 2006; Margulies et al. 2009; Zhang and Li 2012). It is essential to coordinate internal representations integrating egocentric memory with information on spatial, chronological, and social perceptions (Land 2014; Peer et al. 2015). It shows, mostly in association with the posterior cingulate cortex, some correlations with psychometric scores for intelligence (Basten et al. 2015). The fact that such correlations are both positive and negative suggests that its subdivisions may have different roles in this sense.

Although the parietal areas are crucial for all those specific domains, the strong reciprocal connections with the frontal lobes suggest that the actual functional network is represented by the frontoparietal system (Ferraina et al. 1997; Jung and Haier 2007; Caminiti et al. 2015). This is particularly relevant when dealing with complex cognitive processes like those involved in working memory and integrating central executive functions (prefrontal areas) with a phonological loop (lower parietal areas) and with a visuospatial sketchpad (upper parietal areas) (Chafee and Goldman-Rakic 1998; Coolidge and Wynn 2005; Berryhill et al. 2011).

15.2.3 Parietal Lobes and Parietal Bones

The parietal lobes accurately shape the parietal bones during ontogeny and evolution (Moss and Young 1960; Enlow 1990). The parietal bones are formed by direct intramembranous ossification of mesodermal tissues, after interaction with the meninges derived from the neural crest (Jiang et al. 2002; Morriss-Kay and Wilkie 2005). On the endocranial surface, the parietal bones of adult modern humans range between approximately 90 and 120 mm for the sagittal length and between 110 and 150 mm for the maximum biparietal breadth (Bruner et al. 2011). The boundaries and specific areas of the bone and brain counterparts do not show a fixed spatial relationship, although parietal lobe size shows a correlation with parietal bone size (Bruner et al. 2015b). In general, the larger the parietal lobe, the more it approaches the frontal bone and the coronal suture. This suggests that the spatial position of bones and lobes is only partially correlated and there are further factors influencing their respective geometrical organization. Nonetheless, the curvature of the bone is molded directly as a consequence of the underlying brain growth forces, and the endocranial surface is able to reveal sulcal and vascular patterns associated with the parietal cortex (Holloway et al. 2004; Kobayashi et al. 2014). Of course, there is a consistent loss of information when moving from brain anatomy to endocranial morphology, and interpretation of the endocranial surface requires additional experience and expertise (Bruner 2015). In general, it must be considered that interpreting "brain geography" on the endocast surface is more a matter of global considerations than of specific features. The probable position of a specific cortical trait is not due only to the possible expression (visibility) of that character but also to the relative position of the surrounding elements. For example, regarding the never-ending debate on the position of the lunate sulcus in australopiths, Ralph Holloway remarked that although its location may be difficult to establish, this sulcus must necessarily be positioned behind the intraparietal groove. This additional information sets an anterior limit for its position, even in those situations in which a detailed localization cannot be properly identified.

Needless to say, in paleoneurology, cortical features can be only identified as "probable areas." On the one hand, this limit must be seriously considered when making inferences in evolutionary neuroanatomy. At the same time, this information can be useful anyway, and, being the only real evidence associated with brain morphology in extinct species, its unconditioned rejection would be imprudent.

Figure 15.5 shows some anatomical references that can be localized on the endocranial parietal surface. Cranial landmarks (endobregma and endolambda) can be easily localized by following the imprints of the sutures. The central sulcus can be localized by following the precentral and postcentral gyri. The former can be localized as originating

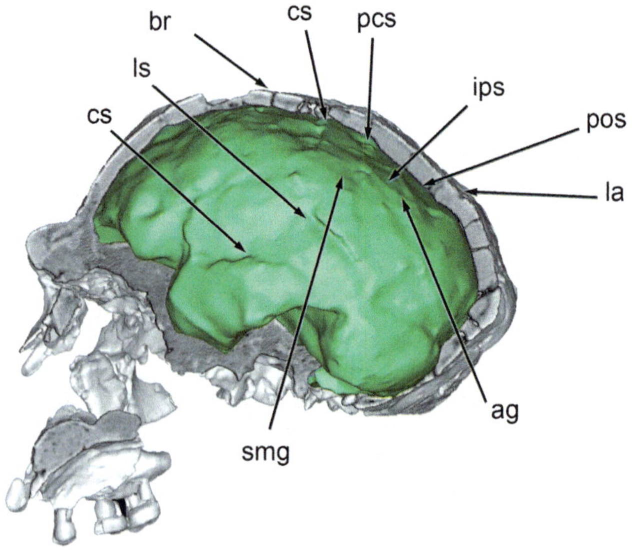

Fig. 15.5 Approximate position of some major parietal landmarks on the digital endocasts of KNM-ER3733 (*Homo ergaster*): angular gyrus (*ag*), bregma (*br*), lower and upper extremes of the central sulcus (*cs*), intraparietal sulcus (*ips*), lambda (*la*); lateral sulcus (*ls*), parieto-occipital sulcus (*pos*), postcentral sulcus (*pcs*), supramarginal gyrus (*smg*)

behind Broca's cap and the latter in front of the intraparietal groove. The lateral sulcus is smooth but generally well visible in most endocasts. The supramarginal and angular gyri can be identified as "bosses" (smooth bulging surfaces, with a recognizable local curvature), the former at the end of the lateral sulcus and the latter anterior to the occipital boundary. The intraparietal sulcus is a very smooth and shallow groove separating the lower bosses and the upper areas. The parieto-occipital boundary is difficult to localize because it corresponds to a region of variable morphology. It is generally positioned more superiorly to lambda, although it can sometimes shift underneath the occipital bone. Once more, it must be stressed that the positions of all these features are better localized by considering all of them at the same time, namely, by taking into account their reciprocal spatial positions. A proper knowledge of the anatomical variations of the sulcal morphology is mandatory. Naturally there is nothing new here: experience, as always, is an issue.

Beyond the cortical pattern, it is worth noting that the parietal bone is the most informative cranial element on the middle meningeal vessels (Bruner and Sherkat 2008) and on the diploic vessels (Rangel de Lázaro et al. 2016). In the case of the middle meningeal vessels, the traces of this vascular system are directly visible on the endocranial surface, and, although correspondence between vessels and imprints is not complete, they generally represent a good proxy for the morphology and distribution of the main vascular branches. The anterior ramus often runs parallel to or within the central sulcus. The middle ramus covers the parietal areas.

15.3 The Paleoneurological Evidence

15.3.1 Human Fossils and Parietal Areas

Paleoanthropologists have directed their attention toward the parietal areas since the early days of the discipline. Back in the early twentieth century, Raymond Dart hypothesized a parietal expansion as the main character separating australopiths from living apes (1925). Franz Weidenreich pointed to the parietal lobes, instead of the frontal areas, as a more apparent source of evolutionary change in the human genus (1936). In 1981, Ralph Holloway, through a pioneering geometric approach to endocast morphology, showed that the parietal surface is a crucial area of variation in humans and apes (see Chap. 9).

The lower parietal gyri, and their association with the upper temporal areas, have generally been the object of functional studies largely because of their involvement in speech comprehension. However, at least taking into account their gross morphology, the fossil record has not provided much useful information on their possible evolutionary changes. The sulcal pattern of those regions is similar in all extinct and extant human groups. An increase of their proportions can be hypothesized in early humans on the basis of a general description (Holloway 1995; Tobias 1995), but the available small and fragmentary sample does not allow any consistent quantitative comparisons. By contrast, the upper parietal districts have proven to be essential when dealing with the morphological differences among different human species (Fig. 15.6).

As soon as spatial analyses were applied to cranial variation in the human genus, the size and shape of the parietal bone were identified as the main factors associated with the modern human globular braincase (Bruner et al. 2004). A multivariate analysis of endocranial chords based on cortical landmarks showed that non-modern humans (namely, all the extinct species belonging to the genus *Homo*) follow a shared allometric trajectory in which brain form differences are largely due to size variation, while modern humans are distinct mainly because of an increase of their parietal diameters (Bruner et al. 2003). Apart from size variations, at present there is no recognized difference in the morphology of the parietal surface among small-brained hominid species, namely, *H. ergaster*, *H. erectus*, and *H. heidelbergensis*. All these taxa display a classic parasagittal depression of the upper parietal lobules, which is supposed to be a plesiomorph condition of the human genus. In Neandertals there is a lateral expansion of the dorsal parietal surface which is absent in their ancestors, and the upper parietal lobules are laterally bulging instead of depressed. In modern humans, this same enlargement is associated with a further increase of the whole upper parietal district, due to a longitudinal expansion of the parietal lobe and recognizable as an evident bulging of the dorsal parietal surface. Regarding this same character, non-modern humans may actually show an inverse tendency: the larger the brain, the shorter the parietal length, in relative terms (Bruner 2004). It has been hypothesized that such an allometric pattern in extinct human species can be related to the position of the parietal lobes, constrained between the frontal and occipital areas, and by geometric limits imposed by the growth of the endocranial cavity. Interestingly, in Neandertals, which represent the extreme degree of such parietal longitudinal compression, the ectocranial surface generally shows supernumerary ossicles, namely, additional ossification centers suggesting some difficulties in the coordination of size and shape changes during morphogenesis (Manzi et al. 1996; Bruner 2014). Parietal differences between modern humans and Neandertals are not only a matter of proportions but also of absolute size, with the former showing larger parietal dimensions (Bruner 2008). The same results were subsequently obtained analyzing the parietal bone instead of the parietal lobe: modern humans are characterized by a species-

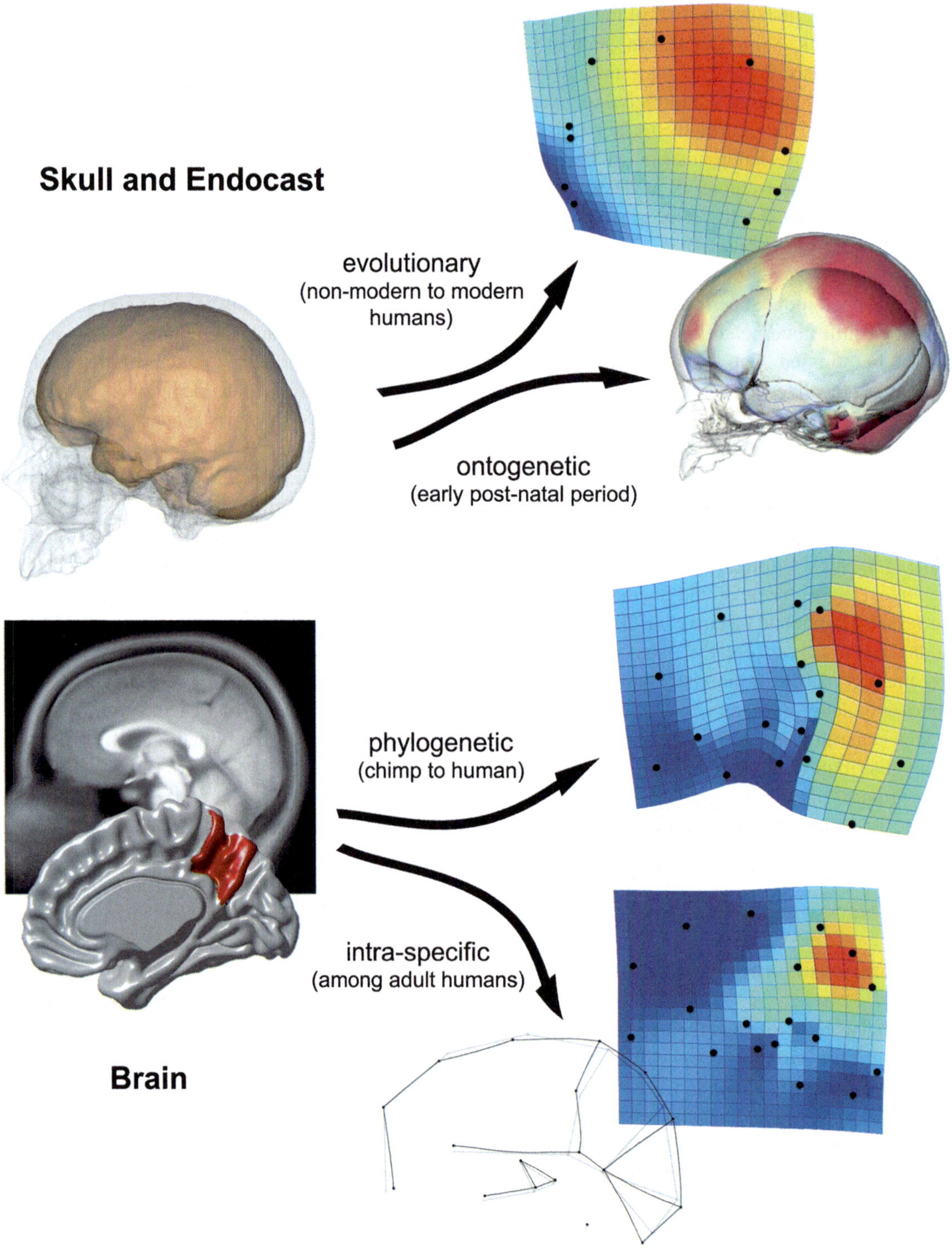

Fig. 15.6 Major shape variation associated with modern human skull (*above*) and brain (*below*). In evolutionary terms, the modern human midsagittal cranial profile is characterized by facial reduction and parietal bulging. In terms of ontogeny, parietal bulging of the braincase is associated with an early postnatal stage specific of our species. The main difference in the midsagittal brain morphology between humans and chimpanzees is an increase of the precuneus dimension. In modern adult individuals, the midsagittal brain shape is mainly influenced by the expansion/reduction of the precuneus. All these spatial changes display very similar spatial patterns (deformation maps: expansion in *red*) (Images modified from Bruner et al. 2004, 2014a, 2016, 2017b and from Neubauer et al. 2009; Gunz et al. 2010, courtesy of Simon Neubauer and Philipp Gunz)

specific lengthening of the parietal bones and increase of their curvature (Bruner et al. 2011).

As previously mentioned, there is a direct relationship between parietal bones and lobes during morphogenesis (Moss and Young 1960), and morphological integration in the brain and in the skull is mostly based on local factors and physical proximity (Bruner and Ripani 2008; Bruner et al. 2010; Gómez-Robles et al. 2013). This suggests that the evolutionary changes at the parietal bone are likely to be due to evolutionary changes in parietal lobe volume, and not to long-range secondary effects associated with other districts (such as variations of the cranial base or the facial block). Most importantly, cranial variation could partially explain parietal bone curvature, but not parietal lobe enlargement. In this second case, volumetric changes are necessarily associated with a variation of specific histological components.

15.3.2 Improving the Database

Recently, computational techniques have been used to perform enhanced digital reconstructions of the skulls and endocasts of several key specimens (e.g., Kondo et al. 2014; Amano et al. 2015). We computed a multivariate analysis of the same sample and variables analyzed in Bruner et al. (2003), adding seven individuals (Fig. 15.7). Mladech 1 and Cro-Magnon 1 represent robust Upper Pleistocene modern humans. Qafzeh 9 and Skhul 5 are essential for investigating early modern origins, because they belong to anatomically modern populations with Mousterian industry. La Chapelle-aux-Saints is a classic Neandertal. The former analysis of this specimen was computed by using the original physical endocast (Grimaud-Hervé 1997). Gibraltar 1 Forbes Quarry Neandertal is interesting because of its morphological similarity to the Italian Neandertal Saccopastore 1. Finally, Amud is extremely relevant because of its large cranial capacity, representing an extreme morphology associated with the Neandertal allometric variation. Figure 15.8 shows the results of the principal component analysis (PCA) computed on the correlation matrix. The addition of these new specimens does not alter the multivariate space, and all the patterns previously described can be confirmed. The first component (71% of the variance) is size related and associated with an increase in all variables. The second component (16%) contrasts the increase of the parietal dimensions (parietal lobe chord and vault heights) against the increase of the rest of the variables. Hemispheric length is confirmed to be a good proxy for size, being almost parallel to PC1 (Bruner 2010b). The successive components are below the standard statistical thresholds of stability. Within the bidimensional space associated with those two multivariate patterns, non-modern humans are positioned along a shared trajectory associated with increasing widths, frontal length, and occipital length. Two specimens from Sima de los Huesos, generally included in the hypodigm of *H. heidelbergensis*, are positioned within the archaic human (SH5) and Neandertal (SH4) ranges of variation, because of their size differences. As previously mentioned, at present no specific endocranial differences are known among archaic humans (namely, *H. ergaster*, *H. erectus*, and *H. heidelbergensis*), except those associated with brain size (Bruner et al. 2015b). Gibraltar 1, a small Neandertal, is extremely similar to Saccopastore 1 also in its endocranial proportions. Apart from this allometric variation, it must be remarked that these smaller early Neandertals already display the lateral parietal bulging typical of this group (Bruner and Manzi 2008) and wider frontal lobes (Bruner and Holloway 2010). The morphology of La Chapelle-aux-Saints is positioned centrally to the Neandertal range. Amud, as expected, lies at the end of the non-modern allometric trajectory. The two early modern humans, Skhul 5 and Qafzeh 9, approach the modern human variation, with the latter halfway between the modern and non-modern group. Mladech 1 also bridges the two distributions. Cro-Magnon 1 lies in the center of the modern variation.

This additional analysis adds to our understanding of later hominid evolution in several ways. It confirms these two main patterns of variation after an increase of 30% of the sample size, and it corroborates the non-modern allometric trajectory. It also confirms the similarity (in this case endocranial) between Saccopastore 1 and Gibraltar 1, which may be the result of shared demographic processes (Bruner and Manzi 2006). It fully acknowledges the expectation of Amud brain morphology as an extreme of the Neandertal allometric variation. It also corroborates the separation of modern humans because of their parietal proportions, even though in this case the intermediate position of Qafzeh 9 and Mladech 1 deserves attention. Mladech 1 is a robust Upper Pleistocene specimen, with a very marked occipital bulging. There is disagreement concerning the possible influence of interbreeding between modern and non-modern populations in Europe as a way to explain the general robusticity of the former. Nonetheless, in this case the extreme occipital bulging may introduce a metric bias, excessively displacing the height of the posterior vault beyond the parietal boundary. In the case of Qafzeh 9, the issue may be more interesting. It belongs to early modern human populations of the Near East, which were in geographical proximity with Neandertal groups, and used tools which were pretty similar to the Neandertal ones. Therefore, beyond the possibility of individual variation or of uncertainty associated with endocast reconstruction or metrics, in this case the morphological proximity with non-modern humans

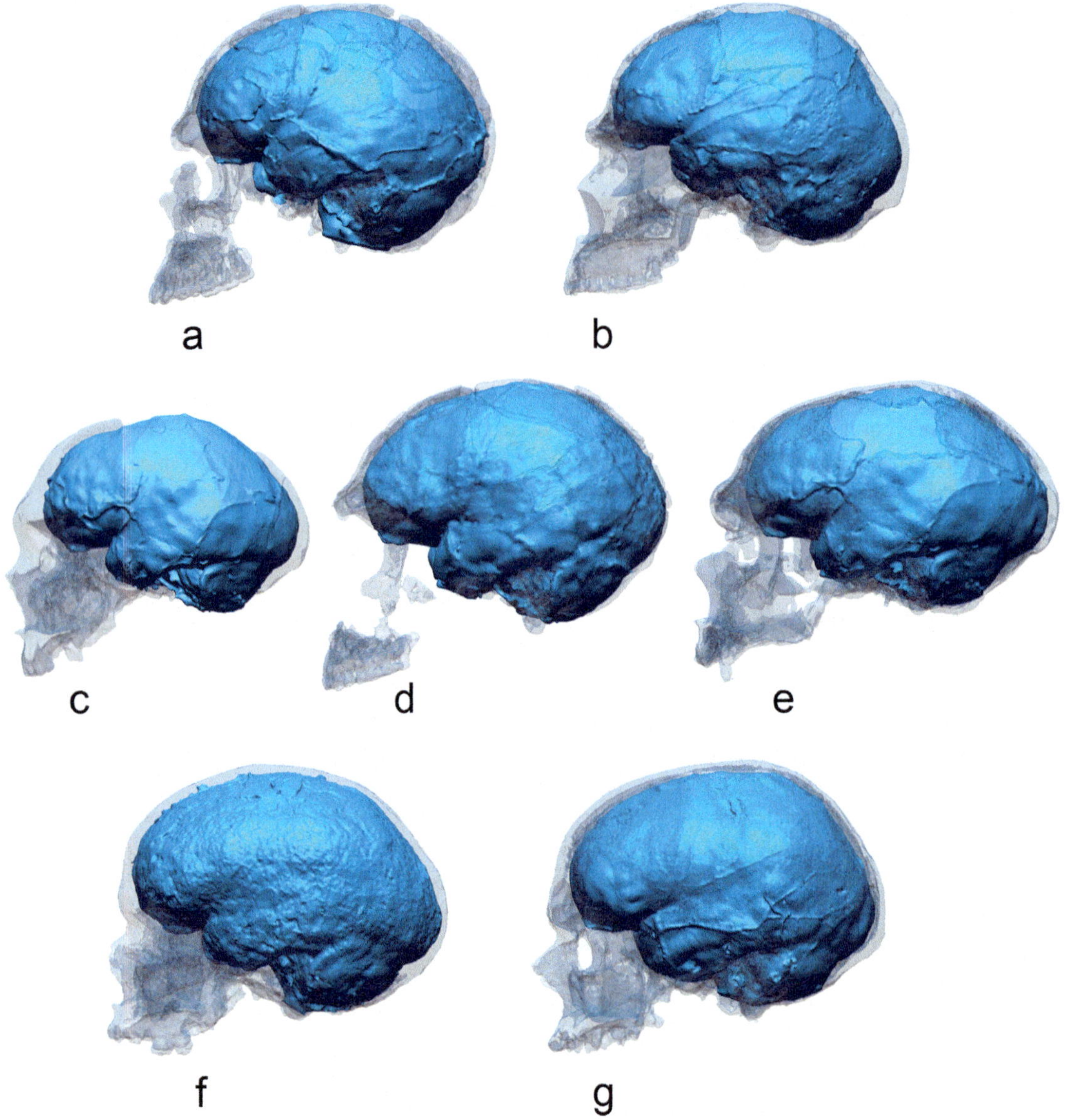

Fig. 15.7 Digital reconstructions of the cranial and endocranial morphology of seven fossil specimens, after computed approaches based on surface extrapolation and shape interpolation of the available fragments: (**a**) Qafzeh 9;(**b**) Skhul 5; (**c**) Gibraltar 1; (**d**) Amud; (**e**) La Chapelle-aux-Saints; (**f**) Mladech 1; (**g**) Cro-Magnon 1 (see Ogihara et al. Chap. 2)

could be the result of an intermediate anatomy, and the specimen could be considered in terms of admixture or, probably more likely, as a "transitional phenotype" (see below).

When considering the large influence of the precuneus in shaping the midsagittal parietal morphology in its longitudinal and vertical aspects, and the lack of influence on the lateral parietal outline (Bruner et al. 2014a; Pereira-Pedro and Bruner 2016), we can tentatively suggest that this element is responsible for the specific brain morphology of modern humans, but not for the lateral bulging described in modern humans and Neandertals. In this case, we should evaluate the possibility that these two encephalized hominid species underwent at least two different parietal changes (Fig. 15.9). Apart from precuneus, the other structure able to influence the upper parietal morphology is the intraparietal sulcus, through its enlargement and consequential pressure on the upper surface or, otherwise, through the enlargement, outfolding, and displacement of the upper parietal surface. It can be therefore hypothesized that a volumetric increase of the intraparietal region can explain the upper parasagittal bulging described in modern humans and Neandertals and a volumetric increase of the precuneus can be associated with the midsagittal parietal bulging described only in the former species.

It is worth noting that the North African skull of Jebel Irhoud, dated to 150–200 ka, is hypothesized to belong the modern lineage because of its facial anatomy but its endocast is noticeably non-modern and similar to the Neandertal

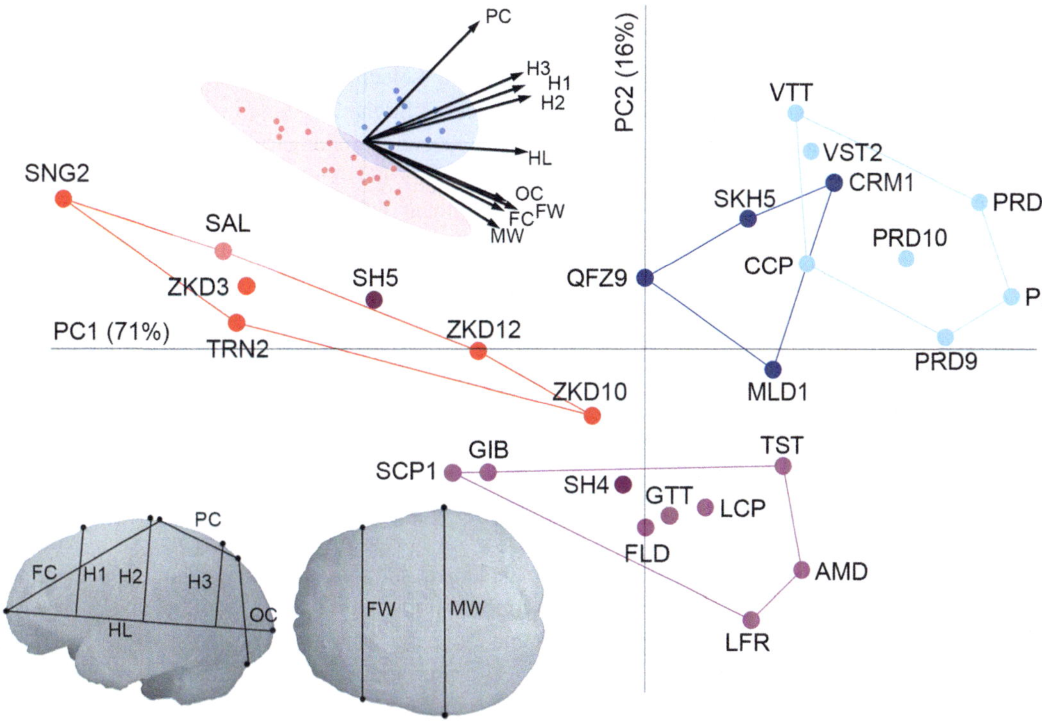

Fig. 15.8 Principal component analysis of endocranial metrics, showing the position of the specimens, the 95% confident ellipses for modern and non-modern groups, and the variable vectors associated with the two components. Variables: *FC* frontal lobe chords, *FW* frontal width, H1, H2, and H3 vault height above the maximum hemispheric length at 25%, 50%, and 75% of the length, *HL* hemispheric length, *MW* maximum endocranial width, *OC* occipital lobe chord, *PC* parietal lobe chord. Specimens: Amud (*AMD*), Combe Capelle (*CCP*), Cro-Magnon (*CRM*), Feldhofer (*FLD*), Gibraltar Forbes' Quarry (*GIB*), Guattari (*GTT*), La Chapelle-aux-Saints (*LCP*), La Ferrassie (*LFR*), Mladech (*MLD*), Predmostí (*PRD*), Qafzeh (*QFZ*); Saccopastore (*SCP*), Salé (*SAL*), Sangiran (*SNG*), Sima de los Huesos (*SH*); Skhul (*SKH*), Tesik Tash (*TST*), Trinil (*TRN*), Vatte di Zambana (*VTT*), Vestonice (*VST*), Zhoukoudian (*ZKD*) (Original data from Bruner et al. 2003)

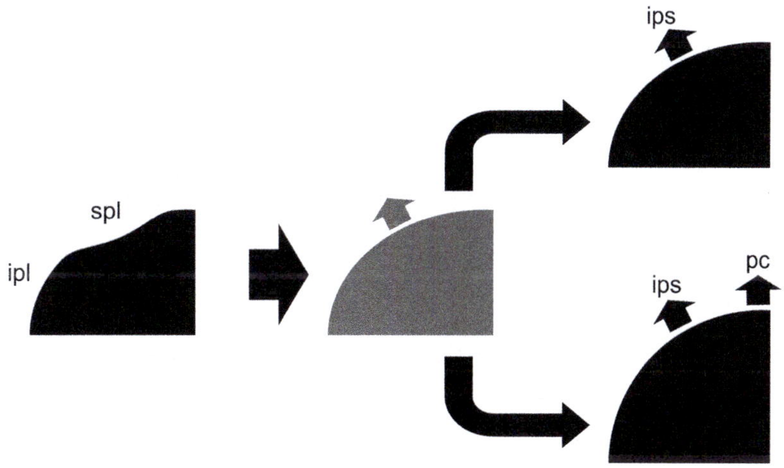

Fig. 15.9 In archaic humans, the inferior parietal lobules (ipl) are bulging, but the superior parietal lobules (spl) are depressed (on the *left*). Neandertals show a parasagittal bulging of the upper parietal surface (*above*) and modern humans a further vertical and longitudinal enlargement of the medial parietal areas (*below*). Taking into account the effect of precuneus dimensions on the parietal outline in adult modern humans, we can tentatively hypothesize that the lateral dilation is associated with changes in the intraparietal sulcus (ips), while the longitudinal and vertical increase is associated with changes in the precuneus (pc). If this scenario is further supported in future studies, it remains to be evaluated whether modern humans evolved both traits together or evolved the lateral enlargement first (*gray* form)

phenotype (Bruner and Pearson 2013; Hublin et al. 2017). This means that the origin of the modern human lineage could have not matched the origin of a modern human brain form, raising two major points. First, it must be determined whether the observed parietal changes in modern humans are the result of a discrete or gradual process. Second, the parietal lobes being highly sensitive to environmental influences, it must be considered whether such changes were based on genetic and selection or on environmental factors and phenotypic responses, including feedback with culture (Bruner and Iriki 2016). The parietal cortex is particularly sensitive to training, and morphological changes can be induced as a physiological response to behavioral practice (Hihara et al. 2006; Quallo et al. 2009). Also, human brain morphology is apparently less constrained by genetic backgrounds and more susceptible to individual influences when compared with living apes (Gómez-Robles et al. 2015). Therefore, although genetic variations can have a role in parietal evolution, the influence and feedback with culture must be taken into account, introducing epigenetic and environmental factors. This is why a specimen such as Qafzeh 9, which belongs to the modern lineage but which used a Mousterian toolkit, can be interpreted as "transitional," not only in genetic/phylogenetic sense but alternatively according to a functional and behavioral perspective. Interestingly, also in the case of Neandertals, there are "intermediate" phenotypes, like the skull from Maba, in China, which apparently display Neandertal traits in the face but archaic braincase, including a parietal parasagittal depression (Wu and Bruner 2016).

15.3.3 Additional Evolutionary Evidence

In the last decade, further studies have provided evidence of the evolutionary differences in the parietal regions of modern humans. Shape analysis has revealed that the bulging of the parietal surface in our species can be associated with a very early postnatal stage (Neubauer et al. 2009). In fact, parietal maturation also occurs very early during ontogeny (Gogtay et al. 2004). This stage of parietal expansion is absent in the growth and developmental patterns of the chimpanzee (Neubauer et al. 2010). Most interestingly, this stage is also absent in Neandertals (Gunz et al. 2010), although the exact timing of this differentiation must be further investigated (Ponce de León et al. 2016). Apart from the "globularization stage" specific to modern humans, all the rest of the postnatal morphogenetic pattern is well conserved among all hominoids (Scott et al. 2014). In general, differences in basic endocranial proportions between hominid species are compatible with random rates of variation, with the exception of general brain size increase for the whole human genus and brain globularity specifically for *H. sapiens* (Gómez-Robles et al. 2017). This further suggests possible selective processes associated with these two aspects of human encephalization.

A recent morphometric analysis on the midsagittal brain morphology in humans and chimpanzees added further and determinant information to this scenario, showing that the size of the precuneus represents the most apparent geometric difference between the two species, being much larger in the former (Bruner et al. 2017b). This pattern of shape difference is the same described as the main intraspecific variation among adult humans. This result further suggests that the precuneus may have undergone recent evolutionary changes in our lineage and that it is a good candidate for the parietal bulging stage specific of *Homo sapiens*, absent in chimps and in Neandertals. The largest cortical area in this region is area 7A according to Scheperjans et al. (2008a), which is the largest and most variable area of the precuneus, extending from the internal medial fold to the external dorsal surface of the brain.

Further additional evidence on parietal evolution comes from indirect analyses based on correlations between bone and brain elements. According to the association between orbit size and occipital lobe size, it has been hypothesized that Neandertals had larger occipital areas when compared with modern humans (Pearce et al. 2013). Taking into account that Neandertals and modern humans share a similar brain size, and that the parietal and occipital bones show an inverse correlation in these two species (Gunz and Harvati 2007), larger occipital lobes in Neandertals are compatible with larger parietal lobes in modern humans. This also fits with the observation that overall parieto-occipital proportions of modern humans fall within the apes' brain scaling patterns (Semendeferi et al. 1997) but with apes having a relatively larger occipital cortex (De Sousa et al. 2010). Such inverse relationships between occipital and parietal lobes, described in living species and supposed in extinct ones, cannot be however confirmed at intraspecific level. In fact, among adult modern humans, there is no inverse relationship between occipital and parietal areas (Allen et al. 2002).

In conclusions, when dealing with the parietal areas, we must evaluate three different aspects in evolutionary neuroanatomy. First, modern humans have larger parietal bones and probably larger parietal lobes, when compared with non-modern human species. Second, the morphology of the precuneus is very variable among adult humans, and it represents a main difference between humans and chimpanzees. Third, these two points may be related, the latter being the reason for the former. These three points must be evaluated independently. The first two issues are but morphometric facts, and their causes must be investigated further. The third is an evolutionary hypothesis.

15.4 Paleoangiology

A final remark on parietal areas and fossils concerns the vascular system. Modern humans display an important increase of the endocranial vascularization, at least as far as we can see taking into account the middle meningeal vessels and the diploic channels (Grimaud-Hervé 1997; Hershkovitz et al. 1999; Bruner and Sherkat 2008; Rangel de Lázaro et al. 2016). Such changes are specific to *H. sapiens*, and apparently not associated with brain size, neurocranial form, or with any gradual phylogenetic trends (Bruner et al. 2005, 2011). This increase of the vascular system is mainly observed on the parietal surface, which suggests an association between parietal size increase, metabolic increase, and vascular complexity (Bruner et al. 2011, 2012). In fact, the precuneus has been reported to have higher thermal and metabolic values than expected (Cavanna and Trimble 2006; Sotero and Iturria-Medina 2011). A simulation modeling of the heat dissipation patterns on modern human and Neandertal endocasts showed that, although brain size is the main factor involved in heat accumulation, also brain shape variations generate differences in the distribution of cortical heat loadings of the two species (Bruner et al. 2014b). In particular, the dorsal parietal surface, being close to the thermal core of the brain geometry in extinct humans, shows large thermal loads. This area is more exposed on the endocranial surface in Neandertals because of their flattened parietals and is more internal in modern humans because of their parietal dilation. The precuneus cortex is mainly vascularized by the anterior cerebral artery, but its boundaries represent the meeting zone of anterior, middle, and posterior cerebral arteries. Therefore, it is a district characterized by a peculiar vascular complexity. The fact that those heat-loaded dorsal parietal surfaces are larger and less superficial in modern humans and that at the same time there is a notable increase of the parietal vascular network in the same districts deserves attention.

This is particularly interesting when considering that metabolic damages at the precuneus have been described in early stages of Alzheimer's disease (Jacobs et al. 2012, 2013; Doré et al. 2013; Huang et al. 2013), a pathology which is associated only with our species. A major feature of Alzheimer's disease is the phosphorylation of the tau protein, a process which is known to occur in mammals in response to thermal changes (Stieler et al. 2011). Main connector hubs like the precuneus represent high-metabolic areas, often sensitive to different kinds of functional impairments (Bertolero et al. 2015). It hence remains to be determined whether or not the evolution of complex parietal lobes may have increased the vulnerability or sensitivity to neurodegenerative processes, introducing or increasing risk factors (Bruner and Jacobs 2013).

15.5 Beyond Fossils: Brains and Bodies

15.5.1 Parietal Lobe Evolution: A Current Synthesis

In the last decade, studies in structural and functional neuroanatomy have dramatically improved our knowledge of the parietal areas, adding relevant information concerning their possible involvement in the evolution of modern humans. The intraparietal sulcus is positioned in a region which has undergone evolutionary changes in terms of both morphology and functions (Vanduffel et al. 2002; Orban et al. 2006; Bruner 2010a). The precuneus shows a notable variation among humans (Bruner et al. 2014a, 2015a, 2017a) and between humans and chimpanzees (Bruner et al. 2017b). The patent similarity between the intraspecific (among adults) and phylogenetic (between humans and apes) pattern of brain variation and the pattern of cranial variation among modern and non-modern human species (in both cases a bulging associated with the dilation/contraction of an area positioned in the middle of the parietal bone and lobe) raises the question of whether the latter can be associated with the former (Bruner et al. 2014a). As previously mentioned, due to the correspondence between brain changes and endocranial changes, it can be hypothesized that lateral changes of the superior parietal areas were associated with variation of the intraparietal sulcus, while midsagittal expansion was associated with the enlargement of the precuneus.

The precuneus is largely involved in visuospatial integration and memory, as well as in consciousness and self-awareness (Fletcher et al. 1995; Cavanna and Trimble 2006; Margulies et al. 2009; Freton et al. 2014). However, it is also a major crossing point between the functional and structural brain networks (Hagmann et al. 2008) and a main hub of the default mode network (Meunier et al. 2010; Utevsky et al. 2014). Connector nodes are essential to bridge and coordinate different brain functional modules, and probably because of this role, they are generally expensive in terms of energy budget and particularly sensitive to functional impairments (Bertolero et al. 2015). It is hence reasonable to hypothesize that an increase of the precuneus volume may have improved the capacity of the brain to integrate its connected modules.

Apart from the importance of the parietal morphology in the evolution of *H. sapiens*, however, it must be taken into consideration that the parietal lobes work largely in tandem with the prefrontal cortex (Jung and Haier 2007; Caminiti et al. 2015). A recent study suggests that, to move from *emulation* (a primate ability to reproduce a given result) to *imitation* (a human ability to reproduce a given process), a specific connection is required, linking the frontal and upper parietal cortex (Hecht et al. 2013). In this sense, it is interesting to note that both modern humans and Neandertals display changes in their frontal and parietal proportions (Bruner and Holloway 2010). Regarding the frontal lobes, in both species, these areas are relatively wider when compared with other hominids. Both species also have a specific spatial position of the prefrontal cortex, which is positioned right above the orbital roof. Hence, we cannot exclude that the frontal widening in modern humans and Neandertals could be a secondary consequence of such a vertical constraint, instead of a neurofunctional change. Nonetheless, it may be no coincidence that such variations involved the prefrontal cortex and in particular Broca's area. Although there is a general disagreement on whether or not modern humans changed their frontal volume proportions (Rilling 2006; Smaers 2013), the absolute size increase is undeniable, and geometrical changes can also influence or be influenced by the underlying schemes of neural connections.

Visuospatial integration is a main function associated with the parietal lobes, and it has been traditionally interpreted in purely mechanical terms (hand coordination, grasping, body orientation, etc.). Recent perspectives in extended cognition challenge this automatic mechanical role (Bruner and Iriki 2016). Following principles like embodiment or brain-artifact interface, theories in extended mind suggest that the cognitive process is the actual result of the interaction between brain, body, and environment (Clark 2007, 2008). The body is the active interface between brain and environment, and material culture is the interface between body and culture (Malafouris 2008, 2010, 2013). In this sense, there is no discontinuity between neural, body, and inorganic elements, and the process we call "mind" is strictly based on their joined association. Following this view, visuospatial integration is not only a mechanical issue but a deeply cognitive one. The eye-hand system is the main port managing inputs and outputs, and cognition is deeply rooted in the body experience (Byrge et al. 2014). In fact, the brain incorporates the external object into the body schemata after visual or physical contact, generating a reciprocal exchange between ecological, neural, and cognitive niches (Iriki and Sakura 2008; Iriki and Taoka 2012). Indeed, the intraparietal sulcus and the precuneus coordinate visuospatial integration and the eye-hand system, and also egocentric memory and self-awareness, so generating a direct link between body and consciousness.

Importantly, visuospatial capacities are also profoundly associated with social aspects. Hand contact (grooming) is the principal behavior associated with social relationships and social group size in primates, and both parameters are correlated to brain size (Dunbar 2010). The precuneus is largely involved in self-centered mental imagery and internal representations (Land 2014), and the capacity of perceiving the other's body is a critical ability for social cognition and integration (Maister et al. 2015). Social relationships are processed following mechanisms which are shared with orientation in space and time, and precuneus is a crucial node for such multiple mapping domains (Peer et al. 2015), all based on the metrics of the body. Social and spatial orientation share patterns aimed at planning a proper exploration and exploitation of the resources (Hills et al. 2015), and some differences have been actually hypothesized between modern humans and Neandertals for their visuospatial capacity, wayfinding ability, and land use (Burke 2012).

With many limitations, visuospatial integration can be investigated in paleoanthropology and archaeology, following the anatomical evidence (paleoneurology and hand evolution) and taking into consideration direct and indirect traces of visuospatial behaviors (Bruner et al. 2016, 2017c). It is intriguing to note that Neandertals lacked the expansion of the upper parietal surface displayed by modern humans and, at the same time, they needed to use their mouth to handle objects much more than any modern human population (Bruner and Lozano 2014, 2015). Recently, it has been hypothesized that language can be embodied too: words stimulate mirror-neuron mechanisms associated with their contextual meaning, suggesting that such a sensorimotor response may be necessary for its codification (Binkofski and Buccino 2004; Buccino et al. 2005; Jirak et al. 2010; Marino et al. 2012). All this is particularly interesting considering that the paleoneurological evidence suggests that the most visible morphological change associated with our brain evolution is precisely localized in the dorsal and medial parietal areas, stressing further the possibility of recent evolutionary changes in the integration between inner and outer cognitive subsystems.

15.5.2 Next Steps

Studies on the parietal lobes are now flourishing, and we have more and more information on their organization and mechanisms. Detailed analysis of their functions will be necessary to disentangle the roles of their many subdivisions. Comparative data on nonhuman primates are necessary to establish the degree of variation and variability of the parietal element and most of all the homology among different taxonomic groups. The remarkable morphological variation among modern humans should be investigated in terms of cells and structure, to localize the histological processes responsible for such macroanatomical differences. These same differences should be also investigated in terms of psychometrics and genetics. It is important to find proper approaches to apply experimental and quantitative methods to the study of extended cognition. In terms of human evolution, it must be determined whether parietal changes have been based on brand new features or else on existing structures and circuits, whether they have been gradual or more punctuated, and most of all to what extent they were and are associated with genetic (adaptations and allelic selection) or environmental (cultural feedback, training, physiological responses) factors. Because of the limits in sample size and in the kind of information provided, it is likely that fossil endocasts can only provide limited information on all these perspectives. Nonetheless, we cannot exclude that accurate geometric models could localize with more precision the spatial origin of the morphological changes, so further orienting the research of the neontological fields. To promote this possibility, it will be necessary to localize and isolate the actual neural information available from the paleoneurological evidence. To do this, first we need to understand what endocranial changes are associated with neural variations and what endocranial changes are secondary cranial adjustments with no neural effect. Second, we have to maintain the field within a robust and consistent quantitative perspective, relying on numerical modeling, multivariate tools, and comparative approaches.

If parietal lobe evolution was really involved in changing the processes of integration between brain, body, and environment, much of our current knowledge on human evolution should be deeply reinterpreted.

Acknowledgments Data presented in this survey were collected within the project "Replacement of Neanderthals by Modern Humans: Testing Evolutionary Models of Learning," funded by the Japanese Government (#22101006). EB is funded by the Spanish Government (CGL2015-65387-C3-3-P) and by the Wenner-Gren Foundation. ASPP is funded by the Atapuerca Foundation, Spain. We are grateful to Todd Preuss, Jim Rilling, Ralph Holloway, Simon Neubauer, Philipp Gunz, Karl Zilles, Osamu Kondo, Gizéh Rangel de Lázaro, Stana Eisová, Hana Píšová, José Manuel de la Cuétara, Giorgio Manzi, Aida Gómez-Robles, Dominique Grimaud-Hervé, Roberto Colom, Manuel Martin-Loeches, Marina Lozano, and Heidi Jacobs for their collaboration and suggestions on the topics presented in this article.

References

Ackerley R, Kavounoudias A (2015) The role of tactile afference in shaping motor behaviour and implications for prosthetic innovation. Neuropsychologia 79:192–205

Allen JS, Damasio H, Grabowski TJ (2002) Normal neuroanatomical variation in the human brain: an MRI-volumetric study. Am J Phys Anthropol 118:341–358

Amano H, Kikuchi T, Morita Y, Kondo O, Suzuki H, Ponce de León MS, Zollikofer C, Bastir M, Stringer C, Ogihara N (2015) Virtual reconstruction of the Neanderthal Amud 1 cranium. Am J Phys Anthropol 158:185–197

Andersen RA, Buneo CA (2002) Intentional maps in posterior parietal cortex. Annu Rev Neurosci 25:189–220

Andersen RA, Snyder LH, Bradley DC, Xing J (1997) Multimodal representation of space in the posterior parietal cortex and its use in planning movements. Annu Rev Neurosci 20:303–330

Ansari D (2008) Effects of development and enculturation on number representation in the brain. Nat Rev Neurosci 9:278–291

Basten U, Hilger K, Fiebach CJ (2015) Where smart brains are different: a quantitative meta-analysis of functional and structural brain imaging studies on intelligence. Intelligence 51:10–27

Battaglia-Mayer A, Ferraina S, Mitsuda T, Marconi B, Genovesio A, Onorati P, Lacquaniti F, Caminiti R (2000) Early coding of reaching in the parietooccipital cortex. J Neurophysiol 83:2374–2391

Battaglia-Mayer A, Caminiti R, Lacquaniti F, Zago M (2003) Multiple levels of representation of reaching in the parieto-frontal network. Cereb Cortex 13:1009–1022

Berryhill ME, Chein J, Olson IR (2011) At the intersection of attention and memory: the mechanistic role of the posterior parietal lobe in working memory. Neuropsychologia 49:1306–1315

Bertolero MA, Yeo BTT, D'Esposito M (2015) The modular and integrative functional architecture of the human brain. Proc Natl Acad Sci U S A 112:E6798–E6807

Binkofski F, Buccino G (2004) Motor functions of the Broca's region. Brain Lang 89:362–369

Bookstein FL (1991) Morphometric tools for landmark data: geometry and biology. Cambridge University Press, Cambridge

Bruner E (2004) Geometric morphometrics and paleoneurology: brain shape evolution in the genus *Homo*. J Hum Evol 47:279–303

Bruner E (2008) Comparing endocranial form and shape differences in modern humans and Neandertal: a geometric approach. PaleoAnthropology 2008:93–106

Bruner E (2010a) Morphological differences in the parietal lobes within the human genus: a neurofunctional perspective. Curr Anthropol 51: S77–S88

Bruner E (2010b) The evolution of the parietal cortical areas in the human genus: between structure and cognition. In: Broadfield D, Yuan M, Schick K, Toth N (eds) The human brain evolving: paleoneurological studies in honor of Ralph L. Holloway. Stone Age Institute, pp 83–96

Bruner E (2014) Functional craniology, human evolution, and anatomical constraints in the Neanderthal braincase. In: Akazawa T, Ogihara N, Tanabe HC, Terashima H (eds) Dynamics of learning in Neanderthals and modern humans (Vol. 2). Springer, Japan, pp 121–129

Bruner E (2015) Functional craniology and brain evolution. In: Bruner E (ed) Human Paleoneurology. Springer, Switzerland, pp 57–94

Bruner E, Holloway RL (2010) A bivariate approach to the widening of the frontal lobes in the genus *Homo*. J Hum Evol 58:138–146

Bruner E, Iriki A (2016) Extending mind, visuospatial integration, and the evolution of the parietal lobes in the human genus. Quat Int 405:98–110

Bruner E, Jacobs HIL (2013) Alzheimer's disease: the downside of a highly evolved parietal lobe? J Alzheimers Dis 35:227–240

Bruner E, Lozano M (2014) Extended mind and visuo-spatial integration: three hands for the Neandertal lineage. J Anthropol Sci 92:273–280

Bruner E, Lozano M (2015) Three hands: one year later. J Anthropol Sci:191–195

Bruner E, Manzi G (2006) Saccopastore 1: the earliest Neanderthal? In: Harvati K, Harrison T (eds) Neanderthals revisited. New approaches and perspectives. Vertebrate paleobiology and paleoanthropology series. Springer, New York

Bruner E, Manzi G (2008) Paleoneurology of an "early" Neandertal: endocranial size, shape, and features of Saccopastore 1. J Hum Evol 54:729–742

Bruner E, Pearson O (2013) Neurocranial evolution in modern humans: the case of Jebel Irhoud 1. Anthropol Sci 121:31–41

Bruner E, Ripani M (2008) A quantitative and descriptive approach to morphological variation of the endocranial base in modern humans. Am J Phys Anthropol 137:30–40

Bruner E, Sherkat S (2008) The middle meningeal artery: from clinics to fossils. Childs Nerv Syst 24:1289–1298

Bruner E, Manzi G, Arsuaga JL (2003) Encephalization and allometric trajectories in the genus *Homo*: evidence from the Neandertal and modern lineages. Proc Natl Acad Sci U S A 100:15335–15340

Bruner E, Saracino B, Ricci F, Tafuri M, Passarello P, Manzi G (2004) Midsagittal cranial shape variation in the genus *Homo* by geometric morphometrics. Coll Antropol 28:99–112
Bruner E, Mantini S, Perna A, Maffei C, Manzi G (2005) Fractal dimension of the middle meningeal vessels: variation and evolution in *Homo erectus*, Neanderthals, and modern humans. Eur J Morphol 42:217–224
Bruner E, Martin-Loeches M, Colom R (2010) Human midsagittal brain shape variation: patterns, allometry and integration. J Anat 216:589–599
Bruner E, Mantini S, Musso F, De la Cuétara JM, Ripani M, Sherkat S (2011) The evolution of the meningeal vascular system in the human genus: from brain shape to thermoregulation. Am J Hum Biol 23:35–43
Bruner E, De La Cuétara JM, Musso F (2012) Quantifying patterns of endocranial heat distribution: brain geometry and thermoregulation. Am J Hum Biol 24:753–762
Bruner E, De la Cuétara JM, Masters M, Amano H, Ogihara N (2014a) Functional craniology and brain evolution: from paleontology to biomedicine. Front Neuroanat 8:19
Bruner E, Rangel de Lázaro G, de la Cuétara JM, Martín-Loeches M, Colóm R, Jacobs HIL (2014b) Midsagittal brain variation and MRI shape analysis of the precuneus in adult individuals. J Anat 224: 367–376
Bruner E, Amano H, de la Cuétara JM, Ogihara N (2015a) The brain and the braincase: a spatial analysis on the midsagittal profile in adult humans. J Anat 227:268–276
Bruner E, Grimaud-Hervé D, Wu X, De la Cuétara JM, Holloway R (2015b) A paleoneurological survey of *Homo erectus* endocranial metrics. Quat Int 368:80–87
Bruner E, Román FJ, de la Cuétara JM, Martín-Loeches M, Colóm R (2015c) Cortical surface area and cortical thickness in the precuneus of adult humans. Neuroscience 286:345–352
Bruner E, Lozano M, Lorenzo C (2016) Visuospatial integration and human evolution: the fossil evidence. J Anthropol 94:81–97
Bruner E, Pereira-Pedro AS, Chen X, Rilling JK (2017a) Precuneus proportions and cortical folding: a morphometric evaluation on a racially diverse human sample. Ann Anat 211:120–128
Bruner E, Preuss T, Chen X, Rilling J (2017b) Evidence for expansion of the precuneus in human evolution. Brain Struct Funct 222: 1053–1060
Bruner E, Spinapolice E, Burke A, Overmann K (2017c) Visuospatial integration: paleoanthropological and archaeological perspectives. In: Di Paolo LD, Di Vincenzo F, D'Almeida AF (eds) Evolution of primate social cognition. Springer, Cham. (in press)
Buccino G, Riggio L, Melli G, Binkofski F, Gallese V, Rizzolatti G (2005) Listening to action-related sentences modulates the activity of the motor system: a combined TMS and behavioral study. Brain Res Cogn Brain Res 24:355–363
Burke A (2012) Spatial abilities, cognition and the pattern of Neanderthal and modern human dispersals. Quat Int 247:230–235
Byrge L, Sporns O, Smith LB (2014) Developmental process emerges from extended brain–body–behavior networks. Trends Cogn Sci 18: 395–403
Caminiti R, Innocenti GM, Battaglia-Mayer A (2015) Organization and evolution of parieto-frontal processing streams in macaque monkeys and humans. Neurosci Biobehav Rev 56:73–96
Cantlon JF, Brannon EM, Carter EJ, Pelphrey KA (2006) Functional imaging of numerical processing in adults and 4-y-old children. PLoS Biol 4:e125
Caspers S, Geyer S, Schleicher A, Mohlberg H, Amunts K, Zilles K (2006) The human inferior parietal cortex: cytoarchitectonic parcellation and interindividual variability. NeuroImage 33:430–448
Caspers S, Eickhoff SB, Rick T, von Kapri A, Kuhlen T, Huang R, Shah NJ, Zilles K (2011) Probabilistic fibre tract analysis of cytoarchitectonically defined human inferior parietal lobule areas reveals similarities to macaques. NeuroImage 58:362–380
Cavanna AE, Trimble MR (2006) The precuneus: a review of its functional anatomy and behavioural correlates. Brain 129:564–583
Chafee MV, Goldman-Rakic PS (1998) Matching patterns of activity in primate prefrontal area 8a and parietal area 7ip neurons during a spatial working memory task. J Neurophysiol 79:2919–2940
Choi H-J, Zilles K, Mohlberg H, Schleicher A, Fink GR, Armstrong E, Amunts K (2006) Cytoarchitectonic identification and probabilistic mapping of two distinct areas within the anterior ventral bank of the human intraparietal sulcus. J Comp Neurol 495:53–69
Clark A (2007) Re-inventing ourselves: the plasticity of embodiment, sensing, and mind. J Med Philos 32:263–282
Clark A (2008) Supersizing the mind. Embodiment, action and cognitive extension. Oxford University Press, Oxford
Coolidge FL, Wynn T (2005) Working memory, its executive functions, and the emergence of modern thinking. Camb Archaeol J15: 5–26
Culham JC, Kanwisher NG (2001) Neuroimaging of cognitive functions in human parietal cortex. Curr Opin Neurobiol 11:157–163
Dart RA (1925) *Australopithecus africanus*: the man-ape of South Africa. Nature 2884:195–199
De Sousa AA, Sherwood CC, Mohlberg H, Amunts K, Schleicher A, MacLeod CE, Hof PR, Frahm H, Zilles K (2010) Hominoid visual brain structure volumes and the position of the lunate sulcus. J Hum Evol 58:281–292
Deschamps I, Baum SR, Gracco VL (2014) On the role of the supramarginal gyrus in phonological processing and verbal working memory: evidence from rTMS studies. Neuropsychologia 53:39–46
Doré V, Villemagne VL, Bourgeat P, Fripp J, Acosta O, Chetélat G, Zhou L, Martins R, Ellis KA, Masters CL, Ames D, Salvado O, Rowe CC (2013) Cross-sectional and longitudinal analysis of the relationship between aβ deposition, cortical thickness, and memory in cognitively unimpaired individuals and in Alzheimer's disease. JAMA Neurol 70:903–911
Dunbar RIM (2010) The social role of touch in humans and primates: behavioural function and neurobiological mechanisms. Neurosci Biobehav Rev 34:260–268
Ebeling U, Steinmetz H (1995) Anatomy of the parietal lobe: mapping the individual pattern. Acta Neurochir 136:8–11
Eidelberg D, Galaburda AM (1984) Inferior parietal lobule: divergent architectonic asymmetries in the human brain. Arch Neurol 41:843–852
Enlow DH (1990) Facial growth. Sounders, Philadelphia
Ferraina S, Garasto MR, Battaglia-Mayer A, Ferraresi P, Johnson PB, Lacquaniti F, Caminiti R (1997) Visual control of hand-reaching movement: activity in parietal area 7m. Eur J Neurosci 9:1090–1095
Fletcher PC, Frith CD, Baker SC, Shallice T, Frackowiak RSJ, Dolan RJ (1995) The mind's eye—precuneus activation in memory-related imagery. NeuroImage 2:195–200
Freton M, Lemogne C, Bergouignan L, Delaveau P, Lehéricy S, Fossati P (2014) The eye of the self: precuneus volume and visual perspective during autobiographical memory retrieval. Brain Struct Funct 219:959–968
Gogtay N, Giedd JN, Lusk L, Hayashi KM, Greenstein D, Vaituzis AC, Nugent TF III, Herman DH, Clasen LS, Toga AW, Rapoport JL, Thompson PM (2004) Dynamic mapping of human cortical development during childhood through early adulthood. Proc Natl Acad Sci U S A 101:8174–8179
Gómez-Robles A, Hopkins WD, Sherwood CC (2013) Increased morphological asymmetry, evolvability and plasticity in human brain evolution. Proc R Soc B 280:20130575
Gómez-Robles A, Hopkins WD, Schapiro SJ, Sherwood CC (2015) Relaxed genetic control of cortical organization in human brains

compared with chimpanzees. Proc Natl Acad Sci U S A 112: 14799–14804
Gómez-Robles A, Smaers JB, Holloway RL, Polly PD, Wood BA (2017) Brain enlargement and dental reduction were not linked in hominin evolution. Proc Natl Acad Sci U S A 114:468–473
Grefkes C, Fink GR (2005) The functional organization of the intraparietal sulcus in humans and monkeys. J Anat 207:3–17
Grimaud-Hervé D (1997) L'évolution de l'encéphale chez l'*Homo erectus* et l'*Homo sapiens*. CNRS Editions, Paris
Gunz P, Harvati K (2007) The Neanderthal "chignon": variation, integration, and homology. J Hum Evol 52:262–274
Gunz P, Neubauer S, Maureille B, Hublin J-J (2010) Brain development after birth differs between Neanderthals and modern humans. Curr Biol 20:R921–R922
Hagmann P, Cammoun L, Gigandet X, Meuli R, Honey CJ, Wedeen VJ, Sporns O (2008) Mapping the structural core of human cerebral cortex. PLoS Biol 6:e159
Hammer Ø, Ryan P, Harper D (2001) PAST: paleontological statistics software package for education and data analysis. Palaeontol Electron 4:9
Hauk O, Johnsrude I, Pulvermüller F (2004) Somatotopic representation of action words in human motor and premotor cortex. Neuron 41:301–307
Hecht EE, Gutman DA, Preuss TM, Sanchez MM, Parr LA, Rilling JK (2013) Process versus product in social learning: comparative diffusion tensor imaging of neural systems for action execution–observation matching in macaques, chimpanzees, and humans. Cereb Cortex 23:1014–1024
Hershkovitz I, Greenwald C, Rothschild BM, Latimer B, Dutour O, Wish-Baratz S, Pap I, Leonetti G (1999) The elusive diploic veins: anthropological and anatomical perspective. Am J Phys Anthropol 108:345–358
Hihara S, Notoya T, Tanaka M, Ichinose S, Ojima H, Obayashi S, Fujii N, Iriki A (2006) Extension of corticocortical afferents into the anterior bank of the intraparietal sulcus by tool-use training in adult monkeys. Neuropsychologia 44:2636–2646
Hills TT, Todd PM, Lazer D, Redish AD, Couzin ID (2015) Exploration versus exploitation in space, mind, and society. Trends Cogn Sci 19:46–54
Holloway RL (1995) Toward a synthetic theory of human brain evolution. In: Changeux JP, Chavaillon J (eds) Origins of the human brain. Clarendon Press, Oxford, pp 42–54
Holloway RL, Broadfield DC, Yuan MS (2004) Brain endocasts: the paleoneurological evidence. The human fossil record, vol. III. Wiley-Liss, Hoboken
Hublin JJ, Ben-Ncer A, Bailey SE, Freidline SE, Neubauer S, Skinner MM, Bergmann I, Le Cabec A, Benazzi S, Harvati K, Gunz P (2017) New fossils from Jebel Irhoud, Morocco and the pan-African origin of *Homo sapiens*. Nature 546:289–292
Huang K-L, Lin K-J, Hsiao I-T, Kuo HC, Hsu WC, Chuang WL, Kung MP, Wey SP, Hsieh CJ, Wai YY, Yen TC, Huang CC (2013) Regional amyloid deposition in amnestic mild cognitive impairment and Alzheimer's disease evaluated by [18F]AV-45 positron emission tomography in Chinese population. PLoS One 8:e58974
Im K, Lee J-M, Lee J, Shin Y, Kim IY, Kwon JS, Kim SI (2006) Gender difference analysis of cortical thickness in healthy young adults with surface-based methods. NeuroImage 31:31–38
Iriki A, Sakura O (2008) The neuroscience of primate intellectual evolution: natural selection and passive and intentional niche construction. Philos Trans R Soc B 363:2229–2241
Iriki A, Taoka M (2012) Triadic (ecological, neural, cognitive) niche construction: a scenario of human brain evolution extrapolating tool use and language from the control of reaching actions. Philos Trans R Soc B 367:10–23
Jacobs HIL, Van Boxtel MPJ, Jolles J, Verhey FRJ, Uylings HBM (2012) Parietal cortex matters in Alzheimer's disease: an overview of structural, functional and metabolic findings. Neurosci Biobehav Rev 36:297–309
Jacobs HIL, Radua J, Lückmann HC, Sack AT (2013) Meta-analysis of functional network alterations in Alzheimer's disease: toward a network biomarker. Neurosci Biobehav Rev 37:753–765
Jiang X, Iseki S, Maxson RE, Sucov HM, Morriss-Kay GM (2002) Tissue origins and interactions in the mammalian skull vault. Dev Biol 241:106–116
Jirak D, Menz MM, Buccino G, Borghi AM, Binkofski F (2010) Grasping language – a short story on embodiment. Conscious Cogn 19:711–720
Jung RE, Haier RJ (2007) The Parieto-frontal integration theory (P-FIT) of intelligence: converging neuroimaging evidence. Behav Brain Sci 30:135–154
Kacar E, Nas OF, Okeer E, Hakyemez B (2015) Pattern, variability, and hemispheric differences of the subparietal sulcus on multiplanar reconstructed MR images. Surg Radiol Anat 38:89–96
Klepp A, Niccolai V, Buccino G, Schnitzler A, Biermann-Ruben K (2015) Language–motor interference reflected in MEG beta oscillations. NeuroImage 109:438–448
Klingenberg CP (2011) Morphoj: an integrated software package for geometric morphometrics. Mol Ecol Resour 11:353–357
Kobayashi Y, Matsui T, Haizuka Y, Hirai N, Matsumura G (2014) Cerebral sulci and gyri observed on macaque endocasts. In: Akazawa T, Ogihara N, Tanabe HC, Terashima H (eds) Dynamics of learning in Neanderthals and modern humans volume 2. Springer, Japan, pp 131–137
Kondo O, Kubo D, Suzuki H, Ogihara N (2014) Virtual Endocast of Qafzeh 9: a preliminary assessment of right-left asymmetry. In: Akazawa T, Ogihara NC, Tanabe H, Terashima H (eds) Dynamics of learning in Neanderthals and modern humans volume 2. Springer, Japan, pp 183–190
Land MF (2014) Do we have an internal model of the outside world? Philos Trans R Soc B 369:20130045–20130045
Leshinskaya A, Caramazza A (2016) Abstract categories of functions in anterior parietal lobe. Neuropsychologia 76:27–40
Luders E, Narr KI, Thompson PM, Rex DE, Jancke I, Toga AW (2006) Hemispheric asymmetries in cortical thickness. Cereb Cortex 16: 1232–1238
Lundstrom BN, Ingvar M, Petersson KM (2005) The role of precuneus and left inferior frontal cortex during source memory episodic retrieval. NeuroImage 27:824–834
Maister L, Slater M, Sanchez-Vives MV, Tsakiris M (2015) Changing bodies changes minds: owning another body affects social cognition. Trends Cogn Sci 19:6–12
Malafouris L (2008) Between brains, bodies and things: tectonoetic awareness and the extended self. Philos Trans R Soc Lond Ser B Biol Sci 363:1993–2002
Malafouris L (2010) The brain–artefact interface (BAI): a challenge for archaeology and cultural neuroscience. Soc Cogn Affect Neurosci 5:264–273
Malafouris L (2013) How things shape the mind: a theory of material engagement. MIT Press, Cambridge
Manzi G, Vienna A, Hauser G (1996) Developmental stress and cranial hypostosis by epigenetic trait occurrence and distribution: an exploratory study on the Italian Neandertals. J Hum Evol 30: 511–527
Marcus DS, Wang TH, Parker J, Csernansky JG, Morris JC, Buckner RL (2007) Open access series of imaging studies (OASIS): cross-sectional MRI data in young, middle aged, nondemented, and demented older adults. J Cogn Neurosci 19: 1498–1507
Margulies DS, Vincent JL, Kelly C, Lohmann G, Uddin LQ, Biswal BB, Villringer A, Castellanos FX, Milham MP, Petrides M (2009) Precuneus shares intrinsic functional architecture in humans and monkeys. Proc Natl Acad Sci U S A 106:20069–20074

Marino BFM, Gallese V, Buccino G, Riggio L (2012) Language sensorimotor specificity modulates the motor system. Cortex 48: 849–856

Mars RB, Jbabdi S, Sallet J, O'Reilly JX, Croxson PL, Olivier E, Noonan MP, Bergmann C, Mitchell AS, Baxter MG, Behrens TE, Johansen-Berg H, Tomassini V, Miller KL, Rushworth MF (2011) Diffusion-weighted imaging tractography-based parcellation of the human parietal cortex and comparison with human and macaque resting-state functional connectivity. J Neurosci 31:4087–4100

Meunier D, Lambiotte R, Bullmore ET (2010) Modular and hierarchically modular organization of brain networks. Front Neurosci 4

Morriss-Kay GM, Wilkie AOM (2005) Growth of the normal skull vault and its alteration in craniosynostosis: insights from human genetics and experimental studies. J Anat 207:637–653

Moss ML, Young RW (1960) A functional approach to craniology. Am J Phys Anthropol 18:281–292

Mountcastle VB (1995) The parietal system and some higher brain functions. Cereb Cortex 5:377–390

Nejad KK, Sugiura M, Nozawa T, Kotozaki Y, Furusawa Y, Nishino K, Nukiwa T, Kawashima R (2015) Supramarginal activity in interoceptive attention tasks. Neurosci Lett 589:42–46

Neubauer S, Gunz P, Hublin J-J (2009) The pattern of endocranial ontogenetic shape changes in humans. J Anat 215:240–255

Neubauer S, Gunz P, Hublin J-J (2010) Endocranial shape changes during growth in chimpanzees and humans: a morphometric analysis of unique and shared aspects. J Hum Evol 59:555–566

Orban GA, Claeys K, Nelissen K, Smans R, Sunaert S, Todd JT, Wardak C, Durand JB, Vanduffel W (2006) Mapping the parietal cortex of human and non-human primates. Neuropsychologia 44:2647–2667

Pearce E, Stringer C, Dunbar RIM (2013) New insights into differences in brain organization between Neanderthals and anatomically modern humans. Proc R Soc Lond B Biol Sci 280:1758

Peer M, Salomon R, Goldberg I, Blanke O, Arzy S (2015) Brain system for mental orientation in space, time, and person. Proc Natl Acad Sci U S A 112:11072–11077

Pereira-Pedro AS, Bruner E (2016) Sulcal pattern, extension, and morphology of the precuneus in adult humans. Ann Anat 208:85–93

Ponce de León MS, Bienvenu T, Akazawa T, Zollikofer CPE (2016) Brain development is similar in Neanderthals and modern humans. Curr Biol 26:R665–R666

Posner MI, Walker JA, Friedrich FJ, Rafal RD (1984) Effects of parietal injury on covert orienting of attention. J Neurosci 4: 1863–1874

Preuss TM (2011) The human brain: rewired and running hot: the human brain: rewired and running hot. Ann N Y Acad Sci 1225: E182–E191

Quallo MM, Price CJ, Ueno K, Asamizuya T, Cheng K, Lemon RN, Iriki A (2009) Gray and white matter changes associated with tool-use learning in macaque monkeys. Proc Natl Acad Sci U S A 106: 18379–18384

Rangel de Lázaro G, de la Cuétara JM, Píšová H, Lorenzo C, Bruner E (2016) Diploic vessels and computed tomography: segmentation and comparison in modern humans and fossil hominids. Am J Phys Anthropol 159:313–324

Rilling JK (2006) Human and non-human primate brains: are they allometrically scaled versions of the same design? Evol Anthropol 15: 65–77

Rilling JK (2008) Neuroscientific approaches and applications within anthropology. Am J Phys Anthropol 137:2–32

Runer E, Spinapolice E, Burke A, Overmann K (2017) Visuospatial integration: paleoanthropological and archaeological perspectives. In: Di Paolo LD, Di Vincenzo F, D'Almeida AF (eds) Evolution of primate social cognition. Springer, Cham. (in press)

Rushworth MFS, Paus T, Sipila PK (2001) Attention systems and the organization of the human parietal cortex. J Neurosci 21:5262–5271

Sakata H, Taira M, Kusunoki M, Murata A, Tanaka Y (1997) The parietal association cortex in depth perception and visual control of hand action. Trends Neurosci 20:350–357

Salinas J, Mills ED, Conrad AL, Koscik T, Andreasen NC, Nopoulos P (2012) Sex differences in parietal lobe structure and development. Gend Med 9:44–55

Scheperjans F, Eickhoff SB, Hömke L, Mohlberg H, Hermann K, Amunts K, Zilles K (2008a) Probabilistic maps, morphometry, and variability of cytoarchitectonic areas in the human superior parietal cortex. Cereb Cortex 18:2141–2157

Scheperjans F, Hermann K, Eickhoff SB, Amunts K, Schleicher A, Zilles K (2008b) Observer-independent cytoarchitectonic mapping of the human superior parietal cortex. Cereb Cortex 18:846–867

Scott N, Neubauer S, Hublin JJ, Gunz P (2014) A shared pattern of postnatal endocranial development in extant hominoids. Evol Biol 41:572–594

Seghier ML (2013) The angular gyrus multiple functions and multiple subdivisions. Neuroscientist 19:43–61

Semendeferi K, Damasio H (2000) The brain and its main anatomical subdivisions in living hominoids using magnetic resonance imaging. J Hum Evol 38:317–332

Semendeferi K, Damasio H, Frank R, Van Hoesen GW (1997) The evolution of the frontal lobes: a volumetric analysis based on three-dimensional reconstructions of magnetic resonance scans of human and ape brains. J Hum Evol 32:375–388

Sereno MI, Pitzalis S, Martinez A (2001) Mapping of contralateral space in retinotopic coordinates by a parietal cortical area in humans. Science 294:1350–1354

Sherwood CC, Smaers JB (2013) What's the fuss over human frontal lobe evolution? Trends Cogn Sci 17:432–433

Smaers JB (2013) How humans stand out in frontal lobe scaling. Proc Natl Acad Sci U S A 110:E3682–E3682

Sotero RC, Iturria-Medina Y (2011) From blood oxygenation level dependent (BOLD) signals to brain temperature maps. Bull Math Biol 73:2731–2747

Stieler JT, Bullmann T, Kohl F, Tøien Ø, Brückner MK, Härtig W, Barnes BM, Arendt T (2011) The physiological link between metabolic rate depression and tau phosphorylation in mammalian hibernation. PLoS One 6:e14530

Stout D, Chaminade T (2007) The evolutionary neuroscience of tool making. Neuropsychologia 45:1091–1100

Stout D, Toth N, Schick K, Stout J, Hutchins G (2000) Stone tool-making and brain activation: position emission tomography (PET) studies. J Archaeol Sci 27:1215–1223

Studer B, Cen D, Walsh V (2014) The angular gyrus and visuospatial attention in decision-making under risk. NeuroImage 103: 75–80

Tobias PV (1995) The brain of the first hominids. In: Changeaux JP, Chavaillon J (eds) Origins of the human brain. Clarendon Press, Oxford, pp 61–83

Tunik E, Rice NJ, Hamilton A, Grafton ST (2007) Beyond grasping: representation of action in human anterior intraparietal sulcus. NeuroImage 36:T77–T86

Utevsky AV, Smith DV, Huettel SA (2014) Precuneus is a functional core of the default-mode network. J Neurosci 34:932–940

Vanduffel W, Fize D, Peuskens H, Denys K, Sunaert S, Todd JT, Orban GA (2002) Extracting 3D from motion: differences in human and monkey intraparietal cortex. Science 298:413–415

Wardak C, Hamed SB, Duhamel JR (2005) Parietal mechanism of selective attention in monkeys and humans. In: Dehaene S, Duhamel JR, Hauser MD, Rizzolatti G (eds) From monkey brain to human brain. MIT Press, Cambridge, MA

Weidenreich F (1936) Observations on the form and proportions of the endocranial casts of *Sinanthropus pekinensis*, other hominids and the great apes: a comparative study of brain size. Palaeontol Sin B VII:1–150

Wild HM, Heckemann RA, Studholme C, Hammers A (2017) Gyri of the human parietal lobe: volumes, spatial extents, automatic labelling, and probabilistic atlases. PLoS One 12:e0180866

Wise SP, Boussaoud D, Johnson PB, Caminiti R (1997) Premotor and parietal cortex: corticocortical connectivity and combinatorial computations. Annu Rev Neurosci 20:25–42

Wu X, Bruner E (2016) The endocranial anatomy of Maba 1. Am J Phys Anthropol 160:633–643

Yantis S, Schwarzbach J, Serences JT, Carlson RL, Steinmetz MA, Pekar JJ, Courtney SM (2002) Transient neural activity in human parietal cortex during spatial attention shifts. Nat Neurosci 5:995–1002

Zhang S, Li CR (2012) Functional connectivity mapping of the human precuneus by resting state fMRI. NeuroImage 59: 3548–3562

A Comparative Perspective on the Human Temporal Lobe

16

Katherine L. Bryant and Todd M. Preuss

Abstract

The temporal lobe is a morphological specialization of primates resulting from an expansion of higher-order visual cortex that is a hallmark of the primate brain. Among primates, humans possess a temporal lobe that has significantly expanded. Several uniquely human cognitive abilities, including language comprehension, semantic memory, and aspects of conceptual processing, are represented in the temporal lobe. Understanding how the temporal lobe has been modified and reorganized in the human lineage is crucial to understanding how it supports human cognitive specializations. Identifying these structural modifications requires a direct comparison with other primates, with special attention to our closest relatives, the chimpanzees. Comparative examination of data from architectonics, tract tracing, and newer imaging methodologies suggests modifications to external morphology (gyri and sulci), preferential expansion of association areas, and elaboration of white matter fasciculi, distinguishing the human temporal lobe from those of Old World monkeys. Chimpanzees and humans share some of these features of cortical expansion, although more research is needed in order to elucidate whether humans possess simply a large hominoid temporal lobe or whether important reorganization has happened since our divergence from chimpanzees.

Keywords

Cortical expansion • Association cortex • Evolution • Multimodal • Hominoid • Fasciculus • Perisylvian

16.1 Introduction

The temporal lobe, as a morphological feature of the cerebrum set off from the rest of the cortex by the Sylvian (lateral) fissure, is arguably unique to Primates, because only primates possess a Sylvian fissure (Preuss 2007) and it appears early in primate evolution (Allman 1982). The majority of mammalian orders completely lack a temporal lobe. In the few non-primate species that have laterally expanded cortices which appear similar to the primate temporal lobe, these expansions lack the anterior projection that is developed as in the Primates. Further, these specializations are found in three different orders (Proboscidea, Cetacea, and Carnivora), indicating that they evolved independently from the primate temporal lobe and from each other. Within Carnivora, canids alone evolved a sulcal analog to the Sylvian fissure three separate times (Lyras 2009), and these analogous "temporal lobes" vary considerably. In the domestic dog, *Canis familiaris*, the two major gyri that make up the temporal region wrap around the canid Sylvian fissure and extend into the parietal regions (Datta et al. 2012; Hecht et al. 2016), quite unlike

K.L. Bryant
Donders Institute for Brain, Cognition and Behaviour, Radboud University Nijmegen, Kapittelweg 29, 6525 EN Nijmegen, The Netherlands
e-mail: k.bryant@donders.ru.nl

T.M. Preuss (✉)
Yerkes National Primate Research Center, Emory University, 954 Gatewood Rd, Atlanta, GA 30329, USA
e-mail: tpreuss@emory.edu

E. Bruner et al. (eds.), *Digital Endocasts*, Replacement of Neanderthals by Modern Humans Series,
DOI 10.1007/978-4-431-56582-6_16

Primates. Another carnivoran species, the sea lion (*Zalophus californianus*), has such unusual sulcal and gyral patterns in the temporal cortex that analogous gyri and sulci with other carnivorans are not apparent (Sawyer et al. 2016). In cetaceans, the temporal region is markedly different in Primates, in gross morphology, sulcal patterning, and functional organization (Hof and Van Der Gucht 2007; Marino 2002; Morgane et al. 1980). These are some of the reasons that we argue that the anterolateral expansion of cortex, which is found in all primate species, is a uniquely primate specialization.

These observations lead to several important questions: Why did Primates evolve a temporal lobe? What morphological features distinguish the primate temporal lobe, and what behaviors does it support? Finally, what structural and functional changes have occurred in the lobe of ape and human lineages? And how do the unique structural adaptations of the human temporal lobe permit the complex cognitive abilities that are the hallmark of our species?

Let us begin with the first question: Why did Primates evolve a temporal lobe? Among mammals, primates distinguish themselves by their specialization in the processing of visual information. The significance of this primate neuroanatomical adaptation may be linked to the evolution of the dual visual pathway system and, specifically, to the evolution of the areas of the ventral pathway (Kaas 2006). The expansion of higher-order visual areas in the ventral stream, which handle object recognition and semantic information, may have been a driving force behind the early appearance of the temporal lobe in primate evolution (Barton 1998; Preuss 1993, 2007). In the most well-studied nonhuman primate, the macaque monkey, the temporal cortices are important for housing both primary and secondary auditory cortices, higher-order visual processing areas, and multimodal areas.

What features distinguish the primate temporal lobe? Fossil evidence suggests that temporal neocortical expansion occurred early in the primate lineage, in the early Eocene period (Gurche 1982), while frontal lobes remained small. Nearly all modern primates possess distinct temporal lobes, and haplorrhine primates (Old World monkeys) possess a distinctive sulcus that extends in the anteroposterior direction along the lobe. The development of the temporal lobe in Primates probably reflects the visual specialization of the order (Allman 1982).

Understanding how the temporal lobe was further modified in the hominoid (ape and human) lineage is more challenging, but we have several clues. When we compare human brains to the primate group from which Hominoidea diverged, the Old World monkeys, we can observe important modifications to the location of visual motion area MT+. This cortical area, which may be used as a landmark for locating the border between occipital and temporal cortices, has been displaced posteriorly and inferiorly when compared to macaques (Fig. 16.2). This displacement suggests that in addition to temporal expansion, reorganization – in the form of disproportionate expansion of more anterior temporal cortex, at least – has occurred. Another clue to the structural modifications in the hominoid lineage lies in the gyral and sulcal morphology of the temporal lobe. Although ape temporal lobes do not differ in proportionate size when compared to monkeys (Rilling and Seligman 2002), visual inspection reveals that two novel temporal gyri, the middle temporal gyrus and fusiform gyrus, have appeared in the hominoid lineage (Figs. 16.1 and 16.2). This could be a passive consequence of cortical expansion – larger cortices possess more folds – but it could also reflect changes in cortical organization.

In order to understand how the human temporal lobe supports human-unique cognitive abilities like language and conceptual processing, a more thorough characterization of human temporal lobe organization is necessary. Comparable cortical maps of humans, chimpanzees, and Old World monkeys are necessary to understand what features of human brain organization are uniquely human and which we share with our great ape relatives. Building comparative cortical maps has been challenging because of the lack of neuroscientific methods that are applicable across species and, thus, directly comparable (Preuss 2010); currently, the most frequently cited cortical maps are over a century old (Brodmann 1905).

The problem of human temporal cortex mapping is large, and the information currently available to us is very incomplete, but there are several inroads that can guide our inquiry. First, we know that human temporal lobes have expanded when compared to other primates, including our closest relatives, the great apes (Rilling and Seligman 2002). Another potentially important source of information on human temporal specializations is architectonics – the structural parcellation of cortex based on regional differences in the distribution of cell bodies (cytoarchitecture) or the distribution and density of myelin in the gray matter (myeloarchitecture). Parcellation schemes from comparative architectonic studies suggest important differences between humans and nonhuman primates with respect to the proportion of association cortices when compared to primary areas (Orban et al. 2004; Glasser et al. 2014). However, the majority of architectonic studies focus on macaques, with less work in humans and even sparser data from our closest relatives, the chimpanzees.

The advent of neuroimaging technologies, especially structural magnetic resonance (MR) imaging and diffusion tensor imaging (DTI), offers a new opportunity to directly compare cortical organization in humans, chimpanzees, and

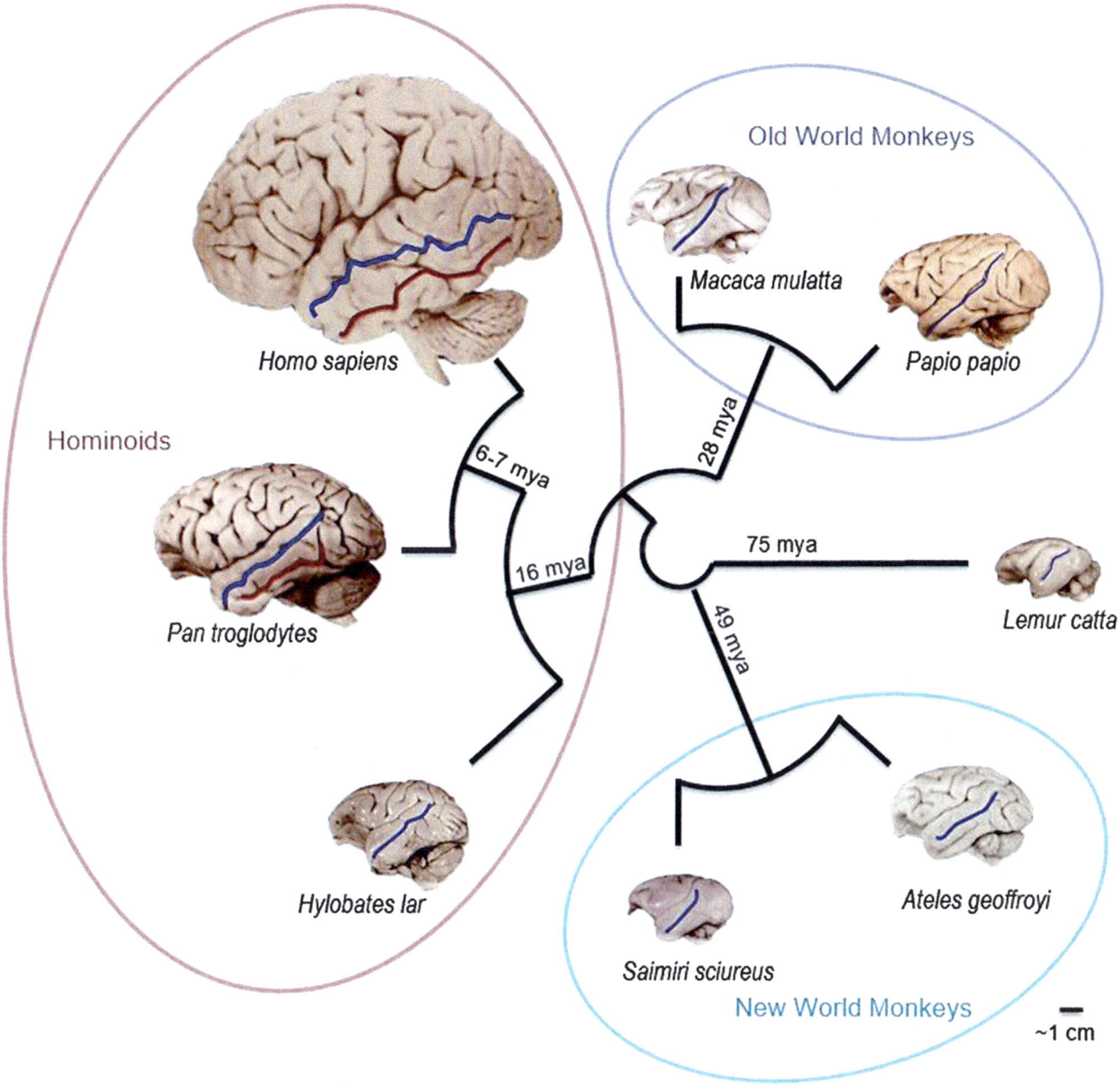

Fig. 16.1 The evolutionary relationship of extant primates. *Blue line* highlights the superior temporal sulcus, present in the majority of primate species. *Red line* highlights the inferior temporal sulcus, present in hominids (chimpanzees and humans) only. Brains are roughly to scale (all brain images adapted from the University of Wisconsin and Michigan State Comparative Mammalian Brain Collections and the National Museum of Health and Medicine, on their website neurosciencelibrary.org, funded by NSF and NIH. Divergence date estimates are from Steiper and Seiffert 2012)

macaques. This review will cover some of the current knowledge on the structural features of the human temporal lobe from comparative studies of morphometry, architectonics, and neuroimaging. Structural data, available across many primate species, will be examined alongside functional data (chiefly from macaque and human studies) in order to infer how human structural specializations may contribute to uniquely human cognitive abilities.

16.2 Morphological Modifications to Temporal Lobes in the Old World Monkey Lineage

Among primates, the human temporal lobe has a number of distinctive structural features (Fig. 16.1). The most easily observable modification to human temporal cortex is its

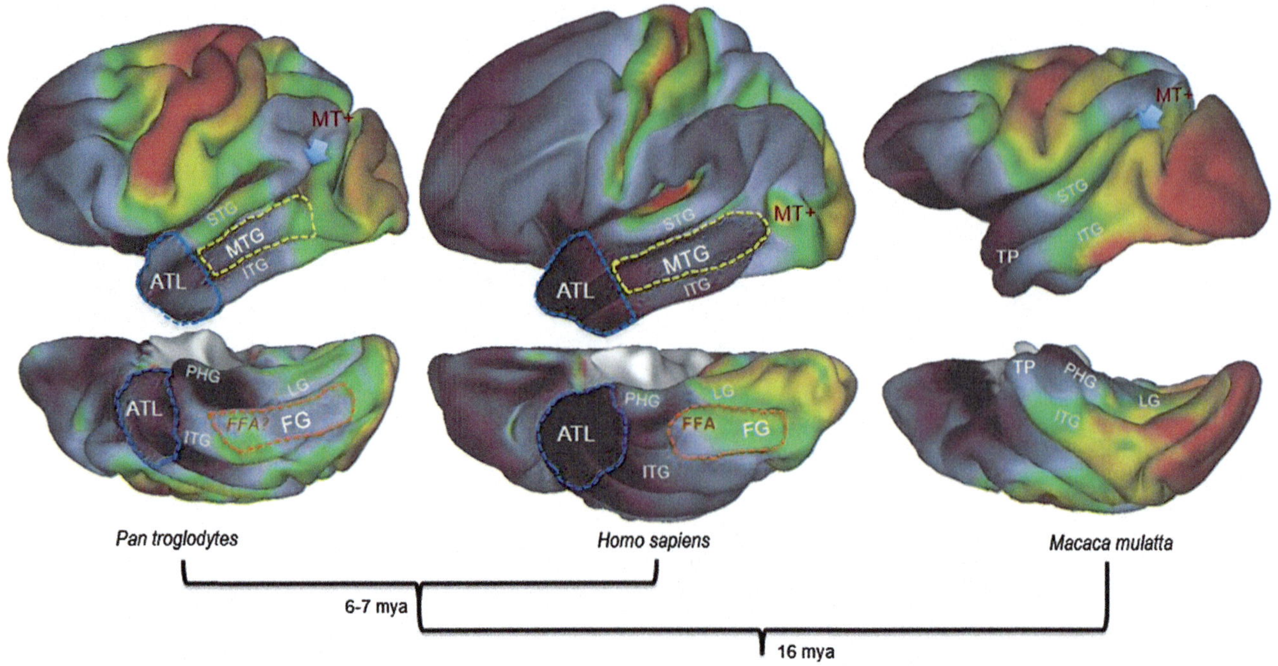

Fig. 16.2 Major temporal multimodal association areas in humans, chimpanzees, and macaques. Brains are not to scale. *ATL* anterior temporal lobe, *FFA* fusiform face area, *FG* fusiform gyrus, *ITG* inferior temporal gyrus, *LG* lingual gyrus, *MT+* visual motion area MT complex, *MTG* middle temporal gyrus, *PHG* parahippocampal gyrus, *STG* superior temporal gyrus, *TP* temporal pole. Divergence date estimates are from Steiper and Seiffert (2012). Cortical myelin map reconstructions made available by the Human Connectome Project, WU-Minn Consortium (Principal Investigators: David Van Essen and Kamil Ugurbil; 1U54MH091657), funded by the 16 NIH Institutes and Centers that support the NIH Blueprint for Neuroscience Research and by the McDonnell Center for Systems Neuroscience at Washington University

absolute and relative expansion when compared to nonhuman primates. In addition to overall non-allometric scaling of human brain (Rilling 2006; Schoenemann 1997), the temporal lobe is larger than would be predicted based on anthropoid allometry (Rilling and Seligman 2002; but see Semendeferi and Damasio 2000). However, humans do share features of temporal lobe anatomy with apes that distinguish them from other primate species.

Compared to macaques, the primary auditory cortex of humans and chimpanzees appears to have been displaced posteriorly and occupies a smaller proportion of cortical surface area in the superior temporal plane (Hackett et al. 2001). The relative location of visual areas has also been modified: humans display not only a posterior displacement of visual motion area MT (Ungerleider and Desimone 1986; Watson et al. 1993), as noted above, but a posterior displacement of other higher-order visual areas (Orban et al. 2004; Glasser and Van Essen 2011) and expansion of cortical areas intervening between auditory core and area MT (Orban et al. 2004).

Visual inspection of the temporal lobe of Old World monkeys, such as macaques, shows they lack a deep inferior temporal sulcus and therefore lack the discrete middle temporal gyrus present in humans and chimpanzees (Fig. 16.1). Moving to the ventral aspect of the temporal lobe, the fusiform gyrus, which houses the fusiform facial area (FFA), is found in humans and chimpanzees, but there is no gyral counterpart in macaques (Fig. 16.2; Nasr et al. 2011; Weiner and Zilles 2015). If there are differences in the internal organization of temporal cortex between monkeys and hominoids, or differential expansion of particular cortical regions, these landmarks could provide information where evolutionary change occurred.

The advent of neuroimaging provides new tools for understanding the cortical organization of humans and nonhuman primates. When gyral and sulcal landmarks are compared in humans and macaques, expansions in the superior, middle, and inferior temporal cortices are evident (Hill et al. 2010; Fjell et al. 2015). These territories, including the anterior temporal lobe, show low cortical myelin density in both humans and chimpanzees when compared to macaques (Glasser and Van Essen 2011). Cortical expansion, coupled with lower myelin content, a feature of association cortex, is indicative of expansion of temporal association cortices in humans (Preuss 2011).

Taken together, these structural differences strongly suggest that important modifications to the sizes and organization of cortical areas within the temporal lobe have occurred since the divergence of hominoids from old-world monkeys. The hominoid temporal lobe has expanded and rewired over the course of evolution, and the association regions within this lobe have disproportionately expanded with respect to sensory regions (Glasser et al. 2014; Preuss 2011). In order to determine what features of temporal lobe are human evolutionary specializations, and which are shared with chimpanzees, further comparative work is needed to characterize the territories within these expanded association areas.

16.3 The Significance of Multimodal Association Cortices

16.3.1 Overview

The human temporal lobe anatomy possesses a large proportion of association cortex relative to total cortical surface area when compared to our primate relatives (Fig. 16.2). Association cortices are classically defined as regions that receive projections from multiple lower-order sensory areas (Flechsig 1901). If we examine cortex for myelin content, we can easily distinguish association cortices from primary areas; association cortex contains less myelin than primary areas and myelinates later, as well. Morphometric, cytoarchitectonic, myeloarchitectonic, and imaging studies suggest greater expansion of association areas relative to primary areas in humans and, to a lesser degree, chimpanzees, compared to old-world monkeys (Schoenemann 1997; Glasser et al. 2011; Passingham and Smaers 2014; Rilling 2006; Rilling and Seligman 2002; reviewed in Orban et al. 2004; Preuss 2011). Less is known, however, about how association areas have been modified during this expansion.

Both structural and functional data suggest that humans are using the expanded cortical regions in the temporal lobe to perform novel, human-specific or hominoid-specific functions such as language, configural processing for object and face recognition, theory of mind representation, and understanding the identity and functional properties of tools. Broadly, these cognitive functions may be referred to as semantic or conceptual representations, and their construction arises from the synthesis of information originating from multiple primary sensory modalities. In the following section, three temporal regions that are involved in the construction of conceptual representations in humans – middle temporal gyrus, fusiform gyrus, and anterior temporal lobe – will be discussed in terms of their structural modifications and putative functional adaptations in the human and hominoid lineage.

16.3.2 The Fusiform Gyrus

Of the face-responsive cortical areas in humans, the fusiform gyrus is the most robust in its face-specific activation (Kanwisher et al. 1997), with fMRI activations for faces over both scenes (Epstein and Kanwisher 1999) and objects (Allison et al. 1994; Kanwisher et al. 1999). The face-selective activation of the FG is more reliably observed in the right hemisphere of humans (Kanwisher et al. 1997; McCarthy et al. 1997). In both hemispheres, this territory, termed the fusiform face area (FFA) has been localized in the middle portion of FG (Allison et al. 1994; Saygin et al. 2012), just anterior to areas responsible for color perception (Clarke and Miklossy 1990; Allison et al. 1993, 1994). Face recognition in the FFA is dependent on expertise (Gauthier et al. 1999) and affective judgments (Pizzagalli et al. 2002). Evidence for dissociation of whole-face versus face-component processing has been observed, with whole-face processing correlated with activation in the right FG (Rhodes 1993; Hillger and Koenig 1991) and left FG activation correlated with the processing of face components (Rossion et al. 2000).

The processing of face components as a "whole," whose recognition depends on the relative spatial relationship of component facial features, is termed configural face processing. Configural face processing has been argued to also be a part of humans' and chimpanzees' cognitive repertoire, but not macaques (Parr et al. 1998, 2006, 2008), and a chimpanzee homolog of FFA in the FG has been localized using positron emission tomography (PET; Parr et al. 2009). Other territories within FG have been implicated in expertise in object recognition beyond face processing. The left FG houses the visual word form area (VWFA) in the middle portion of the gyrus, approximately in Brodmann's area (BA) 37 (Cohen et al. 2000; McCandliss et al. 2003). VWFA has shown activation for both words and pictures, specifically to the abstract, orthographic properties of words (Polk and Farah 2002; Binder et al. 2006). Starrfelt and Gerlach (2007) propose the VWFA is specialized for letter and word recognition as a configural processing task. The exact nature of the function of the VWFA with regard to reading and word processing – i.e., is it operating at the lexical or pre-lexical level? – is still up for debate (see Devlin et al. 2006). Further complicating the issue, an area posterior and medial to VWFA has been shown to activate during grapheme-to-phoneme sound conversion tasks, another important component of reading (Dietz et al. 2005). However, it seems clear that that territories within the middle and posterior portions of the left FG play a role that interfaces with both auditory and visual sensory modalities, as well as abstracted or "supramodal" representations.

Within the FFA, it is possible that subregions may be distinguishable as imaging techniques become more sophisticated. Localization of the FFA with evoked potentials had previously demonstrated significant individual variation in the location of activation within the FG (Allison et al. 1994).

A region of the FG responsive to bodies was reported by Peelen and Downing (2005). Subsequent examination of this area with high-resolution fMRI by Schwarzlose and co-workers (2005) has provided evidence that this body-selective territory is anatomically distinct from the FFA. This "fusiform body area" (FBA) was localized anterior and lateral to FFA (Schwarzlose et al. 2005). More recent work by Saygin et al. (2012), using a novel structure-function connectivity fingerprint approach, found that for many individual subjects, two adjacent but discrete areas of face-selective activation were discernable within FG. Clearly, additional work is needed to determine whether the FFA and associated configural processing areas can be subdivided further, functionally as well as anatomically.

In summary, the right fusiform gyrus houses the FFA and FBA, and the left FG contains the VWFA and associated grapheme and phoneme processing areas. These functional areas are responsible for expertise-based recognition and configural face processing in humans, with the latter being right lateralized. Evidence for configural face processing abilities in chimpanzees but not macaques, along with the lack of FG as a discrete convolution in old-world monkeys, suggests FG, a morphological specialization of hominoids, houses cortical areas that are specialized in, and possibly unique to, apes and humans.

16.3.3 The Middle Temporal Gyrus

Human MTG is well documented in humans as a multimodal cortical region (Binder et al. 2009) that functions as an important language center. Evidence has been mounting for the role of MTG in mapping sounds to meanings (Hickok and Poeppel 2004, 2007), perhaps being essential to semantic comprehension, based on data from patients with lesions in that region (Dronkers et al. 2004; Bates et al. 2003). Based on structural and functional data, the posterior portion of the MTG (pMTG), in particular, has been proposed to constitute a "semantic hub" (Turken and Dronkers 2011). The pMTG in humans has been implicated in naming and retrieving information about tools (Martin et al. 1996; Mummery et al. 1998; Chao et al. 1999; Martin and Chao 2001), generating action words (Wise et al. 1991; Martin et al. 1995; Fiez et al. 1996), and participating as part of a structural network supporting tool semantics and motor behaviors (Ramayya et al. 2010) and, further, has been suggested to be the site of storing information about non-biological object motion more generally (Martin et al. 1996). Chao et al. (1999) have speculated that this is possibly related to its anatomical position, close to visual motion processing areas of the MT+ complex. The MT+ complex itself is a visual motion area which is not novel to the hominoid lineage but in fact is well documented in several primate species. Area MT was first identified in owl monkeys (Allman and Kaas 1971), later in rhesus macaques (Rockland and Pandya 1979), and in galagos (Allman et al. 1973) indicating that visual specializations for motion processing occurred early in the primate lineage.

The anterior half of MTG (aMTG) is also implicated to be involved in a semantic processing network (Copland et al. 2003; Schwartz et al. 2009; Butler et al. 2014). Human imaging studies suggest the aMTG is involved in lexical decision-making, for example, reading words with atypical spelling-to-sound correspondences or "exception words" (Wilson et al. 2012), in visual word recognition (Pammer et al. 2004), and in spoken word recognition (Roxbury et al. 2014). In one of the few studies examining the different role of anterior vs. posterior MTG, Vandenberghe et al. (1996) found aMTG had stronger activation in semantic tasks involving processing images of words rather than pictures when compared to pMTG. In contrast, Visser and co-workers (2012) found pMTG specialized for semantic processing of words, while aMTG responds equivalently to both words and pictures. Both findings are consistent with a recent meta-analysis suggesting the full anterior-posterior axis of the MTG acts as a multimodal convergence zone (Binder and Desai 2011). However, unlike pMTG, there is less evidence for aMTG handling semantic and action knowledge related to tools. Anterior MTG appears to be recruited for recognition of famous faces (Leveroni et al. 2000) and proper names of famous individuals (Gorno-Tempini et al. 1998), tasks that may be considered as tapping into semantic "meaningfulness" (Binder et al. 2009). The latter two findings are similar to functions that have been localized in the anterior temporal lobe (ATL) broadly, perhaps reflecting conflicting interpretations regarding the location and extent of the ATL as it encroaches posteriorly (reviewed in Bonner and Price 2013).

16.3.4 The Anterior Temporal Lobe

The ATL, a large swathe of cortex encompassing the most anterior parts of the STG, MTG, ITG, and temporal pole, is a multimodal association region that, as modern functional neuroimaging studies make clear, plays an important role in both semantic memory and affective cognition in humans. The semantic and affective cognitive functions that have been localized to the ATL in humans include the production and comprehension of spoken and written words and pictures (Coccia et al. 2004; Pobric et al. 2007); taste recognition (Small et al. 1997); olfactory memory (Rausch et al. 1977; Eskenazi et al. 1986); stimulus-invariant perception of emotional facial expressions (Schmolck and Squire 2001; Cancelliere and Kertesz 1990); generation of emotions in response to visual cues (Reiman et al. 1997); a storage site for unique, socially relevant entities, such as familiar people and landmarks (Damasio et al. 2004; Frith 2007; Kriegeskorte et al. 2007); comprehension of social concepts (Zahn et al. 2007; Zahn et al. 2009; Ross and Olson 2010); emotional memory retrieval (Dolan et al. 2000); and coherent conceptual categorization of objects (Rogers et al. 2004; Lambon Ralph et al. 2010).

The conceptual processing that occurs in the ATL has been argued to be transmodal, or perhaps amodal (Pobric et al. 2010), in that conceptual information is computed regardless of the sensory modality of the stimulus, as auditory, visual motion, olfactory, and gustatory processing streams converge at the temporal pole (Binder and Desai 2011). Lambon Ralph and Patterson (2008) observed undergeneralization and overgeneralization of concepts in patients with bilateral degeneration of the temporal lobes and suggest that the ATL plays a crucial role in binding perceptual features across stimulus categories to form modality-invariant conceptual information that links back to modality-specific association cortices. On this view, the ATL constitutes a modality-invariant semantic hub (Lambon Ralph et al. 2010; Visser et al. 2012). Others have argued that the ATL binds multimodal inputs with visceral emotional responses while maintaining segregation of perceptual modalities (Olson et al. 2007). These sometimes conflicting reports can be categorized into three separate accounts of the role of the ATL in semantic memory: (1) as a supramodal/transmodal/amodal semantic hub, (2) as a storage site for unique entities (e.g., famous names and faces), and (3) as a center for social conceptual knowledge (Simmons and Martin 2009; Simmons et al. 2009).

Human ATL has been implicated in the comprehension and expression of social knowledge, including theory of mind (Gallagher and Frith 2003, but see Shaw et al. 2007). Imaging studies in humans support the role of TP in inferring deceit (Grezes et al. 2004), ethical decision-making (Heekeren et al. 2003), and moral and social judgments (Moll et al. 2001, 2002). While few studies have attempted to subdivide the ATL into functional divisions, the superior ATL has been linked to processing of abstract social concepts (Zahn et al. 2007), in contrast to inferior ATL, which has been found to be a hotspot for semantic memory (Visser et al. 2010).

Despite its evident importance in higher-order aspects of human cognition, the anatomical components and boundaries of the ATL in humans are not currently well understood. The region includes the anterior parts of Brodmann's areas 22 (STG), 21 (MTG), 20 (ITG), and 38 (temporal pole also known as area TG) (Economo and Koskinas 1925; Bailey and Bonin 1951). In both chimpanzees and macaques, the same architectonic identifiers have been applied to the cortex of the anterior temporal lobe (e.g., Bailey et al. 1950; Bonin and Bailey 1947). There is little evidence of semantic representation in the anterior temporal cortex of nonhuman primates, which suggests that the region was modified in hominoid or hominin evolution. Unfortunately, detailed comparative anatomical studies of this region with modern methods are lacking, so at present it is not possible to give a comprehensive account of those features that are shared among catarrhine primates and which, if any, are specializations of hominoids or hominins. The least arguable component is the cortex at the tip of the temporal lobe, BA38 (also known as area TG), which appears to be present in all catarrhines that have been examined. In fact, it is a composite of bands of different types of cortex – periallocortex, proisocortex, and true isocortex – distinguishable based upon the number of cortical layers present. This arrangement of bands, which is continuous with bands of insular cortical tissue, makes area TG identifiable across species (e.g., Mesulam and Mufson 1982; Blaizot et al. 2010; Insausti 2013). However, TG makes up only a portion of the ATL, as identified functionally.

Some functional commonalities between the anterior temporal cortex of humans and macaques have been proposed. Olson et al. (2012) propose, based partially on a review of human and nonhuman primate studies, that connectivity with the amygdala and orbitofrontal areas underpins the ATL's social cognitive function, which they argue is a form of semantic or conceptual knowledge processing that privileges emotionally salient information. Another model, based on PET data in humans, includes ATL as part of an extensive neural network with other cortical regions, including medial and superior prefrontal cortices and cingulate cortices (Goel et al. 1995; Calarge et al. 2003). Evidence for vocal and facial identity discrimination extending into anterior portions of macaque STS and IT cortex (Perrett et al. 1992) is consistent with a role for this region in social cognition. On the other hand, Mars et al. (2013) have identified a region at the anterior end of the human STG with a pattern of functional connectivity similar to that of the human temporoparietal junction cortex (a region considered essential for higher-order social cognition; Saxe and Kanwisher 2003), whereas they could identify no corresponding region in macaques using similar methods.

Thus, much work remains to be done in this important region in order determine whether the human anterior temporal lobe has simply expanded, or also been reorganized, and what features, if any, are unique to humans. Another important piece of the puzzle is the organization of chimpanzee temporal cortex, which has been virtually unexamined. Adding to this problem, the temporal pole's location at the sinusoidal interface of the skull creates problematic artifacts for many neuroimaging modalities, potentially reducing the sensitivity of DTI and fMRI measures (Glasser and Van Essen 2011).

16.4 Major Fasciculi of the Temporal Lobe

16.4.1 Overview

Connectivity between distant association areas relies on fasciculi – large, coherent fiber bundles that travel long distances through white matter. These structures have been studied traditionally with blunt dissection and more recently via structural MR imaging, including diffusion tensor imaging and diffusion spectrum imaging. This section will cover the structure and putative functional contributions of fasciculi that traverse the temporal lobe in humans, with some discussion of differences with nonhuman primates.

There are six fasciculi that are usually recognized in macaques that connect to the temporal lobe and seven in humans (Fig. 16.3). These numbers are not universally agreed upon but are a reasonable comparative representation of human and macaque fascicular organization from a survey of the literature. Fasciculi that have been documented in the temporal lobe of both species include the arcuate fasciculus (AF; sometimes considered a branch of the superior longitudinal fasciculus system, SLF III), the middle longitudinal fasciculus (MdLF), the inferior longitudinal fasciculus (ILF), the uncinate fasciculus (UF), the ventral pathway or extreme/external capsule (VP, EmC/EC), and the cingulum bundle (CB). In humans, an additional fasciculus, adjacent to the ILF, is usually referred to as the inferior fronto-occipital fasciculus (IFOF).

One of the more frequently mentioned fasciculi in the human literature is the arcuate, perhaps due to the attention Geschwind's model of language and speech production has received. Our understanding of the neural substrate of human language has been strongly influenced by Geschwind's description of disconnection syndrome (Geschwind 1965), a form of language impairment resulting from interruption of communication between association areas. At the time, virtual reconstructions of white matter via DTI were unavailable, and therefore this model was based on an understanding of white matter bundles based on older blunt dissections (e.g., Dejerine and Dejerine-Klumpke 1901) and Geschwind's own studies of monkeys and of human aphasic patients (Geschwind 1965, 1970).

There is considerable disagreement regarding the naming conventions of individual fasciculi, as well as the grouping of white matter bundles into individual fasciculi. Of the complex white matter bundles that make up the connectional anatomy of language in humans, Dick and Tremblay (2012) review and describe approximately six different interpretations of organization, ranging from one to seven fasciculi. This raises an important epistemic problem: How do we define and differentiate white matter bundles into individual fasciculi? In humans, chimpanzees, and, to a lesser degree, macaques, broad cortical connections have been described on both ends of major fasciculi. Does a coherent white matter bundle, as visible in a map derived from DTI, constitute a fasciculus? Or should a fasciculus be defined by common connectivity points between cortical areas? How do the limits of DTI resolution impact our identification of fasciculi? It is possible, for example, that bundles that are discrete upon examination in DTI may in fact be composed of multiple fasciculi that are packed and run together with other fasciculi with different cortical terminations. Keeping these concerns in mind, this section will outline the state of our understanding of the presence of major fasciculi in humans and our primate relatives.

16.4.2 Arcuate Fasciculus

The arcuate is unique among major fasciculi in having important connections across frontal, parietal, and temporal association cortices, as it arches around the Sylvian fissure (Fig. 16.3). In humans, AF terminations reach STG, MTG, and ITG in the temporal lobe, linking them to Broca's area (BA 45 and BA44) as well as ventral premotor cortex and middle frontal gyrus (Fig. 16.3a; Glasser and Rilling 2008; Powell et al. 2006). The current conception of the AF has been heavily influenced by Norman Geschwind's model (Geschwind 1970; Dick and Tremblay 2012), which conceptualized the arcuate as the major connection between Broca's and Wernicke's areas. Disruption of the left AF has been implicated in the classical conception of aphasia as a disconnection syndrome (Geschwind 1970) resulting in conduction aphasia (impairment of repetition function; Damasio and Damasio 1980; Catani and ffytche 2005; but see Bernal and Ardila 2009). Because of the importance of the AF in speech production (e.g., Marchina et al. 2011; Yeatman et al. 2011), it is sometimes described as containing the "phonological pathway," particularly the portion that interconnects the posterior superior temporal gyrus with the fronto-opercular cortex (Duffau 2008; Glasser and Rilling 2008). Leftward asymmetry of the arcuate in humans (Nucifora et al. 2005; Glasser and Rilling 2008; Rilling et al. 2008) supports the model of the arcuate as crucial for the development human language. Less well studied, the right arcuate may play a role in processing music, particularly vocal-based music (Halwani et al. 2011).

Of the major fiber bundles in primates, the arcuate has arguably been studied most carefully in a comparative light (Rilling et al. 2008, 2011). Rilling and colleagues, using DTI, reported that human AF connects the inferior frontal gyrus not only with the posterior superior temporal gyrus

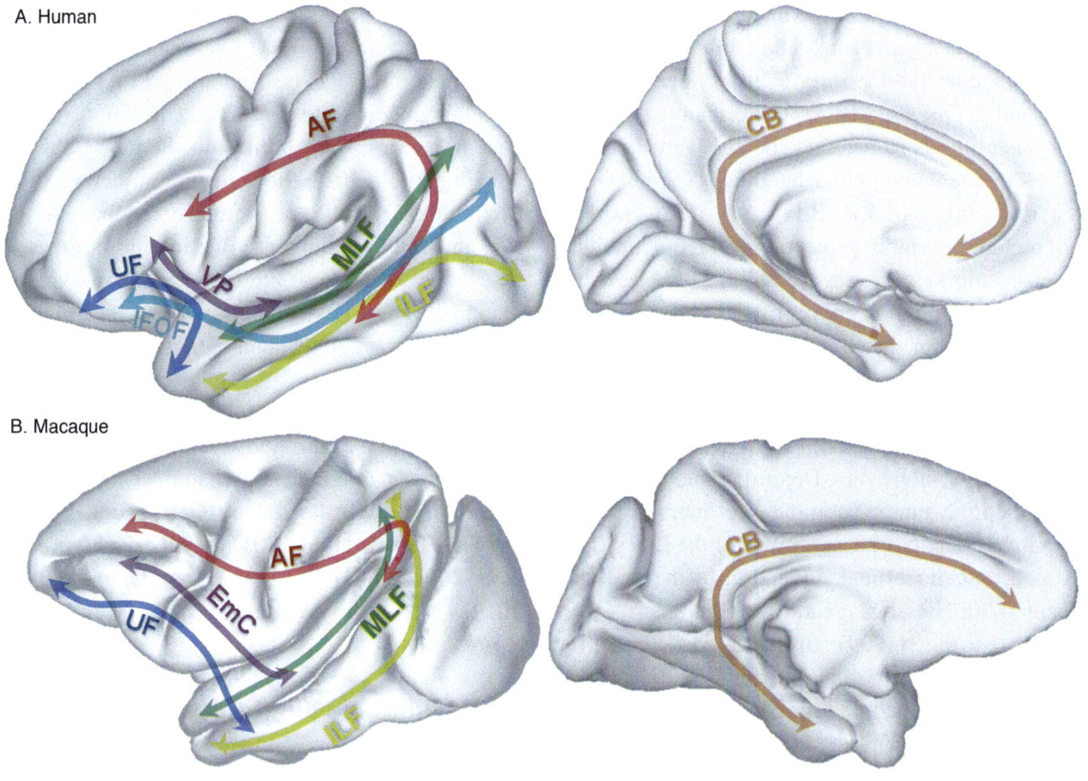

Fig. 16.3 A simplified diagram of major fasciculi that connect with and/or course through the temporal lobe. Human fascicular trajectories are based on Catani and Thiebaut de Schotten (2008), Rilling et al. (2008), Saur et al. (2008), Frey et al. (2008), Makris et al. (2009), and Turken and Dronkers (2011). Macaque fascicular trajectories are based on Schmahmann and Pandya (2007) and Schmahmann et al. (2007). *AF* arcuate fasciculus, *CB* cingulum bundle, *EmC* extreme capsule, *IFOF* inferior fronto-occipital fasciculus, *ILF* inferior longitudinal fasciculus, *MLF* middle longitudinal fasciculus, *UF* uncinate fasciculus, *VP* ventral pathway. Inflated cortical reconstructions were generated using the Freesurfer image analysis suite (http://surfer.nmr.mgh.harvard.edu; Fischl et al. 1999)

(which includes the *planum temporale*, often equated with Wernicke's area) but also with cortex in the middle temporal gyrus involved in semantic representation. In chimpanzees, AF connections are largely restricted to the superior temporal gyrus, and the AF appears to be very poorly developed in macaques. These results point to important language-related evolutionary changes in the human temporal lobe; moreover, the changes in connectivity detected with DTI may reflect changes in the organization and function of cortex occupying the MTG. Interestingly, Rilling et al. (2011) report that while the AF is larger on the left than the right in humans, it is even more strongly left lateralized in chimpanzees. As counterintuitive as this might seem, it is consistent with results indicating humanlike asymmetries of white matter volume in the perisylvian region (Cantalupo et al. 2009; Hopkins et al. 2010) and especially the finding that the *planum temporale* is even more asymmetrical (left > right) in chimpanzees than in humans (Gannon et al. 1998). It is therefore likely that at least some of the asymmetries of cortical regions and fascicles involved in human language evolved prior to the emergence of language (Preuss 2011).

16.4.3 Cingulum Bundle

The cingulum is a medial associative bundle that runs along the cingulate gyrus, medial to the corpus callosum (Fig. 16.3), with the longest fibers extending from the orbitofrontal cortex to the anterior medial temporal lobe by way of the parahippocampal gyrus (Catani and Thiebaut de Schotten 2008). It has been described in several primate species, including the marmoset (Beevor 1891) and rhesus macaque (Baleydier and Mauguiere 1980; Goldman-Rakic et al. 1984; Schmahmann and Pandya 2007). Like the arcuate, it also bridges frontal, parietal, occipital, and temporal cortices, although perhaps less extensively than the arcuate. The cingulum is implicated in memory, emotions, and attention functions (Rudrauf et al. 2008; Catani 2006).

16.4.4 Inferior Longitudinal Fasciculus and Inferior Fronto-occipital Fasciculus

In humans, there are fibers that reach from the inferior frontal lobe and extend through nearly the entire length of the inferior temporal lobe. The white matter tracts that make up these connections have been called both the inferior longitudinal fasciculus and the inferior fronto-occipital fasciculus (ILF and IFOF; Fig. 16.3). Historically, there has been some contention as to whether the ILF and the IFOF exist as fully separate white matter bundles.

The first description of the ILF came from Burdach (1822) and included ventral projections via the uncinate and extreme capsule system. Sachs (1892) later described what became a more common conceptualization of the ILF, as a fiber bundle system that travels from occipitoparietal regions, passing through the MTG and STG and terminating in the temporal lobe. Dejerine and Dejerine-Klumpke's (1895) description is similar to both Burdach, in that it also continues ventrally; however Dejerine and Dejerine-Klumpke included the anterior commissure fibers as part of this system. Other concurrent descriptions of the ILF system describe it as a thalamocortical projection to the occipital lobe (Flechsig 1896; Niessl-Mayendorf 1903).

Soon after, Curran (1909), using blunt dissection, provided the earliest description of a bundle adjacent but separate from the ILF, which he termed the fronto-occipital fasciculus inferior (FOFI). This description indicated the two run concurrently from occipitotemporal areas to the inferior frontal lobe, with the only difference being that the fronto-occipital fasciculus is inferior to the ILF. Later, Davis (1921) reviewed the literature on these two bundles and performed his own blunt dissections. Davis argued that Curran's FOFI is not distinct from the ILF sensu Dejerine. However, Davis identified a bundle discrete from the ILF which courses ventrally into the frontal lobes, in close apposition to and medial/superior to the uncinated fasciculus, which he named the fronto-occipital fasciculus. This system is frequently termed the "IFOF"; however some anatomists refer to it as the "IOFF" (Kier et al. 2004; Turken and Dronkers 2011).

The ILF and the IFOF continue to be recognized as separate fasciculi in neuroimaging literature. Catani et al. (2003) and Catani and Thiebaut de Schotten (2008) track the IFOF from two inputs, the posterior parietal and the occipital lobes, to the length of the temporal lobe and extending ventrally through the extreme/external capsule to both orbitofrontal and lateral prefrontal cortices, dorsal to the uncinate. In humans, Catani and Thiebaut de Schotten (2008) describe the ILF as a ventral associative bundle connecting the occipital and temporal lobes, coursing through the middle temporal gyrus, with short fibers extending to the amygdala and hippocampus (Catani et al. 2003). These workers acknowledge the difficulty of distinguishing ILF and IFOF fibers in DTI scans. Menjot de Champfleur (2012) localizes IFOF medial to ILF with an additional ventral extension.

In macaques, although a radiographic study found no evidence of a discrete ILF (Tusa and Ungerleider 1985), a more recent neuroimaging study identified the ILF as coursing through the ITG (Schmahmann et al. 2007). The chimpanzee ILF has been identified in a comparative study with humans (Rilling et al. 2011), as having a similar course and trajectory, but unlike in humans, the chimpanzee ILF does not about the inferior projection of the arcuate fasciculus. Overall, the literature supports the existence of ILF and IFOF in humans, ILF in chimpanzees, and ILF in macaques. Because of the current lack of evidence for non-human IFOF, it has been argued that the IFOF is unique to the human brain (Catani 2006; Thiebaut de Schotten et al. 2012).

Separating the two bundles in functional studies is more challenging. We do know that the IFOF and ILF are implicated in reading (Epelbaum et al. 2008; Catani and Mesulam 2008), language (Catani and Mesulam 2008; DeWitt-Hamer et al. 2011), and visual processing (Fox et al. 2008; Rudrauf et al. 2008; ffytche 2008; ffytche et al. 2010).

16.4.5 Middle Longitudinal Fasciculus

The middle longitudinal fasciculus (Fig. 16.3; MdLF, occasionally MLF) is not universally recognized by neuroanatomists as a fasciculus discrete from IFOF and ILF in humans. Originally described in macaques as a bundle connecting the inferior parietal lobule with the white matter of the STG (Seltzer and Pandya 1984), later as coursing through the STG to the temporal pole (Schmahmann and Pandya 2007), it was first identified in humans using DTI (Makris et al. 2009). In humans, the MdLF connects the inferior parietal lobule/angular gyrus, with some occipital branching, to STG (DeWitt-Hamer et al. 2011), while the ILF runs through MTG/ITG (Menjot de Champfleur, 2012). Makris and colleagues (Makris et al. 2009; Makris and Pandya 2009) also chart the MdLF superior to the IFOF and ILF and inferior and medial to the AF/SLF III. Other workers have proposed that the MdLF is constitutive of the vertical segment of the AF (Catani et al. 2005; Frey et al. 2008; Makris et al. 2009).

The MdLF's connectivity between the angular gyrus, which houses a recently localized part of the perisylvian language network called Geschwind's territory (Catani et al. 2005), and the STG, containing Wernicke's territory, suggests a critical role for language comprehension. In addition, the bundle shows leftward asymmetry, and its location supports a role in the ventral language pathway, linked to sentence comprehension (Menjot de Champfleur et al. 2013). However, there is still little data on the function of MdLF, as it is difficult to disambiguate from other ventral fasciculi. Electrostimulation of the MdLF in the left dominant hemisphere of patients failed to elicit semantic deficits;

instead, these authors report semantic paraphasias only during stimulation of the lateral border of the IFOF, adjacent to the MdLF (DeWitt-Hamer et al. 2011), indicating the function of the MdLF in humans is still unknown.

16.4.6 Ventral Pathways

A fiber bundle connecting Broca's area with STG, MTG, and Wernicke's area has been described in humans (Parker et al. 2005; Saur et al. 2008) and other primates (Kaas and Hackett 1999; Romanski et al. 1999; Schmahmann et al. 2007). This pathway is problematic for a number of reasons; first, the terms extreme capsule pathway, external capsule pathway, and ventral pathway are used interchangeably by some authors. In macaques, the uncinate has been described as a pathway independent from the extreme capsule (EmC) or external capsule (ExtC) (Petrides and Pandya 2007; Schmahmann et al. 2007) or as coursing through the extreme/external capsule (Anwander et al. 2007). Another interpretation involves fascicular extensions of the IFOF or ILF through the extreme/external capsule to the prefrontal cortex (PFC). One way to deal with this confusion is to recognize three possible interpretations: (1) the ventral pathway as a white matter bundle in its own right which passes through either or both the extreme and external capsules (from orbitofrontal cortex/lateral PFC to temporal pole) and the uncinate as a separate pathways; (2) the ventral pathway and uncinate as described in 1, except the ventral pathway is an extension of the IFOF or ILF fascicular bundle; or (3) the uncinate as a pathway which passes through the external and/or extreme capsule.

Catani and Thiebaut de Schotten (2008) localize the uncinate as distinct from the IFOF, passing through the external capsule ventrally to the IFOF. However, their atlas uses the same IFOF waypoint mask in the extreme/external capsule, which suggests that the IFOF and ventral pathways are not mutually exclusive but rather two segments along the same fasciculus. Other workers (Anwander et al. 2007) found evidence for separate ventral and uncinate pathways in some but not all of their subjects.

Using a combination of fMRI and DTI tractography, Saur and co-workers (2008) identify the ventral pathway as passing between ventrolateral prefrontal cortex through the extreme capsule to the temporal lobe. High angular resolution diffusion fiber tractography also provides evidence for the extreme capsule as part of the ventral pathway, connecting auditory association areas with ventral prefrontal cortex (Frey et al. 2008). Functionally, the ventral pathway appears to be involved in comprehension of sentences rather than individual words (Saur et al. 2008), sound-to-word learning tasks (Wong et al. 2011), naming (Ueno et al. 2011), and the syntactic components of language (Weiller et al. 2011), although relatively little is known about it at the present time (Friederici 2009).

16.5 Methodological Approaches to Understanding Temporal Lobe Evolution

16.5.1 Overview

The study of endocasts from extinct hominids has advanced our understanding of the differences in brain size and structure in the hominid lineage. Data from these disciplines have supported the notion that human evolution is characterized by increases in petalial lateralization (Holloway and De La Costelareymondie 1982) and have helped pinpoint when gross morphological changes may have occurred in our ancestry (e.g., Falk et al. 2000). Brains do not fossilize, however, putting limits on the structural information we can infer from cranial impressions. The temporal lobe, in particular, poses a special challenge to the paleoneurologist, because of its anatomical placement within the cranium. However, advances in this field, including a move from studying endocasts to studying the endocranial cavity itself (Bruner 2015), have permitted more detailed observations on fossil crania, including advancements in our understanding of morphological changes to the temporal lobe in human evolution (Bastir et al. 2008).

In addition to paleoneurology, knowledge about our closest relatives, the chimpanzees, can contribute to our understanding of human brain evolution. Although this has long been appreciated, there is, despite many years of intensive research, relatively little consensus in the field of comparative cognition about the cognitive abilities of chimpanzees vis-à-vis humans. If we focus on cognitive functions that are housed in the temporal lobe, a review of the literature shows cognitive researchers have both supported (Gardner and Gardner 1969, 1975, 1980; Savage-Rumbaugh et al. 1977, 1985) and disputed (Terrace et al. 1979; Wallman 1992; Rivas 2005) the idea that apes possess language or language-like abilities. Also, there is also substantial disagreement on whether chimpanzees possess theory of mind (e.g., Povinelli and Preuss 1995; Call and Tomasello 1999; Penn and Povinelli 2007; Matsuzawa 1996; De Waal 1991; Whiten et al. 1996; Hare et al. 2001; Karg et al. 2016; Krachun et al. 2010; reviewed in Call and Tomasello 2008).

Paleoneurology and comparative cognitive research share a common goal: to understand how evolution has shaped human brains. Comparative anatomical research can provide further insight into this question. We know a good deal about the differences between humans' and macaques' temporal structure, but much less is known about chimpanzees. Evolutionary neuroanatomists need a concrete plan for evaluating structural differences in the temporal lobe. First, homologous areas between humans and other primates need to be identified so that putative sites of expansion can be identified. But how do we identify a discrete cortical "area"?

16.5.2 Paleoneurology

Let's first examine what we can learn from the fossil record about temporal lobes. Fossil endocasts of primates show expansion in temporal areas, both laterally and ventrally, beginning in the early Eocene (Gurche 1982). The Sylvian fissure is observable in the lemuriform ancestor *Smilodectes gracilis* (Gurche 1982) and in stem strepshirrhine *Tetonius homunculus* (Radinsky 1979), suggesting that the temporal lobe of primates was established very early in primate evolution. Fossil endocasts of early primates also give us clues about the evolution of sulcal patterning within the temporal lobe, with Miocene primate *Aegyptopithecus* as an early example of a fossil primate with an observable superior temporal sulcus (Radinsky 1974). However, the location of the temporal lobe and the temporal pole, in particular, in relationship to the surrounding bony structures, makes it difficult to study using endocasts.

The cranial bones surrounding the temporal lobe are fragile, particularly the sphenoid bone, which interfaces with the temporal pole. The sphenoid is a component of the middle cranial fossa (MCF), a large depression in the cranium which interacts and encompasses the temporal lobes. The MCF, in turn, makes up part of the cranial base whose shape is influenced not only by brain structure but also by craniofacial morphology (Bastir et al. 2004, 2006; Bastir and Rosas 2005). Lieberman (2000) and Lieberman et al. (2002) suggest that the lateral widening of the middle cranial fossa that is observed in hominoids may be partially responsible for the facial retraction of modern humans. However, it is not possible to determine whether the MCF widened in response to facial retraction or if the widening of the MCF (and presumably, the temporal lobes) led the way for facial retraction. An examination of modern human, chimpanzee, and fossil hominin MCFs suggests that the MCF underwent important modifications approximately 130 kya (Bastir et al. 2008). Although Neanderthals possessed brain volumes comparable to modern *Homo sapiens*, Bastir and co-workers (2008) found their MCF anatomy more consistent with other fossil hominins in the study (*H. heidelbergensis*, *H. ergaster*, middle Pleistocene *H. sapiens*). An investigation of temporal sulcal patterning (Bastir et al. 2014) further supports the hypothesis that a lateral displacement, combined with a more forward projection of the anterior pole of the MCF, is a human specialization. Additionally, Neanderthals appear to have a more prominent ITG and MTG at the temporal pole, whereas human temporal pole is mainly comprised of MTG and STG (Bastir et al. 2014).

The implications of changes to cranial fossa for the evolution of brain function are difficult to evaluate and raise several questions, described by Bruner (2015): Are these modifications driven by changes in brain size and shape or the cranium itself? If the change originates in the crania, does the change in shape change the organization and function of the brain markedly, or does the brain simply adapt to these external physical forces? Although we cannot definitively establish the direction of causality when it comes to temporal lobe evolution, we can still make a few reasonable observations from paleoneurological studies of the human temporal lobe. Differences between chimpanzees and humans are large enough to indicate that the temporal lobes changed in size as well as orientation. Temporal lobe expansion in human lineage is a hominoid specialization, and further modifications have occurred in the human lineage, particularly in the anterior temporal lobe. Human-specific modifications in the temporal lobes may hold clues to human-unique cognitive abilities.

16.5.3 Histological Approaches

In principle, one could gain a great deal of information about the organization of the temporal lobe and about the differences between species, by examination of histological sections obtained from postmortem specimens. This, of course, was the approach employed in the classical cyto- and myeloarchitectonic studies. Interestingly, the comparative work of Brodmann (1909), who compared humans and *Cercopithecus* (an Old World monkey closely related to macaques), and of von Bonin and Bailey and colleagues (von Bonin and Bailey 1947; Bailey et al. 1950; Bailey and von Bonin 1951), who studied humans, chimpanzees, and macaques, provides little indication of significant species differences in temporal lobe structure, unlike the recent MRI-based myeloarchitecture results obtained using MRI (Glasser et al. 2014), as illustrated in Fig. 16.2. Certainly, there are additional opportunities for architectonic/histological analysis using, for example, the variety of receptor-selective ligands currently available in combination with quantitative densitometric analysis, as has been employed in different regions of human and macaque cortex by Zilles and colleagues (e.g., Zilles et al. 1995; Caspers et al. 2013). A receptorarchitectonic study on temporal areas in humans (Morosan et al. 2005) provided evidence for a novel temporal area, "Te3" on the lateral bulge of the STG in humans. This group has also pioneered with neurotransmitter receptor fingerprinting, where the density of receptor molecules in cortex was measured across cortical areas (Zilles et al. 2015). These workers found that posterior STG/STS and the inferior frontal sulcus, both involved in language, showed highly similar receptor fingerprints. Replicating this method across primate species could offer a way to both map cortical areas in greater detail, while also providing evidence for the function of these areas.

16.5.4 Structural and Functional Imaging

Can we identify homologous temporal cortical areas across primate species? In macaques, it is possible to acquire fMRI data in macaques in some cases, but it is not possible (as far as is known) to train chimpanzees to cooperate for an fMRI scan, not to mention the risks for MRI operators and equipment. Another possibility is diffusion tensor imaging (DTI), which permits the reconstruction of white matter pathways from the motion of water molecules in the brain. In addition, structural connectivity is being probed with higher resolution and greater breadth than before with DTI. A more recent methodology, functional connectivity MRI (fcMRI), uses coincident activations across the brain to infer structural connectivity (Hampson et al. 2002). The two fields appear to largely corroborate each other (Damoiseaux and Greicius 2009), suggesting that the major white matter pathways that are trackable in DTI represent important functional groupings of cortical areas. For example, Greicius et al. (2009) found that bilateral cingulum bundles interconnected the posterior cingulate cortex (PCC), the medial temporal lobes (MTLs) and the medial prefrontal cortex (MPFC), regions known to activate as part of the default mode network. When followed up by an fcMRI analysis, the same functional connectivity appeared (albeit with an extra connection between the PCC and the MPFC which did not appear in the DTI analysis). A related approach is to cluster cortical territories based on correlated individual variations in cortical thickness, which yields networks similar to those derived from structural and functional connectivity (Chen et al. 2008).

When connectivity data are available for large expanses of cortex, graph theory provides bases for inferring the large-scale network organization of the cortex (e.g., van den Heuvel and Sporns 2013). Li et al. (2013) employed these methods to compare the cortical networks of humans, chimpanzees, and macaques. Here, putative hubs were found in association areas, including medial parietal cortex, inferior parietal cortex, insular cortex, medial prefrontal cortex, and ventrolateral prefrontal cortex in both macaques and chimpanzees, while humans showed differences in the location of parietal hubs (one additional hub in the intraparietal sulcus and superior parietal cortex), as well as a novel hub in retrosplenial cortex.

Recent diffusion tractography work from our lab suggests modifications to temporal connectivity in humans and chimpanzees when compared to macaques. Apparent connectivity between primary visual areas and primary auditory areas to anterior temporal regions was observed in humans and chimpanzees, but not macaques, likely mediated by modifications to ILF or MdLF in the hominoid lineage (Bryant et al. 2016). Detailed structural information about network organization of the temporal lobe is needed in chimpanzees in order for us to infer whether these networks are expansions of conserved hominoid cortical areas or new forms of cortico-cortical organization.

The recent US federal mandate to retire the majority of research chimpanzees (Chimpanzee Health Improvement, Maintenance, and Protection Act, 2013) means that it is extremely unlikely that new in vivo structural or functional scans of chimpanzees will be acquired in the future. Future studies may rely on either comparisons of human structural scans with macaques or postmortem chimpanzee scans.

16.5.5 Summary

Several methodological challenges present themselves in the current and future endeavors to map the human temporal lobe. In order to understand the significance of human brain structure from an evolutionary perspective, it will be important to take into account the relationship between structure and function at the level of cortical organization; the inherent interconnectedness of cortical function, which will necessitate the consideration of the role of cortical areas in networks; the importance of highly interconnected hubs as part of association cortex organization; and the possible significance of interindividual variation. Lastly, practical considerations require some methodological workarounds when it comes to directly comparing humans with chimpanzees. Without chimpanzee functional neuroimaging data, it is necessary to examine human and macaque functional data in a comparative manner, in concert with structure, in order to infer the cognitive function of putative homologs in the chimpanzee temporal lobe.

16.6 Future Directions

Future work on temporal lobe evolution will require additional data on structural and functional homology. To gather more evidence for structural homology between humans, chimpanzees, and macaques, there are several options. Improving our maps of human and hominoid temporal lobes, including the major temporal association areas, will require further tractographic analyses.

Part of association cortex functionality is the participation in long-distance networks. We might expect the temporal association areas in humans and chimpanzees to have long-distance connections with prefrontal cortex, posterior temporal lobe, and parietal areas, as part of a language network. These connections could be investigated using graph-theoretic analyses of diffusion tractography, similar to those conducted by Li et al. (2013), or, even more simply, by analyzing the global connectivity of discrete territories, such as the ATL, by seeding a region of interest based on other data, for example, cortico-cortical tractography.

Another possible source of data for understanding human temporal cortex organization would be a fasciculo-cortical tractographic analysis, similar to the work by Turken and Dronkers (2011) used to identify a putative novel cortical area in human pMTG. Here, evidence for crucial linguistic function (in the form of devastating loss of function after injury to the area) correlated with local strong interconnectivity with major fasciculi. The anterior expansion of arcuate fibers in humans and, to a lesser degree, chimpanzees (Rilling et al. 2008) suggests modifications to fascicular architecture may accompany the evolution of novel functional areas.

In humans, we might expect to see modifications to fascicular connectivity to other temporal association areas, like the ATL as compared to neighboring regions – in addition to the AF, we might expect to see MdLF and IFOF, as in the pMTG (Turken and Dronkers 2011). We might also expect to see stronger streamline connectivity with the ventral pathway, another fasciculus implicated in language function. Given the cingulum's role in emotional processing, and the possible role for human ATL in processing emotional saliency of semantic information, we might expect to see stronger streamline connectivity with this fiber bundle as well. The same analyses, performed in chimpanzees and macaques, would provide insight into the role of fascicular connectivity in primate temporal cortex organization.

It is important to note that there is structural variation among individuals that may obscure finer anatomical and functional parcellations. A standard practice within the field of neuroimaging is the use of template brain atlases, such as the Montreal Neurological Institute's ICBM 452, to analyze neuroimaging data from individuals. These templates are formed by "warping" the scans of hundreds of human brains into a single aligned space and averaging the values. One important problem with the use of averaged template brains is the distortion of scan data, which necessarily occurs during warping of individual brains to a brain template. Each individual brain's volumetric and/or topological data is shifted to conform to the template, and, as a result, noise is introduced. A second problem is the common practice of analyzing MR data at a population level. While boosting p-values, this method arguably obscures details of organization by blurring them into structural trends rather than individual maps. Saygin and co-workers (2012) probed this issue by developing a novel connectivity fingerprint approach that combined structural and functional neuroimaging. These authors report that for many subjects, instead of seeing a single area of face-selective activation in the FG, two adjacent but discrete areas were observed. It is possible that this information had previously been collapsed and obscured into one single area in traditional group-based analyses. Creating individual structural and functional maps is possible and indeed may be critical for accurate cortical mapping.

Association cortices have expanded in the human lineage. In the temporal lobe, association areas combine lower level unimodal perceptual inputs into multimodal and subsequently supramodal or amodal conceptual representations. It is reasonable to surmise that these expanded temporal areas are crucial to human-unique abilities in manipulating conceptual representations. Language, abstract concepts, and the use and manufacture of tools are prominent examples of our ability to construct conceptual representations by integrating sensory information from multiple modalities. Producing detailed comparative structural neuroanatomical maps in chimpanzees, and examining these data alongside functional data in humans and macaques, will permit us to develop a better understanding of how the temporal lobe was modified in hominoid evolution. In turn, this improved neuroanatomical understanding will illuminate how the human temporal lobe has evolved to produce the highly complex semantic and conceptual representations that are the hallmark of human cognition. Knowledge about the structural uniqueness of the human temporal lobe is part of a larger project of comparative human brain mapping which has the potential to lay a foundation for understanding of the physical instantiation of semantic knowledge in the brain.

Acknowledgments Brain images in Fig. 16.1 were adapted from the University of Wisconsin and Michigan State Comparative Mammalian Brain Collections and the National Museum of Health and Medicine, on their website neurosciencelibrary.org, funded by NSF and NIH.
Figures 16.2 and 16.3 used data provided in part by the Human Connectome Project, WU-Minn Consortium (Principal Investigators: David Van Essen and Kamil Ugurbil; 1U54MH091657) funded by the 16 NIH Institutes and Centers that support the NIH Blueprint for Neuroscience Research and by the McDonnell Center for Systems Neuroscience at Washington University.
The authors would like to acknowledge the John Templeton Foundation (Award 40463) and the NIH Office of Infrastructure Programs (OD P51OD11132) for supporting this work.

References

Allison T, Begleiter A, McCarthy G, Roessler E, Nobre AC, Spencer DD (1993) Electrophysiological studies of color processing in human visual cortex. Electroencephalogr Clin Neurophysiol 88:343–355

Allison T, Ginter H, McCarthy G, Nobre AC, Puce A, Luby M, Spencer DD (1994) Face recognition in human extrastriate cortex. J Neuro-Oncol 71:821–825

Allman J (1982) Reconstructing the evolution of the brain in primates through the use of comparative neurophysiological and

neuroanatomical data. In: Armstrong E, Falk D (eds) Primate brain evolution. Springer, New York, pp 13–28
Allman JM, Kaas JH (1971) A representation of the visual field in the caudal third of the middle temporal gyrus of the owl monkey (Aotus trivirgatus). Brain Res 31(1):85–105
Allman JM, Kaas JH, Lane RH (1973) The middle temporal visual area (MT) in the bushbaby, Galago senegalensis. Brain Res 57(1):197–202
Anwander A, Tittgemeyer M, von Cramon DY, Friederici AD, Knosche TR (2007) Connectivity-based parcellation of Broca's area. Cereb Cortex 17:816–825
Bailey P, von Bonin G (1951) The isocortex of man. University of Illinois Press, Urbana
Bailey P, von Bonin G, Mcculloch WS (1950) The isocortex of the chimpanzee. University of Illinois Press, Urbana
Baleydier C, Mauguiere F (1980) The duality of the cingulate gyrus in monkey. Neuroanatomical study and functional hypothesis. Brain 103:525–554
Barton JJ (1998) Higher cortical visual function. Curr Opin Ophthalmol 9:40–45
Bastir M, Rosas A (2005) Hierarchical nature of morphological integration and modularity in the human posterior face. Am J Phys Anthropol 128:26–34
Bastir M, Rosas A, Kuroe K (2004) Petrosal orientation and mandibular ramus breadth: evidence for an integrated petroso-mandibular developmental unit. Am J Phys Anthropol 123:340–350
Bastir M, Rosas A, O'Higgins P (2006) Craniofacial levels and the morphological maturation of the human skull. J Anat 209:637–654
Bastir M, Rosas A, Lieberman DE, O'Higgins P (2008) Middle cranial fossa anatomy and the origin of modern humans. Anat Rec 291:130–140
Bastir M, Böhme M, Sanchiz B (2014) Middle Miocene remains of Alytes (Anura, Alytidae) as an example of the unrecognized value of fossil fragments for evolutionary morphology studies. J Vertebr Paleontol 34:69–79
Bates E, Wilson SM, Saygin AP, Dick F, Sereno MI, Knight RT, Dronkers NF (2003) Voxel-based lesion-symptom mapping. Nat Neurosci 6:448–450
Beevor CE (1891) On the course of the fibres of the cingulum and the posterior parts of the corpus callosum and fornix in the marmoset monkey. Philos Trans R Soc London B 182:135–199
Bernal B, Ardila A (2009) The role of the arcuate fasciculus in conduction aphasia. Brain 132:2309–2316
Binder JR, Desai RH (2011) The neurobiology of semantic memory. Trends Cogn Sci 15:527–536
Binder JR, Medler DA, Westbury CF, Liebenthal E, Buchanan L (2006) Tuning of the human left fusiform gyrus to sublexical orthographic structure. NeuroImage 33:739–738
Binder JR, Desai RH, Graves WW, Conant LL (2009) Where is the semantic system? A critical review and meta-analysis of 120 functional neuroimaging studies. Cereb Cortex 19:2767–2796
Blaizot XF, Mansilla F, Insausti AM, Constans JM, Salinas-Alaman A, Pro-Sistiaga P, Mohedano-Moriano A, Insausti R (2010) The human parahippocampal region: I. Temporal pole cytoarchitectonic and MRI correlation. Cereb Cortex 20:2198–2212
Bonner MF, Price AR (2013) Where is the anterior temporal lobe and what does it do? J Neurosci 33:4213–4215
Brodmann K (1905) Beitrage zur histologischen Localisation der Grosshirnrinde. Dritte Mitteilung. Die Rindefelder der niederen Affen. J Psychol Neurol 4:177–226
Brodmann K (1909) Vergleichende Lokalisationslehre der Grosshirnrinde in ihren Prinzipien dargestellt auf Grund des Zellenbaues. Barth, Leipzig
Bruner E (2015) Functional craniology and brain evolution. In: Bruner E (ed) Human paleoneurology. Springer, Cham, pp 57–94
Bryant KL, Li L, Preuss TM (2016) Reorganization of temporal association cortico-cortical connectivity in hominoids. The American Association of Physical Anthropologists' 85th Annual Meeting. Atlanta
Burdach KF (1822) Vom Baue und Leben des Gehirns. Dyk, Leipzig
Butler RA, Lambon Ralph MA, Woollams AM (2014) Capturing multidimensionality in stroke aphasia: mapping principal behavioral components to neural structures. Brain 137:3248–3266
Calarge C, Andreasen NC, O'Leary DS (2003) Visualizing how one brain understands another: a PET study of theory of mind. Am J Psychiatry 160:1954–1964
Call J, Tomasello M (1999) A nonverbal false belief task: the performance of children and great apes. Child Dev 70:381–395
Call J, Tomasello M (2008) Does the chimpanzee have a theory of mind? 30 years later. Trends Cogn Sci 12:187–192
Cancelliere AEB, Kertesz A (1990) Lesion localization in acquired deficits of emotional expression and comprehension. Brain Cogn 13:133–147
Cantalupo C, Oliver J, Smith J, Nir T, Taglialatela JP, Hopkins WD (2009) The chimpanzee brain shows human-like perisylvian asymmetries in white matter. Eur J Neurosci 30:431–438
Caspers S, Schleicher A, Bacha-Trams M, Palomero-Gallagher N, Amunts K, Zilles K (2013) Organization of the human inferior parietal lobule based on receptor architectonics. Cereb Cortex 23:615–628
Catani M (2006) Diffusion tensor magnetic resonance imaging tractography in cognitive disorders. Curr Opin Neurol 19:599–606
Catani M, ffytche DH (2005) The rises and falls of disconnection syndromes. Brain 128:2224–2239
Catani M, Mesulam M (2008) The arcuate fasciculus and the disconnection theme in language and aphasia: history and current state. Cortex 44:953–961
Catani M, Thiebaut de Schotten M (2008) A diffusion tensor imaging tractography atlas for virtual in vivo dissections. Cortex 44:1105–1132
Catani M, Jones DK, Donato R, ffytche DH (2003) Occipito-temporal connections in the human brain. Brain 126:2093–2107
Catani M, Jones DK, ffytche DH (2005) Perisylvian language networks of the human brain. Ann Neurol 57:8–16
Chao LL, Haxby JV, Martin A (1999) Attribute-based neural substrates in posterior temporal cortex for perceiving and knowing about objects. Nat Neurosci 2:913–919
Chen ZJ, He Y, Rosa-Neto P, Germann J, Evans AC (2008) Revealing modular architecture chimpanzee health improvement, protection, and maintenance act. 42 U.S.C. § 283m
Clarke S, Miklossy J (1990) Occipital cortex in man: organization of callosal connections, related myelo-and cytoarchitecture, and putative boundaries of functional visual areas. J Comp Neurol 298:188–214
Coccia M, Bartolini M, Luzzi S, Provinciali L, Lambon Ralph MA (2004) Semantic memory is an amodal, dynamic system: evidence from the interaction of naming and object use in semantic dementia. Cogn Neuropsychol 21:513–527
Cohen L, Dehaene S, Naccache L, Lehéricy S, Dehaene-Lambertz G, Hénaff MA, Michel F (2000) The visual word form area. Brain 123:291–307
Copland DA, de Zubicaray GI, McMahon K, Wilson SJ, Eastburn M, Chenery HJ (2003) Brain activity during automatic semantic priming revealed by event-related functional magnetic resonance imaging. NeuroImage 20:302–310
Curran EJ (1909) A new association fiber tract in the cerebrum with remarks on the fiber tract dissection method of studying the brain. J Comp Neurol 19:645–656
Damasio H, Damasio AR (1980) The anatomical basis of conduction aphasia. Brain 103:337–350

Damasio H, Tranel D, Grabowski T, Adolphs R, Damasio A (2004) Neural systems behind word and concept retrieval. Cognition 92:179–229

Damoiseaux JS, Greicius MD (2009) Greater than the sum of its parts: a review of studies combining structural connectivity and resting-state functional connectivity. Brain Struct Funct 213:525–533

Datta R, Lee J, Duda J, Avants BB, Vite CH, Tseng B, Aguirre GK (2012) A digital atlas of the dog brain. PLoS One 7:e52140

Davis LE (1921) An anatomic study of the inferior longitudinal fasciculus. Arch Neurol Psychiatr 5:370–381

De Waal FB (1991) Complementary methods and convergent evidence in the study of primate social cognition. Behaviour 118:297–320

Dejerine JJ, Dejerine-Klumpke A (1895) Anatomie des centres nerveux, vol 1. Rueff, Paris

Dejerine JJ, Dejerine-Klumpke A (1901) Anatomie des centres nerveux, vol 2. Rueff, Paris

Devlin JT, Jamison HL, Gonnerman LM, Matthews PM (2006) The role of the posterior fusiform gyrus in reading. J Cogn Neurosci 18:911–922

DeWitt-Hamer PC, Moritz-Gasser S, Gatignol P, Duffau H (2011) Is the human left middle longitudinal fascicle essential for language? A brain electrostimulation study. Hum Brain Mapp 32:962–973

Dick AS, Tremblay P (2012) Beyond the arcuate fasciculus: consensus and controversy in the connectional anatomy of language. Brain 135:3529–3550

Dietz NAE, Jones KM, Gareau L, Zeffiro TA, Eden GF (2005) Phonological decoding involves left posterior fusiform gyrus. Hum Brain Mapp 26:81–93

Dolan RJ, Lane R, Chua P, Fletcher P (2000) Dissociable temporal lobe activations during emotional memory retrieval. NeuroImage 11:203–209

Dronkers NF, Wilkins DP, Van Valin RD Jr, Redfern BB, Jaeger JJ (2004) Lesion analysis of the brain areas involved in language comprehension. Cognition 92:145–177

Duffau H (2008) The anato-functional connectivity of language revisited: new insights provided by electrostimulation and tractography. Neuropsychologia 46:927–934

Epelbaum S, Pinel P, Gaillard R, Delmaire C, Perrin M, Dupont S, Dehaene S, Cohen L (2008) Pure alexia as a disconnection syndrome: new diffusion imaging evidence for an old concept. Cortex 44:962–974

Epstein R, Kanwisher N (1999) Repetition blindness for locations: evidence for automatic spatial coding in an RSVP task. J Exp Psychol Hum Percept Perform 25:1855

Eskenazi B, Cain WS, Novelly RA, Mattson R (1986) Odor perception in temporal lobe epilepsy patients with and without temporal lobectomy. Neuropsychologia 24:553–562

Falk D, Redmond JC Jr, Guyer J, Conroy C, Recheis W, Weber GW, Seidler H (2000) J Hum Evol 38:695–717

ffytche DH (2008) The hodology of hallucinations. Cortex 44:1067–1083

ffytche DH, Blom JD, Catani M (2010) Neuropsychiatry review series: disorders of visual perception. J Neurol Neurosurg Psychiatry 81:1280–1287

Fiez JA, Raichle ME, Balota DA, Tallal P, Petersen SE (1996) PET activation of posterior temporal regions during auditory word presentation and verb generation. Cereb Cortex 6:1–10

Fischl B, Sereno M, Dale A (1999) Cortical surface-based analysis. II: inflation, flattening, and a surface-based coordinate system. NeuroImage 9:195–207

Fjell AM, Westlye LT, Amlien I, Tamnes CK, Grydeland H, Engvig A, Espeseth T, Reinvang I, Lundervold AJ, Lundervold A, Walhovd KB (2015) High-expanding cortical regions in human development and evolution are related to higher intellectual abilities. Cereb Cortex 25:26–34

Flechsig PE (1896) Die Localisation der geistigen Vorgänge insbesondere der Sinnesempfindungen des Menschen: Vortrag, gehalten auf der 68. Versammlung Deutscher Naturforscher und Ärtze. Veit, Frankfurt

Flechsig PE (1901) Developmental (myelogenetic) localization of the cerebral cortex in the human subject. Lancet 1898:1027–1029

Fox CJ, Iaria G, Barton JJS (2008) Disconnection in prosopagnosia and face processing. Cortex 44:996–1009

Frey S, Campbell JSW, Pike GB, Petrides M (2008) Dissociating the human language pathways with high angular resolution diffusion fiber tractography. J Neurosci 28:11435–11444

Friederici AD (2009) Pathways to language: fiber tracts in the human brain. Trends Cogn Sci 13(4):175–181

Frith CD (2007) The social brain? Phil Trans R Soc Lond B 362:671–678

Gallagher HL, Frith CD (2003) Functional imaging of 'theory of mind'. Trends Cogn Sci 7:77–83

Gannon PJ, Holloway RL, Broadfield DC, Braun AR (1998) Asymmetry of chimpanzee planum temporale: humanlike pattern of Wernicke's brain language area homolog. Science 279(5348):220–222

Gardner RA, Gardner BT (1969) Teaching sign language to a chimpanzee. Science 165:664–672

Gardner BT, Gardner RA (1975) Evidence for sentence constituents in the early utterances of child and chimpanzee. J Exp Biol 104:244–267

Gardner RA, Gardner BT (1980) Comparative psychology and language acquisition. In: Sebeok T, Umiker-Sebeok J (eds) Speaking of apes – topics in contemporary semiotics. Springer, Boston MA

Gauthier I, Tarr MJ, Anderson AW, Skudlarski P, Gore JC (1999) Activation of the middle fusiform 'face area' increases with expertise in recognizing novel objects. Nat Neurosci 2:569–573

Geschwind N (1965) Disconnection syndromes in animals and man, part 1. Brain 88:237–294

Geschwind N (1970) The organization of language and brain. Science 170:940–944

Glasser MF, Rilling JK (2008) DTI tractography of the human brain's language pathways. Cereb Cortex 18:2471–2482

Glasser MF, Van Essen DC (2011) Mapping human cortical areas in vivo based on myelin content as revealed by T1- and T2-weighted MRI. J Neurosci 31:11597–11616

Glasser M, Preuss T, Snyder L, Nair G, Rilling J, Zhang X, Li L, Van Essen D (2011) comparative mapping of cortical myelin content in humans, chimpanzees, and macaques using T1-weighted and T2-weighted MRI. Society for Neuroscience, Washington, DC

Glasser MF, Goyal MS, Preuss TM, Raichle ME, Van Essen DC (2014) Trends and properties of human cerebral cortex: correlations with cortical myelin content. NeuroImage 93:165–175

Goel V, Grafman J, Sadato N, Hallett M (1995) Modeling other minds. Neuroreport 6:1741–1746

Goldman-Rakic PS, Selemon LD, Schwartz ML (1984) Dual pathways connecting the dorsolateral prefrontal cortex with the hippocampal formation and parahippocampal cortex in the rhesus monkey. Neuroscience 12:719–743

Gorno-Tempini ML, Price CJ, Josephs O, Vandeberghe R, Cappa SF, Kapur N, Frackowiak RSJ (1998) The neural systems sustaining face and proper name processing. Brain 121:2103–2118

Greicius MD, Supekar K, Menon V, Dougherty RF (2009) Resting-state functional connectivity reflects structural connectivity in the default mode network. Cereb Cortex 19:72–78

Grezes J, Frith C, Passingham RE (2004) Brain mechanisms for inferring deceit in the actions of others. J Neurosci 24:5500–5505

Gurche JA (1982) Early primate brain evolution. In: Armstrong E, Falk D (eds) Primate brain evolution. Springer, New York, pp 227–246

Hackett TA, Preuss TM, Kaas JH (2001) Architectonic identification of the core region in auditory cortex of macaques, chimpanzees, and humans. J Comp Neurol 441:197–222

Halwani GF, Loui P, Rueber T, Schlaug G (2011) Effects of practice and experience on the arcuate fasciculus: comparing singers, instrumentalists, and non-musicians. Front Psychol 2:156

Hampson M, Peterson BS, Skudlarski P, Gatenby JC, Gore JC (2002) Detection of functional connectivity using temporal correlations in MR images. Hum Brain Mapp 15:247–262

Hare B, Call J, Tomasello M (2001) Do chimpanzees know what conspecifics know? Anim Behav 61:139–151

Hecht EE, Gutman DA, Dunn W, Keifer OP Jr, Sakai S, Kent M, Preuss T (2016) Neuroanatomical variation in domestic dog breeds. Program No. 834.13/III15. Society for Neuroscience, San Diego

Heekeren HR, Wartenburger I, Schmidt H, Schwintowski HP, Villringer A (2003) An fMRI study of simple ethical decision-making. Neuroreport 14:1215–1219

Hickok G, Poeppel D (2004) Dorsal and ventral streams: a framework for understanding aspects of the functional anatomy of language. Cognition 92:67–99

Hickok G, Poeppel D (2007) The cortical organization of speech processing. Nat Rev Neurosci 8:393–402

Hill J, Inder T, Neil J, Dierker D, Harwell J, Van Essen D (2010) Similar patterns of cortical expansion during human development and evolution. Proc Natl Acad Sci 107:13135–13140

Hillger LA, Koenig O (1991) Separable mechanisms in face processing: evidence from hemispheric specialization. J Cogn Neurosci 3:42–58

Hof PR, Van Der Gucht E (2007) Structure of the cerebral cortex of the humpback whale, *Megaptera novaeangliae* (Cetacea, Mysticeti, Balaenopteridae). Anat Rec 290:1–31

Holloway RL, De La Costelareymondie MC (1982) Brain endocast asymmetry in pongids and hominids: some preliminary findings on the paleontology of cerebral dominance. Am J Phys Anthropol 58:101–110

Hopkins WD, Taglialatela JP, Nir T, Schenker NM, Sherwood CC (2010) A voxel-based morphometry analysis of white matter asymmetries in chimpanzees (*Pan troglodytes*). Brain Behav Evol 76:93–100

Insausti R (2013) Comparative neuroanatomical parcellation of the human and nonhuman primate temporal pole. J Comp Neurol 521:4163–4176

Kaas JH (2006) Evolution of the neocortex. Curr Biol 16:910–914

Kaas JH, Hackett TA (1999) 'What' and 'where' processing in auditory cortex. Nat Neurosci 2:1045–1047

Kanwisher N, McDermott J, Chun MM (1997) The fusiform face area: a module in human extrastriate cortex specialized for face perception. J Neurosci 17:4302–4311

Kanwisher N, Stanley D, Harris A (1999) The fusiform face area is selective for faces not animals. Neuroreport 10:183–187

Karg K, Schmelz M, Call J, Tomasello M (2016) Differing views: can chimpanzees do level 2 perspective-taking? Anim Cogn 19:555–564

Kier EL, Staib LH, Davis LM, Bronen RA (2004) MR imaging of the temporal stem: anatomic dissection tractography of the uncinate fasciculus, inferior occipitofrontal fasciculus, and Meyer's loop of the optic radiation. Am J Neuroradiol 25:677–691

Krachun C, Call J, Tomasello M (2010) A new change-of-contents false belief test: children and chimpanzees compared. Int J Comp Psychol 23:145–165

Kriegeskorte N, Formisano E, Sorger B, Goebel R (2007) Individual faces elicit distinct response patterns in human anterior temporal cortex. PNAS 104:20600–20605

Lambon Ralph MA, Patterson K (2008) Generalization and differentiation in semantic memory: insights from semantic dementia. Ann N Y Acad Sci 1124:61–76

Lambon Ralph MA, Sage K, Jones RW, Mayberry EJ (2010) Coherent concepts are computed in the anterior temporal lobes. PNAS 107:2717–2722

Leveroni CL, Seidenberg M, Mayer AR, Mead LA, Binder JR, Rao SM (2000) Neural systems underlying the recognition of familiar and newly learned faces. J Neurosci 20:878–886

Li L, Hu X, Preuss TM, Glasser MF, Damen FW, Qiu Y, Rilling J (2013) Mapping putative hubs in human, chimpanzee and rhesus macaque connectomes via tractography. NeuroImage 80:462–474

Lieberman DE (2000) Ontogeny, homology and phylogeny in the hominid craniofacial skeleton: the problem of the brow ridge. In: O'Higgins P, Cohn M (eds) Development, growth, and evolution. Linnean Society Symposium Series 20. pp 85–122

Lieberman DE, McBratney BM, Krovitz G (2002) The evolution and development of cranial form in *Homo sapiens*. Proc Natl Acad Sci 99:1134–1139

Lyras GA (2009) The evolution of the brain in Canidae (Mammalia: Carnivora). Scripta Geol 139:1–93

Makris N, Pandya DN (2009) The extreme capsule in humans and rethinking of the language circuitry. Brain Struct Funct 213:343–358

Makris N, Papadimitriou GM, Kaiser JR, Sorg S, Kennedy DN, Pandya DN (2009) Delineation of the middle longitudinal fascicle in humans: a quantitative, in vivo, DT-MRI study. Cereb Cortex 19:777–785

Marchina S, Zhu LL, Norton A, Zipse L, Wan CY, Schlaug G (2011) Impairment of speech production predicted by lesion load of the left arcuate fasciculus. Stroke 42:2251–2256

Marino L (2002) Convergence of complex cognitive abilities in cetaceans and primates. Brain Behav Evol 59:21–32

Mars RB, Sallet J, Neubert FX, Rushworth MFS (2013) Connectivity profiles reveal the relationship between brain areas for social cognition in human and monkey temporoparietal cortex. Proc Natl Acad Sci USA 110:10806–10811

Martin A, Chao LL (2001) Semantic memory and the brain: structure and processes. Curr Opin Neurobiol 11:194–201

Martin A, Haxby JV, Lalonde FM, Wiggs CL, Ungerleider LG (1995) Discrete cortical regions associated with knowledge of color and knowledge of action. Science 270:102–105

Martin A, Wiggs CL, Ungerleider LG, Haxby JV (1996) Neural correlates of category-specific knowledge. Nature 379:649–652

Matsuzawa T (1996) Chimpanzee intelligence in nature and in captivity: isomorphism of symbol use and tool use. In: Marchant LF, Nishida T (eds) Great ape societies. Cambridge University Press, Cambridge

McCandliss BD, Cohen L, Dehaene S (2003) The visual word form area: expertise for reading in the fusiform gyrus. Trends Cogn Sci 7:293–299

McCarthy G, Puce A, Gore JC, Allison T (1997) Face-specific processing in the human fusiform gyrus. J Cogn Neurosci 9:605–610

Menjot de Champfleur N (2012) La voie ventrale sémantique du langage: une étude de connectivité anatomique, de connectivité fonctionnelle et de sa plasticité périopératoire. Unpublished dissertation, Montpellier 1

Menjot de Champfleur NM, Maldonado IL, Moritz-Gasser S, Machi P, Le Bars E, Bonafé A, Duffau H (2013) Middle longitudinal fasciculus delineation within language pathways: a diffusion tensor imaging study in human. Eur J Radiol 82:151–157

Mesulam M, Mufson EJ (1982) Insula of the old world monkey. I: architectonics in the insulo-orbito-temporal component of the paralimbic brain. J Comp Neurol 212:1–22

Moll J, Eslinger PJ, de Oliveira-Souza R (2001) Frontopolar and anterior temporal cortex activation in a moral judgement task: preliminary functional MRI results in normal subjects. Arq Neuropsiquiatr 59:657–664

Moll J, de Oliveira-Souza R, Eslinger PJ, Bramati IE, Grafman J (2002) Functional networks in emotional moral and nonmoral social judgments. NeuroImage 16:696–703

Morgane PJ, Jacobs MS, McFarland WL (1980) The anatomy of the brain of the bottlenose dolphin (*Tursiops truncatus*). Surface configurations of the telencephalon of the bottlenose dolphin with comparative anatomical observations in four other cetacean species. Brain Res Bull 5:1–107

Morosan P, Schleicher A, Amunts K, Zilles K (2005) Multimodal architectonic mapping of human superior temporal gyrus. Anat Embryol 210:401–406

Mummery CJ, Patterson K, Hodges JR, Price CJ (1998) Functional neuroanatomy of the semantic system: divisible by what? J Cogn Neurosci 10:766–777

Nasr S, Liu N, Devaney KJ, Yue X, Rajimehr R, Ungerleider LG, Tootell RB (2011) Scene-selective cortical regions in human and non-human primates. J Neurosci 31:13771–13785

Niessl-Mayendorf V (1903) Vom fasciculus longitudinalis inferior. Eur Arch Psychiatry Clin Neurosci 37:537–563

Nucifora PGP, Verma R, Melhem ER, Gur RE, Gur RC (2005) Left-ward asymmetry in relative fiber density in the arcuate fasciculus. Neuroreport 16:791–794

Olson IR, Plotzker A, Exxyat Y (2007) The enigmatic temporal pole: a review of findings on social and emotional processing. Brain 130:1718–1731

Olson IR, McCoy D, Klobusicky E, Ross LA (2012) Social cognition and the anterior temporal lobes: a review and theoretical framework. Soc Cogn Affect Neurosci 10:123–133

Orban GA, Van Essen D, Vanduffel W (2004) Comparative mapping of higher visual areas in monkeys and humans. Trends Cogn Sci 8:315–324

Pammer K, Hansen PC, Kringelback ML, Holliday I, Barnes G, Hillebrand A, Singh KD, Cornelissen PL (2004) Visual word recognition: the first half second. NeuroImage 22:1819–1825

Parker GJM, Luzzi S, Alexander DC, Wheeler-Kingshott CA, Ciccarelli O, Ralph MAL (2005) Lateralization of ventral and dorsal auditory-language pathways in the human brain. NeuroImage 24:656–666

Parr LA, Dove T, Hopkins WD (1998) Why faces may be special: evidence of the inversion effect in chimpanzees. J Cogn Neurosci 10:615–622

Parr LA, Heintz M, Akamagwuna U (2006) Three studies on configural face processing by chimpanzees. Brain Cogn 62:30–42

Parr LA, Heintz M, Pradhan G (2008) Rhesus monkeys (*Macaca mulatta*) lack expertise in face processing. J Comp Psychol 122:390–402

Parr LA, Hecht E, Barks SK, Preuss SK, Preuss TM, Votaw JR (2009) Face processing in the chimpanzee brain. Curr Biol 19:50–53

Passingham RE, Smaers JB (2014) Is the prefrontal cortex especially enlarged in the human brain allometric relations and remapping factors. Brain Behav Evol 84:156–166

Peelen MV, Downing PE (2005) Cortical representation of faces, bodies and their parts. J Vis 5:825–825

Penn DC, Povinelli DJ (2007) On the lack of evidence that non-human animals possess anything remotely resembling a 'theory of mind'. Philos Trans R Soc Lond Ser B Biol Sci 362:731–744

Perrett DI, Hietanen JK, Oram MW, Benson PJ (1992) Organization and functions of cells responsive to faces in the temporal cortex. Philos Trans R Soc Lond Ser B Biol Sci 335:23–30

Petrides M, Pandya DN (2007) Efferent association pathways from the rostral prefrontal cortex in the macaque monkey. J Neurosci 27:11573–11586

Pizzagalli DA, Lehmann D, Hendrick AM, Regard M, Pascual-Marqui RD, Davidson RJ (2002) Affective judgments of faces modulate early activity (~160 ms) within the fusiform gyri. NeuroImage 16:663–677

Pobric G, Jefferies E, Ralph MA (2007) Anterior temporal lobes mediate semantic representation: mimicking semantic dementia by using rTMS in normal participants. PNAS 104:20137–20141

Pobric G, Jefferies E, Ralph MA (2010) Amodal semantic representations depend on both anterior temporal lobes: evidence from transcranial magnetic stimulation. Neuropsychologia 48:1336–1342

Polk TA, Farah MJ (2002) Functional MRI evidence for an abstract, not perceptual, word-form area. J Exp Psychol Gen 131:65–72

Povinelli DJ, Preuss TM (1995) Theory of mind: evolutionary history of a cognitive specialization. Trends Neurosci 18:418–424

Powell HR, Parker GJ, Alexander DC, Symms MR, Boulby PA, Wheeler-Kingshott CA, Barker GJ, Noppeney U, Koepp MJ, Duncan JS (2006) Hemispheric asymmetries in language-related pathways: a combined functional MRI and tractography study. NeuroImage 32:388–399

Preuss TM (1993) The role of the neurosciences in primate evolutionary biology: historical commentary and prospectus. In: MacPhee RDE (ed) Primates and their relatives in phylogenetic perspective. Springer, New York, pp 333–362

Preuss TM (2007) Primate brain evolution in phylogenetic context. In: Kaas JH, Preuss TM (eds) Evolution of nervous systems, Primates, vol 4. Elsevier, Oxford, pp 1–34

Preuss TM (2010) Reinventing primate neuroscience for the twenty-first century. Primate Neuroethol 1:422–454

Preuss TM (2011) The human brain: rewired and running hot. Annu NY Acad Sci 1225(Suppl 1):182–191

Radinsky L (1974) The fossil evidence of anthropoid brain evolution. Am J Phys Anthropol 41:15–27

Radinsky LB (1979) The fossil record of primate brain evolution. American Museum of Natural History, New York, pp 1–27

Ramayya AG, Glasser MF, Rilling JK (2010) A DTI investigation of neural substrates supporting tool use. Cereb Cortex 20:507–516

Rausch R, Serafetinides EA, Crandall PH (1977) Olfactory memory in patients with anterior temporal lobectomy. Cortex 13:445–452

Reiman EM, Lane RD, Ahern GL, Schwartz GE, Davidson RJ, Friston KJ, Yun LS, Chen K (1997) Neuroanatomical correlates of externally and internally generated human emotion. Am J Psychiatry 154:918–925

Rhodes G (1993) Configural coding, expertise, and right hemisphere advantage for face recognition. Brain Cogn 22:19–41

Rilling JK (2006) Human and non-human primate brains: are they allometrically scaled versions of the same design? Evol Anthropol 15:65–77

Rilling JK, Seligman RA (2002) A quantitative morphometric comparative analysis of the primate temporal lobe. J Hum Evol 42:505–533

Rilling JK, Glasser MF, Preuss TM, Ma X, Zhao T, Hu X, Behrens TEJ (2008) The evolution of the arcuate fasciculus revealed with comparative DTI. Nat Neurosci 11:426–428

Rilling JK, Glasser MF, Jbabdi S, Andersson J, Preuss TM (2011) Continuity, divergence, and the evolution of brain language pathways. Front Evol Neurosci 3:11

Rivas E (2005) Recent use of signs by chimpanzees (*Pan troglodytes*) in interactions with humans. J Comp Psychol 119:404–417

Rockland KS, Pandya DN (1979) Laminar origins and terminations of cortical connections of the occipital lobe in the rhesus monkey. Brain Res 179:3–20

Rogers TT, Lambon Ralph MA, Garrard P, Bozeat S, McClelland JL, Hodges JR (2004) Structure and deterioration of semantic memory: a neuropsychological and computational investigation. Psychol Rev 111:205–235

Romanski LM, Tian B, Fritz JB, Mishkin M, Goldman-Rakic PS, Rauschecker JP (1999) Dual streams of auditory afferents target multiple domains in the primate prefrontal cortex. Nat Neurosci 2:1131–1136

Ross LA, Olson IR (2010) Social cognition and the anterior temporal lobes. NeuroImage 49:3452–3462

Rossion B, Dricot L, Devolder A, Bodart JM, Crommelinck M, De Gelder B, Zoontjes R (2000) Hemispheric asymmetries for whole-based and part-based face processing in the human fusiform gyrus. J Cogn Neurosci 12:793–802
Roxbury T, McMahon K, Copland DA (2014) An fMRI study of concreteness effects in spoken word recognition. Behav Brain Funct 10:34
Rudrauf D, Mehta S, Grabowski TJ (2008) Disconnection's renaissance takes shape: formal incorporation in group-level lesion studies. Cortex 44:1084–1096
Sachs H (1892) Das Hemisphärenmark des menschlichen Grosshirns. Der Hinterhauptlappen. Breslau Universität psychiatrische und Nervenklinik. Arbeiten. Thieme, Leipzig
Saur D, Kreher BW, Schnell S, Kümmerer D, Kellmeyer P, Vry MS, Umarova R, Musso M, Glauche V, Abel S, Huber W (2008) Ventral and dorsal pathways for language. Proc Natl Acad Sci 105:18035–18040
Savage-Rumbaugh ES, Wilkerson BJ, Bakeman R (1977) Spontaneous gestural communication among conspecifics in the pygmy chimpanzee (*Pan paniscus*). In: Bourne G (ed) Progress in ape research. Elsevier, Amsterdam, pp 97–116
Savage-Rumbaugh ES, Sevcik RA, Rumbaugh DM, Rubert E (1985) The capacity of animals to acquire language: do species differences have anything to say to us? Phil Trans R Soc B Biol Sci 308:177–185
Sawyer EK, Turner EC, Kaas JH (2016) Somatosensory brainstem, thalamus, and cortex of the California sea lion (*Zalophus californianus*). J Comp Neurol 524:1957–1975
Saxe R, Kanwisher N (2003) People thinking about thinking people: the role of the temporo-parietal junction in "theory of mind". NeuroImage 19:1835–1842
Saygin ZM, Osher DE, Koldewyn K, Reynolds G, Gabrieli JDE, Saxe RR (2012) Anatomical connectivity patterns predict face selectivity in the fusiform gyrus. Nat Neurosci 15:321–327
Schmahmann JD, Pandya DN (2007) The complex history of the fronto-occipital fasciculus. J Hist Neurosci 16(4):362–377
Schmahmann JD, Pandya DN, Wang R, Dai G, D'Arceuil HE, de Crespigny AJ, Wedeen VJ (2007) Association fibre pathways of the brain: parallel observations from diffusion spectrum imaging and autoradiography. Brain 130:630–653
Schmolck H, Squire LR (2001) Impaired perception of facial emotions following bilateral damage to the anterior temporal lobe. Neuropsychology 15:30–38
Schoenemann PT (1997) An MRI study of the relationship between human neuroanatomy and behavioral ability. Unpublished dissertation. University of California, Berkeley
Schwartz MF, Kimberg DY, Walker GM, Faseyitan O, Brecher A, Dell GS, Coslett HB (2009) Anterior temporal involvement in semantic word retrieval: voxel-based lesion-symptom mapping evidence from aphasia. Brain 132:3411–3427
Schwarzlose RF, Baker CI, Kanwisher N (2005) Separate face and body selectivity on the fusiform gyrus. J Neurosci 25:11055–11059
Seltzer B, Pandya DN (1984) Further observations on parieto-temporal connections in the rhesus monkey. Exp Brain Res 55:301–312
Semendeferi K, Damasio H (2000) The brain and its main anatomical subdivisions in living hominoids using magnetic resonance imaging. J Hum Evol 38:317–332
Shaw P, Lawrence E, Bramham J, Brierley B, Radbourne C, David AS (2007) A prospective study of the effects of anterior temporal lobectomy on emotion recognition and theory of mind. Neuropsychologia 45:2783–2790
Simmons W, Martin A (2009) The anterior temporal lobes and the functional architecture of semantic memory. J Int Neuropsychol Soc 15:645–649
Simmons WK, Reddish M, Bellgowan PSF, Martin A (2009) The selectivity and functional connectivity of the anterior temporal lobes. Cereb Cortex 25:813–825
Small DM, Jones-Gotman M, Zatorre RJ, Petrides M, Evans AC (1997) A role for the right anterior temporal lobe in taste quality recognition. J Neurosci 17:5136–5142
Starrfelt R, Gerlach C (2007) The visual what for area: words and pictures in the left fusiform gyrus. NeuroImage 35:334–342
Steiper ME, Seiffert ER (2012) Evidence for a convergent slowdown in primate molecular rates and its implications for the timing of early primate evolution. Proc Natl Acad Sci 109:6006–6011
Terrace HS, Petitto LA, Sanders RJ, Bever TG (1979) Can an ape create a sentence? Science 206:891–902
Thiebaut de Schotten MT, Dell'Acqua F, Valabregue R, Catani M (2012) Monkey to human comparative anatomy of the frontal lobe association tracts. Cortex 48(1):82–96
Turken U, Dronkers NF (2011) The neural architecture of the language comprehension network: converging evidence from lesion and connectivity analyses. Front Syst Neurosci 5:1–20
Tusa RJ, Ungerleider LG (1985) The inferior longitudinal fasciculus: a reexamination in humans and monkeys. Ann Neurol 18:583–591
Ueno T, Saito S, Rogers TT, Lambon Ralph MA (2011) Lichtheim synthesizing aphasia and the neural basis of language in a neurocomputational model of the dual dorsal-ventral language pathways. Neuron 72:385–396
Ungerleider LG, Desimone R (1986) Cortical connections of visual area MT in the macaque. J Comp Neurol 248:190–222
van den Heuvel MP, Sporns O (2013) Network hubs in the human brain. Trends Cogn Sci 17:683–696
Vandenberghe R, Price C, Wise R, Josephs O, Frackowiak RSJ (1996) Functional anatomy of a common semantic system for words and pictures. Nature 383:254–256
Visser M, Jefferies E, Ralph ML (2010) Semantic processing in the anterior temporal lobes: a meta-analysis of the functional neuroimaging literature. J Cogn Neurosci 22:1083–1094
Visser M, Jefferies E, Embleton KV, Lambon Ralph MA (2012) Both the middle temporal gyrus and the ventral anterior temporal area are crucial for multimodal semantic processing: distortion-corrected fMRI evidence for a double gradient of information convergence in the temporal lobes. J Cogn Neurosci 24:1766–1778
von Bonin G, Bailey P (1947) The neocortex of *Macaca mulatta*. University of Illinois Press, Urbana
Von Economo C, Koskinas GN (1925) Die cytoarchitektonik der hirnrinde des erwachsenen menschen. J. Springer, Wien
Wallman J (1992) Aping language. Cambridge University Press, New York
Watson JD, Myers R, Frackowiak RS, Hajnal JV, Woods RP, Mazziotta JC, Shipp S, Zeki S (1993) Area V5 of the human brain: evidence from a combined study using positron emission tomography and magnetic resonance imaging. Cereb Cortex 3:79–94
Weiller C, Bormann T, Saur D, Musso M, Rijntjes M (2011) How the ventral pathway got lost – and what its recovery might mean. Brain Lang 118:29–39
Weiner KS, Zilles K (2015) The anatomical and functional specialization of the fusiform gyrus. Neuropsychologia 83:48–62
Whiten A, Custance DM, Gomez JC, Teixidor P, Bard KA (1996) Imitative learning of artificial fruit processing in children (*Homo sapiens*) and chimpanzees (*Pan troglodytes*). J Comp Psychol 110:3–14
Wilson MA, Joubert S, Ferre P, Belleville S, ANsaldo AI, Joanette Y, Rouleau I, Brambati SM (2012) The role of the left anterior temporal lobe in exception word reading: reconciling patient and neuroimaging findings. NeuroImage 60:2000–2007

Wise R, Chollet F, Hadar U, Friston K, Hoffner E, Frackowiak R (1991) Distribution of cortical neural networks involved in word comprehension and word retrieval. Brain 114:1803–1817

Wong FCK, Chandrasekaran B, Garibaldi K, Wong PCM (2011) White matter anisotropy in the ventral language pathway predicts sound-to-word learning success. J Neurosci 31:8780–8785

Yeatman JD, Dougherty RF, Rykhlevskaia E, Sherbondy AJ, Deutsch GK, Wandell BA, Ben-Shachar M (2011) Anatomical properties of the arcuate fasciculus predict phonological and reading skills and children. J Cogn Neurosci 23:3304–3317

Zahn R, Moll J, Krueger F, Huey ED, Garrido G, Grafman J (2007) Social concepts are represented in the superior anterior temporal cortex. PNAS 104:6430–6435

Zahn R, Moll J, Iyengar V, Huey ED, Tierney M, Krueger F, Grafman J (2009) Social conceptual impairments in frontotemporal lobar degeneration with right anterior temporal hypometabolism. Brain 132:604–616

Zilles K, Schlaug G, Matelli M, Luppino G, Schleicher A, Qu M, Dabringhaus A, Seitz R, Roland PE (1995) Mapping of human and macaque sensorimotor areas by integrating architectonic, transmitter receptor, MRI and PET data. J Anat 187:515–537

Zilles K, Bacha-Trams M, Palomero-Gallagher N, Amunts K, Friederici AD (2015) Common molecular basis of sentence comprehension network revealed by neurotransmitter receptor fingerprints. Cortex 63:79–89

Evolution of the Occipital Lobe

17

Orlin S. Todorov and Alexandra A. de Sousa

Abstract

In this chapter, we review and summarize the current body of knowledge on the anatomy, function, and evolution of the occipital lobes in humans, with reference to the brains of other key species. The anatomical landmarks that can be used to delineate the occipital lobe have been defined and explored in detail, and its functional significance in regard to visual processing has been elucidated. We give an overview of the current understanding about the evolution of the occipital lobe in primates from comparative perspective and present findings related to cortical reorganization, reduction, folding, and gyrification in the primate lineage over evolutionary time. Implications for further directions of inquiry that might shed light on less clear issues are also suggested.

Keywords

Occipital lobe • Human evolution • Vision • Paleoneurology • Comparative psychology • Neuroanatomy

17.1 The Primate Occipital Lobe

17.1.1 Defining the Occipital Lobe

The human occipital lobe is the best defined and best studied region of the cerebral cortex. It has been crucial to our understanding of brain evolution in general, and in particular, it has shed light on how brain structure is related to function. The occipital lobe is the most posterior lobe in the cerebrum, and its main function is visual processing. It was first described as an anatomical unit by Gratiolet, who named the brain's lobes after the cranial bones which surrounded them (Gratiolet 1854; Pearce 2006). In 1856 Gratiolet traced the optic radiations, a pathway from the optic nerve terminus in the thalamus to the occipital cortex, and described the anatomy of every structure along this pathway. Therefore, the occipital lobe is a structure that can be defined according to its morphology and position in the cerebrum and/or its relative role within the optic pathway. The morphological definition of the human occipital cortex is the cerebral cortex posterior to an anterior limit: a line that connects the parieto-occipital sulcus (from where it emerges on the superomedial edge of the cerebral hemisphere) to the preoccipital notch. However, gross morphological landmarks may be of limited reliability, particularly in comparisons across species (Allen et al. 2006; Malikovic et al. 2012; Filimonoff 1932, 1933a) {Smith 1907 #5355}. Alternatively, the occipital cortex can be defined as the part of the cerebral cortex that is primarily responsible for processing information from the optic pathway. In humans the occipital lobe is comprised entirely of early visual areas (mostly the primary visual cortex, Brodmann area 17). Therefore, the term occipital cortex is sometimes used in other nonhuman species to describe early visual regions

O.S. Todorov
School of Biological Sciences, The University of Queensland, St. Lucia 4072, QLD, Australia
e-mail: o.s.todorov@uq.net.au

A.A. de Sousa (✉)
Psychology, Culture and Environment, Bath Spa University, Bath, Newton Park BA2 9BN, UK
e-mail: a.desousa@bathspa.ac.uk

E. Bruner et al. (eds.), *Digital Endocasts*, Replacement of Neanderthals by Modern Humans Series,
DOI 10.1007/978-4-431-56582-6_17

which correspond histologically and/or functionally to human occipital areas 17, 18, and 19 (Brodmann 1909).

The position of a given neuron in the visual cortical areas is related to properties of its receptive field; that is, the region of space in which the occurrence of a visual stimulus will evoke action potentials. However, visual topography represents space in the visual field in a distorted and flipped-over fashion. Due to the crossing over of half of the retinal fibers at the optic chasm, the left visual field is represented in the right cerebral hemisphere, and the right visual field is represented in the left cerebral hemisphere. Speculation about the crossing over of half of the optical nerves fibers at the optic chiasm had been brewing since the twenty-seventh century BC, proposed by the ancient Egyptian polymath Imhotep (Glaser 2008), and the role of this in binocular vision was predicted as early as 1966 by Newton (1966). Finally, Vater and Heinicke (1723) were the first to clinically demonstrate this hemidecussation.

The occipital lobe is comprised of multiple visual areas that are based on findings from functional studies, and it is also divided histologically according to several different schemes (Zilles and Clarke 1997; Amunts and Zilles 2015). Brodmann divided the occipital lobe into three histological areas; Von Economo and Koskinas (1925) also described three areas but further delineated these into subareas. These parcellation schemes are difficult to replicate as they provide little information about the cortex within sulci and are based on qualitative cytoarchitectonic criteria. Studies by Amunts and others (Kujovic et al. 2013; Amunts et al. 2000) elaborate on these parcellations using a quantitative cytoarchitectonic approach (Schleicher and Zilles 1990) and refer to more recent information about functional regions. Most clearly there is a division between the striate area and the extrastriate areas. The striate cortex is comprised of the largest functional area, the primary visual area (V1), and was first recognized as a distinct unit according to histology. The term V1 is a functional definition that corresponds to the histological definition of Brodmann's area (BA) 17, and area OC of Von Economo and Koskinas (1925), and thus can be equally well partitioned from the adjacent cortex according to either functional and histological definitions. The term extrastriate cortex designates the other regions within the occipital lobe including Von Economo's parastriate OB which is BA 18 and part of Von Economo's parastriate OA which is BA 19. BA 18 seems to correspond roughly to the secondary visual area (V2), which borders V1 (Clarke 1993; Amunts et al. 2000). BA 19 is a single large area of the occipital lobe covering both the ventral and dorsal aspects from the occipital-temporal gyrus to the parieto-occipital sulcus. Due to a lack of sulcal detail provided by Brodmann about BA 19, it is difficult to demonstrate correspondence to other anatomical parcellations, let alone functional ones. According to quantitative cytoarchitectonic parcellations, area 19 actually includes 7 distinct histological areas (and putative corresponding functional areas): hOc3d (V3d), hOc4d (V3A), hOc3v (VP/V3v), hOc4v (V4), hOc5 (V5/MT+), hOc4la (LO-2), and hOc4lp (LO-1) (Rottschy et al. 2007; Kujovic et al. 2013; Malikovic et al. 2007, 2016).

It would be very handy to have measurements of the occipital lobe given the relative functional homogeneity. However, structural studies note that the occipital lobe is not differentiable from the adjacent parietal lobe due to a lack of surface landmarks consistently identifiable across species (Semendeferi and Damasio 2000).

Our fundamental understanding of visual area organization in the human brain has progressed from research on brain-damaged patients. Topography is most easily described in terms of the topographic layout of V1 neurons in respect to the part of the visual field they represent, but similar principles apply to higher areas. The upper visual field projects to inferior V1, and the upper visual field projects to superior V1. This organization within each occipital lobe was first suggested by Haab (1882) who determined from a study of brain damage that the most posterior part of the occipital lobe, the tip, was the location of the representation of central (macular) vision; the central representation of the visual field is the most posterior part. Inouye (1909) mapped the visual field onto the cortex based on patients with gunshot injuries inflicted in war, and it was confirmed in the more widely distributed work of Holmes (1918). These findings have been further revised and have demonstrated an even higher cortical magnification of the central visual field, in brain-damaged patients by Horton and Hoyt (1991), as well as experimental mapping methods, including microelectrode recordings in macaques (Daniel and Whitteridge 1961) and fMRI in humans (Sereno et al. 1995). Central vision is disproportionally represented on the visual cortex surface in humans. The macular representation (corresponding to the central 12° radius of the visual field) occupies 50–60% of the entire surface area of V1 (Horton and Hoyt 1991); the foveal representation (corresponding to the central 2° radius of the visual field) occupies slightly over 10% of V1 (Van Essen et al. 1992).

17.1.2 Visual Areas

The visual cortex is the portion of the cerebral cortex concerned mainly with processing of visual information. Visual areas are involved in segregating and analyzing the features (e.g., color, orientation) of visual images. The visual cortex can be further subdivided, or parcellated, into several individual visual areas. Visual areas are distinguished when reliable differences can be demonstrated in one or more of the following criteria: retinotopy, function, histology, and connections. Some visual areas are recognized in all parcellation schemes, while others remain controversial. Areas V1 (primary visual area), V2 (secondary visual area), and MT (medial temporal area) can be identified in all primates and are the least contested visual areas (Kaas 1993; Rosa and Tweedale 2005). These are the only visual areas that have been fully mapped across primate species and have a complete retinotopic representation of the visual hemifield (Van Essen 1985). In addition to these, many other visual areas can be located in primates that are less precisely defined in terms of function, location, and homology across species. Also, areas that are disputed because they do not comprise a complete retinotopic representation of the contralateral visual hemifield are termed "improbable areas" (Kaas 1993; Zeki 2003), and it has been argued that the best known of these, VP, should be considered part of V3 (Lyon and Kaas 2002). V3A (V3 accessory), on the other hand, is distinct from its neighbor, V3, and fits all the criteria for an independent visual area (Tootell et al. 1997). However, as mentioned below, human and macaque analogues of V3 and V3A show physiological differences.

17.1.3 Parcellation of the Visual Cortex

Originally, it was thought that the cerebral cortex was a uniform sheet of tissue. However, the parcellation of the cerebral cortex into discrete cortical areas is now supported by convergent anatomical and physiological evidence. The earliest brain maps were based on either cyto- or myeloarchitectonics alone. In fact the first cortical area ever discovered was the primary visual cortical area (V1), which on an unstained brain stood out from the adjacent cortex due to a white stripe of myelinated fibers corresponding to layer 4b – Gennari's stripe – from which the "striate" cortex received its name (Glickstein and Rizzolatti 1984). Further investigation of the cerebral cortex has led to the naming of numerous architectonically defined cortical areas.

Visual areas are no longer distinguished solely on the basis of histological evidence. Convergent evidence from different experimental techniques has added definition to maps, particularly in humans, macaques, and several other experimental species (Kaas 2006). However, in several key species only cytoarchitectural maps of the visual cortex are available – these include the apes, which are more closely related to humans than macaques, as well as a range of Old World monkey species relatively equally related to humans and macaques. Data from more species contribute to a phylogenetic context necessary to determine whether findings from macaques can be extrapolated to humans (de Sousa et al. 2013). Criteria for cytoarchitectural parcellations include (1) thickness of the cortex, (2) thickness of individual cortical layers, (3) number of layers, (4) staining intensity of neurons or of ground substance, (5) vertical or radial arrangement of neurons, (6) the packing density of neuronal cell bodies, (7) neuronal cell body size (8), the presence of special cell types, and (9) peculiarities unique to a specific cortical regions (Lashley and Clark 1946; Zilles et al. 2002). Usually, only the cytoarchitectonic parcellation of Brodmann (1909) and occasionally that of Von Economo (1929) are referenced in the neurosciences. Brodmann identified three visual areas – 17, 18, and 19 – in the human occipital lobe on the basis of cytoarchitecture. These roughly correspond to areas OC, OB, and OA, respectively, in the terminology of Von Economo. Von Economo also mentions some subregions but does not give much indication about the borders between them (von Economo 1929). It had long been the goal to ascribe function to structure, and early brain mappers such as Brodmann and the Vogts saw the purpose of their work to be "the development of a comparative organology of the cerebral surface, based upon anatomic criteria"(Brodmann 1925). However, recent comparisons of early cytoarchitectonic maps to functional studies indicate that these tripartite divisions of the human occipital lobe are oversimplifications.

In contrast, physiological data are used to distinguish over 25 distinct visual areas in macaques (Felleman and Van Essen 1991; Van Essen 2004), many of which are outside the occipital lobe and are not included within the extent of areas 17, 18, and 19. In response, a new nomenclature has arisen to describe these functionally relevant monkey visual areas (this terminology has also become accepted in human neuroimaging studies) (e.g., Tootell et al. 1997; Sereno et al. 1995; DeYoe et al. 1996; McKeefry and Zeki 1997; Vanduffel et al. 2001).

Two publications have mapped extrastriate cytoarchitectonic areas in great ape species, both of which are based on tripartite parcellations of the occipital lobe. The first is a parcellation of the occipital lobe by Filimonoff (1933b), which uses the nomenclature of Brodmann to parcellate the orangutan occipital lobe. The second is a parcellation of the entire chimpanzee cerebral cortex by Bailey et al. (1950), which is based on Von Economo's descriptions. These maps provide an overview of the tripartite organization of the visual cortex in individuals of these species but alone are not sufficient to reproduce these cytoarchitectonic areas. For example, Filimonoff (1933b) left sulci for which he was unable to find homologues in other species unlabeled. Bailey et al. acknowledge that "area OA resembles OB so closely, and the transition between the two areas is so gradual, that it is impossible to draw a line between them" (p. 48; 1950). Although they are able to indicate the position of some cortical areas in superficial maps, borders located deep in sulci are not indicated.

Recently, neuroimaging has confirmed that humans have more visual areas than classical cytoarchitecture suggests.

De Yoe et al. (1996) demonstrated that on the basis of physiological data, area 19 can be subdivided in humans, as it is in macaques. In addition, human homologues have been found for several visual areas located outside of the occipital lobe (Van Essen 2004). Of the human cytoarchitectonic areas, only Brodmann area 17 has been shown to be directly identical to a single functional area (V1) on the basis of a high-field MRI study which permitted histological identification and functional imaging (Bridge et al. 2005). Area 17 also happens to be the easiest visual area to identify on histological sections, as its layer four is divided into three sublayers, of which 4B corresponds to the macroscopically visible stripe of Gennari.

Cytoarchitectonic-based parcellations of the cerebral cortex have been criticized for being incongruent with each other, even within species (Lashley and Clark 1946). Different authors have divided the human brain into different numbers of cortical regions – ranging from just four main types described by Bailey and von Bonin (1951) to over 150 fields identified by the Vogts (Gerhardt 1940; Riegele 1931). Several variables have led to this, including the use of different types of staining, differences in cytoarchitectonic criteria, and insufficient consideration of intraspecific variability. It has been argued that many of the criteria used in parcellating cortex may have nothing to do with function. For example, the gyrification of the cortex itself leads the changes in cortical thickness, laminar density, and columnarity (von Economo 1929).

It has been stated that cytoarchitectonic parcellations depend largely on the intuition of the observer, that the basis of differentiation of cortical areas is not always obvious, and that individual estimates of cell size and density were not confirmed by actual measurements (Lashley and Clark 1946). In recent publications, this problem has been addressed by the observer-independent (OI) method of cortical parcellation of Schleicher and others (1999). The OI method requires the cerebral cortex to be photographed and converted into gray-level images, which provide quantitative data about variation in neuronal volume density. These data are used to estimate laminar pattern across vertical cortical columns, and these patterns are compared for statistically significant differences. Abrupt changes in laminar pattern, which in theory correspond to cortical area borders, are in this way justified on the basis of quantitative data (Schleicher et al. 1999).

Cytoarchitectonic parcellations are mainly based on Nissl-stained material, and the input is limited to the variability visible in the total population of cell bodies. The use of different staining methods and neuronal markers may provide better, or at least complementary, information about cortical area patterns and borders. Architectonic maps have been further refined using different kinds of histological (and functional) data (Toga et al. 2006). Visual areas have also been defined in a computer-aided observer-independent histological manner using myeloarchitecture (Annese et al. 2004, 2005). Braak (1977) subdivided the human occipital lobe into ten different areas on the basis of pigment architecture, which reveals additional details about laminar pattern that are not visible in adjacent Nissl-stained sections. Also, extrastriate cortex heterogeneity has been suggested on the basis of the patterns of termination of axons passing through the corpus callosum. This method involves visualizing callosal axons that have degenerated – either due to unilateral occipital infarctions or, in the case of experimental animals, transection of the posterior corpus callosum – prior to death. In rhesus monkeys, a direct comparison between callosal axon termination pattern and physiological organization has demonstrated that callosal projections characterize boundaries between V1 and V2, V3 and V3A, V3 and V4, and the anterior border of VP (Van Essen et al. 1982; Van Essen and Zeki 1978). Using a combination of callosal afferent organization and myeloarchitecture, Clarke and Miklossy (1990) have proposed functional human analogues to macaque areas V1, V2, V3d, VP, V4, and MT. In two species of Old World monkeys (*Cercopithecus aethiops* and *Macaca mulatta*) and humans, cytochrome oxidase, an endogenous mitochondrial enzyme revealing what would have been metabolically highly active zones in postmortem brain tissue, indicates regional distinctions for several extrastriate visual areas including V1, V2, MT, and possibly V3 (Tootell and Taylor 1995). Autoradiographic labeling of multiple transmitter receptors in humans and macaques had revealed regional and laminar cortical patterns that are consistent with known myelo- and cytoarchitectonic borders and additionally delineate regions that are not detectable using cytoarchitecture alone (Zilles and Palomero-Gallagher 2001; Zilles et al. 1995, 2002).

Immunohistochemical techniques may offer the most thorough and readily applicable histological method for parcellating the visual cortex. Immunohistochemistry localizes tissue constituents in situ by means of a specific antigen-antibody interaction, using a labeled antibody. This allows visual areas to be identified and defined in terms of laminar patterns of neuronal populations containing a specific protein. In addition, immunohistochemical markers reveal differences between taxonomic groups (Hof 2000; Sherwood et al. 2007). SMI-32 is an antibody marker that reacts with non-phosphorylated epitopes on neurofilament H (high molecular weight neurofilament; Sternberger and Sternberger 1983). Neurofilament proteins are assembled into neurofilaments, which are the main cytoskeletal components of axons and dendrites (Lacoste-Royal et al. 1990). In crab-eating macaques (*Macaca fascicularis*), non-phosphorylated neurofilament protein (NPNFP) staining with the SMI-32 antibody has been used to identify 28 visual areas (Hof and Morrison 1995). SMI-32 primarily visualizes neuronal cell bodies and dendrites of a subset of pyramidal neurons which are defined by large soma size and thick, heavily myelinated axons. Certain SMI32-immunoreactive (SMI-32-ir) neurons in human V1 which in monkeys have long projections from lower to higher cortical areas (Hof et al. 1996) are reduced in number in Alzheimer's disease cases – possibly related to visual deficiencies observed in Alzheimer's patients (Hof and Morrison 1990). A comparison of diverse cortical regions, including V1 and area 4, demonstrates that, overall, hominoids have a greater proportion of SMI-32-ir pyramidal neurons than do cercopithecoids (Sherwood et al. 2004; Campbell and Morrison 1989). NPNFP labeling has been qualitatively investigated in V1 of several anthropoid including humans, some apes, and some cercopithecines (Preuss et al. 1999). Also, V1 and V2 interneurons labeled by calcium-binding proteins have been quantified in anthropoids, and it has been found that hominoids had relatively fewer calbindin-immunoreactive interneurons than did monkeys (Sherwood et al. 2007).

17.1.4 Visual Pathways

The organization of the visual cortex is often described in terms of the pathways that it is comprised of. As visual inputs travel from the retina to the lateral geniculate nucleus (LGN) to V1 and to extrastriate cortex, the same inputs are processed by neurons with different receptive field properties. This is organized hierarchically, such that as information travels from the retina to the LGN to V1 and then to extrastriate cortex, the receptive field size of individual neurons increases. Within the cerebral cortex, structures early in the hierarchy are sometimes termed "lower order" or "early," and those later in the hierarchy are termed "higher order" or "late"; however, there actually exists a complexity of inputs with reentrant feedback from higher-order to lower-order structures. As visual information ascends the hierarchy, neurons preferentially respond to increasingly complex stimuli. For example, in V1 neurons exist which respond best to bars in a particular location or orientation (Hubel and Wiesel 1968). In area V4, many neurons are tuned for contours, i.e. angles and curves (Pasupathy and Connor 2002). In area TE, the final exclusively visual area, a large proportion of neurons respond exclusively to faces (Perrett et al. 1985).

Although originally two major pathways were recognized (Livingstone and Hubel 1988), it is thought that multiple parallel pathways carry visual sensory cues from the retina to the visual cortex (Nassi and Callaway 2009). Three of these, the magnocellular (M), parvocellular (P), and koniocellular (K) pathways, are each comprised of a distinct group of nerve fibers originating from retinal ganglion cells and terminating in the lateral geniculate nucleus (LGN) of the thalamus, where they were first recognized according to their specific features and location. The M pathway originates in the large, sensitive parasol ganglion cells of the retina, which primarily receive inputs from rods and which synapse in the magnocellular (i.e. large-celled) layers of the LGN and then project to layer 4Cα of cortical area V1 and carry high-contrast visual information, including information about motion. The P pathway originates in the small, numerous midget ganglion cells of the retina, which primarily get inputs from cones (see below) and which synapse on the parvocellular (i.e. small-celled) layers of the LGN, which then project to layer 4Cβ of V1 and carry information about color and fine structure (Rodiek 1988; Leventhal et al. 1981). M and P pathway organization is maintained in V1's primary target, V2, as the P-I (P-interblob), P-B (P-blob), and M streams (DeYoe and Van Essen 1988). V1 and V2 have feed-forward projections to late visual areas, in which visual streams and related functions are further segregated. Areas V3 (Felleman and Van Essen 1987), V3A (Tootell et al. 1997), and MT (Albright et al. 1984) are involved in motion detection and are associated with the M stream (Deyoe et al. 1990). Area V4, dubbed the "color center" (Lueck et al. 1989; McKeefry and Zeki 1997; Zeki 2004), is associated with two P sub-streams, P-interblob and P-blob (DeYoe et al. 1994; Van Essen et al. 1992). The less well-studied K pathway originates in both small and large bistratified ganglion cells. In the LGN, K cells are scattered throughout intercalated layers and throughout the M and P layers and are defined according to the expression of positive immunohistochemical staining for the alpha subunit of calmodulin-dependent protein kinase 2 (αCAMKII; Hendry and Yoshioka 1994; Yoshioka and Hendry 1995), the calcium-binding protein calbindin (Jones and Hendry 1989; Goodchild and Martin 1998), and the gamma subunit of protein kinase C (Fukuda et al. 1994).

Since most LGN inputs to extrastriate areas pass through V1, it was originally suggested that the M and P pathway divisions correspond to two previously described divisions of the extrastriate visual association areas, the dorsal and ventral streams (Livingstone and Hubel 1988). The ventral stream (the "what" pathway) is involved in object identification, includes area V4, and terminates in the inferior temporal cortex; whereas the dorsal stream (the "where" pathway) is involved in the spatial localization of objects (and in action), includes area V5/MT, and terminates in the posterior parietal cortex (Ungerleider and Mishkin 1982; Gattass et al. 1990; Goodale and Milner 1992). The M pathway predominates the input to MT, which itself provides major inputs to inferior temporal cortex, and the P pathway predominates the inputs to V4, which itself provides major inputs to posterior parietal cortex. However, the correspondence between the M and P pathways and the dorsal and ventral stream is not clear-cut

(Nassi and Callaway 2009). The discovery of the K pathway (and others), as well as details about cortical connections, has indicated the organization is much more complex, for example, V4 receives strong inputs from both M and P pathways (Ferrera et al. 1994).

Both the dorsal and ventral pathways interact with other cortical regions and seem to play important roles in both nonconscious and conscious vision (Breitmeyer 2014). According to such models and empirical findings (Breitmeyer 2014), the ventral (P pathway) has a subpathway processing an object's contour features and another one processing the texture or surface attributes. On the other hand, the dorsal (M pathway) processes motion and is crucial for conscious vision by activating a feed-forward loop by projecting to the prefrontal cortex and then the ventral P pathway (Breitmeyer 2014). Additional projections to the premotor cortex and the medial temporal lobe passing through posterior cingulate and retrosplenial cortices, together with the "motion" loop, contribute to conscious and nonconscious visuospatial processing functions such as spatial working memory, visually guided action, and navigation (Kravitz et al. 2011).

17.1.5 Organization of Visual Areas in Other Mammals

At least 13 visual cortical areas have been identified in rodents (Sereno and Allman 1991) although the size and number of areas vary across species (Krubitzer et al. 2011). In each rodent group (murines, sciurumorphs, and cavies), V1 is the largest, but its binocular portion accounts for less than half of its total area due to the lateral position of the eyes. In rats, big parts of the visual field are re-represented in extrastriate areas (Thurlow and Cooper 1988). V1 is encircled laterally by six areas – rostrolateral (RL), anterolateral (AL), lateromedial (LM), posterolateral (PR), and posterior 1 and 2 (P1 and P2) – and medially by three areas: anterior (A), anteromedial (AM), and medial (M). There is also a second tier of lateral areas – laterointermediate (LI), laterolateral (LL), and laterolateral anterior (LLa). There also appears to be hierarchy of visual areas based on laminar origins and corticocortical projection targets, similarly to that in primates, and also all areas seem to be retinotopic (Coogan and Burkhalter 1993). Many of these extrastriate areas appear to be visually driven without the role of V1 – visual thalamic areas in the pulvinar project extensively to these areas, and moreover, lesions in the lateral extrastriate areas produce much larger deficits in visuospatial tasks compared to lesions in V1 (McDaniel and Wall 2013).

The organization of the visual areas in lagomorphs is similar to that in rodents – V1 has a binocular strip and is surrounded by multiple extrastriate areas. These animals (and also rodents, sheep, and other mammals), unlike primates and carnivores, seem to have two functionally distinctive viewing strategies. When in "freeze" position, the portion of the visual field seen by one of the eyes does not overlap with the portion seen by the other one. In the "forward fixation" position, the binocular cells overlap (Hughes and Vaney 1982).

The visual organization in carnivores is slightly different to that in lagomorphs and rodents – the visual areas in cats, for example, are at least 19. Extrastriate areas also occupy larger area of the neocortex proportionally. Differences in the organization of the upper- and lower-field representations (area 18 and 19) and their asymmetry in carnivores suggest different evolutionary origins to that in rodents. Additionally, rostral to area 19, there are numerous small visual areas – one at the extreme lower-field representation, one onto the posterior part of the suprasylvian gyrus, one next to the center of the gaze, eight into the suprasylvian sulcus, four onto the small inferotemporal region, and one located in the anterior ectosylvian region near the face representation in primary somatosensory cortex. The last visual cortical area, unlike all the rest, is entirely surrounded by nonvisual cortical areas. In cats, receptive field size increases in the more rostral areas, and there are only few non-retinotopic areas. Similarly to rodents, lagomorphs, and primates, there is also a hierarchy of areas (area 17 to 18 to 19). Similarly to rats, large portions of the extrastriate cortex are visually driven in the absence of V1 activation (Raczkowski and Rosenquist 1983; Tusa et al. 1978).

It is important to mention that laying the foundations of visual research in animals were Hubel and Wiesel (1959) who, inspired by the experiments of Kuffler (1953), were among the first to do single-neuron recordings in the striatal cortex of the domestic cat in attempts to delineate the receptive fields and their structure and function. They were able to demonstrate that not only different neurons in the visual areas respond to slits of light with different orientation but, moreover, different neurons in the striate cortex are hierarchically organized as to process input sequentially. In doing so, it can be said that they were the first to pave the road for further studies of visual areas' organization in humans, primates, and other animals.

17.2 Human-Specific Features of the Visual Cortex

It is becoming increasingly clear that humans differ from macaques (and other primates) in some features of the visual cortex (Preuss 2005). The evolution of the human brain is specifically adapted to the socioecological problems the species are facing (Ghazanfar and Santos 2004), so neuroanatomical comparisons are interpreted in the context of species-specific behaviors and cognitive abilities. Some evidence suggests that humans have anatomical specializations of the occipital lobe related to the detection of motion. Motion detection is important in visual learning and therefore could be especially relevant in important human behaviors such as social learning and tool use. Humans possess a unique meshwork arrangement of M pathway fibers in layer 4A of V1 (Preuss and Coleman 2002; Preuss et al. 1999). This specialization may give humans increased sensitivity to motion and luminance contrast. If humans are specialized for motion detection, then it is likely that other aspects of their neuroanatomy are also derived for this skill. Specifically, the magnocellular layers of the LGN and cerebral cortical areas V3, V3A, and, especially, MT have important roles in motion detection, so related specializations may be detected in the volumes and microstructural organization of these extrastriate areas. However, note that Meynert cells, which are involved in motion detection, show a size increase in terrestrial cercopithecines, but not in humans (Sherwood et al. 2003).

Cognitive and behavioral studies demonstrate variation in aspects of visual perception within catarrhines and within hominoids. For example, when shown a compound visual pattern, humans perceive the global form before the local forms, whereas baboons show a local precedence (Fagot and Deruelle 1997). Chimpanzees seem to have either a local precedence or no consistent bias at all (Fagot et al. 2001). Physiological studies highlight certain cortical areas as participating in the aforementioned cognitive task for some species. Specifically, the dorsal inferotemporal cortex (TEd) appears to be involved in the identification of local, but not global, features by crab-eating macaques (*Macaca fascicularis*; Horel 1994). This is consistent with an fMRI study of humans and anesthetized rhesus monkeys, demonstrating that early visual areas (e.g., V1 and V2) may respond to global rather than simple local features (Kourtzi et al. 2003). Humans differ from chimpanzees when perceiving the shape of an object based on shading cues (Tomonaga 2001), an attribute which may suggest ecologically relevant differences in visually processing objects, with a bias in humans for seeing objects as if they were lit from above and in chimpanzees as if they were lit from the side. Further, the neural systems involved in visual processing may differ between species. The precise anatomy involved could differ between the species as well: an fMRI investigation of humans found a role for lower visual areas V1, V2, and V3 during shape from shading tasks (Humphrey et al. 1997), whereas in snow monkeys, V4 is involved in related texture segregation tasks (Hanazawa and Komatsu 2001).

Further studies suggest anatomical differences between cercopithecoids and all hominoids in aspects of visual system neuroanatomy. Compared to cercopithecoids, hominoids show a decrease in neurons, and specifically GABAergic interneurons, in V1 and V2 (Sherwood et al. 2007). It had been found that in cercopithecoids, increases in the volumes of V1 and LGN accompanying encephalization scale with a steeper slope than in hominoids (de Sousa et al. 2010a). Hominoids also differ from cercopithecoids in that they lack a band of cytochrome oxidase activity in layer 4A of V1, indicating a loss of P inputs to this layer (Preuss et al. 1999). This difference between these hominoids and monkeys may be related to differences in color perception. Evidence has suggested that folivorous diet has maintained routine trichromatic color vision in catarrhines (Dominy and Lucas 2001; Lucas et al. 1998, 2003).

17.3 The Fossil Evidence

17.3.1 Cortical Folding and Gyrification

The gyration pattern of the occipital lobe, as its size and shape, is determined as much by genetic factors as it is by physical constraints from the skull shape and volume. Several studies highlight these distinct processes. De Juan Romero et al. (2015) conducted a large-scale transcriptomic analysis of individual germinal layers in the developing cortex of the gyrencephalic (i.e. defined by convoluted cerebral cortex) ferret and compared the expression pattern between different regions and to that of the lissencephalic (i.e. defined by smooth cerebral cortex) mouse. They were able to show that proto-maps of gene expression within germinal layers around the splenial gyrus and its adjacent lateral sulcus (located in visual areas A17 and A19) distinguish the development of gyrencephalic cortices, and these may contribute to define cortical folds or functional areas. Through microarray analysis of the developing ferret cerebral cortex, they were able to pinpoint to transcriptomic differences between prospective folds and fissures around the area of interest. Such differentially expressed genes included 80% of those mutated in human cortical malformations, implying that similar genetic mechanisms are at play during gyrification in different gyrencephalic taxa. They were able to show that expression levels of certain genes change drastically and repeatedly across the cortex and that pattern of change might come to distinguish multiple domains or modules with differential gene expression. That modular expression pattern was observed only in the developing cortex of gyrencephalic species (ferret and human) but not in the lissencephalic cortex of mouse.

Nonaka-Kinoshita et al. (2013) were able to identify some of the neural stem and progenitor cell types involved differentially in cortical size expansion and gyrification. They were able to show that expansion in unipotent basal progenitors in mouse embryos, leading to megalencephaly, led only to increase in cerebral cortex surface area but not to cortical folding. On the other hand, in the gyrencephalic ferret brain, expansion of multipotent basal progenitors was sufficient for formation of additional folds and fissures. Thus, there seem to be distinct progenitor and neural stem cell types with specific role in regulation of cerebral cortex size and folding during development.

Tallinen et al. (2016) made a step toward understanding the biophysical causes of gyrification. They had shown that although many molecular and genetic determinants are at play during cortical expansion, the size, shape, placement, and orientation of folds arise in big part due to mechanical properties governed by physical laws. Using a combination of numerical model, observations on a growing fetal brain, and a 3D gel brain model, they were able to show that elementary mechanical instability, modulated by early fetal brain geometry, was contributing largely to the observed pater of brain gyrification. They also show that lateral asymmetry can be partially understood as a product of variations in the initial geometry of the fetal brain, which might also explain the variation in the exact locations of sulci.

17.3.2 Reorganization in Hominin Visual Brain Structures

The human primary visual cortex is about half the size it would be expected to be for a primate of human brain size (de Sousa et al. 2010a). A relative reduction of the occipital lobe means that other parts of the brain must be relatively larger in humans. Compared to monkeys, humans have proportionally more cortical surface dedicated to processing higher functions including the temporoparietal junction and the dorsolateral prefrontal cortex, whereas the occipital lobe is more similar in surface area (Van Essen and Dierker 2007). Thus, higher-order cortical regions are implicated as beneficiaries in a size trade-off, although precisely which areas of the brain enlarged, and when, is less clear (Bruner et al. 2016).

It has previously been proposed that human V1 is relatively reduced in comparison with the size of the rest of the brain as a result of expansion of higher-order cortical regions of the posterior parietal cortex involved in complex functions such as toolmaking and language (Holloway 1966, 1968). That is, it was hypothesized that brain evolution acts on a mosaic of functionally specific units reacting differentially to selection pressures. In the case of humans, it was hypothesized that higher-order areas expanded at an unusually fast pace, while primary areas maintained a slower rate of expansion, resulting in a large brain that is mostly devoted to higher-order processing.

Organizational changes involving V1 size have been suggested to be linked to changes in parietal lobe volumes (Holloway 1966). The arealization of the cerebral cortex is affected by both signaling molecule expression and neuronal inputs. Changes in the expression of transcription factors as well as differences in neuronal inputs can cause cortical areas to take on aspects of adjacent cortical regions. For example, fetal enucleation causes the cortex normally destined to be V1 to resemble adjacent area V2 instead (Dehay et al. 1996). In *Emx2* mutant neonatal mice, the border between visual areas and somatosensory areas is shifted caudally (Bishop et al. 2000). Similarly, it is speculated that subtle genetic, epigenetic, and developmental differences could alter the relative proportions of V1 and nearby cortical areas in closely related species.

However, within humans, the size of the occipital is inversely correlated with size aspects of the frontal and temporal lobes, but not the adjacent parietal lobe (Allen et al. 2002; Schwarzkopf et al. 2011). If there is in fact a tighter relationship between the size of the occipital and parietal lobes, this may be because the occipital lobe is the first place in which visuospatial information is mapped and processed in the cerebral cortex, and such information is then further processed in the parietal lobe. Although these regions are further away, dramatic changes in sensory area function affecting the size and organization of many cortical areas do occur with the early deprivation of visual experience (Karlen and Krubitzer 2009).

17.3.3 System-Level Brain Organization

Although gross structure volumes tend to scale to overall brain size, the microanatomical details of brain organization have more specific scaling relationships. The volume fraction of cortical tissue occupied by cell bodies in striate and extrastriate visual areas was found not to be correlated with overall brain size and yet to scale to the volumes of visual system structures, particularly V1 volume (de Sousa et al. 2010b). This implies that the properties governing the density of neurons may depend specifically on the size of the visual system or whichever other system in which they are found. This suggests that there are local scaling relationships which predominate, versus global scaling constraints on aspects of neuronal connections. Therefore, caution should be taken in interpreting results produced by pooling brain tissue across multiple regions, as in the isotropic fractionator method (see, e.g., Herculano-Houzel et al. 2007). Previous reports of visual cortex neuron number scaling to brain weight (e.g., Sherwood et al. 2007; Cragg 1967) may actually be demonstrating indirectly that neuron densities scale to visual area volumes and that these volumes, in turn, scale to brain weight. In area 9L, neuron density does not scale to brain weight in anthropoids (Sherwood et al. 2006). Among hominoids, neither neuron density nor GLI values of areas 10, 13, and 4 scale to brain weight (Sherwood and Hof 2007), but this is likely to be due to the low taxonomic level of comparison and/or small sample size. It would be interesting to investigate whether visual area neuron densities and GLI values more specifically, scale to the area in which they are located – indicating a standard of total neuron number per cortical area – or whether they systematically scale to early and subcortical structures volumes which are indicative of total neuronal input.

17.3.4 Left Occipital Right Frontal (LORF) Petalia

A petalia is a protrusion of one cerebral hemisphere relative to the other. The left-occipital right-frontal (LORF) petalia is an asymmetrical pattern in which there is a wider and more posteriorly protruding left occipital pole and a wider right frontal lobe. The LORF petalia is typical of modern humans and is statistically related to right-handedness – i.e. left-handed and ambidextrous people are more likely to be symmetrical or have the opposite pattern (Le May 1976). It is not clear whether apes exhibit modern humanlike petalias. Le May and colleagues (1976, 1982) found that petalias are also common in great apes. However, Holloway and de Lacoste-Lareymondie (1982) found them to be less frequent in apes than in modern humans and rarely involved both the frontal and the occipital lobes but noted a high incidence of left occipital petalias in gorillas. In a more recent MR study, however, Hopkins and Marino (2000) found that great apes display modern humanlike right-frontal and left-occipital petalias.

In a study evaluating cranial asymmetry in great apes, fossil hominins, and AMHs from quantitative endocast models, Balzeau et al. (2012) show that the observed pattern of cerebral asymmetry is quite similar between AMH and great apes but with a lower variation and a lower degree of fluctuating asymmetry in the latter. Even though variations in the position of the frontal and occipital poles on the right and left hemispheres would be expected to show some degree of antisymmetry considering population distribution, the authors show that the observed pattern of variation among the samples is more closely related to fluctuating asymmetry for most of the components of the petalias. They also show a significant rightward asymmetry for the anteroposterior and lateral component of the frontal petalia in both samples (AMH and great apes). The anteroposterior and lateral component of the occipital petalia were also shown to exhibit similar directional asymmetries, either leftward or rightward, while the vertical component of the frontal and occipital petalias was shown to exhibit leftward asymmetry in AMH only.

This evidence suggests that the observed traits were most probably inherited from a last common ancestor of the extant great apes and *Homo sapiens* and also has implications for possible relationships between endocranial shape asymmetries and cognitive ability in hominins.

17.3.5 Lunate Sulcus and Primary Visual Cortex Reduction

The lunate sulcus is within the secondary visual area, close to the anterior border of the primary visual cortex. Modern humans have a more posteriorly located lunate sulcus than do the great apes. A regression of striate cortex volumes against mean brain volumes from small samples of diverse primate species suggests that modern humans have substantially less (−121 %) primary visual cortex than expected for a nonhuman primate of similar brain size (Holloway 1992), a finding which is supported even when phylogeny is controlled for (de Sousa et al. 2010a). Although chimpanzees typically have a relatively larger primary visual cortex than do modern humans, a minority of chimpanzees show repositioning of the lunate sulcus to a more modern human-like posterior position (Holloway et al. 2003). Holloway et al. (2003) use this point to argue that the hypothetical panin-hominin common ancestor must also have had within its population individuals with reduced primary visual cortices, so one would expect this condition in early hominins such as *Australopithecus afarensis*. The lunate sulcus may be unique among the cortical sulci visible on endocasts in that it may provide information about the proportion of cortex allocated to distinct functional categories and provides an estimate of the aforementioned ratio of association to the sensory cortex (Holloway 1966, 1968). Gross anatomical markers used to estimate relative position of the lunate sulcus give an indication of histologically defined relative V1 volume in apes (Fig. 17.1; de Sousa et al. 2010a).

17.3.6 Variability in the Organization of Fossil Hominin Brains

Predictions about the neural organization of fossil species based on brain size alone have led to the conclusion that observations of a posterior lunate sulcus on small-brained early hominins *must* have been misinterpretations (Armstrong et al. 1991; Jerison 1975). But hominin encephalization and reorganization is not a unilinear process, and other examples of fossil hominin species diverging from a linear evolutionary model from the *Pan-Homo* common ancestor to modern humans have been recognized.

Fig. 17.1 3D surface reconstruction of the left cerebral hemisphere of adult female chimpanzee from magnetic resonance images, on which gross anatomical landmarks have been identified for distance measurements

Differences in hominoid visual brain structures point to issues in understanding the variation that exists in the hominin fossil record. The fossil record indicates that hominin species vary in brain shape, brain mass, and cognitive abilities. In general, the species with the largest, most human-looking brains are associated with the most sophisticated technology, but exceptions exist. Primate scaling relationships have been used to predict the size of brain components, including V1, in fossil hominin species (Conroy and Smith 2007), based on assumption that all brain component volumes are closely related to brain size due to developmental constraints (Finlay et al. 2001). However, it has also been indicated that brains evolve as a mosaic of systems, in which brain component volumes are better predicted by the size of functionally related brain structures than by overall brain size (Barton and Harvey 2000). Chimpanzees and bonobos provide an example that closely related species with similarly sized brains can differ in V1 volume, V1 and extrastriate volume density, and perhaps, other aspects of brain organization. One recent paleoanthropological puzzle is how *Homo floresiensis*, with its chimp-sized brain, became associated with sophisticated stone tools. Notably, *Homo floresiensis* has a posteriorly positioned lunate sulcus, which may be indicative of the increased size of posterior parietal cortical areas involved in toolmaking as compared to chimpanzees. *Australopithecus africanus* also has a similarly sized brain and a posteriorly positioned lunate sulcus but lacks the association with stone tools. The degree to which fossil species like *Australopithecus africanus* and *Homo floresiensis* differ in brain organization will never be known, because endocasts only provide information about gross morphology.

According to the mosaic brain evolution hypothesis, certain brain areas evolve independently from the rest of the brain and thus follow different enlargement patterns to that of the whole brain or that of unrelated processing systems. It has been shown that in primates all the components of the visual system scale with each other – orbits and eyes, optical tract, LGN, primary cortical area (V1), and downstream visual areas (V2, V3, and V5/MT) (Schultz 1940; Pearce et al. 2013; Barton and Harvey 2000; Barton 2007; Andrews et al. 1997; Yan et al. 2009).

17.3.7 Neanderthals Versus Modern Humans

Using a method for estimating visual area size in the occipital lobes of Neanderthals from endocast volumes by applying corrections and transformation due to allometry of body and eye size, and using known scaling coefficients between orbital size and visual areas size, ratio of visual to nonvisual cortex surface area, and ratio of gray to white matter and assuming regression slopes from primate models, Pearce and colleagues propose that Neanderthal's visual areas (and thus occipital lobes) must have been larger proportionally to the rest of the brain, compared to AMH (Pearce et al. 2013). They attribute that enlargement not only to body size and brain size differences but also to larger orbital size in Neanderthals, possibly due to the fact that the species inhabited higher latitudes.

However, more details about the behavior and phylogenetic relationships of these species will broaden the context for comparison. Also, it is encouraging that some information about brain organization can be derived from the fossil endocasts: of the great apes, the bonobo specimen has the most anteriorly positioned lunate sulcus. And there is a correlation between lunate sulcus position and V1 volume across apes.

17.4 Implications for Future Studies

Visual aspects of human behaviors such as toolmaking, art, symbolic activity, language, and social complexity are predicted to have a species-specific neuroanatomical basis. However, such complex functions are extremely difficult to relate to specific neuroanatomical variables. There remains much to be explored in human and nonhuman hominoid brain evolution. In fact, the meshwork arrangement of V1 layer 4A is the only qualitative neuroanatomical characteristic of any brain system specific to humans. Although the visual system is the most studied sensory system, many aspects of visual system organization, function, and morphology are still coming to light. For example, in recent years much is being revealed about the koniocellular pathway and other lesser known visual pathways. Comparative neuroimaging is a new field, implicating that, for example, the human and macaque extrastriate cortices are not functionally identical (Orban et al. 2004).

The relationship of the lunate sulcus to V1 volume can be further addressed as the current studies have small sample sizes, and lunate sulcus arc distances can be compared as well. Eventually, this approach can be expanded to investigate post-lunate surface area measurements, and to indicate the reliability of other cerebral surface landmarks for determining cortical area volumes in ape species. If robust relationships are found between external landmarks and functional regions, these data can be used to test predictions made about the sizes of V1 and other brain components in fossil taxa (Conroy and Smith 2007).

In humans, Brodmann's area 7, a superior parietal association area involved in somatosensory and visuomotor integration as well as visuospatial attention and memory shows differential activation during toolmaking by skilled toolmakers (Stout et al. 2000). Although area 7 and its subdivisions have recently been mapped in humans using the observer-independent method (Scheperjans et al. 2008), little is known about the organization of the posterior parietal lobe in great ape species. Mapping of posterior parietal cytoarchitectonic areas in apes would indicate whether V1 volume reduction is directly related to the volumetric or numerical expansion of functionally distinct posterior parietal areas. This would also allow for comparative studies of aspects of occipital and parietal lobe microanatomical organization, including tests of overall neuron volume and numerical density scaling relationships, and examination of specific neuron populations within these areas. Further details about the anatomy and function of visual pathways in hominoid brains will contribute to, and open up new, questions about human evolution.

Acknowledgments This review is based in part on work done in partial completion of a PhD by Alexandra de Sousa at the George Washington University.

References

Albright TD, Desimone R, Gross CG (1984) Columnar organization of directionally selective cells in visual area MT of the macaque. J Neurophysiol 51:16–31

Allen JS, Damasio H, Grabowski TJ (2002) Normal neuroanatomical variation in the human brain: an MRI-volumetric study. Am J Phys Anthropol 118:341–358

Allen JS, Bruss J, Damasio H (2006) Looking for the lunate sulcus: a magnetic resonance imaging study in modern humans. Anat Rec (Hoboken) 288:867–876

Amunts K, Zilles K (2015) Architectonic mapping of the human brain beyond Brodmann. Neuron 88:1086–1107

Amunts K, Malikovic A, Mohlberg H, Schormann T, Zilles K (2000) Brodmann's areas 17 and 18 brought into stereotaxic space-where and how variable? NeuroImage 11:66–84

Andrews TJ, Halpern SD, Purves D (1997) Correlated size variations in human visual cortex, lateral geniculate nucleus, and optic tract. J Neurosci 17:2859–2868

Annese J, Pitiot A, Dinov ID, Toga AW (2004) A myelo-architectonic method for the structural classification of cortical areas. NeuroImage 21:15–26

Annese J, Gazzaniga MS, Toga AW (2005) Localization of the human cortical visual area MT based on computer aided histological analysis. Cereb Cortex 15:1044–1053

Armstrong E, Zilles K, Curtis M, Schleicher A (1991) Cortical folding, the lunate sulcus and the evolution of the human brain. J Hum Evol 20:341

Bailey P, Von Bonin G (1951) The Isocortex of man. University of Illinois Press, Urbana

Bailey P, Von Bonin G, McCulloch WS (1950) The isocortex of chimpanzee. University of Illinois Press, Urbana

Balzeau A, Gilissen E, Grimaud-Herve D (2012) Shared pattern of endocranial shape asymmetries among great apes, anatomically modern humans, and fossil hominins. PLoS One 7:e29581

Barton RA (2007) Evolutionary specialization in mammalian cortical structure. J Evol Biol 20:1504–1511

Barton RA, Harvey PH (2000) Mosaic evolution of brain structure in mammals. Nature 405:1055–1058

Bishop KM, Goudreau G, O'leary DD (2000) Regulation of area identity in the mammalian neocortex by Emx2 and Pax6. Science 288:344–349

Braak H (1977) The pigment architecture of the human occipital lobe. Anat Embryol (Berl) 150:229–250

Breitmeyer BG (2014) Contributions of magno- and parvocellular channels to conscious and non-conscious vision. Philos Trans R Soc Lond Ser B Biol Sci 369:20130213

Bridge H, Clare S, Jenkinson M, Jezzard P, Parker AJ, Matthews PM (2005) Independent anatomical and functional measures of the V1/V2 boundary in human visual cortex. J Vis 5:93–102

Brodmann K (1909) Vergleichende Lokalisationslehre der Grosshirnrinde in ihren Prinzipien dargestellt auf Grund des Zellenbaues. Barth, Leipzig

Brodmann K (1925) Vergleiehende Lokalisationslehre der Grosshirnrinde. J. A. Barth, Leipzig

Bruner E, Lozano M, Lorenzo C (2016) Visuospatial integration and human evolution: the fossil evidence. J Anthropol Sci 94:81–97
Campbell MJ, Morrison JH (1989) Monoclonal antibody to neurofilament protein (SMI-32) labels a subpopulation of pyramidal neurons in the human and monkey neocortex. J Comp Neurol 282:191–205
Clarke S (1993) Callosal connections and functional subdivision of the human occipital cortex. In: Gulyas B, Ottoson D, Roland PE (eds) Functional organization of the human visual cortex, 1st edn. Perhamon Press, Oxford
Clarke S, Miklossy J (1990) Occipital cortex in man: organization of callosal connections, related myelo- and cytoarchitecture, and putative boundaries of functional visual areas. J Comp Neurol 298:188–214
Conroy GC, Smith RJ (2007) The size of scalable brain components in the human evolutionary lineage: with a comment on the paradox of Homo floresiensis. HOMO – J Comp Hum Biol 58:1–12
Coogan TA, Burkhalter A (1993) Hierarchical organization of areas in rat visual cortex. J Neurosci 13:3749–3772
Cragg BG (1967) The density of synapses and neurones in the motor and visual areas of the cerebral cortex. J Anat 101:639–654
Daniel PM, Whitteridge D (1961) The representation of the visual field on the cerebral cortex in monkeys. J Physiol 159:203–221
De Juan Romero C, Bruder C, Tomasello U, Sanz-Anquela JM, Borrell V (2015) Discrete domains of gene expression in germinal layers distinguish the development of gyrencephaly. EMBO J 34:1859–1874
De Sousa AA, Sherwood CC, Mohlberg H, Amunts K, Schleicher A, Macleod CE, Hof PR, Frahm H, Zilles K (2010a) Hominoid visual brain structure volumes and the position of the lunate sulcus. J Hum Evol 58:281–292
De Sousa AA, Sherwood CC, Schleicher A, Amunts K, Macleod CE, Hof PR, Zilles K (2010b) Comparative cytoarchitectural analyses of striate and extrastriate areas in hominoids. Cereb Cortex 20:966–981
De Sousa AA, Sherwood CC, Hof PR, Zilles K (2013) Lamination of the lateral geniculate nucleus of catarrhine primates. Brain Behav Evol 81:93–108
Dehay C, Giroud P, Berland M, Killackey H, Kennedy H (1996) Contribution of thalamic input to the specification of cytoarchitectonic cortical fields in the primate: effects of bilateral enucleation in the fetal monkey on the boundaries, dimensions, and gyrification of striate and extrastriate cortex. J Comp Neurol 367:70–89
Deyoe EA, Van Essen DC (1988) Concurrent processing streams in monkey visual cortex. Trends Neurosci 11:219–226
Deyoe EA, Hockfield S, Garren H, Van Essen DC (1990) Antibody labeling of functional subdivisions in visual cortex: cat-301 immunoreactivity in striate and extrastriate cortex of the macaque monkey. Vis Neurosci 5:67–81
Deyoe EA, Felleman DJ, Van Essen DC, McClendon E (1994) Multiple processing streams in occipitotemporal visual cortex. Nature 371:151–154
Deyoe EA, Carman GJ, Bandettini P, Glickman S, Wieser J, Cox R, Miller D, Neitz J (1996) Mapping striate and extrastriate visual areas in human cerebral cortex. Proc Natl Acad Sci U S A 93:2382–2386
Dominy NJ, Lucas PW (2001) Ecological importance of trichromatic vision to primates. Nature 410:363–366
Fagot J, Deruelle C (1997) Processing of global and local visual information and hemispheric specialization in humans (Homo sapiens) and baboons (Papio papio). J Exp Psychol Hum Percept Perform 23:429–442
Fagot J, Tomonaga M, Deruelle C (2001) Processing of the global and local dimensions of visual hierarchical stimuli by humans (*Homo sapiens*), chimpanzees (*Pan troglodytes*), and baboons (*Papio papio*). In: Matsuzawa T (ed) Primate origins of human cognition and behavior. Springer, New York
Felleman DJ, Van Essen DC (1987) Receptive field properties of neurons in area V3 of macaque monkey extrastriate cortex. J Neurophysiol 57:889–920
Felleman DJ, Van Essen DC (1991) Distributed hierarchical processing in the primate cerebral cortex. Cereb Cortex 1:1–47
Ferrera VP, Nealey TA, Maunsell JH (1994) Responses in macaque visual area V4 following inactivation of the parvocellular and magnocellular LGN pathways. J Neurosci 14:2080–2088
Filimonoff IN (1932) Über die Variabilität der Großhirnrindenstruktur. Mitteilung II Regio occipitalis beim erwachsenen Menschen. J Psychol Neurol 44:1–96
Filimonoff IN (1933a) Über die Variabilität der Großhirnrindenstruktur. Mitteilung III Regio occipitalis bei der höheren und niederen Affen. J Psychol Neurol 45:69–137
Filimonoff IN (1933b) Über die Variabilität der Großhirnrindenstruktur. Mitteilung III Regio occipitalis bei der höheren und niederen Affen. J Psychol Neurol 45:69–137
Finlay BL, Darlington RB, Nicastro N (2001) Developmental structure in brain evolution. Behav Brain Sci 24:263–278. discussion 278–308
Fukuda K, Saito N, Yamamoto M, Tanaka C (1994) Immunocytochemical localization of the alpha-, beta I-, beta II- and gamma-subspecies of protein kinase C in the monkey visual pathway. Brain Res 658:155–162
Gattass R, Rosa MG, Sousa AP, Pinon MC, Fiorani Junior M, Neuenschwander S (1990) Cortical streams of visual information processing in primates. Braz J Med Biol Res 23:375–393
Gerhardt E (1940) Die Cytoarchitektonik des Isocortex parietalis beim Menschen. J Psychol Neurol 49:367–419
Ghazanfar AA, Santos LR (2004) Primate brains in the wild: the sensory bases for social interactions. Nat Rev Neurosci 5:603–616
Glaser JS (2008) Romancing the chiasm: vision, vocalization, and virtuosity. J Neuroophthalmol 28:131–143
Glickstein M, Rizzolatti G (1984) Francesco Gennari and the structure of the cerebral cortex. Trends Neurosci 7:464–467
Goodale MA, Milner AD (1992) Separate visual pathways for perception and action. Trends Neurosci 15:20–25
Goodchild AK, Martin PR (1998) The distribution of calcium-binding proteins in the lateral geniculate nucleus and visual cortex of a New World monkey, the marmoset, Callithrix jacchus. Vis Neurosci 15:625–642
Gratiolet PL (1854) Mémoire sur les plis cérébraux de l'homme et des primates. A. Bertrand, Paris
Haab O (1882) Ueber cortex – Hemianopie. Monatsbl f Augenhlkde 20:141–153
Hanazawa A, Komatsu H (2001) Influence of the direction of elemental luminance gradients on the responses of V4 cells to textured surfaces. J Neurosci 21:4490–4497
Hendry SH, Yoshioka T (1994) A neurochemically distinct third channel in the macaque dorsal lateral geniculate nucleus. Science 264:575–577
Herculano-Houzel S, Collins CE, Wong PY, Kaas JH (2007) Cellular scaling rules for primate brains. Proc Nat Acad Sci U S Am 104:3562–3567
Hof, PR (2000) Neurochemical and cellular specializations in the mammalian neocortex reflect phylogenetic relationships: evidence from primates, cetaceans, and artiodactyls. Brain Behav Evol 300
Hof PR, Morrison JH (1990) Quantitative analysis of a vulnerable subset of pyramidal neurons in Alzheimer's disease: II. Primary and secondary visual cortex. J Comp Neurol 301:55–64
Hof PR, Morrison JH (1995) Neurofilament protein defines regional patterns of cortical organization in the macaque monkey visual

system: a quantitative immunohistochemical analysis. J Comp Neurol 352:161–186
Hof PR, Ungerleider LG, Webster MJ, Gattass R, Adams MM, Sailstad CA, Morrison JH (1996) Neurofilament protein is differentially distributed in subpopulations of corticocortical projection neurons in the macaque monkey visual pathways. J Comp Neurol 376:112–127
Holloway RL (1966) Cranial capacity, neural reorganization, and hominid evolution – search for more suitable parameters. Am Anthropol 68:103–121
Holloway RL (1968) The evolution of the primate brain: some aspects of quantitative relations. Brain Res 7:121–172
Holloway RL (1992) The failure of the Gyrification index (Gi) to account for volumetric reorganization in the evolution of the human brain. J Hum Evol 22:163–170
Holloway RL, De Lacoste-lareymondie MC (1982) Brain endocast asymmetry in pongids and hominids: some preliminary findings on the paleontology of cerebral dominance. Am J Phys Anthropol 58:101–110
Holloway RL, Broadfield DC, Yuan MS (2003) Morphology and histology of chimpanzee primary visual striate cortex indicate that brain reorganization predated brain expansion in early hominid evolution. Anat Rec 273A:594–602
Holmes G (1918) Disturbances of vision by cerebral lesions. Br J Ophthalmol 2:353–384
Hopkins WD, Marino L (2000) Asymmetries in cerebral width in nonhuman primate brains as revealed by magnetic resonance imaging (MRI). Neuropsychologia 38:493–499
Horel JA (1994) Local and global perception examined by reversible suppression of temporal cortex with cold. Behav Brain Res 65:157–164
Horton JC, Hoyt WF (1991) The representation of the visual field in human striate cortex. A revision of the classic Holmes map. Arch Ophthalmol 109:816–824
Hubel DH, Wiesel TN (1959) Receptive fields of single neurones in the cat's striate cortex. J Physiol 148:574–591
Hubel DH, Wiesel TN (1968) Receptive fields and functional architecture of monkey striate cortex. J Physiol 195:215–243
Hughes A, Vaney DI (1982) The organization of binocular cortex in the primary visual area of the rabbit. J Comp Neurol 204:151–164
Humphrey GK, Goodale MA, Bowen CV, Gati JS, Vilis T, Rutt BK, Menon RS (1997) Differences in perceived shape from shading correlate with activity in early visual areas. Curr Biol 7:144–147
Inouye T (1909) Die Sehstörungen bei Schussverletzungen der kortikalen Sehsphäre: nach Beobachtungen an Verwundeten der letzten japanischen Kriege. W. Engelmann, Leipzig
Jerison HJ (1975) Fossil evidence of evolution of the human brain. Annu Rev Anthropol 4:27–58
Jones EG, Hendry SH (1989) Differential calcium binding protein immunoreactivity distinguishes classes of relay neurons in monkey thalamic nuclei. Eur J Neurosci 1:222–246
Kaas JH (1993) The organization of visual cortex in primates: problems, conclusions, and the use of comparative studies in understanding brain evolution. In: Gulyas B, Ottoson D, Roland PE (eds) Functional organization of the human visual cortex, 1st edn. Perhamon Press, Oxford
Kaas JH (2006) The evolution of visual cortex in primates. In: Kremers J (ed) The primate visual system. John Wiley & Sons, Ltd., Chichester, UK, pp 267–283
Karlen SJ, Krubitzer L (2009) Effects of bilateral enucleation on the size of visual and nonvisual areas of the brain. Cereb Cortex 19:1360–1371
Kourtzi Z, Tolias AS, Altmann CF, Augath M, Logothetis NK (2003) Integration of local features into global shapes: monkey and human FMRI studies. Neuron 37:333–346
Kravitz DJ, Saleem KS, Baker CI, Mishkin M (2011) A new neural framework for visuospatial processing. Nat Rev Neurosci 12:217–230
Krubitzer L, Campi KL, Cooke DF (2011) All rodents are not the same: a modern synthesis of cortical organization. Brain Behav Evol 78:51–93
Kuffler SW (1953) Discharge patterns and functional organization of mammalian retina. J Neurophysiol 16:37–68
Kujovic M, Zilles K, Malikovic A, Schleicher A, Mohlberg H, Rottschy C, Eickhoff SB, Amunts K (2013) Cytoarchitectonic mapping of the human dorsal extrastriate cortex. Brain Struct Funct 218:157–172
Lacoste-Royal G, Mathieu M, Nalbantoglu J, Julien JP, Gauthier S, Gauvreau D (1990) Lack of association between two restriction fragment length polymorphisms in the genes for the light and heavy neurofilament proteins and Alzheimer's disease. Can J Neurol Sci 17:302–305
Lashley KS, Clark G (1946) The cytoarchitecture of the cerebral cortex of Ateles – a critical examination of architectonic studies. J Comp Neurol 85:223–305
Le May M (1976) Morphological cerebral asymmetries of modern man, fossil man and nonhuman primates. Ann N Y Acad Sci 280:349–366
Le May M, Billig MS, Geschwind N (1982) Asymmetries in the brains and skulls of nonhuman primates. In: Armstrong E, Falk D (eds) Primate brain evolution: methods and concepts. Plenum Press, New York
Leventhal AG, Rodieck RW, Dreher B (1981) Retinal ganglion cell classes in the old world monkey: morphology and central projections. Science 213:1139–1142
Livingstone M, Hubel D (1988) Segregation of form, color, movement, and depth: anatomy, physiology, and perception. Science 240:740–749
Lucas PW, Darvell BW, Lee PK, Yuen TD, Choong MF (1998) Colour cues for leaf food selection by long-tailed macaques (*Macaca fascicularis*) with a new suggestion for the evolution of trichromatic colour vision. Folia Primatol (Basel) 69:139–152
Lucas PW, Dominy NJ, Riba-Hernandez P, Stoner KE, Yamashita N, Loria-Calderon E, Petersen-Pereira W, Rojas-Duran Y, Salas-Pena R, Solis-Madrigal S, Osorio D, Darvell BW (2003) Evolution and function of routine trichromatic vision in primates. Evolution Int J Org Evolution 57:2636–2643
Lueck CJ, Zeki S, Friston KJ, Deiber MP, Cope P, Cunningham VJ, Lammertsma AA, Kennard C, Frackowiak RS (1989) The colour centre in the cerebral cortex of man. Nature 340:386–389
Lyon DC, Kaas JH (2002) Evidence for a modified V3 with dorsal and ventral halves in macaque monkeys. Neuron 33:453–461
Malikovic A, Amunts K, Schleicher A, Mohlberg H, Eickhoff SB, Wilms M, Palomero-Gallagher N, Armstrong E, Zilles K (2007) Cytoarchitectonic analysis of the human extrastriate cortex in the region of V5/MT+: a probabilistic, stereotaxic map of area hOc5. Cereb Cortex 17:562–574
Malikovic A, Vucetic B, Milisavljevic M, Tosevski J, Sazdanovic P, Milojevic B, Malobabic S (2012) Occipital sulci of the human brain: variability and morphometry. Anat Sci Int 87:61–70
Malikovic A, Amunts K, Schleicher A, Mohlberg H, Kujovic M, Palomero-gallagher N, Eickhoff SB, Zilles K (2016) Cytoarchitecture of the human lateral occipital cortex: mapping of two extrastriate areas hOc4la and hOc4lp. Brain Struct Funct 221:1877–1897
McDaniel WF, Wall TT (2013) Visuospatial functions in the rat following injuries to striate, peristriate, and parietal neocortical sites. Psychobiology 16:251–260
McKeefry DJ, Zeki S (1997) The position and topography of the human colour centre as revealed by functional magnetic resonance imaging. Brain 120(Pt 12):2229–2242

Nassi JJ, Callaway EM (2009) Parallel processing strategies of the primate visual system. Nat Rev Neurosci 10:360–372
Newton I (1966) Opticks: or a treatise of the reflexions, refractions, inflexions and colours of light. Also two treatises of the species and magnitude of curvilinear figures. Sam. Smith and Benj. Walford, London. 1704, Culture et civilisation
Nonaka-Kinoshita M, Reillo I, Artegiani B, Martinez-Martinez MA, Nelson M, Borrell V, Calegari F (2013) Regulation of cerebral cortex size and folding by expansion of basal progenitors. EMBO J 32:1817–1828
Orban GA, Van Essen D, Van Duffel W (2004) Comparative mapping of higher visual areas in monkeys and humans. Trends Cogn Sci 8:315–324
Pasupathy A, Connor CE (2002) Population coding of shape in area V4. Nat Neurosci 5:1332–1338
Pearce JM (2006) Louis Pierre Gratiolet (1815–1865): the cerebral lobes and fissures. Eur Neurol 56:262–264
Pearce E, Stringer C, Dunbar RI (2013) New insights into differences in brain organization between Neanderthals and anatomically modern humans. Proc Biol Sci 280:20130168
Perrett DI, Smith PA, Potter DD, Mistlin AJ, Head AS, Milner AD, Jeeves MA (1985) Visual cells in the temporal cortex sensitive to face view and gaze direction. Proc R Soc Lond B Biol Sci 223:293–317
Preuss TM (2005) Evolutionary specializations of primate brain systems. In: Ravoso MJ, Dagosto M (eds) Primate origins and adaptations. Kluwer Academic/Plenum Press, New York
Preuss TM, Coleman GQ (2002) Human-specific organization of primary visual cortex: alternating compartments of dense Cat-301 and calbindin immunoreactivity in layer 4A. Cereb Cortex 12:671–691
Preuss TM, Qi H, Kaas JH (1999) Distinctive compartmental organization of human primary visual cortex. Proc Natl Acad Sci U S A 96:11601–11606
Raczkowski D, Rosenquist AC (1983) Connections of the multiple visual cortical areas with the lateral posterior-pulvinar complex and adjacent thalamic nuclei in the cat. J Neurosci 3:1912–1942
Riegele L (1931) Die Cytoarchitektonik der Felder der Broca'schen Region. J Psychol Neurol 42:496–514
Rodiek RW (1988) The primate retina. In: Stelis HD, Erwin J (eds) Comparative primate biology, Neurosciences, vol 4. Liss, New York
Rosa MG, Tweedale R (2005) Brain maps, great and small: lessons from comparative studies of primate visual cortical organization. Philos Trans R Soc Lond B 360:665–691
Rottschy C, Eickhoff SB, Schleicher A, Mohlberg H, Kujovic M, Zilles K, Amunts K (2007) Ventral visual cortex in humans: cytoarchitectonic mapping of two extrastriate areas. Hum Brain Mapp 28:1045–1059
Scheperjans F, Hermann K, Eickhoff SB, Amunts K, Schleicher A, Zilles K (2008) Observer-independent cytoarchitectonic mapping of the human superior parietal cortex. Cereb Cortex 18:846–867
Schleicher A, Zilles K (1990) A quantitative approach to cytoarchitectonics: analysis of structural inhomogeneities in nervous tissue using an image analyser. J Microsc 157:367–381
Schleicher A, Amunts K, Geyer S, Morosan P, Zilles K (1999) Observer-independent method for microstructural parcellation of cerebral cortex: a quantitative approach to cytoarchitectonics. NeuroImage 9:165–177
Schultz AH (1940) The size of the orbit and of the eye in primates. Am J Phys Anthropol 26:389–408
Schwarzkopf DS, Song C, Rees G (2011) The surface area of human V1 predicts the subjective experience of object size. Nat Neurosci 14:28–30
Semendeferi K, Damasio H (2000) The brain and its main anatomical subdivisions in living hominoids using magnetic resonance imaging. J Hum Evol 38:317–332
Sereno M, Allman JM (1991) Cortical visual areas in mammals. In: Leventhal AG (ed) The neural basis of visual function. Macmillan, London
Sereno MI, Dale AM, Reppas JB, Kwong KK, Belliveau JW, Brady TJ, Rosen BR, Tootell RB (1995) Borders of multiple visual areas in humans revealed by functional magnetic resonance imaging. Science 268:889–893
Sherwood CC, Hof PR (2007) The evolution of neuron types and cortical histology in apes and humans. In: Kaas JH, Preuss TM (eds) Evolution of nervous systems, The Evolution of Primate Nervous Systems, vol 4. Academic Press, Oxford
Sherwood CC, Lee PW, Rivara CB, Holloway RL, Gilissen EP, Simmons RM, Hakeem A, Allman JM, Erwin JM, Hof PR (2003) Evolution of specialized pyramidal neurons in primate visual and motor cortex. Brain Behav Evol 61:28–44
Sherwood CC, Holloway RL, Erwin JM, Hof PR (2004) Cortical orofacial motor representation in old world monkeys, great apes, and humans. II. Stereologic analysis of chemoarchitecture. Brain Behav Evol 63:82–106
Sherwood CC, Stimpson CD, Raghanti MA, Wildman DE, Uddin M, Grossman LI, Goodman M, Redmond JC, Bonar CJ, Erwin JM, Hof PR (2006) Evolution of increased glia-neuron ratios in the human frontal cortex. Proc Natl Acad Sci U S A 103:13606–13611
Sherwood CC, Raghanti MA, Stimpson CD, Bonar CJ, De Sousa AA, Preuss TM, Hof PR (2007) Scaling of inhibitory interneurons in areas V1 and V2 of anthropoid primates as revealed by calcium-binding protein immunohistochemistry. Brain Behav Evol 69:176–195
Stout D, Toth N, Schick K, Stout J, Hutchins G (2000) Stone toolmaking and brain activation: position emission tomography (PET) studies. J Archaeol Sci 27:1215–1223
Tallinen T, Chung JY, Rousseau F, Girard N, Lefevre J, Mahadevan L (2016) On the growth and form of cortical convolutions. Nat Phys, advance online publication
Thurlow GA, Cooper RM (1988) Metabolic activity in striate and extrastriate cortex in the hooded rat: contralateral and ipsilateral eye input. J Comp Neurol 274:595–607
Toga AW, Thompson PM, Mori S, Amunts K, Zilles K (2006) Towards multimodal atlases of the human brain. Nat Rev Neurosci 7:952–966
Tomonaga M (2001) Investigating visual perception and cognition in chimpanzees (Pan troglodytes) through visual search and related tasks: from basic to complex processes. In: Matsuzawa T (ed) Primate origins of human cognition and behavior. Springer, Tokyo, pp 55–86
Tootell RB, Taylor JB (1995) Anatomical evidence for MT and additional cortical visual areas in humans. Cereb Cortex 5:39–55
Tootell RB, Mendola JD, Hadjikhani NK, Ledden PJ, Liu AK, Reppas JB, Sereno MI, Dale AM (1997) Functional analysis of V3A and related areas in human visual cortex. J Neurosci 17: 7060–7078
Tusa RJ, Palmer LA, Rosenquist AC (1978) The retinotopic organization of area 17 (striate cortex) in the cat. J Comp Neurol 177:213–235
Ungerleider LG, Mishkin M (1982) Two cortical visual systems. In: Ingle DJ, Goodale MA, Mansfield RJW (eds) Analysis of visual behavior. MIT Press, Cambridge, MA
Van Duffel W, Fize D, Mandeville JB, Nelissen K, Van Hecke P, Rosen BR, Tootell RB, Orban GA (2001) Visual motion processing investigated using contrast agent-enhanced fMRI in awake behaving monkeys. Neuron 32:565–577
Van Essen DC (1985) Functional organization of the primate visual cortex. In: Peters A, Jones EG (eds) Cerebral cortex. Plenum Press, New York
Van Essen DC (2004) Organization of visual areas in macaque and human cerebral cortex. In: Chalupa L, Werner J (eds) The visual neurosciences. MIT Press, Cambridge, MA

Van Essen DC, Dierker DL (2007) Surface-based and probabilistic atlases of primate cerebral cortex. Neuron 56:209–225
Van Essen DC, Zeki SM (1978) The topographic organization of rhesus monkey prestriate cortex. J Physiol 277:193–226
Van Essen DC, Newsome WT, Bixby JL (1982) The pattern of interhemispheric connections and its relationship to extrastriate visual areas in the macaque monkey. J Neurosci 2:265–283
Van Essen DC, Anderson CH, Felleman DJ (1992) Information processing in the primate visual system: an integrated systems perspective. Science 255:419–423
Vater A, Heinicke JC (1723) Dissertatio qua visus vitia duo rarissima: alterum duplicati, alterum dimidiati physiologice et pathologice considerata exponuntur
Von Economo C (1929) The cytoarchitectonics of the human cortex. Oxford University Press, Oxford
Von Economo C, Koskinas GN (1925) Die Cytoarchitektonik der Hirnrinde des erwachsenen Menschen. Springer, Vienna
Yan T, Jin F, Wu J (2009) Correlated size variations measured in human visual cortex V1/V2/V3 with functional MRI. In: Zhong N, Li K, Lu S, Chen L (eds) Brain informatics: international conference, BI 2009 Beijing, China, October 22–24, 2009 proceedings. Springer Berlin Heidelberg, Berlin
Yoshioka T, Hendry SH (1995) Compartmental organization of layer IVA in human primary visual cortex. J Comp Neurol 359:213–220
Zeki S (2003) Improbable areas in the visual brain. Trends Neurosci 26:23–26
Zeki SM (2004) Improbable areas in color vision. In: Chalupa LM, Werner JS (eds) The visual neurosciences. MIT Press, Cambridge, MA
Zilles K, Clarke S (1997) Architecture, connectivity, and transmitter receptors of human extrastraite visual cortex. In: Rockland KS, Kaas JH, Peters A (eds) Extrastriate cortex in primates. Plenum Press, New York
Zilles K, Palomero-Gallagher N (2001) Cyto-, myelo-, and receptor architectonics of the human parietal cortex. NeuroImage 14:S8–20
Zilles K, Schlaug G, Matelli M, Luppino G, Schleicher A, Qu M, Dabringhaus A, Seitz R, Roland PE (1995) Mapping of human and macaque sensorimotor areas by integrating architectonic, transmitter receptor, MRI and PET data. J Anat 187(Pt 3):515–537
Zilles K, Schleicher A, Palomero-Gallagher N, Amunts K (2002) Quantitative analysis of cyto- and receptor architecture of the human brain. In: Toga AW, Mazziotta JC (eds) Brain mapping: the methods, 2nd edn. Elsevier Science, Atlanta

Cerebellum: Anatomy, Physiology, Function, and Evolution

18

Hiroki C. Tanabe, Daisuke Kubo, Kunihiro Hasegawa, Takanori Kochiyama, and Osamu Kondo

Abstract

Until recently, few studies have focused on the evolution of the cerebellum in human lineage. While the major functional role of the cerebellum was traditionally thought to be fine motor control, recent neuroimaging and neurological evidence suggests that the cerebellum is deeply involved in a variety of cognitive and social functions. Moreover, the cerebellum has been found to have a unique gross anatomy and microstructure. In contrast to the cerebrum, the cortex of the cerebellum is structured as a homogeneous sheet with a similar internal structure throughout. This cortex contains cerebellar neural circuitry that functions as a learning system capable of constructing and storing internal models of the external environment. Converging evidence suggests that the greater the volume of the cerebellar cortex, the more internal models it is able to store. This neuroanatomical organization may affect innate learning, cognitive ability, and the human capacity to innovate. In this chapter, we review the relationship between cerebellar volume and various cognitive abilities in modern humans and then discuss the evolutionary changes of cerebellar size based on the comparative anatomy of extant primates and the evidence from fossil specimens with our recent findings.

Keywords

Cerebellar hemisphere • Cognitive function • Cerebellar circuit • Internal model

H.C. Tanabe (✉)
Department of Cognitive and Psychological Sciences, Graduate School of Informatics, Nagoya University, Furo-cho, Chikusa-ku, Nagoya, Aichi 464-8601, Japan
e-mail: htanabe@i.nagoya-u.ac.jp

D. Kubo
Division of Human Evolution Studies, Graduate School of Medicine, Hokkaido University, Sapporo, Japan
e-mail: dkubo@med.hokudai.ac.jp

K. Hasegawa
Automotive Human Factors Research Center, National Institute of Advanced Industrial Science and Technology, Tsukuba, Japan
e-mail: hasegawa.kunihiro@aist.go.jp

T. Kochiyama
Department of Cognitive Neuroscience, Advanced Telecommunications Research Institute International, Kyoto, Japan
e-mail: kochiyam@atr.jp

O. Kondo
Graduate School of Science, The University of Tokyo, Tokyo, Japan
e-mail: kondo-o@bs.s.u-tokyo.ac.jp

18.1 Introduction

The cerebellum is an important part of the vertebrate central nervous system. In humans, the cerebellum is located in the posterior of the brain stem and pons and is separated from the overlaying cerebrum by the cerebellar tentorium (Fig. 18.1, shown in pink). The cerebellum accounts for around 10–11% of the volume of the whole brain, but contains over half of the total number of neurons. In the present chapter, we first describe the gross anatomy, microanatomy, neural circuitry, and internal models of the cerebellum. Second, we discuss the function of the cerebellum, particularly its role in higher cognitive functions. In this section, we also review the relationship between cerebellar volume and various cognitive abilities using data from modern humans. Finally, we discuss the evolutionary changes of the cerebellar size in hominins as

E. Bruner et al. (eds.), *Digital Endocasts*, Replacement of Neanderthals by Modern Humans Series,
DOI 10.1007/978-4-431-56582-6_18

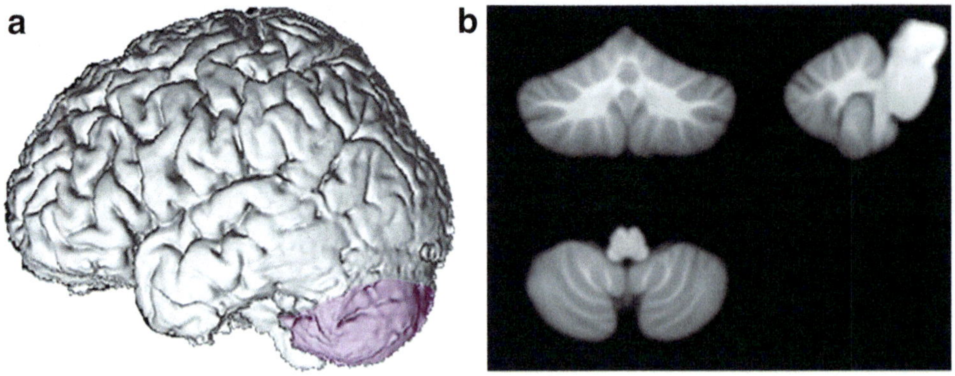

Fig. 18.1 (**a**) MRI scan of a modern human brain (lateral view of the left hemisphere). The *pink* region indicates the location of the cerebellum. (**b**) Multislice view of the cerebellum. The nonlinear normalization to the unbiased infratentorial template with respect to the affine registration (SUIT) image was used to display (Diedrichsen 2006). *Upper left*, coronal view; *upper right*, sagittal view; *lower left*, axial view

well as the technical issues related to the cerebellar volume estimation from fossil specimens.

18.2 Anatomical Structure of the Modern Human Cerebellum

18.2.1 Gross Anatomy

The basic structure of the cerebellum resembles a piece of cauliflower and is not obviously split down the middle. A bump called the vermis is located on the midline of the cerebellum and separates the two lateral hemispheres. The surface of the cerebellum is organized like a repeatedly folded thin sheet, containing several types of neurons. The cerebellum is divided into ten segments by the deep transverse fissures, and each segment is known as a lobule. According to the Montréal Neurological Institute (MNI) 152 template, the bilateral hemisphere parts of lobule VII are additionally separated into three segments: Crus I, Crus II, and lobule VIIb (Fig. 18.2). Neurons also exist in the white matter of the cerebellum, forming the deep cerebellar nuclei, which relay most of the cerebellar cortical output to some brain stem nuclei, the reticular formation, and the thalamus. The vermis and the hemispheres have different functional roles: the vermis sends output signal to the ventromedial descending spinal pathways which control the axial musculature, whereas the hemispheres connect various parts of the contralateral cerebral cortex via the cerebello-thalamic pathway.

The hemispheres of the cerebellum, also known as the neocerebellum, are well developed in humans compared with other species (MacLeod et al. 2003). MacLeod et al. (2003) reported that hominoids have disproportionally large cerebellar hemispheres as compared to monkeys when the hemispheres were regressed against the vermis. This finding, in accord with several other reports, indicates that there may have been a transformation of the lateral cerebellum at some point in human evolution (MacLeod et al. 2003; Weaver 2005; Whiting and Barton 2003).

18.2.2 Microanatomy and Neural Circuits

The cerebellar cortex has a relatively uniform thickness of 1 mm. Histological analysis indicates that the cerebellar cortex comprises three layers: the molecular layer, which is a surface layer containing mostly axons, the Purkinje layer, and the granular cell layer, which is densely packed with granule neurons in humans (approximately 10^{10}–10^{11}; Braitenberg and Atwood 1958). In addition, five types of neurons have been found in the cortex of the cerebellum: superficial stellate cells, basket cells, Purkinje cells, granule cells, and Golgi cells. Both granule cells and Purkinje cells receive afferent input, whereas the other three neuron types are inhibitory interneurons. Granule cells are innervated from mossy fibers originating in the pontine and vestibular nuclei and the spinal cord, while Purkinje cells have direct inputs from climbing fibers whose cell bodies are in the inferior olivary nuclei of the brain stem. Purkinje cells are the only output cells in the cerebellar cortex (Apps and Garwicz 2005; Ito 2011).

The neural circuitry of the cerebellum, known as the cerebellar circuit, is structured around Purkinje cells. Axons of granule cells extend from the granule cell layer, through the Purkinje cell layer, and into the molecular layer, where they bifurcate. These thin unmyelinated branches are called parallel fibers and terminate on the Purkinje cell's dendrites. One Purkinje cell connects $1–2 \times 10^6$ parallel fibers. Purkinje cells also receive input from climbing fibers. In contrast to the input that granule cells receive via parallel fibers, each Purkinje cell receives input from only one

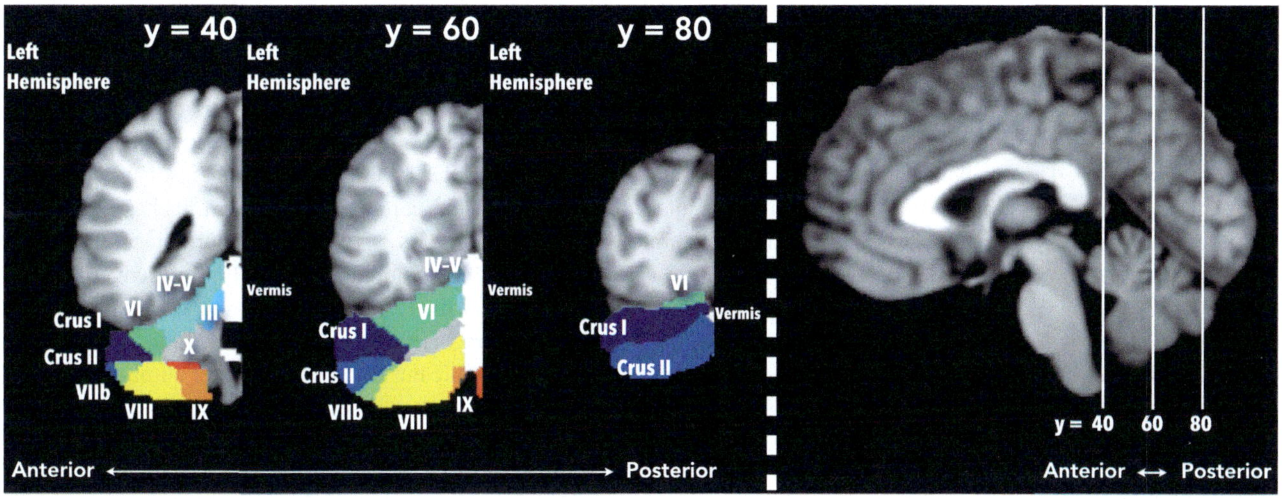

Fig. 18.2 The MNI152 anatomical atlas of the modern human cerebellum

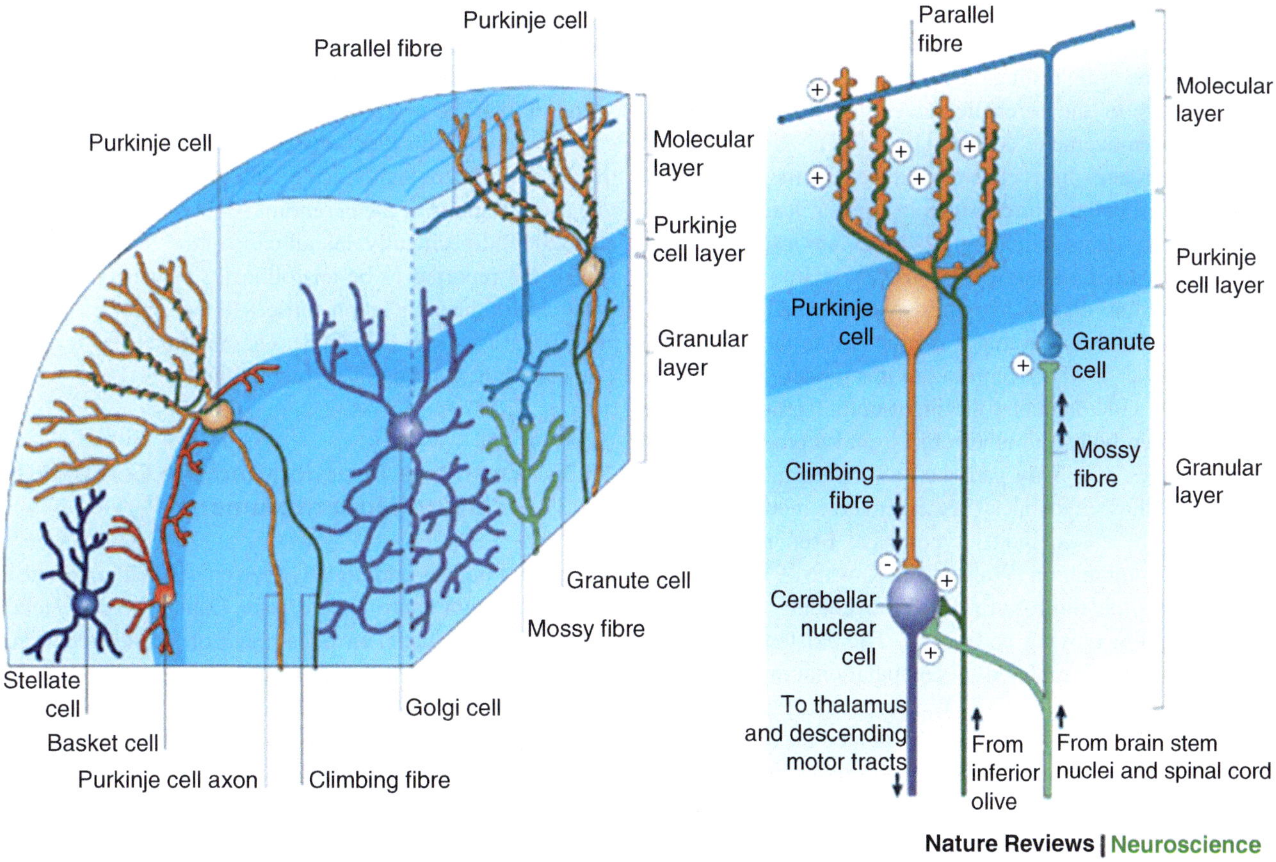

Fig. 18.3 Structure of the human cerebellar cortex (Apps and Garwicz 2005) (Reprinted by permission from Nature Publishing Group: Nature Reviews Neuroscience Apps and Garwicz 2005, copyright 2005)

climbing fiber. The cerebellar cortex contains inhibitory interneurons, which form another link between the granule cells and Purkinje cells. A schematic illustration of the cerebellar cortical structure is shown in Fig. 18.3 (Apps and Garwicz 2005).

Unlike the cerebral cortex, the cerebellar cortical structure is a homogeneous sheet of tissue. All parts of this tissue have the same internal structure such as the cerebellar circuits. The cerebellar circuits are compartmentalized into 1 × 10 mm regions, referred to as microzones or microcomplexes

(Apps and Garwicz 2005). Each microcomplex contains 500–1,000 Purkinje cells and operates as a functional unit, somewhat like a computer chip. In humans, the cerebellar cortex contains a large number of microcomplexes (Porrill et al. 2013). The microcomplex is thought to be involved in learning (Ito 2011). The cerebellar circuits have been found to exhibit long-term synaptic plasticity, indicating that experience-dependent learning processes are a salient feature of cerebellar function (Apps and Garwicz 2005).

18.2.3 Internal Model and Function of the Cerebellum

As described above, the cerebellum is able to build and store "internal models" which neurally represent the external world (Ito 2008). The internal model in the cerebellum is defined as neural circuits which simulates or emulates input-output properties of the outside of the brain. Internal models are formed and stored various input-output properties of the outside of the brain through learning. The internal modeling functionality of the cerebellum was first described in the context of movement control (Ito 2011). Around 1970, three researchers independently advocated the hypothesis that the cerebellar cortex supports motor learning using input signals from climbing fibers as an instructor (the Marr-Albus-Ito hypothesis) (Albus 1971; Ito 1970; Marr 1969). Building on this hypothesis, a computational model of cerebellar motor learning was proposed in the late 1980s (Kawato et al. 1987). According to this model, the cerebellum forms and maintains the input-output properties of a person's own body and other objects and reproduces either the dynamics of a body part (i.e., a forward model) or the inverse of those dynamics (i.e., an inverse model) through learning. These two model types have been proposed to operate in combination (Wolpert and Kawato 1998). Internal models help the control of precise movements without any feedback from moving body parts. The existence of these models has been supported by computational modeling and robotics researches (Ito 2008; Wolpert et al. 1998). Furthermore, Imamizu and Kawato have examined the neural representation of internal models in the cerebellum (Imamizu et al. 2000; Imamizu and Kawato 2009 for review) (Fig. 18.4, Imamizu and Kawato 2009).

Van Overwalle and colleagues recently suggested that such internal models are able to encode essential properties of mental representation in the cerebrum for various cognitive functions (Van Overwalle and Mariën 2016). Although the exact nature of input-output relationships remains unclear, internal models are considered to represent various features of the external environment, including the mental states of others (Imamizu and Kawato 2009). Furthermore, Ito (2011) proposed that the implicit process underlying the manipulation of thoughts involves internal models in the cerebellum, while the explicit component of the process is undertaken in the cerebral cortex. Ito (2011) suggests that this implicit process may be an important component of creativity and innovation. Importantly, this converging evidence suggests that internal models are stored in the cerebellar circuitry. Therefore, a greater volume of cerebellar cortex may enable the storage of a greater number of internal models.

18.2.4 Anatomical and Functional Connections Between Cerebellum and Cerebrum

Although the anatomical characteristics of the cerebellar cortex appear to be homogenous, the functions of the cerebellar hemisphere differ according to location. It has been proposed that the cerebellum is able to support a variety of cognitive and motor functions because the different parts of the cerebellum are anatomically and functionally connected to various regions of the cerebrum (Buckner 2013; O'Reilly et al. 2009). Specifically, lateral parts of the cerebellar hemisphere are reported to be anatomically connected with the other side of the association cortices in the cerebrum (Bostan et al. 2013; Jissendi et al. 2008). These regions play a significant role in higher cognitive functioning.

18.3 Cognitive Function and the Cerebellum in Living Modern Humans

The cerebellum has traditionally been considered to operate as a center of motor control in the brain. Damage to the cerebellum can trigger ataxia, an impairment of balance, coordination, gait, extremity movement, and eye movements. However, mounting evidence indicates that the cerebellum is not just involved in motor function. Several recent neuroimaging studies revealed anatomical and functional connections between the cerebral cortex and the cerebellum (Buckner et al. 2011; Krienen and Buckner 2009; O'Reilly et al. 2009; Yeo et al. 2010). This evidence indicates that the posterior parts of the cerebellum are linked to the association cortex of the cerebrum. These cerebro-cerebellar connections provide a neural substrate by which the cerebellum could contribute to higher cognitive functions, including language, working

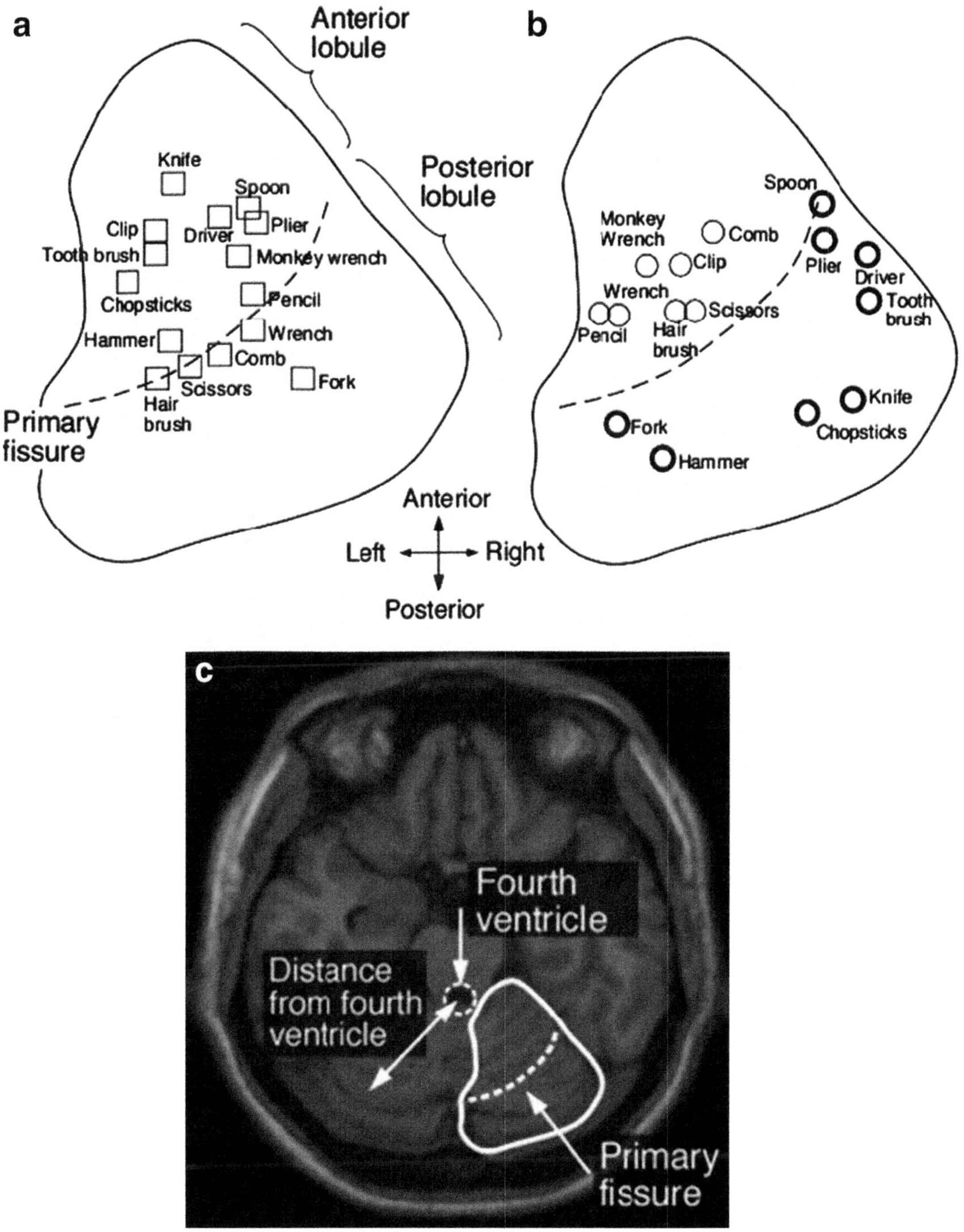

Fig. 18.4 Neural substrates of internal models of tool use (Adapted from Imamizu and Kawato 2009). Internal model of each tool was stored in different regions of the cerebellum. (**a**) Centroid of activation coordinates (*rectangles*) when participants really used tools. (**b**) Centroid of activation coordinates (*circles*) when participants imagined to use tools. (**c**) Axial anatomical image of the human cerebellum. Anatomical location of **a** and **b** is shown (a region surrounded by a *white line*) (Reprinted by permission from Psychological Research Psychologische Forshung: Psychological Research Imamizu and Kawato 2009, copyright 2009)

memory, and executive functions (Fig. 18.5) (Desmond and Fiez 1998; Ito 2008; Marvel and Desmond 2010).

Converging neuroimaging and neurological evidence from studies of people with brain injury also suggests that the cerebellar hemisphere is involved in the variety of cognitive functions (Keren-Happuch et al. 2014). To be more specific, the right cerebellar hemisphere is related to language processing, whereas left cerebellar damage leads to dysfunction of attention and visuospatial cognition (Schmahmann 1996, 2004; Schmahmann and Scherman

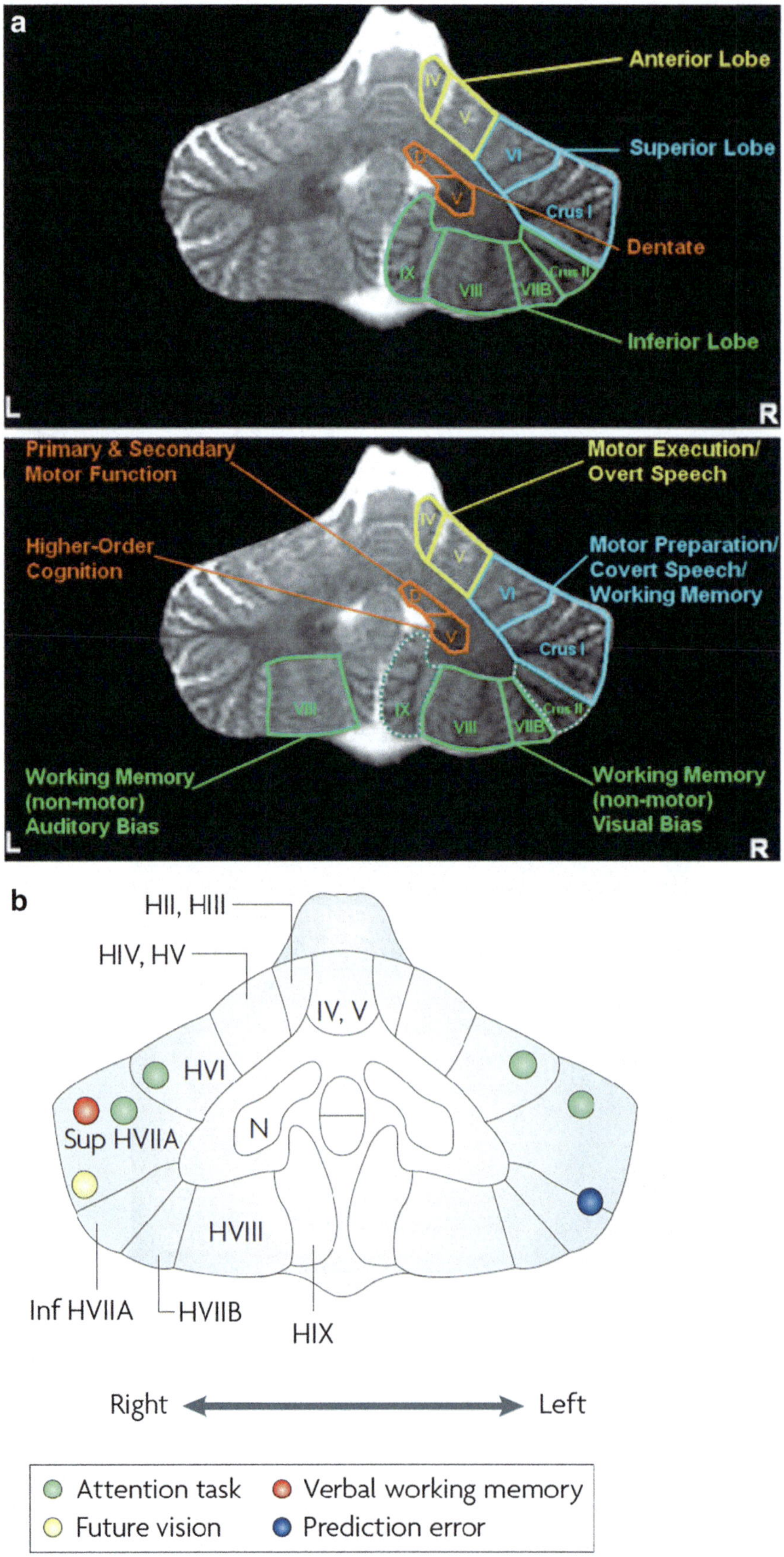

Fig. 18.5 Anatomical subregions in the cerebellum and cognitive functions corresponding to these regions. (**a**) Figure from Marvel and Desmond (2010) (Reprinted by permission from Springer: Neuropsychology Review Marvel and Desmond 2010, copyright 2010). (**b**) Figure from Ito (2008) (Reprinted by permission from Nature Publishing Group: Nature Reviews Neuroscience Ito 2008, copyright 2008)

1998; De Smet et al. 2013). However, we have to mention that human can live with complete absence of the cerebellum or cerebellar agenesis. For example, Yu et al. (2015) reported that a 24-year-old female patient showed symptoms of mild mental retardation and cerebellar ataxia. They also surveyed nine published living cases with primary cerebellar agenesis and showed that eight of nine patients showed symptoms of motor, language, and mental development (Yu et al. 2015). Taken this into consideration, the cerebellum seems to be not essential for motor and cognitive function, but be necessary for these "normal" functions.

Next, we discuss the role of the cerebellum in the major higher cognitive abilities including language, working memory, and executive functions.

18.3.1 Language

Language refers to the mental processes that enable communication between individuals through the sharing of any symbols (e.g., sounds, letters, and gestures). The importance of language in human evolution is beyond doubt, but many details remain to be clarified. For example, the question of whether Neanderthals possessed any language ability remains contentious (Lieberman 2009; Dediu and Levinson 2013). Klein and Edgar's "neural hypothesis" suggests that linguistic and symbolic abilities had a dramatic influence on the divergence between modern humans and other hominids (Klein and Edgar 2002). A recent study reported that Neanderthal hyoid bones are consistent with a capacity for speech (D'Anastasio et al. 2013). In addition, the Neanderthal genome project revealed that Neanderthals and *Homo sapiens* exhibit the same mutations in the language-related gene *FOXP2* (Krause et al. 2007). Evidence from studies of the neural underpinnings of language has important implications for furthering our understanding of human evolution.

In modern humans, language ability is divided into two major functions: speech production, which is strongly associated with activity in Broca's area, and speech comprehension, which involves Wernicke's area. Broca's area is in the inferior frontal gyrus (Brodmann area 44 and 45) of the dominant (left in most cases) cerebral hemisphere (shown as a red circle in Fig. 18.6). In contrast, Wernicke's area is in the posterior part of the superior temporal gyrus (Brodmann area 22) of the dominant (left in most cases) hemisphere (shown as a cyan circle in Fig. 18.6) and is strongly

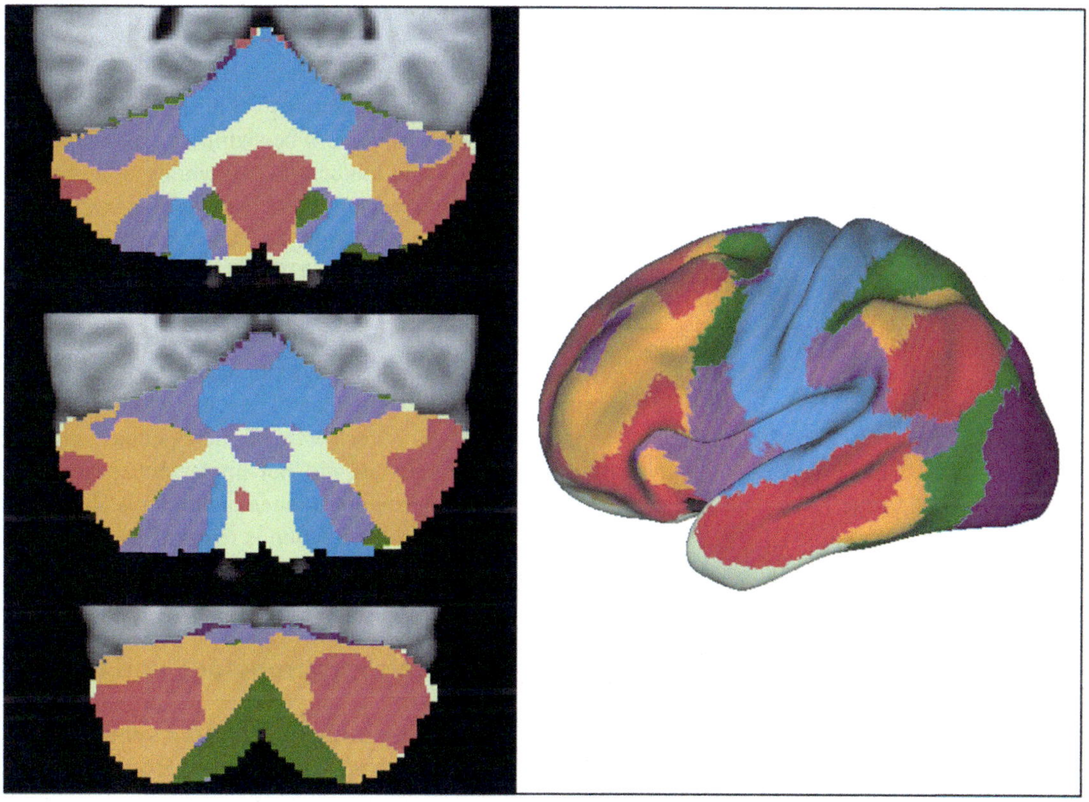

Fig. 18.6 The major networks between the cerebellum and cerebral cortex (Buckner 2013). Regions of the same *color* indicate a functional connection (e.g., the temporal fluctuation pattern of the resting-state brain activity is similar between cerebellar and cerebral cortices) (Reprinted by permission from Elsevier: Neuron Buckner 2013, copyright 2013)

connected with the arcuate fasciculus. Importantly, previous studies suggest that there is functional connectivity between the language-related areas of the cerebral cortex and the cerebellum (Buckner et al. 2011; Fig. 18.6), suggests that at least some parts of the cerebellum might play a role in language processing.

In accord with this functional connectivity, many neuropsychological studies have reported language impairment resulting from cerebellar lesions. The most commonly reported impairments are dysfunction of verbal fluency, grammatical/syntactic ability, and semantic access (Baillieux et al. 2006; Fabbro et al. 2000; Fiez et al. 1992; Hassid 1995; Leggio et al. 2000; Silveri et al. 1994; Zettin et al. 1997). Moreover, a small number of studies have reported hemispheric laterality in language-related cerebellar function (Schmahmann and Scharman 1998). Generally, the left hemisphere of the cerebral cortex plays a critical role in language (Frost et al. 1999), and language impairment is most often associated with lesions of the right hemisphere of the cerebellum, consistent with crossed cerebro-cerebellar connectivity (Hassid 1995).

18.3.2 Working Memory and Executive Functions

Executive functions play a critical role in environmental adaptation through the control of cognitive processes. Working memory is a particularly influential model of higher cognitive functioning in cognitive psychology, acting both as a temporary memory storage system and an executive information processing system (Baddeley 1992, 2003; Baddeley and Hitch 1974). The dorsolateral prefrontal cortex (DLPFC; shown as a green circle in Fig. 18.6) is well established as a critical brain region for working memory (Braver et al. 1997; Cohen et al. 1997). In the cerebellum, the posterior lateral region has been found to play a significant role in working memory (E et al. 2014). This cerebellar region is connected to the DLPFC both functionally and anatomically (Buckner et al. 2011; Kelly and Strick 2003). In addition, several studies have reported that cerebellar lesions are associated with impairments of working memory (Baier et al. 2014; Justus and Ivry 2009).

Barkley proposed that working memory has exerted a strong influence on human evolution, enabling the development of social self-defense against resource theft and social exchanges such as reciprocal altruism, selfish cooperation, and imitating or learning from others (Barkley 2001). The enhanced working memory (EWM) hypothesis (Coolidge and Wynn 2005; Wynn and Coolidge 2003) seeks to explain the replacement of Neanderthals by modern humans. They employed several indirect methods such as cognitive neuropsychology, cognitive anthropology, and cognitive archaeology, because it is impossible to access living Neanderthals (Wynn and Coolidge 2003). They combined the results from these approaches and infer Neanderthal's cognition. According to this hypothesis, Neanderthal's extinction was caused by a difference in working memory ability. The lateral prefrontal cortex is a particularly important brain region in the context of the EWM hypothesis, playing a critical role in working memory ability (D'Esposito et al. 1995). However, no anatomical differences in the lateral prefrontal cortex between Neanderthals and modern humans have been reported (Bruner and Holloway 2010; Jerison 2006).

18.3.3 Relationships Between Cerebellar Size and Cognitive Abilities

There is a long-running debate about the possible existence of a relationship between the size and function of the human brain (Jerison 1973). In the field of cognitive neuroscience, a number of studies have reported a correlation between global and local volume of the brain and psychological ability (Haier et al. 2004; Kanai and Rees 2011; Lange et al. 2010, McDaniel 2005; Sousa and Proulx 2014). However, the relationship between cerebellar volume and cognitive ability is currently not well understood.

In an early neuroimaging study, Paradiso et al. (1997) examined whether cerebellar size was associated with general intelligence, verbal memory, nonverbal memory, and motor dexterity (Paradiso et al. 1997). The entire cerebellar volume was found to be significantly correlated with verbal memory and motor dexterity ability (Paradiso et al. 1997). With the development of brain imaging techniques, the size-function relationship has been investigated using voxel-based morphometry (VBM) (Ashburner and Friston 2000). Studies using this method have reported that the gray matter volume of the cerebellum is associated with grammatical processes in a second language (Pliatsikas et al. 2013) and with the performance of executive function tasks (Lin et al. 2012). However, both of these studies involved a small sample size and did not implement a comprehensive test battery to examine a range of cognitive abilities. More comprehensive investigation is required to better understand the relationship between cerebellar size and cognitive ability.

18.4 Paleoanthropology and the Cerebellum

18.4.1 Cerebellar Size and Asymmetry in Extant Primates and Fossil Hominins

Evolutionary changes in the cerebellum in the hominins have been examined using the comparative anatomy of extant primates, as well as the paleoneurology of fossil hominins. The data obtained from extant primates provides relevant evidence to aid interpretation of the paleoneurological data.

A number of studies within the last two decades (Rilling and Insel 1998; Rilling 2006; MacLeod et al. 2003: MacLeod 2012; Barton and Venditti 2014) have revealed that apes have a disproportionally large cerebellum compared with non-hominoid anthropoids of the same body weight or brain mass. According to MacLeod et al. (2003), the cerebellum accounts for 11% of the brain in modern humans, 13.4% in lesser and great apes, and less than 10% in monkeys, on average. Their regression analysis showed that the volume of the cerebellum and cerebellar hemispheres of the ape were approximately 1.7 and 1.5 times larger than those expected for a monkey of the same brain mass, respectively, and the cerebellar hemispheres were 2.4 times larger in an ape compared with a monkey of the same body weight (MacLeod et al. 2003). In addition, they reported that the cerebellar hemispheres of hominoids (including modern humans) were 2.7 times larger than those of a monkey with an equivalent sized cerebellar vermis (MacLeod et al. 2003).

Among the extant primates, the modern human has the largest cerebellum, which is 2.9 times larger than would be expected in an ape of the same body weight (MacLeod et al. 2003). However, the modern human cerebellum appears to be smaller than would be expected in a hypothetical ape of the same brain mass (Rilling and Insel 1998; Semendeferi and Damasio 2000; MacLeod et al. 2003). Rilling and Insel (1998) noted that the relatively large cerebellum shared among extant apes and the relatively small cerebellum in humans are best explained by a grade shift of cerebellar expansion in the common ancestor of hominoids, followed by a disproportionate cerebral expansion in hominids. Barton and Venditti (2014) provided an alternative proposal: based on the rates of cerebellar volume change relative to the (cerebral) neocortex volume change for each branch of the phylogenetic tree of the primates, they suggested that the volume of the cerebellum increased disproportionately relative to that of the cerebral neocortex in the hominin branch as well as in the chimpanzee/bonobo branches. If this is the case, the last common ancestor of the chimpanzees and humans (and possibly the early hominins) may have had smaller cerebella compared to that of extant apes with the same brain mass. Further paleoneurological investigation is required to test these competing hypotheses.

Paleoanthropologists have examined the brain anatomy of fossil hominins based on the endocast, or the mold of the endocranial cavity (Holloway et al. 2004; Falk 2014). Endocasts roughly reflect the external form of the brain because brain growth is accompanied by the expansion of the endocranial cavity in normal ontogeny (Moss and Young 1960; Friede 1981; Peña-Melian 2000). A large portion of the cerebellum is housed in the posterior cranial fossa (PCF) of the endocranial cavity, and the external form of the cerebellum can be partially modeled from PCF morphology. Some researchers (Weil 1929; Schepers 1950; Kochetkova 1978; Holloway and Yuan 2004) have attempted to evaluate the development of the cerebellum in fossil hominins based on the dimensions of the PCF. These studies have indicated that the relative cerebellar size of australopithecines was likely smaller than that of the extant apes. However, there is disagreement over the relative cerebellar size of Neanderthals, the late Paleolithic *Homo sapiens*, and recent modern humans.

Weil (1929) conducted the earliest study of the PCF size of fossil hominins. The cerebrum/cerebellum ratio was calculated based on the surface areas of the endocast of extant lesser and great apes, five recent modern humans from different ethnic groups, and four fossil *Homo* specimens (Trinil 2, Kabwe, La Chapelle-aux-Saints, and Predmostí 9). According to Weil's (1929) measurements, the cerebrum/cerebellum ratio was smaller in apes (4.53–5.16) than modern humans (5.90–6.40), consistent with the observations from recent brain volumetric studies (e.g., MacLeod et al. 2003). The ratio was reported to be 6.28 in the late Paleolithic modern human (Predmostí 9), falling within the range of recent modern humans. Moreover, the reported ratio was 7.67 in an African archaic *Homo* specimen (Kabwe) and 7.86 in a Neanderthal specimen (La Chapelle-aux-Saints), indicating that they had relatively small cerebella compared with the extant and fossil modern humans.

In another early study, Schepers (1950) evaluated the ratios of the forebrain with the midbrain and hindbrain of extant great apes; australopithecine specimens from Taung, Sterkfontein (I, II, V, VII, and VIII), and Kromdraai (TM1517); and several representatives of the genus *Homo*. However, the data are somewhat limited, because Schepers (1950) did not describe the way these volumes were obtained from the endocast and did not record the names of the examined *Homo* specimens. Despite these limitations, Schepers (1950) reported that the mid- and hindbrain occupy 16–17% of the brain in great apes and 11–14% in australopithecines, which appears to be compatible with the predictions of Barton and Venditti (2014). As a caveat, it should be noted that some of the examined australopithecine crania (particularly Sterkfontein II and TM1517) are fragmentary specimens (Broom and Schepers 1946; Broom et al. 1950; Neubauer et al. 2012). In contrast to Weil's (1929) findings, the measurements of Schepers (1950) indicated that Neanderthals had relatively large cerebella compared with modern humans; the mid- and hindbrain ratio was found to be 16% in *H. erectus*, 15% in Neanderthals, and 11% in modern humans.

Kochetkova (1978) assessed evolutionary trends in the length, width, and height of the PCF. The comparative sample included chimpanzees, more than 20 fossil hominin specimens, and 40 recent modern humans (Kochetkova 1978).

The results revealed a marked increase in the size of the cerebellum between the australopithecines and early *Homo*, followed by a decrease between late Paleolithic *H. sapiens* and recent modern humans. It should be noted that the difference between the australopithecines and early *Homo* is likely to have been exaggerated due to the underestimation of the PCF width of the Taung endocast (Schepers 1950; Falk and Clarke 2007). Kochetkova (1978) also reported that the ratio of cerebellar height to brain height reached a peak in early *Homo*, then decreased gradually. In addition, the data suggested that the PCF width of the late Paleolithic and Mesolithic *H. sapiens* ($n = 8$) tended to be absolutely and relatively larger than those of Neanderthals ($n = 7$) or recent modern humans.

Holloway and Yuan (2004) examined the cerebellar size of *Australopithecus afarensis* and the other australopithecines, in which they calculated the "estimated cerebellar volume of the endocast" by multiplying three diameters obtained from the cerebellar hemisphere (Holloway and Yuan 2004). They reported that the ratio of the estimated cerebellar volume to the total endocranial volume was 10.8 in australopithecines ($n = 9$), 13.3 in modern humans ($n = 14$), 14.4 in gorillas ($n = 37$), 13.7 in chimpanzees ($n = 15$), and 14.7 in bonobos ($n = 37$). In accord with Schepers' (1950) report, Holloway and Yuan's (2004) findings suggest that australopithecines had small cerebella compared to extant apes.

Weaver (2005) estimated cerebellar volume of fossil hominins based on the PCF volume obtained from the endocast, then evaluated the relative size by calculating a cerebellar quotient (CQ) as the ratio of the estimated cerebellar volume to the predicted volume from a regression based on the extant mammals. The fossil sample included two australopithecines, three *H. habilis/rudolfensis*, seven *H. erectus*, three Neanderthals, two other archaic *Homo* (Kabwe and Swanscombe), and a fossil *H. sapiens* (Cro-Magnon; Weaver 2005). The results revealed that the CQs of early hominins (australopithecines, *H. habilis/rudolfensis*, and *H. erectus*) were smaller than those of extant apes, and the CQs of Neanderthals and Cro-Magnon were significantly smaller than those of both the earlier hominins and the recent modern humans (Weaver 2005).

Recently, using reconstructed skulls of fossil hominins, we developed a method to extrapolate the brain morphology from the head MRI of living humans (Kochiyama et al. 2014; Ogihara et al. 2015). This method is based on the spatial transformation and standardization algorithm used in functional brain imaging research, which allows direct comparison of brain morphology among different populations (Kochiyama et al. 2014). Applying the method to a few reconstructed Neanderthal skulls, we extrapolated their brains and compared them to living human brains. The preliminary results showed that the cerebellar hemispheres were larger in modern humans than in Neanderthals.

There appears to be little consensus regarding the asymmetry of the cerebellum in primates. Phillips and Hopkins (2007) reported a rightward predominance in the posterior part of the cerebellum in capuchins ($n = 11$) and a leftward predominance in the anterior part of the cerebellum in chimpanzees ($n = 16$). Moreover, this asymmetry was related to the handedness in the capuchins while not in the chimpanzees (Phillips and Hopkins 2007). Another study reported no significant asymmetry in the volume of the cerebellar hemispheres in chimpanzees ($n = 53$) on average, while the asymmetry found at the individual level was related to handedness (Cantalupo et al. 2008). In addition, Smaers et al. (2013) reported no significant left-right asymmetry in scaling trends between the volume of the posterior cerebellar lobes and "the brain volume minus those of the frontal lobes and posterior cerebellar lobes" across 16 anthropoid species including apes and humans.

Regarding the cerebellar volume asymmetry in modern humans, Allen et al. (2002) reported that there was no directional asymmetry between the cerebellar hemispheres. However, some studies have reported a leftward predominance of 0.45% (Raz et al. 2001) and a rightward predominance of 1 cm^3 (Fan et al. 2010). Snyder et al. (1995) found that the anterior part of the cerebellum was significantly larger on the right side, while the posterior part was larger on the left side, in a sex-combined human sample. However, Szeszko et al. (2003) found the opposite asymmetry (left-anterior and right-posterior predominance) in another human sample. Based on a large MRI dataset, our recent analysis confirmed that the degree of volume asymmetry was not distinct in humans (data not shown). Few studies have examined the asymmetry of the cerebellum in fossil hominins. Broadfield et al. (2001) described the endocranial morphology of a Javanese *H. erectus* specimen (Sambungmacan 3) and noted that the left cerebellar hemisphere was smaller than the right hemisphere. White (2005) evaluated the asymmetry of the PCF using a geometric morphometric approach and reported that there was a significant shape asymmetry in a macaque sample, but not in a composite catarrhine sample that included fossil hominins. In a preliminary investigation, we examined the asymmetry of the PCF in a set of Neanderthal endocasts ($n = 3$) reconstructed by Ogihara and colleagues (Amano et al. 2015) and found that the left half of the PCF was larger than the right half in all of the Neanderthal endocasts studied. However, it should be noted that the results are small sample size and uncertainty about the volumetric relationship between the PCF and the cerebellum.

Aside from the matter of cerebellar size and asymmetry, some researchers have discussed the position of the cerebellum and its evolutionary change. According to Grimaud-Hervé (1997), the cerebellum was positioned under the occipital lobes in *H. erectus* and under the occipitotemporal area in Neanderthals, while it is under the temporal lobes in modern humans. Such positional change of the cerebellum or PCF can be interpreted as a result from biomechanical interactions with basicranial and posterior braincase structures, irrelevant to any change in brain functions (Kyriacou and Bruner 2011; Bruner 2015; Bruner et al. 2015).

18.4.2 Are Endocasts Useful for Estimating Cerebral/Cerebellar Volume?

As discussed above, it has traditionally been assumed that the PCF dimensions in hominin endocasts reflect cerebellar size. This assumption appears to be reasonable, considering that the cerebellum occupies a large portion of the PCF, and that the tentorium cerebella overlying the cerebellum attaches to the vicinity of the superior border of the PCF. However, the anterior and superior parts of the cerebellum are far away from the surface of the endocranial cavity. Moreover, some intervening structures (e.g., the meninges, vascular structures, and cerebrospinal fluid) obscure the anatomical correspondence between the endocast and the brain (Connolly 1950; Holloway et al. 2004; Falk 2014; Kobayashi et al. 2014). Thus, the relationship between PCF size and cerebellar volume needs to be examined in more detail.

The first attempt to quantify the relationship between the cerebellum and the PCF was performed by Weaver (2001). She examined the volumetric relationship based on magnetic resonance imaging (MRI) data from a sample of 34 extant hominoid specimens (4 gibbons, 4 orangutans, 2 gorillas, 4 chimpanzees, 3 bonobos, and 17 humans), reporting a strong correlation between cerebellar and PCF volumes (coefficient of determination $r^2 = 0.89$). However, it is likely that the strong correlation was at least partly due to interspecies differences in body and/or brain size. We reexamined Weaver's (2001) dataset and found that the correlation within the human sample of the dataset was very weak ($r^2 = 0.16$) and not statistically significant (Kubo et al. 2011). This finding raises doubt about reliably estimating cerebellar volume from PCF size.

To resolve this issue, we examined the correlation between cerebellar volume and PCF metrics obtained from the MRI data of the whole head of 32 living modern human subjects (Kubo et al. 2014b). The PCF metrics examined included PCF volume, PCF width, PCF height, the maximum chord length of the cerebellar hemisphere, and the geometric mean of these linear measurements. The accuracy of PCF volumes taken from MRI data was validated by comparing with the PCF volumes taken from CT data of the same individuals ($n = 3$). The results indicated that PCF volume was a valid measure for estimating the cerebellar volume, showing a strong correlation with cerebellar volume ($r^2 = 0.78$). Moreover, the error in the cerebellar volume values estimated by regression from PCF volume was approximately ± 12 cc in terms of the 95% prediction interval in the reference population. However, the precision of cerebellar volume estimated by the regressions based on PCF linear measurements was not sufficient. For example, the error rose to approximately ± 18 cc when geometric mean-based estimation was implemented.

Another limitation is related to the estimation of the whole brain or cerebral volume. While Weaver (2005) estimated cerebellar volume from PCF volumes directly, she determined the whole brain volumes of a fossil sample by multiplying their endocranial volumes by 0.88, the coefficient representing the occupation ratio of the brain in the endocranial cavity (Pickering 1930). This means that the coefficient affected the brain volume other than the cerebellum, causing a systematic error in relative cerebellar size of fossil specimens. This error hampers comparisons with volumetric measures of extant species obtained from MRI and/or cadavers. To avoid this problem, we devised the reduced major axis (RMA) equations to estimate the cerebral and cerebellar volumes from the volumes of the corresponding endocranial regions (i.e., the supratentorial and PCF regions) based on the MRI data from living modern humans discussed above (Kubo et al. 2014a).

We then estimated cerebral and cerebellar volumes from the newly reconstructed endocasts of four Neanderthals, two Middle Paleolithic modern humans from Levant (Qafzeh 9 and Skhul V), and an Upper Paleolithic modern human from Cro-Magnon 1 and calculated the "volume fraction of the cerebellum" (F_{CBL}), which is the ratio of the cerebellar volume to the whole brain volume (Kubo et al. 2014a). For comparison, F_{CBL} values of living humans were obtained using MRI-based measurements. A Mann-Whitney U test (with Holm correction) was performed to examine intergroup differences. The data revealed that F_{CBL} estimates of the Neanderthal specimens were significantly smaller than those of the living human sample, while those of the Skhul/Qafzeh crania were similar to those of the living human sample. This result indicates that Middle Paleolithic modern humans already had cerebro-cerebellar systems comparable to that of present-day humans, while Neanderthals did not. These results were partially compatible with those of Weaver (2001, 2005), indicating that the cerebella of Neanderthals were smaller than the average cerebellar volume of recent modern humans. However, our F_{CBL} estimates were not as small as the F_{CBL} values calculated using the measurements reported by Weaver (2001). The latter F_{CBL} estimates for the Neanderthal specimens were approximately 0.8 on average, which is notably small among mammalian species (Clark et al. 2001). As detailed in Kubo et al. (2014b), and as noted above in terms of the occupation ratio of the brain in the endocranial cavity, the volume estimates of Weaver (2001, 2005) likely contain systematic errors. Therefore, the ratio of the cerebellar volume to the whole brain volume in the Neanderthals might have not been as small as suggested by Weaver's (2005) findings. However, it must be noted that the results of our analysis are also based on a limited number of incomplete fossil crania.

18.5 Conclusion

In this chapter, we introduced the anatomical and physiological properties of the modern human cerebellum. Notably, unlike the cerebral cortex, the cerebellar cortex appears to have a similar internal structure throughout and is organized into simple neural circuits. The cerebellum is compartmentalized into microcomplexes, which operate as functional units. Moreover, the cerebellar neural circuits function as a learning system, building internal models that are important for body movement, tool use, imitation of other's movement and thought, and other features of the external environment. Taken together, these findings indicate that the larger volume of the cerebellar cortex, the more internal models it can contain. This functionality differentiates the cerebellum from the cerebral cortex and its neural circuits.

The cerebellar hemispheres play significant roles not only in fine motor control but also in a variety of cognitive functions. In particular, language processing and working memory have significant positive correlations with cerebellar volume, indicating that the cerebellum plays a special role in these cognitive functions. Our studies based on fossil specimens suggest that the Neanderthal cerebellum, particularly the right cerebellar hemisphere, might have been smaller than that of living and ancient modern humans. This indicates that there may have been a difference in cognitive abilities such as language and working memory between Neanderthals and modern humans.

The role of the cerebellum has traditionally been underestimated in the context of cognitive functioning. However mounting evidence implicates the cerebellum in a variety of higher cognitive abilities, in conjunction with the cerebral association cortex. Future research should focus on the cerebellum as an avenue for progressing our understanding of human evolution.

Acknowledgments The authors would like to express the deepest gratitude to Prof. Takeru Akazawa of Kochi Institute of Technology for giving the opportunity to participate in the research project "Replacement of Neanderthals by Modern Humans: Testing Evolutionary Models of Learning" and for lending his continuous guidance and support throughout the course of the study. We're also grateful to N. Sadato of the National Institute for Physiological Sciences, N. Ogihara of Keio University, O. Kondo of the University of Tokyo, and H. Ishida, A. Yogi, and S. Murayama of the University of the Ryukyus, for collaborations in this project. We also thank Y. Rak and I. Hershkovitz of Tel Aviv University and C. P. E. Zollikofer and M. Ponce de Leon of the University of Zurich for kindly allowing the use of CT scan data of the Amud 1 and Qafzeh 9; P. Mennecier and A. Froment of Muséum national d'Histoire naturelle for La Chappelle-aux-Saints 1, and Cro-Magnon 1; M. Bastir of Museo Nacional De Ciencias Naturales and C. Stringer of Natural History Museum for Forbes' Quarry 1; and D. Lieberman, O. Herschensohn, and M. Morgan of Harvard University for Skhul 1. The CT scan data of the Mladec 1 were obtained from the digital archive of fossil hominoids, the University of Vienna. This project is supported by Scientific Research on Innovative Areas "Replacement of Neanderthals by Modern Humans: Testing Evolutionary Models of Learning" (#22101001, #22101006, #22101007) and "The Evolutionary Origin and Neural Basis of the Empathetic Systems" (#16H01486) from the Ministry of Education, Culture, Sports, Science and Technology of Japan (MEXT), and by Grant-in-Aid for Scientific Research C#26350987.

References

Albus JS (1971) A theory of cerebellar function. Math Biosci 10:25–61

Allen JS, Damasio H, Grabowski TJ (2002) Normal neuroanatomical variation in the human brain: an MRI-volumetric study. Am J Phys Anthropol 118:341–358

Amano H, Kikuchi T, Morita Y, Kondo O, Suzuki H, Ponce de León MS, Zollikofer CPE, Bastir M, Stringer C, Ogihara N (2015) Virtual reconstruction of the Neanderthal Amud 1 cranium. Am J Phys Anthropol 158:185–197

Apps R, Garwicz M (2005) Anatomical and physiological foundations of cerebellar information processing. Nat Rev Neurosci 6:297–311

Ashburner J, Friston KJ (2000) Voxel-based morphometry: the methods. NeuroImage 11:805–821

Baddeley A (1992) Working memory. Science 255:556–559

Baddeley A (2003) Working memory: looking back and looking forward. Nat Rev Neurosci 4:829–839

Baddeley AD, Hitch G (1974) Working memory. In: Bower G (ed) Psychology of learning and motivation volume 8. Academic Press, New York, pp 47–89

Baier B, Müller NG, Dieterich M (2014) What part of the cerebellum contributes to a visuospatial working memory task? Ann Neourol 76:754–757

Baillieux H, De Smet HJ, Lesage G, Paquier P, De Deyn PP, Mariën P (2006) Neurobehavioral alterations in an adolescent following posterior fossa tumor resection. Cerebellum 5:289–295

Barkley RA (2001) The executive functions and self-regulation: an evolutionary neuropsychological perspective. Neuropsychol Rev 11:1–29

Barton RA, Venditti C (2014) Rapid evolution of the cerebellum in humans and other great apes. Curr Biol 24:2440–2444

Bostan AC, Dum RP, Strick PL (2013) Cerebellar networks with the cerebral cortex and basal ganglia. Trends Cogn Sci 17:241–254

Braitenberg V, Atwood RP (1958) Morphological observations on the cerebellar cortex. J Comp Neurol 109:1–33

Braver TS, Cohen JD, Nystrom LE, Jonides J, Smith EE, Noll DC (1997) A parametric study of prefrontal cortex involvement in human working memory. NeuroImage 5:49–62

Broadfield DC, Holloway RL, Mowbray K, Silvers A, Yuan MS, Màrquez S (2001) Endocast of Sambungmacan 3 (Sm 3): a new Homo Erectus from Indonesia. Anat Rec 262:369–379

Broom R, Schepers GWH (1946) The South African fossil Ape-man: the Australopithecinae. Transv Mus Mem 2:1–272

Broom R, Robinson JT, Schepers GWH (1950) Sterkfontein Ape-man Plesianthropus. Transv Mus Mem 4:1–117

Bruner E (2015) Functional craniology and brain evolution. In: Bruner E (ed) Human paleoneurology. Springer International Publishing, Cham, pp 57–94

Bruner E, Holloway RL (2010) A bivariate approach to the widening of the frontal lobes in the genus homo. J Hum Evol 58:138–146

Bruner E, Grimaud-Hervé D, Wu X, de la Cuétara JM, Holloway R (2015) A paleoneurological survey of *Homo erectus* endocranial metrics. Quat Int 368:80–87

Buckner RL (2013) The cerebellum and cognitive function: 25 years of insight from anatomy and neuroimaging. Neuron 80:807–815

Buckner RL, Krienen FM, Castellanos A, Diaz JC, Yeo BTT (2011) The organization of the human cerebellum estimated by intrinsic functional connectivity. J Neurophysiol 106:2322–2345

Cantalupo C, Freeman H, Rodes W, Hopkins W (2008) Handedness for tool use correlates with cerebellar asymmetries in chimpanzees (Pan troglodytes). Behav Neurosci 122:191–198

Clark DA, Partha PM, Wang SSH (2001) Scalable architecture in mammalian brains. Nature 411:189–193

Cohen JD, Perlstein WM, Braver TS, Nystrom LE, Noll DC, Jonides J, Smith EE (1997) Temporal dynamics of brain activation during a working memory task. Nature 386:604–608

Connolly CJ (1950) External morphology of the primate brain. Thomas, Springfield

Coolidge FL, Wynn T (2005) Working memory, its executive functions, and the emergence of modern thinking. Camb Archaeol J 15:5–26

D'Anastasio R, Wroe S, Tuniz C, Mancini L, Cesana DT, Dreossi D, Ravichandiran M, Attard M, Parr WC, Agur A, Capasso L (2013) Micro-biomechanics of the Kebara 2 hyoid and its implications for speech in Neanderthals. PLoS One 8:e82261

D'Esposito M, Detre JA, Alsop DC, Shin RK, Atlas S, Grossman M (1995) The neural basis of the central executive system of working memory. Nature 378:279–281

De Smet HJ, Paquier P, Verhoeven J, Mariën (2013) The cerebellum: its role in language and related cognitive and affective functions. Brain Lang 127:334–342

Dediu D, Levinson SC (2013) On the antiquity of language: the reinterpretation of Neandertal linguistic capacities and its consequences. Front Psychol 4:397

Desmond JE, Fiez JA (1998) Neuroimaging studies of the cerebellum: language, learning and memory. Trends Cogn Sci 2:355–362

Diedrichsen J (2006) A spatially unbiased atlas template of the human cerebellum. NeuroImage 33:127–138

E K-H, Chen S-HA, Ho M-HR, Desmond JE (2014) A meta-analysis of cerebellar contributions to higher cognition from PET and fMRI studies. Hum Brain Mapp 35:593–615

Fabbro F, Moretti R, Bava A (2000) Language impairments in patients with cerebellar lesions. J Neurolinguistics 13:173–188

Falk D (2014) Interpreting sulci on hominin endocasts: old hypotheses and new findings. Front Hum Neurosci 8:134

Falk D, Clarke R (2007) New reconstruction of the Taung endocast. Am J Phys Anthropol 134:529–534

Fan L, Tang Y, Sun B, Gong G, Chen ZJ, Lin X, Yu T, Li Z, Evans AC, Liu S (2010) Sexual dimorphism and asymmetry in human cerebellum: an MRI-based morphometric study. Brain Res 1353:60–73

Fiez JA, Petersen SE, Cheney MK, Raichle ME (1992) Impaired non-motor learning and error detection associated with cerebellar damage: a single case study. Brain 115:155–178

Friede H (1981) Normal development and growth of the human neurocranium and cranial base. Scand J Plast Reconstr Surg Hand Surg 15:163–169

Frost JA, Binder JR, Springer JA, Hammeke TA, Bellgowan PSF, Rao SM, Cox RW (1999) Language processing is strongly left lateralized in both sexes: evidence from functional MRI. Brain 122: 199–208

Grimaud-Hervé D (1997) L'évolution de l'encéphale chez *l'Homo erectus* et *l'Homo sapiens*. CNRS, Paris

Haier RJ, Jung RE, Yeo RA, Head K, Alkire MT (2004) Structural brain variation and general intelligence. NeuroImage 23:425–433

Hassid EI (1995) A case of language dysfunction associated with cerebellar infarction. Neurorehabil Neural Repair 9:157–160

Holloway RL, Yuan MS (2004) Endocranial morphology of AL 444-2. In: Kimbel WH, Rak Y, Johanson DC (eds) The skull of Australopithecus Afarensis. Oxford University Press, New York, pp 123–135

Holloway RL, Broadfield DC, Yuan MS (2004) The human fossil record, volume 3, brain endocasts: the paleoneurological evidence. Wiley-Liss, New York

Imamizu H, Kawato M (2009) Brain mechanisms for predictive control by switching internal models: implications for higher-order cognitive functions. Psychol Res 73:527–544

Imamizu H, Miyauchi S, Tamada T, Sasaki Y, Takino R, Puts B, Yoshioka T, Kawato M (2000) Human cerebellar activity reflecting an acquired internal model of a new tool. Nature 403:192–195

Ito M (1970) Neurophysiological basis of the cerebellar motor control system. Int J Neurol 7:162–176

Ito M (2008) Control of mental activities by internal models in the cerebellum. Nat Rev Neurosci 9:304–313

Ito M (2011) The cerebellum: brain for an implicit self. FT Press, New Jersey

Jerison HJ (1973) Evolution of the brain and intelligence. Academic Press, New York

Jerison HJ (2006) Evolution of the frontal lobes. In: Miller BL, Cummings JL (eds) The human frontal lobes, second edition: functions and disorders. Gilford Press, New York, pp 107–118

Jissendi P, Baudry S, Balériaux D (2008) Diffusion tensor imaging (DTI) and tractography of the cerebellar projections to prefrontal and posterior parietal cortices: a study at 3T. J Neuroradiol 35:42–50

Justus TC, Ivry RB (2009) The cognitive neuropsychology of the cerebellum. Intern Rev Psychiatry 13:276–282

Kanai R, Rees G (2011) The structural basis of inter-individual differences in human behaviour and cognition. Nat Rev Neurosci 12: 231–242

Kawato M, Furukawa K, Suzuki R (1987) A hierarchical neural-network model for control and learning of voluntary movement. Biol Cybern 57:169–185

Kelly RM, Strick PL (2003) Cerebellar loops with motor cortex and prefrontal cortex of a nonhuman primate. J Neurosci 23:8432–8444

Keren-Happuch E, Chen SHA, Ho MHR, Desmond JE (2014) A meta-analysis of cerebellar contributions to higher cognition from PET and fMRI studies. Hum Brain Mapp 35:593–615

Klein RG, Edgar B (2002) The dawn of human culture. Wiley, New York

Kobayashi Y, Matsui T, Haizuka Y, Ogihara N, Hirai N, Matsumura G (2014) Cerebral sulci and gyri observed on macaque endocasts. In: Akazawa T, Ogihara N, Tanabe HC, Terashima H (eds) Dynamics of learning in Neanderthals and modern humans, Cognitive and physical perspectives, vol 2. Springer, Tokyo, pp 131–137

Kochetkova VI (1978) Paleoneurology. VH Winston and Sons, Washington, DC

Kochiyama T, Tanabe HC, Ogihara N (2014) Reconstruction of the brain from skull fossil using computational anatomy. In: Akazawa T, Ogihara N, Tanabe HC, Terashima H (eds) Dynamics of learning in Neanderthals and modern humans, Cognitive and physical perspectives, vol 2. Springer, Tokyo, pp 191–200

Krause J, Lalueza-Fox C, Orlando L, Enard W, Burbano GRE, HA HJJ, Hänni C, Fortea J, de la Rasilla M, Bertranpetit J, Rosas A, Pääbo S

(2007) The derived FOXP2 variant of modern humans was shared with Neandertals. Curr Biol 17:1908–1912
Krienen FM, Buckner RL (2009) Segregated fronto-cerebellar circuits revealed by intrinsic functional connectivity. Cereb Cortex 19: 2485–2497
Kubo D, Kono RT, Suwa G (2011) A micro-CT based study of the endocranial morphology of the Minatogawa I cranium. Anthropol Sci 119:123–135
Kubo D, Tanabe HC, Kondo O, Amano H, Yogi A, Murayama S, Ishida H, Ogihara N (2014a) Estimating the cerebral and cerebellar volumes of Neanderthals and Middle and Upper Paleolithic Homo sapiens. In: Akazawa T, Nishiaki Y (eds) RNMH Project Series No. 003: Abstracts of RNMH 2014 Second International Conference, pp 116–118
Kubo D, Tanabe HC, Kondo O, Ogihara N, Yogi A, Murayama S, Ishida H (2014b) Cerebellar size estimation from endocranial measurements: an evaluation based on MRI data. In: Akazawa T, Ogihara N, Tanabe HC, Terashima H (eds) Dynamics of learning in Neanderthals and modern humans, Cognitive and physical perspectives, vol 2. Springer, Tokyo, pp 209–215
Kyriacou A, Bruner E (2011) Brain evolution, innovation, and endocranial variations in fossil hominids. PaleoAnthropology 2011: 130–143
Lange N, Froimowitz MP, Bigler ED, Lainhart JE (2010) Associations between IQ, total and regional brain volumes and demography in a large sample of healthy children and adolescents. Dev Neuropsychol 35:296–317
Leggio MG, Silveri MC, Petrosini L, Molinari M (2000) Phonological grouping is specifically affected in cerebellar patients: a verbal fluency study. J Neurol Neurosurg Psychiatry 69:102–106
Lieberman P (2009) The singing Neanderthals: the origins of music, language, mind and body, and: music, language, and the brain (review). Language 85:732–736
Lin WC, Chou KH, Chen HL, Huang CC, Lu CH, Li SH, Wang YL, Cheng YF, Lin CP, Chen CC (2012) Structural deficits in the emotion circuit and cerebellum are associated with depression, anxiety and cognitive dysfunction in methadone maintenance patients: a voxel-based morphometric study. Psychiatry Res 201:89–97
MacLeod C (2012) The missing link: evolution of the primate cerebellum. In: Hofman MA, Falk D (eds) Evolution of the primate brain: from neuron to behavior. Elsevier, Amsterdam, pp 165–187
MacLeod CE, Zilles K, Schleicher A, Rilling JK, Gibson KR (2003) Expansion of the neocerebellum in Hominoidea. J Hum Evol 44: 401–429
Marr D (1969) A theory of cerebellar cortex. J Physiol Lond 202: 437–470
Marvel CL, Desmond JE (2010) Functional topography of the cerebellum in verbal working memory. Neuropsychol Rev 20:271–279
McDaniel MA (2005) Big-brained people are smarter: a meta-analysis of the relationship between in vivo brain volume and intelligence. Intelligence 33:337–346
Moss ML, Young RW (1960) A functional approach to craniology. Am J Phys Anthropol 18:281–292
Neubauer S, Gunz P, Weber GW, Hublin JJ (2012) Endocranial volume of Australopithecus africanus: new CT-based estimates and the effects of missing data and small sample size. J Hum Evol 62: 498–510
O'Reilly JX, Beckmann CF, Tomassini V, Rammani N, Johansen-berg H (2009) Distinct and overlapping functional zones in the cerebellum defined by resting state functional connectivity. Cereb Cortex 20:953–965
Ogihara N, Amano H, Kikuchi T, Morita Y, Hasegawa K, Kochiyama T, Tanabe HC (2015) Towards digital reconstruction of fossil crania and brain morphology. Anthropol Sci 123:57–68
Paradiso S, Andreasen NC, O'Leary DS, Arndt S, Robinson RG (1997) Cerebellar size and cognition: correlations with IQ, verbal memory and motor dexterity. Neuropsychiar Neuropsychol Behav Neurol 10: 1–8
Peña-Melian A (2000) Development of human brain. Hum Evol 15: 99–112
Phillips KA, Hopkins WD (2007) Exploring the relationship between cerebellar asymmetry and handedness in chimpanzees (Pan troglodytes) and capuchins (Cebus apella). Neuropsychologia 45: 2333–2339
Pliatsikas C, Johnstone T, Marinis T (2013) Grey matter volume in the cerebellum is related to the processing of grammatical rules in a second language: a structural voxel-based morphometry study. Cerebellum 13:55–63
Porrill J, Dean P, Anderson SR (2013) Adaptive filters and internal models: multivariate description of cerebellar function. Neural Netw 47: 134–149
Pryor Pickering S (1930) Correlation of brain and head measurements, and relation of brain shape and size to shape and size of the head. Am J Phys Anthropol 15(1):1–52
Raz N, Gunning-Dixon F, Head D, Williamson A, Acker JD (2001) Age and sex differences in the cerebellum and the ventral pons: a prospective MR study of healthy adults. Am J Neuroradiol 22: 1161–1167
Rilling JK (2006) Human and nonhuman primate brains: are they allometrically scaled versions of the same design? Evol Anthropol 15: 65–77
Rilling JK, Insel TR (1998) Evolution of the cerebellum in primates: differences in relative volume among monkeys, apes and humans. Brain Behav Evol 52:308–314
Schepers GWH (1950) The brain casts of the recently discovered Plesianthropus skulls. Transv Mus Mem 2:89–117
Schmahmann JD (1996) From movement to thought: anatomic substrates of the cerebellar contribution to cognitive processing. Hum Brain Mapp 4:174–198
Schmahmann JD (2004) Disorders of the cerebellum: ataxia, dysmetria of thought, and the cerebellar cognitive affective syndrome. J Neuro-Oncol 16:367–378
Schmahmann JD, Sherman JC (1998) The cerebellar cognitive affective syndrome. Brain 121:561–579
Semendeferi K, Damasio H (2000) The brain and its main anatomical subdivisions in living hominoids using magnetic resonance imaging. J Hum Evol 38:317–332
Silveri MC, Leggio MG, Molinari M (1994) The cerebellum contributes to linguistic production: a case of agrammatic speech following a right cerebellar lesion. Neurology 44:2047–2050
Smaers JB, Steele J, Case CR, Amunts K (2013) Laterality and the evolution of the prefronto-cerebellar system in anthropoids. Ann N Y Acad Sci 1288:59–69
Snyder PJ, Bilder RM, Wu H, Bogerts B, Lieberman JA (1995) Cerebellar volume asymmetries are related to handedness: a quantitative MRI study. Neuropsychologia 33:407–419
Sousa AA d, Proulx MJ (2014) What can volumes reveal about human brain evolution? A framework for bridging behavioral, histometric, and volumetric perspectives. Front Neuroanat 8:51
Szeszko PR, Gunning-Dixon F, Ashtari M, Snyder PJ, Lieberman JA, Bilder RM (2003) Reversed cerebellar asymmetry in men with first-episode schizophrenia. Biol Psychiatry 53:450–459

Van Overwalle F, Mariën P (2016) Functional connectivity between the cerebrum and cerebellum in social cognition: a meta-study analysis. NeuroImage 124:248–255
Weaver AGH (2001) The cerebellum and cognitive evolution in Pliocene and Pleistocene hominids. Dissertation, University of New Mexico
Weaver AH (2005) Reciprocal evolution of the cerebellum and neocortex in fossil humans. Proc Natl Acad Sci U S A 102:3576–3580
Weil A (1929) Measurements of cerebral and cerebellar surfaces: comparative studies of the surfaces of endocranial casts of man, prehistoric men, and anthropoid apes. Am J Phys Anthropol 13: 69–90
White DD (2005) Size and shape of the cerebellum in Catarrhine primates and Plio-Pleistocene fossil hominins: a paleoneurological analysis of endocranial casts. Dissertation, State University of New York
Whiting BA, Barton RA (2003) The evolution of the cortico-cerebellar complex in primates: anatomical connections predict patterns of correlated evolution. J Hum Evol 44:3–10
Wolpert DM, Kawato M (1998) Multiple paired forward and inverse models for motor control. Neural Netw 11:1317–1329
Wolpert DM, Miall RC, Kawato M (1998) Internal models in the cerebellum. Trends Cogn Sci 2:338–347
Wynn T, Coolidge FL (2003) The role of working memory in the evolution of managed foraging. Before Farming 2003(2): 1–16
Yeo BTT, Sabuncu MR, Vercauteren T, Holt DJ, Amunts K, Zilles K, Golland P, Fischl B (2010) Learning task-optimal registration cost functions for localizing cytoarchitecture and function in the cerebral cortex. IEEE Trans Med Imaging 29:1424–1441
Yu F, Jiang Q-j, Sun X-y, Zhang R-w (2015) A new case of complete primary cerebellar agenesis: clinical and imaging findings in a living patient. Brain 138(6):e353–e353
Zettin M, Cappa SF, D'amico A, Rago R, Perino C, Perani D, Fazio F (1997) Agrammatic speech production after a right cerebellar haemorrhage. Neurocase 3:375–380

The manufacturer's authorised representative in the EU is Springer Nature Customer Service Centre GmbH, Europaplatz 3, 69115 Heidelberg, Germany. If you have any concerns regarding our products, please contact ProductSafety@springernature.com

Printed and bound by CPI Group (UK) Ltd, Croydon, CR0 4YY
30/09/2026
02225476-0002